Stochastic Geometry
and its Applications

Stochastic Geometry and its Applications

Third Edition

Sung Nok Chiu

Department of Mathematics, Hong Kong Baptist University, Hong Kong

Dietrich Stoyan

Institut für Stochastik, TU Bergakademie Freiberg, Germany

Wilfrid S. Kendall

Department of Statistics, University of Warwick, UK

Joseph Mecke

Institut für Stochastik
Friedrich-Schiller-Universität Jena, Germany

WILEY

This edition first published 2013
© 2013 John Wiley & Sons, Ltd

Registered office
John Wiley & Sons Ltd, The Atrium, Southern Gate, Chichester, West Sussex, PO19 8SQ, United Kingdom

For details of our global editorial offices, for customer services and for information about how to apply for permission to reuse the copyright material in this book please see our website at www.wiley.com.

Library of Congress Cataloging-in-Publication Data

Stoyan, Dietrich.
  [Stochastische Geometrie. English]
  Stochastic geometry and its applications. – Third edition / Sung Nok Chiu, Dietrich Stoyan, Wilfrid S. Kendall, Joseph Mecke.
       pages cm – (Wiley series in probability and statistics)
  Revision of: Stochastic geometry and its applications / Dietrich Stoyan, Wilfrid S. Kendall, Joseph Mecke. – 2nd ed. – ?1995.
     Includes bibliographical references and indexes.
  ISBN 978-0-470-66481-0 (hardback)
 1. Stochastic geometry. I. Chiu, Sung Nok. II. Kendall, W. S. III. Mecke, Joseph. IV. Title.
  QA273.5.S7813 2013
  519.2–dc23
                                                                2013012066

A catalogue record for this book is available from the British Library.

ISBN: 978-0-470-66481-0

Set in 10/12pt Times by Thomson Digital, Noida, India.

1   2013

Contents

Foreword to the first edition

My good friends the authors of this book have kindly invited me to add a few words describing 'how it all began'. At present one can write only an anecdotal history of stochastic geometry, and we must recognise that the anecdotes of others will certainly much extend the account given here, and will supply fresh perspectives. A serious attempt to write a history would be premature. The historian of mathematics looks always to the future rather than to the past. He hopes to find early instances of general concepts later seen to be of fundamental significance, and so if he works at a time of rapid development (such as the present) he will overlook many clues in the early record which point to a future not yet revealed.

My own first contact with *classical geometrical probability* occurred during the war, when the Superintendent of my group (Louis Rosenhead) asked me to investigate the following problem; for the sake of clarity I formulate it in the modern terminology.

> A euclidean plane with a marked origin O carries a Poisson field of unsensed lines with a uniform intensity. Almost surely the point O will lie in the interior of a unique Crofton cell C, with (unlabelled) shape $\sigma(C)$ and area $a(C)$. What probability statements can be made about C which convey information about the strength of a fabric ('a sheet of paper') consisting of the field of lines ('fibres')? Thus it would be useful to be able to calculate the rate of occurrence of splinter-shaped cells C, and the rate of occurrence of cells with large area $a(C)$.

A few moments of the $a(C)$-distribution were already known, and I managed to add one more to these. One would have preferred to be able to say something about the asymptotics of the marginal $a(C)$-distribution valid for large areas, and to throw light one way or the other on my conjecture that the conditional law for $\sigma(C)|a(C)$ converges weakly, as $a(C) \rightarrow \infty$, to the degenerate law concentrated at the circular shape.

Unfortunately nothing substantial is known about either of these questions even today, apart from the limited information that can be derived from a massive series of simulations carried out in Stanford by E. I. George (1982, 1987) in association with Herbert Solomon.

In 1961 Roger Miles wrote his Cambridge PhD thesis on a generalised version of this problem under the direction of Dennis Lindley and Peter Whittle and in consultation with paper technology experts from Wiggins Teape Research and Development Ltd. This initiated a long series of famous papers by Miles which provide a huge volume of information about the problem as a whole without, however, bringing us any nearer to the answer to the two questions above.[*]

All this work was within the classical Croftonian framework, not however without hints that a statistical theory of shape could play a useful rôle. What we now call *stochastic geometry* began for me with three papers by Maurice Bartlett: 'The spectral analysis of point processes' (*J. Roy. Statist. Soc.* B 1963), 'The spectral analysis of two-dimensional point processes' (*Biometrika* 1964), and especially 'The spectral analysis of line processes' (*5th Berkeley Symp.* 1967). These became available just when Rollo Davidson joined me as a research student in October 1965, and the impact of Bartlett's third paper can be felt in the Smith's Prize Essay which he wrote in 1967. As Bartlett's work had focused on the empirical spectral analysis of point and line processes, I encouraged Davidson to set up an appropriate theoretical framework underlying such an empirical analysis, and the second half of his PhD thesis was concerned with this; see Chapter 2.1 of *Stochastic Geometry* (ed. Harding and Kendall) for a reprint of it. The following two quotations give the flavour of the approach:

> ... *we can talk of flat- and line-processes, meaning the point-processes that they induce on the appropriate manifolds,*

> ... *my results showed that it is profitable, when considering point-processes, to observe simply whether certain sets contain points of the process or not; the usual approach is, of course, to look at the number of points in these sets.*

The first quotation, coupled with a general specification for a point process on a manifold M, takes us away from Croftonian geometric probability to the consideration of arbitrary random fields of geometric objects, while the second hints at a theory of random sets of very general character to replace the point process on the representation-manifold. Two such (closely related) theories of random sets very shortly afterwards became available; both were strongly influenced by earlier ideas due to Gustave Choquet. Stochastic Geometry thus became a reality.

When was the phrase 'stochastic geometry' first used? Klaus Krickeberg, who was to play a leading rôle in its development, thinks that perhaps he and I may have used the phrase informally in the Spring of 1969 when he was in Cambridge and we were planning the

---

[*]David Kendall died in 2007 at age 89 years. So it is the authors' duty to inform the reader about the fate of his conjecture. David Kendall posed the problem in the foreword of the first (1987) edition of the present book, but only in its second (1995) edition the problem, known as Kendall's conjecture, attracted more interest. Miles (1995) offered a heuristic proof, and surprisingly, only two years later, in 1997, a solution was given by the Ukrainian I. N. Kovalenko (1997, 1999), known until that time as a queueing theorist. He even found a similar result for large cells of the Poisson-Voronoi tessellation in Kovalenko (1998). Kovalenko's main idea is to give an upper bound for some conditional probability, by enlarging the numerator and reducing the denominator. For the former he used Bonnesen's inequality (a refined form of the planar isoperimetric inequality), for the latter an explicit construction.

In a series of papers, German stochastic geometers (Hug, Reitzner and Schneider) '*treated very general higher-dimensional versions, variants and analogy of Kendall's problem*', see Schneider and Weil (2008, p. 512) for an excellent overview. They considered the problem in d dimensions and could omit the isotropy assumption. For this, they started from Kovalenko's ideas but employed more sophisticated geometrical tools. In the anisotropic case '*the asymptotic shape of such cells was found to be that of the so-called Blaschke body of the hyperplane process. This is (up to a dilatation) the convex body, centrally symmetric with respect to the origin, that has the spherical directional distribution of the hyperplane process as its surface area measure*', see Hug and Schneider (2010). The last paper presents a solution for k-dimensional faces for $2 \leq k \leq d$, that is, for the k-volume weighted typical k-face, for example a polygonal cell face in the three-dimensional case.

Oberwolfach meeting (for June 1969) on *Integral Geometry and Geometrical Probability*. This is very possible, because 'stochastic analysis' was already in common use in the UK as part of the title of the Stochastic Analysis Group set up in December 1961 under the chairmanship of Harry Reuter, and one phrase naturally suggested the other. Certainly there was obviously no other choice for the titles of the two memorial books produced after Davidson's death in 1970. So 1969 was perhaps the year of coining. The initial group of enthusiasts could readily be identified by inspecting the *Tagungsbuch* at Oberwolfach. From the first Ruben Ambartzumian played a very important rôle and he has continued to influence the development of the subject in characteristic ways.

To some extent and to my great satisfaction there has also been a close association between those interested in stochastic geometry, and those interested in geometrical statistics, so that we now have a broad and lively subject area with abstract and empirical edges to it. I trust that this will continue, and the balance of the present book makes that seem likely.

Shape-theory is generally viewed as part of stochastic geometry and I think that this is as it should be. I have already mentioned one early hint at the need for a theory of shape, and here is a much earlier one. *The Ladies' Diary* (1706–1840), *The Gentleman's Diary* (1741–1840), and *The Lady's and Gentleman's Diary* (1841–1871) are mathematical periodicals known now perhaps only to a few specialists, but are worth very serious study because of the frequency with which important ideas first found explicit mention in their pages. In *TLGD* (1861) there is the following challenge by the London-based mathematician Wesley Stoker Woolhouse (1809–1893):

> Problem 1987. *Three points being taken at random in space as the corners of a plane triangle, determine the probability that it shall be acute.*

A solution by Stephen Watson of Haydonbridge, Northumberland, appeared in the 1862 edition of the diary. It begins with the comment:

> '*Space*' *is equivalent to a sphere of infinite radius, and it is obvious that the chance will be the same whatever be the radius of the sphere within which the three points may lie; hence we may suppose them to always lie within a sphere of radius unity.*

He then continues with an integration argument yielding the probability 33/70, and the probability $4/\pi^2 - 1/8$ for three points in a plane is added in a comment by Woolhouse.

Watson's solution provoked a strong reaction from Augustus de Morgan, who argued in *Trans. Cambridge Philos. Soc.* **11** (1871) 145–189 that 'it is very easily shown that the chance of an acute-angled triangle must be infinitely small'. The controversy raged for many years in the journal *Mathematical Questions with their Solutions from the 'Educational Times'*, where Woolhouse's challenge had been reprinted as Question 1333. The article on it by M. W. Crofton in 1867 is particularly interesting. In the pages of these journals we can see classical geometrical probability taking shape under our eyes.

But from our present point of view a most interesting comment is that Watson could very well have approximated to 'infinite space' by expanding an arbitrary compact convex set K about its centroid, and he would then have obtained a different answer based on the shape measure induced on Σ_2^3 (a hemisphere) by i. i. d.-uniform sampling from the interior of K, so that the resulting probability would depend on the shape of K itself, but not at all upon its size. Of course Watson's choice was a very natural one, because one feels that one is required to respect the isotropy of 'infinite space' – though this is a delusion, for the isotropy is only maintained at the centre of the sphere. Meaningless though Woolhouse's problem is, without further specification, any reader of this book will probably find it instructive as well as amusing to read through these old polemics. As the same protagonists occur again and again in the pages of the journals mentioned, and as they express themselves with considerable freedom, one soon becomes familiar with them on a personal basis, and the early history of our subject comes to life in a most vivid way.

It only remains to say that, given the spherical assumption, the numerical answers obtained by Watson and by Woolhouse are identical with those derived from the recent solution to the general problem in n dimensions given by G. R. Hall in *J. Appl. Prob.* **19**, 712–15 (1982).

David Kendall

From the preface to the first edition

Complicated geometrical patterns occur in many areas of science and technology and often require statistical analysis. Examples include the structures studied in geology, sections of porous media, solid bodies, biological tissues, and patterns formed by the distinction between wood and field in a landscape. Analyses of such data sets require suitable mathematical models and appropriate statistical methods. The area of mathematical research that seeks to provide such models and methods is called *Stochastic Geometry*. The oldest part of this subject considers problems concerning a finite number of geometrical objects of fixed form, whose positions are completely random and (in some sense) uniformly distributed. The famous question of Buffon's needle is the prototype of these problems, which form the subject of *Geometrical Probability*. The modern theory of stochastic geometry (initiated by D. G. Kendall, K. Krickeberg, and R. E. Miles) considers random geometrical patterns (which may be infinite in extent) of more complicated distribution. *Stereology* is that branch of stochastic geometry which studies the problem of recovering information on three-dimensional structures when the only information available is two- or one-dimensional, obtained by planar or linear section.

This monograph grew out of a book originally published in German (Stoyan and Mecke, 1983b), but has undergone considerable expansion and reorganisation. Its aim is to make the results and methods of stochastic geometry more generally accessible to applied scientists, but also to provide an exposition which is mathematically exact and general, and which takes into account the current state of research in order to serve as an introduction to stochastic geometry for mathematicians. Of course these aims conflict and the resulting compromises have strongly influenced the form of the book. In most parts of the monograph proofs are omitted. The level of exposition is uneven and the subjects are treated with varying thoroughness: some topics are illustrated by numerical examples, some results are stated without much comment, others are accompanied by heuristic arguments, and sometimes substantial issues are dismissed with only a few remarks and a few references to the literature. Throughout the text attempts are made to explain the plausible nature and the underlying ideas of mathematical concepts, in order to facilitate the reader's understanding and to pave the way for a deeper study of the literature. Our hope is that those readers who do not wish to invest much effort in following mathematical arguments will nevertheless be able to interpret and to use most of the formulae.

There is some redundancy in the exposition but we believe this will help most readers. At some points it has been appropriate to use formulae and notation in anticipation of their introduction. In any case, readers may prefer to turn directly to the chapters concerning the topics that interest them the most, rather than to read though the book consecutively. Generally we have not sought to use the most elegant possible mathematical style but rather to strike a balance between generality and concrete special cases. For example, we use the theory of marked point processes, and this frees us from the need to consider point processes in abstract spaces.

At places we refer to ideas of mathematical physics that are related to the techniques of stochastic geometry. A closer collaboration between mathematical physicists and stochastic geometers might be very fruitful; the two subjects meet at several points but use different languages.

Mathematical terminology is used throughout the book. This exhibits some peculiarities due to historical accident. The word 'process' as in 'point process' and in 'line process' does not imply any dependence on time (with the possible exception of point processes on the real line). A more logical terminology would use the phrases 'random point field' and 'random line field'. . . .

The examples in the book are for the most part concerned with the analysis and description of images by numbers and functions, and are drawn from various branches of science. Generally the theoretical basis of the statistical methods is not discussed. In some cases statistical methods are given and these enable the fitting of models to empirical data. Much work remains to be done on statistical theory for geometrical structures. For example, little is known of the distribution theory for most estimators appearing in this book.

A brief summary of the contents of the book will illustrate the way in which theory and practice are intertwined. The first chapter briefly introduces areas of mathematics with which most scientists and engineers are not familiar. . . . We assume a basic knowledge of probability theory and statistics.

In the remaining chapters the development of the exposition does not proceed from the general to the particular but rather in the reverse direction. Thus Chapters 2 and 3 discuss the Poisson process and the Boolean model, which are simple cases of the random structures to be discussed in the remainder of the text. Chapters 4 and 5 continue the subject of point processes and give a general discussion; Chapter 6 expounds the general theory of random sets. Chapter 7 briefly introduces the important concept of a random measure, which arises throughout the subject at a more theoretical level. The theory of random processes of geometrical objects is introduced in Chapter 8, which leads on to the discussion of fibre processes in Chapter 9[†] and tessellations in Chapter 10. The final Chapter 11 is on stereology, which is of great importance in practice and uses results and ideas from all of the preceding discussion. . . .

[†] In the third edition Chapters 8 and 9 are combined to form a chapter on line, fibre and surface processes.

Preface to the second edition

We the authors present a second edition of our book. The first edition met with a kind reception and has become a standard reference in its field . This has encouraged us to retain its style and conception. As before this book has an applied character, presents the matter in a less than strictly sequential form and admits inhomogeneities in the presentation. Our personal taste and interests played an important rôle in choosing the topics.

We have tried to present many of the new ideas and developments in the fields of stochastic geometry and spatial statistics since 1987. They seem to us particularly prominent in the fields of Boolean models, stereology, random shapes, Gibbs processes, and random tessellations. The progress of these years is also visible in the jacket of this book: a figure in Chapter 10 of the old edition presented a small part of a Johnson–Mehl tessellation (drawn by hand by H. Stoyan); this has been replaced by a computer-generated figure containing many cells, and we have used a similar figure to decorate the cover of the new edition.

We hope very much that our readers will find the style and presentation of the second edition better than that of its predecessor. It was a pleasure to eliminate a series of misprints and (we have to confess) errors; and also the poor texture of the paper of the first edition can now be forgotten. We also hope that the many minor additions will be noticed, which arose from many discussions with colleagues. On the other hand we have to warn our readers that at a few points notation has been changed; we hope that the number of new misprints and errors is small.

As in the first edition, we do not present all which may go under the names 'stochastic geometry' and 'spatial statistics'. This is quite appropriate since there are already specialised books on spatial statistics (Cressie, 1993), fractals (Falconer, 1990; Stoyan and Stoyan, 1994), random shapes (Stoyan and Stoyan, 1994; Barden, Carne, Kendall, and Le, 1996[‡]), and integral geometry (Schneider, 1993).

Producing the manuscript of the second edition was not an easy task for us because of our various other professional duties. It was only possible with the help of many friends and colleagues. They read whole chapters or parts of them and suggested many corrections and additions. We are very grateful to them: S. N. Chiu, L. M. Cruz-Orive, L. Heinrich, D. G. Kendall, M. N. M. van Lieshout, U. Lorz, K. V. Mardia, I. Molchanov, L. Muche, W. Nagel, J. Ohser, R. Schneider, and E. Schüle.

The hard technical work was done by H. Stoyan, assisted by I. Gugel and R. Pohlink. She did this work with incredible care and patience and also suggested many corrections and improvements of a scientific nature.

We have also to thank two collections of electronic software. LaTeX proved to be an excellent tool for the production of our manuscript: in common with very many other mathematical scientists, we owe an almost incalculable debt to D. E. Knuth and L. Lamport. The first edition

---

[‡]This actually refers to D. G. Kendall *et al.* (1999).

was still produced in the classical way using lead type, and so W.S.K. may be one of the last Englishmen to have seen in his proofs a 'Zwiebelfisch' (the German word for a letter standing on the head).[§] The existence of *e-mail* made the correspondence between Warwick, Chichester and Freiberg easy and very fast. (It would have seemed incredible to us ten years ago, but the authors did not have any personal meeting during the work for the manuscript.) We also thank Stuart Gale of John Wiley & Sons Ltd. for his work as an editor; his predecessor of the first edition, Dr R. Höppner, is now Ministerpräsident of the German Bundesland Sachsen-Anhalt.

July 1995   *The authors*

[§] In some sense, this was not true. In the 1995 edition the name 'Hansen' was written as 'Hausen'. The 'u' can be seen as a upside-down 'n' and hence can be regarded as a Zwiebelfisch, but in fact it was a typo.

Preface to the third edition

It is perhaps unusual to make a third edition of a book 18 years after its second. However, the authors remained active in the field of the book and observed with pleasure that the second edition, abbreviated as 'SKM95' in the text, became a standard reference for (applied) stochastic geometry and wished to maintain this status for the future. Finally and crucially, the original authors found a younger new co-author, so being competent to produce a modernised book.

In the years since 1995 many other books on stochastic geometry have been published, but all have been of a nature different from SKM95. There are now excellent books of a high theoretical level such as Schneider and Weil (2008) and Kendall and Molchanov (2010). The present book uses them as references and source for proofs of complicated mathematical facts and by no means aims to compete with them. Then there are now specialised books which present the methods of image analysis and processing of lattice data coming from modern imaging techniques, such as Ohser and Schladitz (2009). On the other hand there are now various books which present ideas of stochastic geometry to physicists, engineers and others, such as Ohser and Mücklich (2000), Torquato (2002) and Buryachenko (2007). However, none of these books plays the rôle of SKM95, as a book which is accessible for a broad readership of applied mathematicians, physicists and engineers, but also presents mathematical foundations and in some cases mathematical proofs. By the way, the selection of the statements which are proved was made according to the following considerations: proofs are included when they show how the mathematical tools work, where the argument is not too complex and where somehow unexpected results are derived. The book by Schneider and Weil (2008) clearly demonstrates how large a book may become if it aims to give nearly 'all' proofs.

In the process of modernising (and correcting errors in) SKM95, which started in 2010, the authors learned which areas of stochastic geometry have been particularly active. The first such area is the theory of random sets, where new books such as Molchanov (2005) and Nguyen (2006) were published and many new models have been developed, for example in Baccelli and Błaszczyszyn (2009a,b). The second is the theory of tessellations. The corresponding Chapter 9 of this book was enlarged by new sections on networks and random graphs since these areas are becoming more and more important. Of course, the classical branches such as the theory of point processes also developed new ideas, and so the important theory of point-stationarity and balanced partitions appears in Chapter 4. Unfortunately, in order to limit the volume of the new book the section on random shape theory had to be omitted. A reason for this omission is that there are now excellent books on random shape theory with which the present book could never compete. Nevertheless, this edition is still much thicker than its predecessor and has about 700 new references, though about 300 outdated references have been deleted.

In the years since 1995 stochastic geometry further developed as a mathematical discipline. This has resulted in simplifying and generalising its theories and making its notation more elegant. For example, the Minkowski functionals intensively used by Matheron have been replaced by the intrinsic volumes. The description of tessellations has been refined to include not necessarily face-to-face tessellations in the theory, following R. Cowan and V. Weiß.

In discussing changes in the notation, it may be interesting to add some words about the notation used in this book. Unfortunately, there is no unique notation system in stochastic geometry. Even the various book authors have different personal notations. And the notations used by mathematicians and physicists differ greatly.

This book commits to a consistently mathematical notation as exemplified by the use of $E(X)$ for the mean of X instead of $\langle X \rangle$ as physicists would write. As geometers do, the multidimensional space is denoted by $\mathbb{R}^d$, with d as 'dimension'.

Various traditions come together in the notation of the present book. Queueing theorists played a significant rôle in the early development of point process theory and stochastic geometry. As a consequence, the intensity or density is conventionally denoted by λ. This follows the queueing tradition, which denotes the arrival and service rates of queueing systems by λ and μ. Product densities are denoted by $\varrho^{(n)}$, following an old notation system of physicists, to which, by the way, the symbol ϱ for the intensity belongs. The authors considered replacement of λ by ϱ, but the tradition was stronger and even the youngest author argued in favour of old λ; moreover a capital Λ is needed, whereas the capital counterpart to ϱ is P, which has many other uses in the book. The Lebesgue measure is sometimes denoted by λ, which does not fit into this scheme; and also μ is too often used for means. Thus also in this edition the Lebesgue measure is again denoted by ν.

There is some influence of the now classical book Matheron (1975). The French 'fermé' has led to $\mathbb{F}$ for the system of all closed subsets of $\mathbb{R}^d$ and the related $\mathcal{F}$ and $\mathcal{F}_K$. And b_d with 'b' as 'ball' is used for the volume of the unit ball in $\mathbb{R}^d$, for which other authors use κ_d and ω_d, and the ball with centre x and radius r is $B(x, r)$.

Finally, the use of Φ to denote a point process goes back to Klaus Matthes.

It is a pleasure to thank here all the colleagues who helped us in producing the manuscript of the third edition. The list is so long that we may speak of a 'collective work'. In alphabetic order we name R. Adler, A. Baddeley, F. Baccelli, F. Ballani, S. Bernstein, B. Błaszczyszyn, P. Calka, S. Ciccariello, R. Cowan, D. Dereudre, W. Gille, P. Grabarnik, P. Hall, L. Heinrich, H. Hermann, J. Janáček, S. Kärkkäinen, M. Kiderlen, M. Lang, G. Last, T. Mattfeldt, N. Medvedev, I. Molchanov, J. Møller, L. Muche, W. Nagel, A. Penttinen, P. Ponížil, C. Redenbach (née Lautensack), V. Schmidt, R. Schneider, D. Schuhmacher, V. Weiß, K. Y. Wong and S. Zuyev.

A particular rôle as readers and suppliers of mathematical criticism of whole chapters was played by P. Grabarnik, L. Muche, W. Nagel and J. Ohser. G. Last read Chapter 4 and wrote for us the Sections 4.4.9 and 4.4.10. K. Y. Wong offered technical support. Finally, R. Schneider answered with great patience many questions from D.S.

This book contains an accompanying website. Please visit www.wiley.com/go/cskm

Autumn 2012   *The authors*

Notation

This index contains only the notation used throughout the book. Symbols with localised usage are omitted, as are 'standard' symbols such as e and π.

Symbols

Symbol	Usage	Page
$\| \cdot \|$	Euclidean metric	3
1_A	indicator function of A	17
$\# \{\ldots\}$	number of elements of set $\{\ldots\}$	46

Operations

Symbol	Usage	Page
$\circ$	opening	8
$\bullet$	closing	8
$\oplus$	Minkowski-addition	5
$\ominus$	Minkowski-subtraction	5
$*$	convolution	132
conv	convex hull	11
∂	boundary	4

Greek letters

Symbol	Usage	Page
$\alpha^{(n)}$	factorial moment measure	46, 47, 121
χ	connectivity number	25
$\gamma_K, \overline{\gamma}_K(r)$	set covariance function	17, 18
$\kappa(\mathbf{r}), \kappa(r)$	correlation function	75, 218
λ	intensity, intensity function	41, 51, 113, 280
Λ	intensity measure	45, 51, 112, 118, 280
$\mu^{(n)}$	moment measure	44, 47, 120
ν_d	Lebesgue measure	30
Ω	generic sample space	33
Φ	point process, fibre (surface) process, random measure	109, 280, 315, 336
φ, ϕ	realisation of Φ	108, 280, 315
$\Phi_k(K, B)$	curvature measure of K, where B is a Borel set	290
Ψ	marked point process, marked random measure	116, 283
ψ	realisation of Ψ	283
Φ_x	translated point process	42, 112
$\varrho^{(n)}$	product density	46, 47, 122
$\sum^{\neq}$	summation over all distinct pairs (distinct n-tuples)	46, 121
Ξ	random closed set	65, 206

Blackboard bold letters

Symbol	Usage	Page
$C(\mathbb{K})$	system of compact convex subsets of $\mathbb{R}^d$	11
$\mathbb{F}$	system of closed subsets of $\mathbb{R}^d$	4
$\mathbb{G}, \mathbb{G}_n, \mathbb{G}(n, p)$	random graph	402, 404
$\mathbb{K}$	system of compact subsets of $\mathbb{R}^d$	4
$\mathbb{L}_k$	set of all k-subspaces	14
$\mathbb{M}$	space of marks, set of measures	116, 166, 280
$\mathbb{N}$	system of all simple and locally finite point sequences	108
$\mathbb{R}$	real line	1
$\mathbb{R}^d$	d-dimensional Euclidean space	2
$\mathbb{S}$	extended convex ring	25
$\mathbb{W}$	space of marks	283

Script letters

Symbol	Usage	Page
$\mathcal{A}$	σ-algebra of Ω	33
$\mathcal{B}^d$	Borel σ-algebra of $\mathbb{R}^d$	28
$\mathscr{C}, \mathscr{C}^!$	Campbell measure	123, 281
$\mathcal{F}$	hitting σ-algebra	206
$\mathscr{K}$	reduced second moment measure	140
$\mathcal{M}$	σ-algebra of $\mathbb{M}$	116, 280
$\mathcal{N}$	σ-algebra of $\mathbb{N}$	109
$\mathcal{R}$	convex ring	24
$\mathscr{R}$	rose of directions	307
$\mathcal{V}(\varphi)$	Voronoi tessellation relative to φ	347
$\mathcal{W}$	σ-algebra of $\mathbb{W}$	283

Bold letters

Symbol	Usage	Page
E	expectation	33
P	probability	33
cov	covariance	34
var	variance	34
$\boldsymbol{m}$	Euclidean isometry, rigid motion	9, 112
$\boldsymbol{r}$	rotation of $\mathbb{R}^d$ around o	10

Roman letters

Symbol	Usage	Page
$\check{A}$	reflection $-A$	5
A_A	area fraction	217, 414
$A(K)$	area of K	12
$A_{\oplus r}$	$A \oplus B(o, r)$	6
$A_{\ominus r}$	$A \ominus B(o, r)$	6
A_x	translation $A + x$	5
a_V	integral range	219
$B(a, r)$	ball of radius r centred at a	3
$\bar{b}$	average breadth	12
b_k	volume of unit ball in $\mathbb{R}^k$	14
$C(\mathbf{r}), C(r)$	covariance, two-point probability function	74, 218
$C_0(\Theta), C_0$	zero cell of a tessellation	359
C^o	typical cell of a tessellation	364

Symbol	Usage	Page
$D(r)$	nearest-neighbour distance distribution	132
$g(r)$	pair correlation function of a point process	47, 141
h_k	k-dimensional Hausdorff measure	30
$H_B(r), H_d(r), H_l(r),$ $H_q(r), H_s(r)$	contact distribution function	77, 81, 82, 83, 223
$k_{mm}(r)$	mark correlation function	123
$k(\mathbf{r}), k(r)$	covariance function	160, 218, 259
$K(r)$	Ripley's K-function	51, 141
L_A, L_V	intensity of fibre process, length density	289, 308, 316, 331, 415
$l(K)$	length of K in $\mathbb{R}^1$	12
$L(K)$	perimeter, boundary length of K	12
$L(r)$	L-function of a point process, chord length distribution of a random closed set	57, 141 276, 418
M	integral of mean curvature	14
$M(L)$	mark distribution	119, 283
M_V	specific integral of mean curvature	80
N_A, N_V	specific connectivity number	80, 294
N_A^+, N_V^+	specific convexity number	80, 294
$\overline{n}_{kl}$	mean number of l-faces adjacent to the typical k-face of a tessellation	368
o	origin	3
p	volume fraction	70, 74, 216
P	distribution of point process, random set, etc.	109, 207
$P_o, P_o^!, P_x, P_x^!$	Palm distribution	129, 130, 131, 132
S^{d-1}	unit sphere in $\mathbb{R}^d$	4
$S(K)$	surface area of K	12
$s(K, u)$	support function of K	11
$s_m(K, \ell)$	modified support function of K	11
S_V	intensity of surface process, specific surface area	289, 312, 334, 415
$T_\Xi(K)$	capacity functional of Ξ, where K is a compact set	71
$V(K)$	volume of K	12
v_k	specific k^{th} intrinsic volume	79, 249
$\overline{V}_k$	mean k^{th} intrinsic volume	67, 78
$V_k(K)$	k^{th} intrinsic volume of K	15
V_V	volume fraction	217, 414
W	window of observation	55, 145, 230
$W_k(K)$	k^{th} Minkowski functional of K	15
z_α	quantile of the standard normal distribution	55, 421
$Z(x)$	Gaussian random field	260

1

Mathematical foundations

1.1 Set theory

The language of naïve set theory is ubiquitous in geometry and even more so in stochastic geometry. The reader will find a thorough introduction in specialised textbooks. The following briefly summarises notation and defines important sets and operations which will often be employed later.

A *set* is a collection of mathematical objects taken from a suitable domain of discourse. If x is an *element* of a set S this is written as $x \in S$. All sets appearing in this book are constructed from two fundamental sets, which are the set of the natural numbers $\{1, 2, \ldots\}$ and the set of the real numbers (the *real line*) $\mathbb{R} = (-\infty, \infty)$. All the constructions here are suitably regular, and the more profound aspects of mathematical logic and set theory are ignored.

The notation for the sets of natural and real numbers illustrates two useful conventions for the description of sets. The braces { } in the example above enclose a description of the set of natural numbers by (implicit, infinite) enumeration. The notation (u, v) for two real numbers, perhaps equal to $-\infty$ or $+\infty$, describes the set of all real numbers x such that $u < x < v$. This set (u, v) is an *open interval* of the real line $\mathbb{R}$. *Closed* and *half-open intervals* are given by

$$(u, v] = \{x \in \mathbb{R} : u < x \leq v\} \quad \text{(half-open)},$$
$$[u, v) = \{x \in \mathbb{R} : u \leq x < v\} \quad \text{(half-open)},$$
$$[u, v] = \{x \in \mathbb{R} : u \leq x \leq v\} \quad \text{(closed)}.$$

Here the braces { } enclose a description of a set as the collection of elements of another set satisfying some property. Contraction of this notation is often used, as for example:

$$\{x \in \mathbb{R} : x = y + z \text{ with } 0 < y < 1 \text{ and } 0 < z < 1\}$$
$$= \{x : x = y + z \quad \text{with } 0 < y < 1 \text{ and } 0 < z < 1\}$$
$$= \{y + z : 0 < y < 1 \text{ and } 0 < z < 1\}.$$

Stochastic Geometry and its Applications, Third Edition.
Sung Nok Chiu, Dietrich Stoyan, Wilfrid S. Kendall and Joseph Mecke.
© 2013 John Wiley & Sons, Ltd. Published 2013 by John Wiley & Sons, Ltd.

Note that this set is actually the open interval $(0, 2)$. Call A a *subset* of a set S (and S a *superset* of A), and write $A \subset S$, if all elements of A are also elements of S. (This book does not use the symbol $\subseteq$, thus $\subset$ includes also the case that $A = S$.) If $A, B \subset S$ for some set S then their *union*, *intersection*, and *difference* are

$$\text{union} \qquad A \cup B = \{x \in S : x \in A \text{ or } x \in B\},$$

$$\text{intersection} \qquad A \cap B = \{x \in S : x \in A \text{ and } x \in B\},$$

$$\text{difference} \qquad A \setminus B = \{x \in S : x \in A \text{ and } x \notin B\}.$$

Also define the *complement* A^c of A in S as

$$A^c = \{x \in S : x \notin A\}$$

$$= S \setminus A.$$

Notice that the definition of A^c depends on the – usually implicit – choice of superset S. The empty set $\emptyset$ is the set that contains no elements. Formally, it is

$$\emptyset = S \setminus S = A \setminus A$$

for any A.

Special collections of sets (σ-*algebras*) are considered in Section 1.9.

Two sets A and B can be used to form the *Cartesian product* $A \times B$ given by the ordered pairs (a, b), that is,

$$A \times B = \{(a, b) : a \in A \text{ and } b \in B\}.$$

More generally, the Cartesian product of n sets $A_1, \ldots, A_n$ is

$$A_1 \times \cdots \times A_n = \{(a_1, \ldots, a_n) : a_1 \in A_1, \ldots, a_n \in A_n\}.$$

An important example is given by

$$\mathbb{R}^2 = \mathbb{R} \times \mathbb{R}$$

$$= \{(x_1, x_2) : x_1, x_2 \in \mathbb{R}\}$$

which is the Cartesian plane. The higher-dimensional counterparts are

$$\mathbb{R}^3 = \mathbb{R} \times \mathbb{R} \times \mathbb{R}$$

and

$$\mathbb{R}^d = \{(x_1, \ldots, x_d) : x_1, \ldots, x_d \in \mathbb{R}\}.$$

The spaces $\mathbb{R}^2$ and $\mathbb{R}^3$ are often referred to as the *plane* and *space*, respectively, and $\mathbb{R}^d$ as the *d-dimensional space*. Because of additional structures such as topology (see Section 1.2) and linearity (Section 1.3), the term *Euclidean space* is used. An element $x = (x_1, x_2, \ldots, x_d) \in \mathbb{R}^d$ is usually referred to as a *point* in geometry. However, in stochastic geometry a distinction must be drawn. The study of stochastic geometry frequently concerns random collections of points, referred to as *point processes*. It is convenient to refer to members of such processes as *points of the process*, or simply *points*. Therefore points that are merely locations in $\mathbb{R}^d$ with no membership presumed of some special collection of points are referred to as *location points* or *locations*. A particular case of a location point is the *origin*

$$o = (0, \ldots, 0).$$

Here this book makes a notational difference between a real number (0) and a point of $\mathbb{R}^d$, while otherwise normal italic letters are used both for points and numbers, with one exception:

> r is always a real number and $\mathbf{r}$ a point of $\mathbb{R}^d$ of distance r from the origin o.

1.2 Topology in Euclidean spaces

A concept of distance is associated with Euclidean spaces. The *Euclidean metric* measures the distance between two location points $x = (x_1, x_2, \ldots, x_d)$ and $y = (y_1, y_2, \ldots, y_d)$ as

$$\|x - y\| = \sqrt{(x_1 - y_1)^2 + \cdots + (x_d - y_d)^2}. \tag{1.1}$$

This distance is used to define the *closed ball* $B(a, r)$ of *centre a* and of *radius r*

$$B(a, r) = \{x \in \mathbb{R}^d : \|x - a\| \leq r\}. \tag{1.2}$$

Here r is a positive real number and a is a location point. The *open ball* $B^{\text{int}}(a, r)$ is defined similarly but with strict rather than weak inequality, that is,

$$B^{\text{int}}(a, r) = \{x \in \mathbb{R}^d : \|x - a\| < r\}.$$

This means that the boundary of the ball, that is, the *sphere*, does not belong to the set.

The metric (or equivalently the closed and open balls) can be used to define special properties that might be possessed by subsets A of Euclidean space. The set A is said to be *bounded* if there is a ball $B(a, r)$ such that

$$A \subset B(a, r).$$

A sequence $x_1, x_2, \ldots$ is said to *converge* to x if

$$\lim_{n \to \infty} \|x_n - x\| = 0.$$

A set A is said to be *open* if for each $x \in A$ a positive number ε can be found (depending on x) such that $B(x, \varepsilon) \subset A$. The system of open sets of $\mathbb{R}^d$ is denoted by $\mathcal{O}$. (In this and other similar notations the exponent 'd' is omitted whenever the actual dimension is clear in the context.) Examples of open sets in the one-dimensional case of $d = 1$ are the open intervals (u, v). In the general case of $\mathbb{R}^d$ examples include the open balls $B^{\text{int}}(a, r)$ described above, and the open hypercubes

$$(u_1, v_1) \times \cdots \times (u_d, v_d).$$

Finite intersections and arbitrary unions of open sets are again open.

A set A is said to be *closed* if its complement A^c in $\mathbb{R}^d$ is open. The system of closed sets of $\mathbb{R}^d$ is denoted by $\mathbb{F}$. Examples of closed sets in the case $d = 1$ are the closed intervals $[a, b]$. In the general case of $\mathbb{R}^d$ examples include the closed balls $B(a, r)$ and the closed hypercubes

$$[u_1, v_1] \times \cdots \times [u_d, v_d]$$

and the hyperplanes

$$\left\{ x = (x_1, \ldots, x_d) \in \mathbb{R}^d : \sum_{i=1}^{d} a_i x_i = b \right\}$$

for some constants b and $a_1, \ldots, a_d$ with $a_1, \ldots, a_d$ not all zero. Finite unions of closed sets and arbitrary intersections of closed sets are again closed.

The *interior* A^{int} of a general set A is the union of all open sets contained in A. The *closure* A^{cl} of A is the intersection of all closed sets containing A. Thus A^{int} is the largest open set contained in A, while A^{cl} is the smallest closed set containing A. Hence

$$A^{\mathrm{int}} \subset A \subset A^{\mathrm{cl}}.$$

Moreover $A^{\mathrm{int}} = \left((A^c)^{\mathrm{cl}}\right)^c$. A set A is open precisely when $A^{\mathrm{int}} = A$, and closed precisely when $A^{\mathrm{cl}} = A$. If $A = (A^{\mathrm{int}})^{\mathrm{cl}}$ then A is said to be *regular closed*.

The difference $\partial A = A^{\mathrm{cl}} \setminus A^{\mathrm{int}}$ is the *boundary* of A. An important example is the *sphere* of centre a and radius r, which is the boundary of $B(a, r)$ and is given by

$$\partial B(a, r) = \{x \in \mathbb{R}^d : \|a - x\| = r\}.$$

The particular case $\partial B(o, 1)$ is the *unit sphere* of $\mathbb{R}^d$, and is denoted by S^{d-1}.

A set $K \subset \mathbb{R}^d$ is said to be *compact* if it is both closed and bounded. The system of all compact subsets of $\mathbb{R}^d$ is denoted by $\mathbb{K}$. Moreover denote the system of non-empty compact sets by

$$\mathbb{K}' = \mathbb{K} \setminus \{\emptyset\}.$$

Examples of compact sets include the closed balls and the closed hypercubes.

The *distance between a point x and a closed set F* is denoted by $d(x, F)$ and defined as

$$d(x, F) = \inf\{\|x - y\| : y \in F\}.$$

In image analysis one speaks of the *Euclidean distance function* (or *transform*) which assigns to each $x \in F^c$ the value $d(x, F)$. The *signed* Euclidean distance function is defined for all $x \in \mathbb{R}^d$ and takes for $x \in F$ the value $-d(x, \partial F)$. Ohser and Schladitz (2009) discuss algorithms for efficient calculations of the Euclidean distance function and give examples for its use.

1.3 Operations on subsets of Euclidean space

Euclidean space is a *vector space* since it allows the vector space operations of

$$\text{addition} \qquad x + y = (x_1 + y_1, \ldots, x_d + y_d),$$
$$\text{scalar multiplication} \qquad c \cdot x = cx = (cx_1, \ldots, cx_d)$$

for location points $x = (x_1, \ldots, x_d)$ and $y = (y_1, \ldots, y_d)$ in $\mathbb{R}^d$ and real numbers c. This vector structure allows for the definition of set operations special to Euclidean space as follows:

Multiplication by real numbers

$$cA = \{c \cdot x : x \in A\}$$

for real numbers c and $A \subset \mathbb{R}^d$. The case $c = -1$ leads to the particular case of *reflection*

$$\check{A} = -A = \{-x : x \in A\} \qquad \text{for } A \subset \mathbb{R}^d.$$

If $A = \check{A}$ then A is said to be *symmetric*.

Translation

$$A_x = A + x = \{y + x : y \in A\} \qquad \text{for } x \in \mathbb{R}^d \text{ and } A \subset \mathbb{R}^d.$$

Minkowski-addition

$$A \oplus B = \{x + y : x \in A, \ y \in B\} \qquad \text{for } A, B \subset \mathbb{R}^d. \tag{1.3}$$

Applying to a set A the operation of Minkowski-addition by a set B enlarges, translates, and deforms the set A; see Figure 1.1 on p. 9. Minkowski-addition is both associative and commutative. Other properties are summarised in the following formulae:

$$A_x = A \oplus \{x\}, \tag{1.4}$$

$$A \oplus B = \bigcup_{y \in B} A_y = \bigcup_{x \in A} B_x, \tag{1.5}$$

$$A \oplus B = \{x : A \cap (\check{B})_x \neq \emptyset\}, \tag{1.6}$$

$$A \oplus (B_1 \cup B_2) = (A \oplus B_1) \cup (A \oplus B_2). \tag{1.7}$$

If $A_1 \subset A_2$ then $A_1 \oplus B \subset A_2 \oplus B$. Formula (1.5) shows that Minkowski-addition can be represented as the union of the translates B_x as x runs through A. In the special case of $B = B^{\text{int}}(o, r)$ the Minkowski-sum $A \oplus B$ is the union of all location points that are of distance smaller than r from A.

Minkowski-subtraction

$$A \ominus B = \bigcap_{y \in B} A_y. \tag{1.8}$$

Equivalent forms are

$$A \ominus B = (A^c \oplus \check{B})^c \qquad (1.9)$$

and

$$A \ominus B = \{x : (\check{B})_x \subset A\}, \qquad (1.10)$$

where the complement is with respect to the superset $\mathbb{R}^d$. From the definition it is easy to show that

$$A_x \ominus B = A \ominus B_x \qquad \text{for all } x \in \mathbb{R}^d. \qquad (1.11)$$

In general, the Minkowski-subtraction is not an inverse for Minkowski-addition, but the following relationship holds

$$(A \ominus \check{B}) \oplus B \subset A \subset (A \oplus \check{B}) \ominus B.$$

Note that $\{o\} \subset B \ominus \check{B}$, and equality holds if B is bounded.

If B is the ball $B(o, r)$ then $A \ominus B$ is the union of all location points lying within A and such that balls centred at them and of radius r are completely contained in A.

Sometimes the notation

$$A_{\oplus r} = A \oplus B(o, r)$$

and

$$A_{\ominus r} = A \ominus B(o, r)$$

is used, where r is a positive number.

The family $\mathbb{K}'$ of non-empty compact subsets of $\mathbb{R}^d$ can be made into a metric space by using the *Hausdorff metric*:

$$\delta(K_1, K_2) = \inf\{r : K_1 \subset K_2 \oplus B(o, r) \text{ and } K_2 \subset K_1 \oplus B(o, r)\} \qquad \text{for } K_1, K_2 \in \mathbb{K}'.$$

This δ defines distances between compact sets which are positive even when the sets have a non-empty intersection. It is easy to see that δ is compatible with the Euclidean metric in the sense that

$$\delta(\{x_1\}, \{x_2\}) = \|x_1 - x_2\| \qquad \text{for } x_1, x_2 \in \mathbb{R}^d.$$

And the distance between a point x and a compact K as defined in Section 1.2 satisfies

$$d(x, K) = \delta(\{x\}, K).$$

Furthermore, the family $\mathbb{K}'$ equipped with the metric δ is *complete*: if $K_1, K_2, \ldots \in \mathbb{K}'$ with $\delta(K_m, K_n) \to 0$ as $m, n \to \infty$ then there is a $K_\infty \in \mathbb{K}'$ with $\delta(K_m, K_\infty) \to 0$ as $m \to \infty$. The metric is also *countably separated*: there is a sequence $K_1, K_2, \ldots \in \mathbb{K}'$ such that for any $K \in \mathbb{K}'$ and any $\varepsilon > 0$ there is a K_n with $\delta(K_n, K) < \varepsilon$. (A possible choice of such a sequence $K_1, K_2, \ldots$ are the finite subsets of $\mathbb{Q}^d$, where $\mathbb{Q}$ denotes the set of rational numbers. These finite subsets can be enumerated, as every mathematician knows.) So $\mathbb{K}'$ is a *complete separable metric space* when endowed with the Hausdorff metric: that is to say, it is a *Polish space*.

1.4 Mathematical morphology and image analysis

In many fields of technology and science there is a great need for methods of analysis for large quantities of data in the form of *images*. Examples are satellite photographs, geological maps, microscope images of sections of metals, minerals, cellular tissue, and data coming from computerised tomography. The sheer quantity of data requires the use of automatic and quantitative methods. This section describes briefly some ideas of mathematical morphology applied in this context. In particular, the reader will get an idea how statistical procedures on random sets (such as described throughout the rest of this book) can be performed automatically.

Technical equipment yields image data which usually have the form of two- or three-dimensional arrays of *pixels* with grey values. Such a greyscale image can be reduced to a binary image by operations like thresholding, in which grey-tone values lower than a chosen threshold are set to white, and the others to black. The following discussion considers only such binary images, which correspond to random sets, where for example the black pixels stand for some set and the white for its complement.

Once the image is reduced to such an array of pixels then it is possible (with suitable equipment and software) to determine important image characteristics automatically. These quantities are represented by numbers of pixels in specific subsets of the image. Examples are the areas or volumes of the white and black parts of the image, the length or area content of the boundary between these parts, and the number of components of the black part. If one conceives of the image as a realisation of a random structure then these measurements lead to statistical estimates of various characteristics of the random structure. For example, the area fraction p is estimated by the proportion of area covered by black pixels (see Sections 3.4 and 6.4.2), while the specific length of boundary is estimated by the proportion of length of boundary between black and white to unit area in the image (see Sections 7.3 and 8.3).

However, more is possible. A powerful idea is to subject a given image to repeated transformations and to perform measurements on the transformed images (Serra, 1982). On the one hand, automatic operations can be applied to the image to free it from the inevitable image defects and artifacts of the image-processing procedure. This is often done as a preliminary step before visual inspection. On the other hand, important functions can be measured quickly and easily. Examples of these functions are the set covariance, chord length distribution function and contact distribution functions, which will be explained in Section 6.3.

Important image transformations have their origin in *mathematical morphology*, as introduced by Matheron and Serra. Many of its transformations employ 'structuring elements'. This theory is a direct application of the Euclidean set operations discussed in the previous section and it is a useful technical aid in the analysis of images. For thorough discussions, see Heijmans (1994), Goutsias and Heijmans (2000), Serra (1982) and Soille (1999), and for the three-dimensional case, see Ohser and Schladitz (2009).

In the following A will denote the set which is of interest, also called the *image*: in practical applications this is often the set of black pixels. The structuring element B will often be a ball, a disc, a line segment, or a two-point set. In practice, when images are pixel sets, balls or discs are approximated by also pixel sets and the set operations are approximated by operations on lattices, see for example Goutsias and Heijmans (2000). However, the following description follows Euclidean geometry.

The free R package `spatstat` can perform these operations when B is a disc, whilst the commercial `Image Processing Toolbox` of MATLAB allows users to create and manipulate the structuring element.

Dilation

This is the operation

$$A \mapsto A \oplus \check{B}. \tag{1.12}$$

The set A is enlarged (but not scaled) and, at least in the case where the structuring element B is a ball, smoothed. In particular the action of dilation fills in cavities, repairs fissures, and joins together a fragmented image. If A is a realisation of a random closed set, then dilation by rB followed by measurement of the area of $A \oplus r\check{B}$ allows estimation of the contact distribution function $H_B(r)$ of the random set (see Sections 6.3.3 and 6.4.5).

Erosion

This is in some sense dual to the operation of dilation, and is given by

$$A \mapsto A \ominus \check{B}. \tag{1.13}$$

Erosion shrinks the set A, tending to produce smaller fragments, even separating connected sets into several subsets. This can be helpful in the estimation of the number of particles composing an image, where some particles are in contact. The erosion operation is of great importance for the quantitative estimation of various summary characteristics of random sets. For example, if one takes for the structuring element B a two-point set $\{x, y\}$ with $\|x - y\| = r$, then the normalised area of the eroded set $A \ominus \check{B}$ is an estimate for the set covariance $C(r)$ (see Section 6.3.2).

If A is a union of non-overlapping discs, then an estimate of the diameter distribution can be obtained by means of successive erosions of A by discs $B(o, r_1)$, $B(o, r_2)$, ... with $r_1 < r_2 < \cdots$, and counting the numbers of components of the eroded A.

Ohser and Schladitz (2009, pp. 86–7) show how the coordination number of sinter particles can be determined by erosion.

Opening

The opening of A by B can be viewed as an attempt to reverse an erosion by a dilation. It is given by

$$A \mapsto A \circ B = (A \ominus \check{B}) \oplus B. \tag{1.14}$$

The opening $A \circ B$ of a set A by B has an appearance similar to that of the original set A, but is built only on the portions of the image that survive the initial erosion. Thus $A \circ B \subset A$; small disconnected fragments of the image disappear under opening and this is useful in systematically eliminating possible image defects or noise.

A set A is *(morphologically) B-open* if $A = A \circ B$. For example, in the plane a union of discs with radii larger than or equal to r is $B(o, r)$-open.

Closing

This is dual to opening and can be viewed as an attempt to reverse a dilation by an erosion. It is given by

$$A \mapsto A \bullet B = (A \oplus \check{B}) \ominus B. \tag{1.15}$$

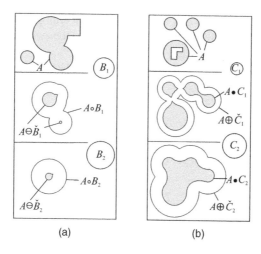

(a)   (b)

Figure 1.1   (a) The operations of erosion and opening with discs applied to a planar set A. Components that overlap are separated while small components and roughnesses vanish or are reduced. (b) The operations of dilation and closing applied to a set. Gaps are closed up, concavities vanish or are reduced, and clusters of small particles are merged.

The same as the opening $A \circ B$, the closing $A \bullet B$ bears an approximate resemblance to A, but now $A \subset A \bullet B$. As in the case of opening, the closing operation is useful in cleaning up an image. The action of closing tends to close up small holes, to join up close but separated subsets, and to smooth out the boundaries of an image. Opening and closing are simple examples of morphological filtration operations.

A set A is *(morphologically) B-closed* if $A = A \bullet B$. For example, in the plane a union of non-intersecting discs with radii smaller than r and with inter-centre distance larger than $4r$ is $B(o, r)$-open.

Figure 1.1 displays typical results of the application of these transformations. The example discussed there shows how repeated application of the transformations of mathematical morphology with perhaps different structuring elements leads to composite image transformations; see Ohser and Schladitz (2009, p. 88) for a three-dimensional example.

1.5 Euclidean isometries

A transformation $x \mapsto mx$ is said to be a Euclidean *isometry* if it leaves invariant the distance between points x and y for all x and y. That is to say,

$$\|x - y\| = \|mx - my\| \qquad \text{for all } x, y \in \mathbb{R}^d.$$

It can be shown that every isometry of Euclidean space can be represented in the form

$$mx_k = b_k + \sum_{l=1}^{d} a_{kl} \cdot x_l \tag{1.16}$$

for $k = 1, \ldots, d$, $b = (b_1, \ldots, b_d) \in \mathbb{R}^d$, and $\mathbf{A} = (a_{kl})$ an orthogonal matrix. Such a matrix has the determinant $\det \mathbf{A} = \pm 1$. The isometry is said to be a *proper isometry* if $\det \mathbf{A} = +1$. So reflections are *not* proper isometries.

If $\mathbf{A}$ is the unit matrix then the isometry is said to be a *translation* and sometimes denoted by

$$T_b x = x - b. \tag{1.17}$$

The set A_x in (1.4) is the *translated set* of A and can be written as

$$T_{-x} A = A_x.$$

Note the composition formula

$$T_x T_y = T_{x+y}. \tag{1.18}$$

Rotations about the origin are proper isometries given by (1.16) for which $b = o$ and are denoted by rx or $\mathbf{A}x$, where $\mathbf{A}$ is an orthogonal matrix with $\det \mathbf{A} = 1$. A further composition formula is

$$r T_x y = T_{rx} r y. \tag{1.19}$$

Formula (1.16) makes it plain that every proper isometry is the composition of a rotation about the origin with a translation.

An alternative term for a proper isometry is *(rigid) motion*.

1.6 Convex sets in Euclidean spaces

A subset K of $\mathbb{R}^d$ is called *convex* if for every pair of points x, y in K the intervening line segment also lies in K. That is to say

$$cx + (1 - c)y \in K \qquad \text{whenever } x, y \in K \text{ and } 0 \leq c \leq 1.$$

Important examples of convex sets include the *affine linear subspaces*, which are subsets L of $\mathbb{R}^d$ with the property that for every x, y in L the whole line $\{cx + (1 - c)y : c \in \mathbb{R}\}$ through x and y also lies in L. The origin o does not necessarily belong to such an L.

The *dimension* of an affine linear subspace is the smallest integer k such that L can be given by the formula

$$L = \left\{ c_1 z_1 + \cdots + c_{k+1} z_{k+1} : c_1, \ldots, c_{k+1} \in \mathbb{R}, \sum_{i=1}^{k+1} c_i = 1 \right\}$$

for some points $z_1, \ldots, z_{k+1}$ in $\mathbb{R}^d$. Affine linear subspaces of dimension k are called *k-flats* or *k-planes*; if a *k-flat* contains o then it is called a *k-subspace*. The $(d-1)$-flats are called *hyperflats* or *hyperplanes*. The 1-flats are called *lines*. In the spatial case $d = 3$ the term *2-flat* is abbreviated to *flat* and the term *2-plane* to *plane*. Note that the hyperflats of the plane $\mathbb{R}^2$ are its lines.

The *k-flats* are closed sets but are not bounded and therefore not compact. Convex sets which are also compact are sometimes called *convex bodies*. Examples are the closed balls, the closed and bounded hypercubes, and the closed discs (intersections of closed balls with 2-flats). The system of all convex bodies is denoted by $C(\mathbb{K})$. In the case $d = 1$ the system $C(\mathbb{K})$ coincides with the system of all closed and bounded intervals. Open balls and open hypercubes are not convex bodies, though they are convex. The sphere

$$\partial B(o, r) = \{x \in \mathbb{R}^d : \|x - a\| = r\}$$

and the torus are of course not convex.

The smallest convex set which contains a given set A is called the *convex hull* of A and denoted by conv A. For example, the convex hull of a sphere is the corresponding closed ball.

An important functional characteristic for a convex body K is the *support function* defined by

$$s(K, u) = \max_{x \in K} \langle u, x \rangle \qquad \text{for } u \in \mathbb{R}^d, \tag{1.20}$$

where $\langle u, x \rangle = u_1 x_1 + \cdots + u_d x_d$ is the scalar product of $u = (u_1, \ldots, u_d)$ and $x = (x_1, \ldots, x_d)$. The support function is convex and positively homogeneous (i.e., $s(K, \alpha u) = \alpha \cdot s(K, u)$ for all $\alpha > 0$). It determines K uniquely.

For $u \in S^{d-1}$, $s(K, u)$ is the signed distance from the origin of the support hyperplane to K with exterior normal vector u; the distance is negative if and only if u points into the open half space containing the origin o. The function $s(K, \cdot)$ is completely determined by its values on S^{d-1} because of positive homogeneity. Therefore in this book sometimes $s(K, u)$ is considered as a function on S^{d-1}, given by (1.20) with $\mathbb{R}^d$ replaced by S^{d-1}. See Schneider (1993, Section 1.7) for more information on the support function.

If K is symmetric (so $\check{K} = K$), then the support function is uniquely determined by its values on one hemisphere of the sphere S^{d-1}. In this case the *modified support function* $s_m(K, \cdot)$ is defined on the set L_1 of all lines through the origin. For $\ell \in L_1$ let $e(\ell)$ be the point on $\ell \cap S^{d-1}$ in the upper hemisphere ($x_d \geq 0$). Then

$$s_m(K, \ell) = s(K, e(\ell)) \qquad \text{for } \ell \in L_1. \tag{1.21}$$

In the planar case ($d = 2$) the line ℓ is uniquely given by the angle α formed by ℓ and the x_1-axis in the upper half-plane, and hence $s_m(K, \cdot)$ becomes a function defined on $(0, \pi]$.

Some set operations preserve the class $C(\mathbb{K})$. In particular if K_1, K_2 belong to $C(\mathbb{K})$ then so do the sets $c \cdot K_1$ for real c, $\check{K}_1$, $K_1 \cap K_2$ and $K_1 \oplus K_2$.

A *convex body functional* $h(K)$, defined on $C(\mathbb{K})$, assigns a real value $h(K)$ to each $K \in C(\mathbb{K})$. Of particular interest are those nonnegative convex body functionals which possess the following properties:

isometry-invariance $h(mK) = h(K)$ if $K \in C(\mathbb{K})$ and m is an isometry,

monotonicity   if $K_1 \subset K_2$ then $h(K_1) \leq h(K_2)$,

C-additivity $h(K_1) + h(K_2) = h(K_1 \cup K_2) + h(K_1 \cap K_2)$

     for $K_1, K_2 \in C(\mathbb{K})$, if $K_1 \cup K_2 \in C(\mathbb{K})$.

Important examples of convex body functionals for the cases $d = 1, 2, 3$ are

the *length* $l(K)$ if $d = 1$,

the *area* $A(K)$ if $d = 2$,

the *volume* $V(K)$ if $d = 3$,

the *boundary length* $L(K)$ if $d = 2$,

the *surface area* $S(K)$ if $d = 3$.

In the case $d = 2$, if K is a line segment then $L(K)$ is defined as *twice* the length of K. Likewise in the case $d = 3$ if K is actually a subset of a flat then $S(K)$ is *twice* the area of K.

The *parallel set* of distance r of a set $A \subset \mathbb{R}^d$ is the set $A_{\oplus r} = A \oplus B(o, r)$. The operation of taking a parallel set preserves the properties of convexity, of compactness, and of being a ball.

Expressing the length ($d = 1$), area ($d = 2$) and volume ($d = 3$) of the parallel set as functions of the distance r is of particular interest. For the case $d = 1$ this is given simply by

$$l\big(K \oplus B(o, r)\big) = l(K) + 2r. \tag{1.22}$$

In the cases $d = 2$ and $d = 3$ the *Steiner formula* holds

$$A\big(K \oplus B(o, r)\big) = A(K) + L(K)\, r + \pi r^2, \tag{1.23}$$

$$V\big(K \oplus B(o, r)\big) = V(K) + S(K)\, r + 2\pi \bar{b}(K) r^2 + \frac{4\pi r^3}{3}. \tag{1.24}$$

Here $\bar{b}(K)$ is yet another convex body functional, called the *average breadth* or *average width*.

The average breadth can be defined as follows. For each line ℓ through the origin let $b_\ell(K)$ be the least distance between two parallel hyperplanes perpendicular to ℓ and enclosing K entirely between them. Then $\bar{b}(K)$ is defined to be the mean value of $b_\ell(K)$ averaging over all lines ℓ through the origin using the uniform direction distribution. This can be given in an explicit formula as

$$\bar{b}(K) = \frac{1}{2\pi} \int_0^{\pi/2} \int_0^{2\pi} l(K|_{S_{\beta,\lambda}}) \sin \beta \, d\lambda \, d\beta, \tag{1.25}$$

where $l(K|_{S_{\beta,\lambda}})$ is the length of the orthogonal projection of K on $S_{\beta,\lambda}$, the line which passes through the origin and through the point $(\sin \beta \cos \lambda, \sin \beta \sin \lambda, \cos \beta)$.

In the special case of polyhedra the following formula holds (see Santaló, 1976, p. 226):

$$\bar{b}(K) = \frac{1}{4\pi} \sum_i l_i \alpha_i,$$
(1.26)

where l_i is the length of the ith edge and α_i is the angle between the normals of the faces which meet at the ith side, where $0 < \alpha_i \leq \pi$.

Table 1.1 displays average breadths, together with volumes and surface areas, for various convex bodies $K \subset \mathbb{R}^3$. Other formulae for average breadths can be found in Hadwiger (1957, p. 215) and Santaló (1976, pp. 226, 229, 230).

Table 1.1 Volumes, surface areas, and average breadths for convex bodies K in $\mathbb{R}^3$.

K	V	S	$\bar{b}$
Ball of radius r	$\frac{4}{3}\pi r^3$	$4\pi r^2$	$2r$
Cylinder, radius r, height h	$\pi r^2 h$	$2\pi r(r+h)$	$\frac{h+\pi r}{2}$
Disc, radius r	0	$2\pi r^2$	$\frac{\pi r}{2}$
Square plate, side a	0	$2a^2$	a
Convex flat (a planar convex subset in $\mathbb{R}^3$), area A and perimeter L	0	$2A$	$\frac{L}{4}$
Segment, length l	0	0	$\frac{l}{2}$
Spheroid, equator radius a, half axes of meridian ellipse a and λa, $\alpha = \sqrt{1-\lambda^2}$, $\beta = \sqrt{\lambda^2-1}$	$\frac{4}{3}\pi\lambda a^3$	$2\pi\lambda a^2(\frac{1}{\lambda}+\frac{\lambda}{\beta}\sin^{-1}\frac{\beta}{\lambda})$, if $\lambda>1$ $2\pi\lambda a^2(\frac{1}{\lambda}-\frac{\lambda}{\alpha}\ln\frac{1-\alpha}{\lambda})$, if $\lambda<1$	$a\{\lambda+\frac{1}{\beta}\ln(\beta+\lambda)\}$, if $\lambda>1$ $a(\lambda+\frac{1}{\alpha}\sin^{-1}\alpha)$, if $\lambda<1$
Cube, side a	a^3	$6a^2$	$\frac{3}{2}a$
Rectangular parallelepiped, sides a, b, c	abc	$2(ab+bc+ca)$	$\frac{a+b+c}{2}$
Tetrahedron, side a	$\frac{\sqrt{2}}{12}a^3$	$\sqrt{3}a^2$	$\frac{3a}{2\pi}\cos^{-1}(-\frac{1}{3})$
Octahedron, side a	$\frac{\sqrt{2}}{3}a^3$	$2\sqrt{3}a^2$	$\frac{3a}{\pi}\cos^{-1}\frac{1}{3}$

The average breadth $\bar{b}$ is closely related to M, the integral of mean curvature, by

$$\bar{b} = \frac{M}{2\pi}. \tag{1.27}$$

The *integral of mean curvature* M is the surface integral over ∂K, the boundary of K, of the mean of the two principal curvatures $1/r_1(x)$ and $1/r_2(x)$ of the surface

$$M = \int_{\partial K} m(x)\,dS,$$

where

$$m(x) = \frac{1}{2}\left(\frac{1}{r_1(x)} + \frac{1}{r_2(x)}\right)$$

is the mean curvature of ∂K in the surface point x and dS is the surface element. For M to be well-defined by this integral the surface ∂K must satisfy suitable regularity conditions, though of course the average breadth makes sense for any convex body.

Formulae (1.23) and (1.24) given above can be generalised to the case of dimension d. Suppose K is in $C(\mathbb{K})$. Then the volume $v_d\big(K \oplus B(o, r)\big)$ of a parallel body for K is given by the *Steiner formula*:

$$v_d\big(K \oplus B(o, r)\big) = \sum_{k=0}^{d} \binom{d}{k} W_k(K)r^k. \tag{1.28}$$

This formula introduces the important *quermassintegrals* or *Minkowski functionals* $W_k(K)$. They are isometry-invariant, monotone, C-additive convex body functionals, defined directly by the formula

$$W_k(K) = \frac{b_d}{b_{d-k}} \int_{\mathbb{L}_k} v_{d-k}(K|_{E^\perp})U_k(dE), \tag{1.29}$$

in which

$\quad v_k \quad$ is the k-dimensional Lebesgue measure (see p. 30),

$\quad \mathbb{L}_k \quad$ is the set of all k-subspaces,

$\quad K|_{E^\perp} \quad$ is the orthogonal projection of K on $E^\perp$,

$\quad E^\perp \quad$ is the $(d-k)$-subspace orthogonal to $E \in \mathbb{L}_k$,

$\quad U_k \quad$ is the uniform probability distribution on $\mathbb{L}_k$.

Furthermore

$$b_k = \frac{\sqrt{\pi^k}}{\Gamma(1 + k/2)} \tag{1.30}$$

is the volume of the unit ball in $\mathbb{R}^k$. Important special cases are

$$b_0 = 1, \qquad b_1 = 2, \qquad b_2 = \pi, \qquad b_3 = 4\pi/3.$$

If K is the unit ball, then

$$W_k\big(B(o, 1)\big) = b_d \qquad \text{for } k = 0, 1, \ldots, d. \tag{1.31}$$

For a general convex body K, the quantity $W_0(K)$ is equal to the volume $v_d(K)$; the quantity $dW_1(K)$ is the $(d-1)$-*dimensional measure of area of the boundary* ∂K; $W_d(K)$ is constant (independent of K) and equal to b_d, and $(2/b_d)W_{d-1}(K)$ is the average breadth. In the particular cases of $d = 1, 2, 3$ this gives

$$d = 1: \ W_0(K) = l(K), \quad W_1(K) = 2, \tag{1.32}$$

$$d = 2: \ W_0(K) = A(K), \quad W_1(K) = \frac{L(K)}{2}, \quad W_2(K) = \pi, \tag{1.33}$$

$$d = 3: \ W_0(K) = V(K), \quad W_1(K) = \frac{S(K)}{3},$$
$$W_2(K) = \frac{2\pi}{3}\bar{b}(K) = \frac{M(K)}{3}, \quad W_3(K) = \frac{4\pi}{3}. \tag{1.34}$$

The Minkowski functionals W_k are closely related to the so-called *intrinsic volumes* V_k:

$$b_{d-k}V_k(K) = \binom{d}{k} W_{d-k}(K) \qquad \text{for } k = 0, 1, \ldots, d. \tag{1.35}$$

Which functional should be used is merely a matter of mathematical convention. The V_k can be considered more natural and depend only on K but not the dimension of its surrounding space (Schneider and Weil, 2008, p. 600). Therefore this book prefers the V_k. Substituting W_0, W_1, W_d and W_{d-1} into Formula (1.35) gives, for example,

$$V_0(K) = 1, \tag{1.36}$$

$$V_{d-1}(K) = \frac{1}{2}S_d(K), \tag{1.37}$$

$$V_d(K) = v_d(K), \tag{1.38}$$

where $S_d(K)$ is the $(d-1)$-dimensional area of ∂K. The relationships corresponding to Formulae (1.28) and (1.31)–(1.34), respectively, are

$$v_d\big(K \oplus B(o, r)\big) = \sum_{k=0}^{d} b_{d-k} V_k(K) r^{d-k}, \qquad \text{(Steiner formula)} \tag{1.39}$$

$$V_k\big(B(o, 1)\big) = \binom{d}{k} \frac{b_d}{b_{d-k}} \qquad \text{for } k = 0, 1, \ldots, d, \tag{1.40}$$

$$d = 1: \ V_0(K) = 1, \quad V_1(K) = l(K), \tag{1.41}$$

$$d = 2: \ V_0(K) = 1, \quad V_1(K) = \frac{L(K)}{2}, \quad V_2(K) = A(K), \tag{1.42}$$

$$d = 3: \quad V_0(K) = 1, \quad V_1(K) = 2\bar{b}(K) = \frac{M(K)}{\pi},$$

$$V_2(K) = \frac{S(K)}{2}, \quad V_3(K) = V(K). \tag{1.43}$$

Hadwiger's characterisation theorem states that every nonnegative, motion-invariant, monotone, C-additive convex body functional h can be written in the form

$$h(K) = \sum_{k=0}^{d} a_k V_k(K), \tag{1.44}$$

where the a_k are nonnegative constants depending on h. Here 'nonnegative monotone' can be replaced by 'continuous' (using Hausdorff metric) if the a_k are allowed to be general real constants. For a modern proof see Schneider and Weil (2008, pp. 628–30), the idea of which can be traced back to Klain (1995).

The volume itself is an intrinsic volume as noted above. Formula (1.39) can be generalised to apply not only to the volume $v_d = V_d$ but to the other intrinsic volumes. The generalised Steiner formula is

$$V_k\big(K \oplus B(o, r)\big) = \sum_{j=0}^{k} \binom{d-j}{d-k} \frac{b_{d-j}}{b_{d-k}} V_j(K) r^{k-j} \qquad \text{for } k = 0, 1, \ldots, d. \tag{1.45}$$

In particular, Formula (1.45) gives

$$d = 2: \qquad L\big(K \oplus B(o, r)\big) = L(K) + 2\pi r \tag{1.46}$$

$$d = 3: \qquad S\big(K \oplus B(o, r)\big) = S(K) + 4\pi \bar{b}(K)\, r + 4\pi r^2, \tag{1.47}$$

$$\bar{b}\big(K \oplus B(o, r)\big) = \bar{b}(K) + 2r. \tag{1.48}$$

Matheron (1978), Miles (1974b) and Weil (1982b) give similar formulae for $V\big(K \ominus B(o, r)\big)$ under suitable smoothness conditions on ∂K.

The intrinsic volumes for intersections of convex bodies with flats satisfy the *Crofton formula*

$$\int_{\mathbb{L}_k} \int_{E^\perp} V_j(K \cap E_x) v_{d-k}(\mathrm{d}x) U_k(\mathrm{d}E) = \frac{\binom{k}{k-j} \frac{b_k}{b_j}}{\binom{d}{k-j} \frac{b_d}{b_{d-k+j}}} V_{d-k+j}(K) \tag{1.49}$$

for $0 \leq j \leq k \leq d - 1$ and $K \in C(\mathbb{K})$. (Recall that E_x is the translate of E by x.) The Lebesgue measure v_{d-k} on $E^\perp$ and the intrinsic volume V_j on E_x are defined by identifying $E^\perp$ and E_x with $\mathbb{R}^{d-k}$ and $\mathbb{R}^k$, respectively.

If $j = 0$ then (1.49) gives

$$V_k(K) = \frac{\binom{d}{d-k} \frac{b_d}{b_k}}{b_{d-k}} \int_{\mathbb{L}_{d-k}} v_k(K|_{E^\perp}) U_{d-k}(\mathrm{d}E), \tag{1.50}$$

which, after substituting into (1.35), is equivalent to Formula (1.29). The intrinsic volumes of intersections of a given convex body with another moved convex body are given by the so-called *principal kinematic formula* (see Schneider and Weil, 2008, Theorems 5.1.3 and 5.1.5).

The results of this and Section 1.8 belong to the fields of *convex geometry* and *integral geometry*. References to these branches of mathematics are Gruber and Wills (1993), Schneider (1993), Klain and Rota (1997) and Schneider and Weil (2008). Modern integral geometry studies generalisations of intrinsic volumes known as *curvature measures*; see Schneider (1993), Schneider and Weil (2008) and Section 7.3.4.

1.7 Functions describing convex sets

1.7.1 General

This section presents some functions that describe size and shape of deterministic convex bodies in $\mathbb{R}^d$, additionally to the support function already mentioned in Section 1.6. This includes the set covariance, the chord length probability density function and the erosion–dilation functions. All these functions serve as descriptors of sets, for example in statistical analysis, where functional data are simpler to handle than set-valued data. They also appear in formulae for other characteristics or can be directly obtained by statistical analysis.

All these functions are independent of the positions and orientations of the sets in space, thus they coincide for congruent sets; it is not necessary to define 'centres' in the sets. That is why these functions are also well suited to shape statistics; see Stoyan and Stoyan (1994). However, they do not uniquely characterise the corresponding convex bodies, that is, there may be different sets with the same function.

In the planar case, both the erosion–dilation function and the chord length probability density function can be easily determined by image analysis.

1.7.2 Set covariance

The *set covariance* $\gamma_K(r)$, *geometric covariogram* or *distance probability*, introduced by Porod (1951), of a convex body K is defined by

$$\gamma_K(\mathbf{r}) = v_d\big(K \cap (K - \mathbf{r})\big), \tag{1.51}$$

where $\mathbf{r}$ is a vector of $\mathbb{R}^d$ with $\|\mathbf{r}\| = r$.

In integral form, the set covariance function $\gamma_K(\mathbf{r})$ can be written as

$$\gamma_K(\mathbf{r}) = \int_{\mathbb{R}^d} \mathbf{1}_K(x)\mathbf{1}_K(x - \mathbf{r})\mathrm{d}x \qquad \text{for } \mathbf{r} \in \mathbb{R}^d, \tag{1.52}$$

where $\mathbf{1}_K(x)$ is the indicator function of K, defined by

$$\mathbf{1}_K(x) = \begin{cases} 1 & \text{for } x \in K \\ 0 & \text{otherwise.} \end{cases} \tag{1.53}$$

A physicist would perhaps write the integral as

$$\int_K dv_1 \int_K dv_2 \delta(\mathbf{r}_1 + \mathbf{r} - \mathbf{r}_2),$$

using the Dirac delta function δ (see Section 1.9).

It is natural to conjecture that the set covariance $\gamma_K(\mathbf{r})$ is able to characterise the set K uniquely, up to translations and reflections. This has been proved for planar convex K (Averkov and Bianchi, 2009) and for convex polyhedra in $\mathbb{R}^3$ (Bianchi, 2009), but in general it is not true for convex polytopes in $\mathbb{R}^d$, $d \geq 4$ (Bianchi, 2005).

Frequently, the *isotropised* set covariance $\overline{\gamma}_K(r)$ is used, the average of $\gamma_K(\mathbf{r})$ over all possible directions of $\mathbf{r}$, assuming uniform directions:

$$\overline{\gamma}_K(r) = \int_{S^{d-1}} \gamma_K(r\boldsymbol{u}) U_1(d\boldsymbol{u}). \tag{1.54}$$

Here r is a positive number, and $\boldsymbol{u}$ is a unit vector representing a point on the unit sphere S^{d-1} and is identified with the line containing the vector $\boldsymbol{u}$, so that with slight abuse of notation, the distribution U_1, defined on p. 14, also denotes the uniform distribution on S^{d-1}.

In physics $\overline{\gamma}_K(r)$ is usually normalised such that it has the value 1 for $r = 0$. This function can be measured directly by scattering methods.

Important analytical properties of the function γ_K are known; see Matheron (1975). In particular, the derivative can be found: if $r > 0$ and $\boldsymbol{u}$ is a unit vector then $\gamma_K(r\boldsymbol{u})$ has

$$\frac{d}{dr}\gamma_K(r\boldsymbol{u}) = -\left(v_{d-1}\left((K \cap (K + r\boldsymbol{u}))|_{\boldsymbol{u}^{\perp}}\right)\right).$$

Here $A|_{\boldsymbol{u}^{\perp}}$ is the orthogonal projection of A on the hyperplane that has $\boldsymbol{u}$ as normal vector. The geometric configuration is illustrated in Figure 1.2.

It is also the case that $\gamma_K(r\boldsymbol{u})$ is convex in r. This follows from the observation that $-d\gamma_K(r\boldsymbol{u})/dr$ is decreasing in r. Thus also $\overline{\gamma}_K(r)$ is decreasing and convex in r.

The isotropised set covariance $\overline{\gamma}_K(r)$ is closely related to the so-called *distance distribution* $P(r)$, which is the probability density function of the random distance between two

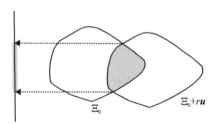

Ξ_0 \hspace{3cm} $\Xi_0 + r\boldsymbol{u}$

Figure 1.2   The orthogonal projection $\left(K \cap (K + r\boldsymbol{u})|_{\boldsymbol{u}^{\perp}}\right)$ in the case $d = 2$. It is the shaded interval on the hyperplane (which in this two-dimensional case is a line) perpendicular to the vector $\boldsymbol{u}$.

independent random points that are uniformly distributed in K. In the spatial case $(d = 3)$

$$P(r) = \frac{4\pi r^2 \overline{\gamma}_K(r)}{\left(V(K)\right)^2},$$ (1.55)

and for general dimension d

$$P(r) = \frac{2\pi^{d/2}}{\Gamma(d/2)} \frac{r^{d-1}\overline{\gamma}_K(r)}{v_d(K)^2} \qquad \text{for } r \geq 0.$$ (1.56)

There is also a relation to the chord length distribution function; see below. The function $P(r)$ can be easily determined based on physical data, but it is not very sensitive with respect to shape variation of K; see Glatter (1979).

In general, $\overline{\gamma}_K(r)$ is not an easy function to calculate. The following are the formulae of the positive part of $\overline{\gamma}_K(r)$ for some particular bodies K:

(a) $K =$ a ball of radius R in $\mathbb{R}^3$,

$$\overline{\gamma}_K(r) = \frac{4}{3}\pi R^3 \left(1 - \frac{3r}{4R} + \frac{r^3}{16R^3}\right) \qquad \text{for } 0 \leq r \leq 2R;$$ (1.57)

(b) $K =$ a disc of radius R in $\mathbb{R}^2$,

$$\overline{\gamma}_K(r) = 2R^2 \cos^{-1}\left(\frac{r}{2R}\right) - \frac{r\sqrt{4R^2 - r^2}}{2} \qquad \text{for } 0 \leq r \leq 2R;$$ (1.58)

(c) $K =$ a rectangle of area A and side-length ratio $\beta\ (\geq 1)$ in $\mathbb{R}^2$,

$$\overline{\gamma}_K(r) = \frac{A}{\pi} \cdot \begin{cases} \pi - 2x - \dfrac{2x}{\beta} + \dfrac{x^2}{\beta} & \text{for } 0 \leq x \leq 1, \\[2mm] 2\sin^{-1}\left(\dfrac{1}{x}\right) - \dfrac{1}{\beta} - 2(x - u) & \text{for } 1 < x \leq \beta, \\[2mm] 2\sin^{-1}\left(\dfrac{\beta - uv}{x^2}\right) + 2u + \dfrac{2v}{\beta} - \beta - \dfrac{1 + x^2}{\beta} & \text{for } \beta < x < \sqrt{\beta^2 + 1}, \\[2mm] 0 & \text{for } x \geq \sqrt{\beta^2 + 1}, \end{cases}$$

(1.59)

where

$$x = \frac{r}{\sqrt{A/\beta}}, \qquad u = \sqrt{x^2 - 1}, \qquad v = \sqrt{x^2 - \beta^2}.$$ (1.60)

The formula for a parallelepiped and formulae for other bodies can be found in Gille (1988) and Gille (2014).

A good approximation for $d = 3$ and for small r is

$$\overline{\gamma}_K(r) \approx V(K) - \frac{S(K)}{4} r, \tag{1.61}$$

and for $d = 2$

$$\overline{\gamma}_K(r) \approx A(K) - \frac{L(K)}{\pi} r. \tag{1.62}$$

Better but more complicated approximations are given in Ciccariello (1995).

The set covariance plays an important rôle in small-angle scattering experiments, since the intensity collected there is the Fourier transform of the autocorrelation function of the scattering density in the sample, see Guinier and Fournet (1995). A particular case is that of a monodisperse, isotropic and dilute 'particulate sample', where the sample can be understood as a system of homogeneous particles of the same shape and size, random orientation and small number density. In this case the sample autocorrelation is proportional to $\overline{\gamma}_K(r)$.

1.7.3   Chord length distribution

Random lines generate chords of random length in intersected convex sets. The standard case is that where the lines are uniform random in the sense of the motion-invariant measure discussed in Section 8.2.2, restricted to the set of lines which actually hit the convex body K. (Note that there are many other ways of defining random chords; see for example Solomon, 1978; Coleman, 1989; Chiu and Larson, 2009, and other literature on the so-called Bertrand paradox (e.g. M. G. Kendall and Moran, 1963, pp. 9–10).) To each random chord there corresponds its length, and its distribution function is called the *chord length distribution function*. For any K with inner points there exists the corresponding probability density function, which is denoted here by $f_K(l)$. Many formulae for $f_K(l)$ and its moments can be found in the literature, in particular in Gille (2014). Some of them will be given here.

Planar case

The mean chord length $\overline{\ell}_K$ is given by

$$\overline{\ell}_K = \frac{\pi A(K)}{L(K)}. \tag{1.63}$$

The third moment of the chord length is

$$\overline{\ell^3}_K = \int_0^\infty l^3 f_K(l) dl = \frac{3\left(A(K)\right)^2}{L(K)} \tag{1.64}$$

(Santaló, 1976).

For the second moment of the chord length there exists no general formula comparable to (1.63) or (1.64). Of course, for many examples of planar convex sets formulae for $\overline{\ell^2}$ are known; see for example Voss (1982).

For a disc of radius R it holds

$$\overline{\ell^n} = \begin{cases} \dfrac{1 \cdot 3 \cdot \ldots \cdot n}{2 \cdot 4 \cdot \ldots \cdot (n+1)} \dfrac{\pi}{2} (2R)^n & \text{for odd } n, \\[3mm] \dfrac{2 \cdot 4 \cdot \ldots \cdot n}{3 \cdot 5 \cdot \ldots \cdot (n+1)} (2R)^n & \text{for even } n. \end{cases} \tag{1.65}$$

The chord length density function $f_K(l)$ is known for many sets K. The following formulae give it for some K:

(a) $K =$ a disc of radius R,

$$f_K(l) = \frac{l}{2R\sqrt{4R^2 - l^2}} \qquad \text{for } 0 \le l \le 2R; \tag{1.66}$$

(b) $K =$ an ellipse with semiaxis lengths a and b $(a \ge b)$,

$$f_K(l) = 3ab\bar{l} \int_{\max\{\ell,\, 2b\}}^{2a} \frac{1}{x^3 \sqrt{x^2 - 4b^2} \sqrt{x^2 - l^2} \sqrt{4a^2 - x^2}} \, dx; \tag{1.67}$$

the integral can be evaluated numerically;

(c) $K =$ a rectangle with side lengths a and b $(a \ge b)$ (Gille, 1988, 2014),

$$f_K(l) = \begin{cases} \dfrac{1}{a+b} & \text{for } 0 \le l \le b, \\[4mm] \dfrac{ab^2}{l^2(a+b)\sqrt{l^2 - b^2}} & \text{for } b < l \le a, \\[4mm] \dfrac{ab}{l^2(a+b)}\left(\dfrac{a}{\sqrt{l^2 - a^2}} + \dfrac{b}{\sqrt{l^2 - b^2}}\right) - \dfrac{1}{a+b} & \text{for } a < l \le \sqrt{a^2 + b^2}; \end{cases} \tag{1.68}$$

(d) $K =$ an equilateral triangle of side length a (Sulanke, 1961, Gille, 2014),

$$f_K(l) = \begin{cases} \dfrac{1}{a}\left(\dfrac{1}{2} + \dfrac{\pi}{3}\sqrt{\dfrac{1}{3}}\right) & \text{for } 0 < l \le \dfrac{\sqrt{3}a}{2}, \\[4mm] -\dfrac{1}{l}\sqrt{1 - \dfrac{3a^2}{4l^2}} + \dfrac{1}{a}\left(\dfrac{1}{2} - \dfrac{2\pi}{3\sqrt{3}} + \dfrac{2}{\sqrt{3}}\sin^{-1}\left(\dfrac{\sqrt{3}a}{2l}\right)\right) & \text{for } \dfrac{\sqrt{3}a}{2} < l \le a. \end{cases} \tag{1.69}$$

The case of a general triangle is considered in Ciccariello (2010).

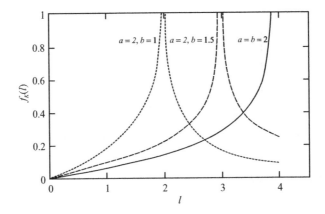

Figure 1.3   Chord length density functions for ellipses with $a = 2$ and $b = 1$, 1.5 and 2.

Figure 1.3 shows chord length density functions for two ellipses and a disc, and Figure 1.4 those for a square and a rectangle.

The form of the chord length density function is related to certain features of the corresponding body K. For example, algebraic singularities of this function correspond to parallel pieces of the contour and the form of $f_K(l)$ for l close to its maximum is essentially related to smaller details of the contour.

Spatial case

The mean chord length $\bar{\ell}_K$ is given by

$$\bar{\ell}_K = \frac{4V(K)}{S(K)}. \tag{1.70}$$

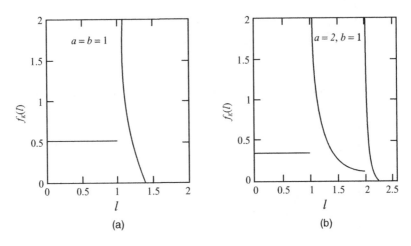

Figure 1.4   Chord length density functions for (a) a square with side length 1 and (b) a rectangle with side lengths 2 and 1.

The fourth moment is

$$\overline{\ell^4}_K = \frac{12\left(V(K)\right)^2}{\pi S(K)}. \tag{1.71}$$

Clearly, for particular bodies also the second and third moment of the chord length are known. For a ball of radius R it holds

$$\overline{\ell^n}_K = \frac{2^{n+1}}{n+2} R^n. \tag{1.72}$$

Note that there is a close relationship between the isotropised set covariance $\overline{\gamma}_K(r)$ and the chord length density function $f_K(l)$: for $d = 2$

$$f_K(l) = \frac{\pi}{L(K)} \frac{\mathrm{d}^2\overline{\gamma}_K(l)}{\mathrm{d}l^2}, \tag{1.73}$$

and for $d = 3$

$$f_K(l) = \frac{4}{S(K)} \frac{\mathrm{d}^2\overline{\gamma}_K(l)}{\mathrm{d}l^2} \quad \text{for } l \geq 0 \tag{1.74}$$

(see Kingman, 1969; Gille, 2014).

These formulae are frequently used to determine chord length density functions. The case of K being a ball is treated in Section 10.4 in the context of stereology and Figure 10.4 shows the chord length density function for a parallelepiped.

In the literature further formulae can be found for $\overline{\gamma}_K(r)$ and $f_K(l)$ for various bodies K. Among the bodies for which calculations have been made are: ellipsoids, cylinders (finite and infinite), hemispheres, tetrahedra, prisms, parallelepipeds and for various isotropic random sets such as the typical cells of the Voronoi, Poisson plane and dead leaves tessellation. The results are scattered in many different journals and collected in Gille (2014). Gille asserts that there are around ten different mathematical techniques that have been applied to determine chord length distributions, which depend on the geometry of the bodies of interest.

Chords also play a rôle for nonconvex sets. There an intersecting line may produce a sequence of segments, which are best analysed together as a totality. Miles (1972a, 1985) shows how the segments can be related together so as to obtain useful formulae, for example, for moments of length. Ciccariello (2009) determines chord length distributions for planar butterfly-shaped objects, considering all section segments as single chords. There $\mathrm{d}^2\overline{\gamma}_K(r)/\mathrm{d}r^2$ can be negative.

Many physical experimental techniques indirectly measure chord length distributions. One of the physical principles is the interference of two monochromatic waves, originating at the endpoints of a chord: scattering investigations of electromagnetic waves (light, X-rays) or neutrons yield a scattering intensity curve, also called 'diffraction pattern'. The scattering intensity I is recorded as a function of the scattering vector q, $I(q)$. Such patterns describe order ranges from nanometers to micrometers. Finally, chord length distributions result from $I(q)$ by integral transformations; see Wu and Schmidt (1973) and Burger and Ruland (2001). The series of formulae for chord length distribution functions above and in Gille (2014) may be of value in finding suitable models when empirical distribution functions are given. Another example of the application of chord length distribution is in the classification of microscopic images of for example starch grains (Tong *et al.*, 2008).

While these applications are related to small-scale objects, chords also play a rôle in the geometrical investigation of very large objects, in astronomy in the context of occultations, for example of asteroids, which yield chord length information, see Hestroffer *et al.* (2002).

1.7.4 Erosion–dilation functions

The *erosion–dilation function* $E_K(r)$ of the convex body K is defined as

$$E_K(r) = \begin{cases} v_d\big(K \oplus B(o, r)\big) & \text{for } r > 0, \\ v_d\big(K \ominus B(o, |r|)\big) & \text{for } r \le 0. \end{cases} \tag{1.75}$$

That is, $E_K(r)$ is the volume of the inner ($r \le 0$) and outer ($r > 0$) parallel set of K. While for $r > 0$ the value of $E_K(r)$ is simply given by the Steiner formula (1.28) or (1.39), the function is more complicated for negative r. The erosion–dilation function can be easily generalised for the case of nonconvex sets — nothing in Formula (1.75) need change. In Section 6.3.3 an analogous function for unbounded sets will be used.

A normalised version of the erosion function is the *spherical erosion function* $Q_K(r)$, which has the nature of a distribution function:

$$Q_K(r) = 1 - \frac{v_d\big(K \ominus B(o, r)\big)}{v_d(K)} \qquad \text{for } r \ge 0. \tag{1.76}$$

It is the distribution function of the distance of a uniform random point in K to the boundary of K.

Like the erosion-based function, also functions that include morphologically 'opened' or 'closed' sets can be used, for example

$$G_K(r) = 1 - \frac{v_d\big(K \circ B(o, r)\big)}{v_d(K)} \qquad \text{for } r \ge 0. \tag{1.77}$$

See Serra (1982, p. 333ff) and Ripley (1988, Chapter 6).

Note that other test sets, for example point pairs or segments, can be used instead of balls $B(o, r)$.

1.8 Polyconvex sets

Some of the results concerning convex sets can be generalised for sets in a wider class known as the *convex ring* $\mathcal{R}$; see Klain and Rota (1997) and Schneider and Weil (2008). The convex ring $\mathcal{R}$ is the system of all subsets A of $\mathbb{R}^d$ which can be expressed as finite unions of convex bodies:

$$A = \bigcup_{i=1}^{n} K_i \qquad \text{for } K_i \in C(\mathbb{K}).$$

If A_1 and A_2 both belong to $\mathcal{R}$ then so do $A_1 \cup A_2$ and $A_1 \cap A_2$. The elements of the convex ring are called *polyconvex* sets.

The *extended convex ring* is the system

$$\mathbb{S} = \{K : K \cap K' \in \mathcal{R} \text{ for all convex bodies } K'\}.$$

Each element of $\mathbb{S}$, which may be a union of infinitely many convex sets, yields a polyconvex set when intersected with a convex body. The elements of $\mathbb{S}$ are called *locally polyconvex*.

An *additive functional h* on $\mathcal{R}$ is a map $h : \mathcal{R} \to \mathbb{R}$ with the properties

$$h(\emptyset) = 0$$

and

$$h(A_1 \cup A_2) + h(A_1 \cap A_2) = h(A_1) + h(A_2)$$

for A_1, $A_2 \in \mathcal{R}$.

Isometry-invariance of functionals on $\mathcal{R}$ is defined as for convex body functionals.

A very important example of an additive and isometry-invariant functional on $\mathcal{R}$ is the *connectivity number* or *Euler–Poincaré characteristic* χ. It can be understood as a generalisation of convex object count. Indeed, for a convex non-empty K it takes the value

$$\chi(K) = 1$$

and

$$\chi(A) = n,$$

if A is the union of n disjoint convex bodies. And it holds $\chi(\emptyset) = 0$. This together with the additivity property defines $\chi(A)$ for general A in $\mathcal{R}$. If A is a union of convex bodies

$$A = \bigcup_{i=1}^{n} K_i \qquad \text{for } K_i \in C(\mathbb{K}),$$

then the additivity property gives the 'inclusion–exclusion' formula

$$\chi(A) = \sum_{i} \chi(K_i) - \sum_{1 \le i_1 < i_2 \le n} \chi(K_{i_1} \cap K_{i_2}) + \cdots + (-1)^{n-1} \chi(K_1 \cap \cdots \cap K_n). \quad (1.78)$$

It can be shown that $\chi(A)$ is independent of the representation of A as a finite union of convex bodies. Figure 1.5 shows four planar sets of connectivity numbers $+1, 0$ and -1 respectively.

The connectivity number can, the same as volume and surface content, be defined also for closed sets outside the extended convex ring; see for example Wilder (1963) and Richeson (2008). For such cases the following informal rules hold:

- In the two-dimensional case $\chi(A)$ equals the number of outer boundaries minus the number of inner boundaries. An outer boundary is a boundary such that an observer moving along it anticlockwise sees the set on his left hand.

- In the three-dimensional case $\chi(A)$ equals the number of connected components of A, minus the number of independent two-dimensional holes or tunnels, and plus the number of three-dimensional holes or voids.

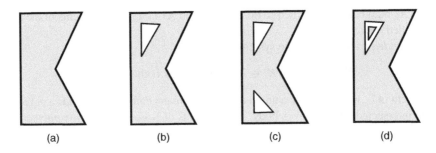

Figure 1.5   Four planar sets with various connectivity numbers: (a) +1, (b) 0, (c) −1, (d) +1.

A reader who wants to get some feeling of the three-dimensional connectivity number should note that the connectivity number of a solid torus as well as a tea cup is 0, that of a closed ball is 1 and that of a sphere as well as the surface of any convex body is 2. The set A formed by the edges of a cube has $\chi(A) = -4$.

The connectivity number makes it possible to generalise the intrinsic volumes, still denoted by V_k, to act on $\mathcal{R}$ by means of additivity, as shown by Schneider (1980) and Schneider and Weil (2008, p. 190). The term $v_k(K|_{E^\perp})$ in Formula (1.50) comes from the inner integral of the left-hand side of Formula (1.49) and must be replaced by

$$\int_{E^\perp} \chi(K \cap E_x) v_k(\mathrm{d}x)$$

so that V_0 coincides with χ on $\mathcal{R}$, and hence for a polyconvex set $A \in \mathcal{R}$, the (generalised) intrinsic volumes are defined as

$$V_k(A) = \frac{\binom{d}{d-k} \frac{b_d}{b_k}}{b_{d-k}} \int_{\mathbb{L}_{d-k}} \int_{E^\perp} V_0(A \cap E_x) v_k(\mathrm{d}x) U_{d-k}(\mathrm{d}E) \tag{1.79}$$

for $k = 1, \ldots, d$. These functionals are additive and isometry-invariant, but not continuous and may be negative for $k \leq d - 1$. The quantity $V_d(A)$ is the d-dimensional volume of A, $V_{d-1}(A)$ is half of the area of ∂A, and $V_0(A)$ is the connectivity number of A.

Minkowski functionals of polyconvex sets, still denoted by $W_k(A)$ and connected with the $V_k(A)$ by Formula (1.35), may also be used.

The intrinsic volumes can be generalised also in another way, to the *positive extension* introduced by Matheron (1975) and Schneider (1980) and referred to in Section 7.3.4. This generalisation yields nonnegative functionals and loses additivity. A quantity that arises by this route of generalisation is $M^+(A)$, the *integral of mean positive curvature*; see also p. 294. A further positive characteristic which is a counterpart to the connectivity number is the *convexity number* χ^+. The following gives an account of its definition after Matheron (1975, pp. 122–3), but using the notation developed in this chapter.

First define the *convexity number of $A \in \mathcal{R}$ with respect to the unit vector $\boldsymbol{u}$*. Let $S_{\boldsymbol{u},r}$ for real r be the hyperplane through the point $r\boldsymbol{u}$ and perpendicular to $\boldsymbol{u}$. For A in $\mathcal{R}$ let $K_1, K_2, \ldots, K_n$ be the connected components of $A \cap S_{\boldsymbol{u},r}$. The connected component K_i is said to be an *entering set* for A (with respect to $\boldsymbol{u}$ and $S_{\boldsymbol{u},r}$) if there is an open and connected set G containing K_i such that $G \cap K_j$ is empty for $j \neq i$ and $A \cap G \cap S_{\boldsymbol{u},r-\varepsilon}$ is empty for all sufficiently small positive ε. Let $n(r)$ denote the number of entering sets associated with $S_{u,r}$. If u is a fixed unit vector then this number $n(r)$ is nonzero for only a finite number of real numbers r, so that it is possible to put

$$\chi^+(A, u) = \sum_r n(r).$$

This is the convexity number of A with respect to the unit vector u. Figure 7.2 on p. 295 demonstrates an application of the convexity number in the planar case with $u = (0, 1)$ and $u = (0, -1)$.

The (mean) convexity number χ^+ is then simply the rotation average of $\chi^+(A, u)$ with respect to the invariant probability measure U_1 on the unit sphere S^{d-1}:

$$\chi^+(A) = \int_{S^{d-1}} \chi^+(A, u)\, U_1(du).$$

One might ask whether the various integral-geometric formulae (the Steiner formula, Hadwiger's characterisation theorem, the principal kinematic formula, and the Crofton formula) hold in the convex ring if the standard intrinsic volumes are systematically replaced by their additive generalisations. This is indeed the case; see Schneider and Weil (2008). In particular,

$$v_d\big(A \oplus B(o, r)\big) = v_d(A) + S(A)r + o(r), \tag{1.80}$$

for regular closed A; see Schneider (1993, Theorem 4.4.1). More general sets are considered in Kiderlen and Rataj (2006).

Still more is possible: the theory can be extended to sets of positive reach and sets which are finite unions of such sets; see Zähle (1984a, 1987a) and Rother and Zähle (1990). For $A \subset \mathbb{R}^d$, reach(A) is defined to be

$$\text{reach}(A) = \sup\{r: \text{ for all } x \in A \oplus B(o, r) \text{ there exists a unique point of } A \text{ nearest to } x\}.$$

If A is of positive reach, then A is closed. If A is convex, then reach(A) $= \infty$. An example of a set which is of positive reach but not in the convex ring is the sphere $\partial B(o, R)$. Its reach is R.

1.9 Measure and integration theory

In modern probability theory and geometry the concept of measure plays a central rôle. Consequently these ideas play a large part in the present book. However, the authors expect that many of the readers of this book are not familiar with this theory, and have therefore written the book so that most of it can be understood without a deeper knowledge of these mathematical theories.

The following introduction is of course too short for a thorough understanding of measure and integration theory. The aim is to remind mathematicians of what they have learned before and to help non-mathematicians understand the notation.

Measures are real-valued functions, defined on families of sets and enjoying properties of additivity and positivity.

To introduce the idea of a measure, the following three simple and linked examples may be helpful. Consider the contents of a body K in space that is made up of some material. Suppose that the mass of this body is not necessarily homogeneously distributed in it. Associated with each portion A of K ($A \subset K$) are the quantities $\alpha_1(A)$, the volume of A, and $\alpha_2(A)$, the mass of K that is contained in A. Suppose that in the body K there are also small grains of another material. Let the number of these grains lying in A be given by $\alpha_3(A)$. Thus there are three set functions α_1, α_2 and α_3 defined on the portions of K. If a portion A is divided into disjoint parts A_1 and A_2 then

$$\alpha_k(A) = \alpha_k(A_1 \cup A_2) = \alpha_k(A_1) + \alpha_k(A_2) \qquad \text{for } k = 1, 2, 3.$$

Set functions with this property are called *additive*.

For mathematical reasons, it is useful to stipulate that such functions should have the stronger property of σ-*additivity*. That is to say, if a set can be divided into a countable disjoint union of subsets then the value of the set function on the whole set should equal the sum of the values of the set function on the subsets. Together with various mathematical technicalities, this consideration leads to the idea that set functions of the kind given above are naturally defined on systems of sets that are closed under the basic set operations of union, intersection, and complementation, and also under the operations of taking countably infinite unions and intersections.

Such systems of sets are called σ-*algebras*. Technically a σ-algebra is a system $\mathcal{X}$ of subsets of some ground set X satisfying three conditions:

(S1) $X \in \mathcal{X}$;

(S2)   if $A \in \mathcal{X}$, then $A^c \in \mathcal{X}$;

(S3)   if $A_1, A_2, \ldots \in \mathcal{X}$, then $\bigcup\limits_{k=1}^{\infty} A_k \in \mathcal{X}$.

From these properties others follow immediately:

$$\emptyset \in \mathcal{X};$$

$$\text{if } A, B \in \mathcal{X}, \text{ then } A \setminus B \in \mathcal{X};$$

$$\text{if } A_1, A_2, \ldots \in \mathcal{X}, \text{ then } \bigcap\limits_{k=1}^{\infty} A_k \in \mathcal{X}.$$

Trivial examples of σ-algebras are given by the 'minimal' σ-algebra $\{\emptyset, X\}$ and the 'maximal' σ-algebra, the *power set* $\mathcal{P}(X)$, that is, the system of all subsets of X.

Often one has to consider the *trace* $\mathcal{X}_A$ of the σ-algebra $\mathcal{X}$ on $A \subset X$, defined by

$$\mathcal{X}_A = \{B \cap A : B \in \mathcal{X}\},$$

which is also a σ-algebra.

A very important example of a σ-algebra is given by the family $\mathcal{B}^d$ of *Borel sets of* $\mathbb{R}^d$. This is the smallest σ-algebra on $\mathbb{R}^d$ that contains all the open subsets of $\mathbb{R}^d$. It can be shown that $\mathcal{B}^d$ is strictly smaller than the power set $\mathcal{P}(\mathbb{R}^d)$, that is, there are subsets of $\mathcal{B}^d$ which are not Borel.

Loosely speaking all 'reasonable' sets are Borel sets; it is difficult to give examples of non-Borel sets. The σ-algebra $\mathcal{B}^d$ contains all the subsets of $\mathbb{R}^d$ that can be constructed from the open subsets by the basic set operations and by limits. Since these operations can be iterated this is an exceedingly large class of sets. It certainly includes all closed sets, and thus all compact sets, and thus all elementary geometrical bodies discussed in this book. Denote the subclass of all bounded Borel sets of $\mathbb{R}^d$ by $\mathcal{B}^d_o$.

The pair $[X, \mathcal{X}]$, formed by a set X and a σ-algebra $\mathcal{X}$ of subsets of X, is called a *measurable space* and the A in $\mathcal{X}$ are called *measurable sets*.

A function $f : X \rightarrow \mathbb{R}$ is said to be $\mathcal{X}$-*measurable* if for each Borel set $B \in \mathcal{B}^1$ the inverse image $f^{-1}(B) = \{x \in X : f(x) \in B\}$ belongs to the σ-algebra $\mathcal{X}$ associated with X. Elementary algebraic operations and operations such as the taking of absolute values, and the taking of limits when applied to measurable functions all yield measurable functions. A particular example of a measurable function is the *indicator function* $\mathbf{1}_A(x)$ of a measurable set A as defined in (1.53) on p. 17.

The coordinate functions $x = (x_1, \ldots, x_d) \mapsto x_i$ are special examples of Borel-measurable functions from $\mathbb{R}^d$ to $\mathbb{R}$.

All continuous functions and hence all differentiable functions on $\mathbb{R}^d$ are Borel-measurable.

If $\mathbb{R}$ is replaced by a general set Y with associated σ-algebra $\mathcal{Y}$ in the definition of a measurable function then a transformation

$$f : X \rightarrow Y$$

is said to be $(\mathcal{X}, \mathcal{Y})$-measurable if $f^{-1}(B) \in \mathcal{X}$ for each $B \in \mathcal{Y}$.

Suppose that $[X, \mathcal{X}]$ is a measurable space. A *measure* on $[X, \mathcal{X}]$ is a function $\mu : \mathcal{X} \rightarrow [0, \infty]$ with the following two properties:

(M1) $\mu(\emptyset) = 0,$

(M2) $\mu \left(\bigcup_{k=1}^{\infty} A_k \right) = \sum_{k=1}^{\infty} \mu(A_k)$

for all $A_1, A_2, \ldots \in \mathcal{X}$ with $A_i \cap A_j = \emptyset$ whenever $i \neq j$. Note that $\mu(A)$ may equal ∞. Property (M2) is referred to as the property of σ-*additivity*. A particular case of this property is finite additivity:

$$\mu(A_1 \cup \cdots \cup A_n) = \mu(A_1) + \cdots + \mu(A_n)$$

if $A_1, \ldots, A_n$ are pairwise disjoint. Another important consequence is that if $A, B \in \mathcal{X}$ and $B \subset A$ then

$$\mu(A \setminus B) = \mu(A) - \mu(B).$$

The measures which occur in this book generally have two other important properties. They are defined on the Borel σ-algebra of $\mathbb{R}^d$ for some d, and they are *locally finite*, that is, finite on bounded sets; such measures are referred to as *Radon measures*. (If $\mu(X) < \infty$ then μ is called *finite*. The set X can be nevertheless an unbounded subset of $\mathbb{R}^d$.) The relationship between measure theory and the topology of the ground space is an important and extensive theory; however, the two definitions above will suffice for the discussions below.

Simple examples of measures on measurable spaces are given by the *Dirac measures* δ_x for x in X. They are defined by

$$\delta_x(A) = \begin{cases} 1 & \text{if } x \in A, \\ 0 & \text{otherwise.} \end{cases} \tag{1.81}$$

On $[\mathbb{R}^d, \mathcal{B}^d]$ the Dirac measures are Radon measures.

A further and most important example is that of *Lebesgue measure* ν_d on $[\mathbb{R}^d, \mathcal{B}^d]$. A fundamental theorem of measure theory has as a special case the result that locally finite measures on $[\mathbb{R}^d, \mathcal{B}^d]$ are unambiguously characterised by their values on hypercubes. Lebesgue measure is characterised by

$$\nu_d(Q) = (v_1 - u_1) \cdot \ldots \cdot (v_d - u_d)$$

if $Q = [u_1, v_1] \times \ldots \times [u_d, v_d]$. This implies that the values of the ν_d-measure of geometrical objects such as balls, cylinders, solid toruses, and so forth, coincide with the volumes in elementary geometry. Indeed, in the case $d = 3$ the Lebesgue measure ν_3 is equal to the volume measure V. Some of such values are given in Table 1.1 on p. 13. In the cases $d = 1$ and $d = 2$, it is $\nu_1 = l$ the length measure and $\nu_2 = A$ the area measure.

If B is a bounded Borel set then $\nu_d(B) < \infty$. This follows from the positivity of ν_d and its finiteness on all hypercubes including those which contain the bounded set B. Thus ν_d is indeed a Radon measure.

Lebesgue measure has the property of being *isometry-invariant*:

$$\nu_d(mA) = \nu_d(A) \tag{1.82}$$

for all isometries m and Borel sets A. Another important theorem states that every isometry-invariant Radon measure μ (indeed every translation-invariant Radon measure) on $[\mathbb{R}^d, \mathcal{B}^d]$ is a constant multiple of the Lebesgue measure,

$$\mu = c\nu_d \qquad \text{for some } c \geq 0. \tag{1.83}$$

Further measures of geometrical interest are the *k-dimensional Hausdorff measures* h_k,

$$h_k(B) = 2^{-k} b_k \lim_{\delta \to 0+} \inf \left\{ \sum_{i=1}^{\infty} (\mathrm{diam}(M_j))^k : B \subset \bigcup_{i=1}^{\infty} M_j, \ \mathrm{diam}(M_j) \leq \delta \right\}$$

$$\text{for } B \in \mathcal{B}^d, \ k = 0, 1, \ldots, d, \tag{1.84}$$

where diam denotes the diameter, the M_i are compact subsets of $\mathbb{R}^d$, and b_k, defined in Formula (1.30), denotes the volume of the k-dimensional unit ball. For $k = d$ the measure coincides with the Lebesgue measure, for $k = 0$ it is the counting measure assigning to a set the number of its elements. Further, $h_k(B)$ for a k-dimensional submanifold B of $\mathbb{R}^d$ coincides with its usual k-dimensional Lebesgue measure: for example, if B is a curve in $\mathbb{R}^d$, then $h_1(B)$ is its length; see Morgan (2009) for more details.

Dirac measures and Lebesgue measure are quite different in that Dirac measures are concentrated on single points while Lebesgue measure gives zero mass to every point. Measures μ on $[\mathbb{R}^d, \mathcal{B}^d]$ that are concentrated on a countably infinite collection of points $x_1, x_2, \ldots$ (so that $\mu(\mathbb{R}^d \setminus \{x_1, x_2, \ldots\}) = 0$) are called *atomic measures*, and are called *counting measures* if in addition they give each point a mass of one or zero. A measure is called a *diffuse measure* if, like Lebesgue measure, it gives zero mass to each single point. Property (M2) then ensures that every countable set has also Lebesgue measure zero. Since the set of all rational numbers in $\mathbb{R}$ is countable, it has Lebesgue measure zero, while the set I of all irrational numbers in $[0, 1]$ has Lebesgue measure 1.

A measure μ on a measurable space $[X, \mathcal{X}]$ defines a triplet $[X, \mathcal{X}, \mu]$, which is called a *measure space*. Associated with each measure space is the *integral with respect to* μ. For a measurable real-valued function $f(x)$, $x \in X$, the integral is written as

$$\int f(x)\,\mu(dx).$$

Such integrals appear at many places in this book, where the variable x is often not a real number or a point of $\mathbb{R}^d$ but, for example, a compact set or a point pattern.

The construction of this general integral starts with the particular case of an indicator function $f(x) = \mathbf{1}_A(x)$ where

$$\int \mathbf{1}_A(x)\,\mu(dx) = \mu(A) \qquad \text{for } A \in \mathcal{X}. \tag{1.85}$$

Remember that for the classical Riemann integral and $A = [a, b]$

$$\int_{-\infty}^{+\infty} \mathbf{1}_A(x)\,dx = \int_a^b dx = b - a,$$

which can be seen as a particular case of (1.85) since $\nu_1(A) = b - a$.

Further steps of the construction are

$$\int \big(c_1 f_1(x) + c_2 f_2(x)\big)\,\mu(dx) = c_1 \int f_1(x)\,\mu(dx) + c_2 \int f_2(x)\,\mu(dx) \qquad \text{for } c_1, c_2 \geq 0 \tag{1.86}$$

and

$$\int \sum_{k=1}^{\infty} f_k(x)\,\mu(dx) = \sum_{k=1}^{\infty} \int f_k(x)\,\mu(dx) \tag{1.87}$$

for nonnegative functions $f_1(x)$, $f_2(x)$, ... For such functions an integral value of $+\infty$ is an accepted convention.

For an arbitrary measurable function $f(x)$ the positive and negative part are considered separately to obtain

$$\int f(x)\,\mu(dx) = \int f(x)\mathbf{1}_B(x)\,\mu(dx) - \int \big(-f(x)\big)\mathbf{1}_{B^c}(x)\,\mu(dx), \tag{1.88}$$

where

$$B = \{x \in X : f(x) > 0\}.$$

Here one says that the integral exists if

$$\int |f(x)|\,\mu(dx) = \int f(x)\mathbf{1}_B(x)\,\mu(dx) + \int \big(-f(x)\big)\mathbf{1}_{B^c}(x)\,\mu(dx) < \infty.$$

Occasionally such integrals are written as

$$\int f \, d\mu = \int f(x) \, \mu(dx) = \int f(t) \, \mu(dt),$$

where the *dummy variables* x and t of integration can be chosen arbitrarily from previously undefined symbols.

The integral derived from the measure space $[\mathbb{R}^d, \mathcal{B}^d, \nu_d]$ is called the *Lebesgue integral*. In the case $d = 2$ this integral can be interpreted graphically. A nonnegative function $f : \mathbb{R}^2 \to [0, \infty)$ can be considered as giving the height $f(x)$ at each point x of a surface lying over the plane $(x_3 = 0)$ in $\mathbb{R}^3$. Then the integral $\int f(x)\nu_2(dx)$ can be interpreted as the volume lying between the surface and the plane $(x_3 = 0)$. A similar interpretation is possible for the case $d = 1$.

The Lebesgue integral can be applied to functions where the Riemann integral does not exist. Consider the case $d = 1$ and $f(x)$ the Dirichlet function, which is the indicator function of the set I of all irrational numbers in $[0, 1]$:

$$\mathbf{1}_I(x) = \begin{cases} 1 & \text{for } x \text{ irrational and } 0 \leq x \leq 1, \\ 0 & \text{otherwise.} \end{cases} \tag{1.89}$$

Here the Riemann integral does not exist, since the upper and lower Riemann sums are always 1 and 0, respectively. In contrast, Formula (1.85) yields the Lebesgue integral value 1 since $\nu_1(I) = 1$.

A frequent contraction of notation involves removing indicator functions from the integrand and inserting the corresponding sets as *domain of integration* or *range of integration*. In the case $d = 1$

$$\int \mathbf{1}_{[a,b]}(x) f(x) \, \nu_1(dx) = \int_{[a,b]} f(x) \, \nu_1(dx),$$

and in the general case

$$\int \mathbf{1}_A(x) f(x) \, \mu(dx) = \int_A f(x) \, \mu(dx).$$

Where no domain of integration is given for the integral it is to be understood that the domain of integration is the whole of the appropriate space.

For elementary functions $f(x)$ on $\mathbb{R}$ the expression $\int_{[a,b]} f(x)\nu_1(dx)$ equals the well-known elementary Riemann integral

$$\int_a^b f(x) \, dx.$$

The Lebesgue integral simply fits better to many theoretical calculations in stochastic geometry.

If two measures μ_1 and μ_2 are given on the same measurable space $[X, \mathcal{X}]$ then sometimes one can be given in terms of the other by an integral formula involving a *density* $g : X \to [0, \infty)$:

$$\mu_2(A) = \int_A g(x) \, \mu_1(dx) \qquad \text{for all } A \in \mathcal{X}.$$

In this case μ_2 is said to be *absolutely continuous with respect to* μ_1, and written as

$$\mu_2 \ll \mu_1.$$

The Radon–Nikodym theorem gives a necessary and sufficient condition for one Radon measure μ_2 to be absolutely continuous with respect to another Radon measure μ_1:

if $\mu_1(A) = 0$ implies $\mu_2(A) = 0$ for all A in $\mathcal{X}$, then $\mu_2 \ll \mu_1$.

In practice the actual calculation of the density function $g(x)$ for given μ_1 and μ_2 may be difficult; see Rao (1993).

Probability theory makes much use of the ideas of measure theory. If a measure space $[\Omega, \mathcal{A}, \mathbf{P}]$ is such that $\mathbf{P}(\Omega) = 1$ then it is called a *probability (measure) space* and the measure $\mathbf{P}$ is called a *probability measure*. Then Ω is called the *sample space* and $\mathcal{A}$ is the σ-*algebra of events*; the elements of Ω are called *sample points*; the subsets of Ω that belong to $\mathcal{A}$ are called *events*. Real-valued $\mathcal{A}$-measurable functions defined on Ω are called *random variables*. By tradition they are usually not denoted by f, g or h but by X, Y or Z. They assign real numbers to sample points. The measurability condition ensures that for a random variable X it is possible to define probabilities such as

$$\mathbf{P}(X \le x) = \mathbf{P}(\{\omega \in \Omega : X(\omega) \le x\}) = F(x). \tag{1.90}$$

The function $F(x)$ is called the *distribution function of* X.

The definition of more general random variables is analogous. They are also measurable mappings, but then the image space is not $\mathbb{R}$ but a suitable other space. In the case of random sets, for example, the image space is $\mathbb{F}$, the space of all closed subsets of $\mathbb{R}^d$; and for defining measurability, it is necessary to introduce a σ-algebra of subsets of $\mathbb{F}$.

If X is a real-valued random variable with $\int |X(\omega)| \mathbf{P}(d\omega) < \infty$ then $\mathbf{E}(X)$ denotes its expectation or mean,

$$\mathbf{E}(X) = \int X(\omega) \mathbf{P}(d\omega) = \int_{-\infty}^{\infty} x \, dF(x). \tag{1.91}$$

where the last expression is a so-called *Riemann–Stieltjes integral*.

If $\mathbf{P}$ is a probability measure on $[\mathbb{R}^1, \mathcal{B}^1]$ and absolutely continuous with respect to ν_1, then X is said to be an absolutely continuous random variable and there exists a nonnegative function $f(x)$, the *probability density function*, satisfying

$$f(x) = F'(x)$$

and the integral given in (1.91) is rewritten as

$$\int_{-\infty}^{\infty} x \, f(x) dx.$$

If $\mathbf{P}$ is atomic, then X is a discrete random variable and the integral in (1.91) is equal to

$$\sum_{i} x_i p_i,$$

where x_i are the atoms with weights p_i. In physics literature, the latter would be written as

$$\int_{-\infty}^{\infty} x \, f(x) \, dx$$

with

$$f(x) = \sum_i \delta_{x_i}(x) = \sum_i \delta(x - x_i)$$

using Dirac delta functions.

Analogously, the notation

$$E\big(g(X)\big) = \int_{-\infty}^{\infty} g(x) \, dF(x) \tag{1.92}$$

is used for a measurable function $g(x)$.

Furthermore, $\mathbf{cov}(X, Y)$ is the *covariance* of two random variables X and Y,

$$\mathbf{cov}(X, Y) = \mathbf{E}\Big(\big(X - \mathbf{E}(X)\big)\big(Y - \mathbf{E}(Y)\big)\Big), \tag{1.93}$$

and in particular

$$\mathbf{var}(X) = \mathbf{cov}(X, X) = \mathbf{E}\Big(\big(X - \mathbf{E}(X)\big)^2\Big) \tag{1.94}$$

is the *variance* of X.

Thorough introductions to the theory of measure, integration and probability can be found in many textbooks; for example, Billingsley (1995), Kallenberg (2002), Pollard (2002), Capiński and Kopp (2004), Rosenthal (2006), Yeh (2006), Bogachev (2007), Athreya and Lahiri (2010), Durrett (2010) and Gut (2013).

2

Point processes I – The Poisson point process

2.1 Introduction

The basic ingredients of practical geometry are points. Similarly random point patterns, point fields or *point processes* (in mathematical terminology) play a fundamental rôle in stochastic geometry and they arise directly as results of investigations into nature and technology. Figures in Chapters 2, 4 and 5 display various point patterns that appear in this fashion.

This chapter begins the task of showing how to study them mathematically. Frequently they occur as systems of points connected with particles, pores, or other geometrical objects. For example, a system of particles gives rise to a point pattern generated by the particle centroids. This leads on to a second, more indirect, fashion in which random point patterns arise. In many problems of stochastic geometry it is helpful to interpret patterns of geometrical objects as systems of 'points' in suitable 'representative spaces'; a 'point' from the system represents a particular geometrical object in the original pattern. This is sketched in Section 4.8 and described in detail for the case of random lines in the plane in Section 8.2.2.

In the current chapter the simplest and most important random point pattern is studied: the Poisson (point) process. This study will be carried out in a heuristic fashion, generally avoiding use of the abstract theory of point processes. It is hoped that readers having less mathematical training will thus be encouraged to read further and study the general theory as presented in Chapter 4 and the discussion in Chapter 5 of further models for random point patterns. A detailed treatment of the Poisson process is given in Kingman (1993, 2006), Daley and Vere-Jones (2003) and Streit (2010).

An appreciation of the possible varieties of random point patterns leads to an understanding of the central position that the Poisson point process holds among theoretical models. A given random point pattern may well exhibit various kinds of interaction between its

Stochastic Geometry and its Applications, Third Edition.
Sung Nok Chiu, Dietrich Stoyan, Wilfrid S. Kendall and Joseph Mecke.

constituent points. For example, the points may occur in clusters (see Figures 5.4 and 5.6) or may exhibit regularity (see Figure 5.7). There may be a *hard-core distance* (a minimum inter-point distance). These features may occur together in the same pattern (see Figure 5.9) and in various scales. The subject of point process statistics aims to detect and to quantify such interactions. Models from the theory of point processes can be used both in comparison to the original point pattern and also in representation of it.

In the absence of any of the above interactions a point pattern can be thought of as completely random. A theoretical model of such a pattern is of importance as a basis for comparison, as a null-model. Calculations can be made of the extent of probable fluctuations in such a model. These fluctuations provide objective grounds for ignoring sufficiently small empirically observed interactions, on the basis that they are insignificant compared with features that might well be observed in a completely random point pattern. If intuitively appealing axioms are imposed (concerning homogeneity and lack of interaction) then completely random processes can be characterised as Poisson point processes. Thus one of the central rôles of the Poisson point process is to serve as a null hypothesis for statistical tests of interaction.

The other main rôle lies in the use of the Poisson process as a basic building block for other more complicated models; see, for example, Sections 3.1, 5.3 and 9.7. Theoretical definitions of varieties of point processes frequently make reference to Poisson point processes. Simulation procedures often include the construction of a Poisson point process, which is then modified into the form required.

A history of the concept of a Poisson process can be found in Guttorp and Thorarinsdottir (2012). They reported that the first recorded use of the Poisson process in spatial statistics appears to be that of Michell (1767), who studied distances of stars, followed by Clausius (1857), who applied it in the kinetic theory of heat, and Abbe (1879), who considered the problem of counting blood particles — of course all these scientists did not know the modern idea of a Poisson process.

2.2 The binomial point process

2.2.1 Introduction

The most elementary example of a point process is one that contains one point only. A random point x *uniformly distributed* in a compact set $W \subset \mathbb{R}^d$ is a random point such that

$$\mathbf{P}(x \in A) = \frac{v_d(A)}{v_d(W)} \tag{2.1}$$

for all Borel sets A contained in W. Many problems in geometric probability arise from the study of uniformly distributed random points and the determination of the volume ratio in (2.1) when A is defined by some geometrical construction. The progress of modern stochastic geometry can to some extent be summarised as systematic replacement of uniformly distributed random points by more general random point patterns.

A uniformly distributed random point is a rather trivial random pattern. However, n independent uniformly distributed random points can be superposed to form a new, more interesting random point pattern, a *binomial point process of n points*. Such a process is formed by n independent points $x_1, \ldots, x_n$ uniformly distributed in the same compact set W. From Formula (2.1)

$$\mathbf{P}(x_1 \in A_1, \ldots, x_n \in A_n) = \mathbf{P}(x_1 \in A_1) \cdot \ldots \cdot \mathbf{P}(x_n \in A_n)$$
$$= \frac{v_d(A_1) \cdot \ldots \cdot v_d(A_n)}{v_d(W)^n} \qquad (2.2)$$

for Borel subsets $A_1, \ldots, A_n$ of W. This leads to an equivalent definition; the points $x_1, \ldots, x_n$ form a binomial point process in W (for $W \subset \mathbb{R}^d$) if the random vector $(x_1, \ldots, x_n)$ is uniformly distributed in W^n. Figure 2.1 (on p. 39) shows the result of a simulation of a binomial point process in $W = [0, 1]^2$ with $n = 100$ points.

If $x_1, \ldots, x_n$ form a binomial point process in W then the random pattern constituted by these points is denoted by $\Phi_{W^{(n)}}$. For most purposes the ordering of the points is irrelevant and $\Phi_{W^{(n)}}$ can be regarded as a random set.

It can also be thought of as a random counting measure; for a Borel set A let $\Phi_{W^{(n)}}(A)$ denote the number of points of $\Phi_{W^{(n)}}$ falling in A. Thus

$$\Phi_{W^{(n)}}(\emptyset) = 0, \qquad \Phi_{W^{(n)}}(W) = n,$$

and

$$\Phi_{W^{(n)}}(A_1 \cup A_2) = \Phi_{W^{(n)}}(A_1) + \Phi_{W^{(n)}}(A_2)$$

whenever A_1 and A_2 are disjoint subsets of W. The σ-additivity property also holds. Hence $\Phi_{W^{(n)}}$ is a random measure.

The binomial point process $\Phi_{W^{(n)}}$ is the first nontrivial example of a point process considered in this book. Notice that it can be viewed either as a point process, or as a random set, or as a random measure; these multiple views are important for the study of all (simple) random point processes. A point process is *simple* if no two points coincide.

2.2.2 Basic properties

The binomial point process $\Phi_{W^{(n)}}$ earns its name from a distributional property. If A is a Borel subset of W then $\Phi_{W^{(n)}}(A)$ has a binomial distribution with parameters $n = \Phi_{W^{(n)}}(W)$ and $p = p(A) = v_d(A)/v_d(W)$. Since the mean of the binomial distribution is np, the mean number of points per unit volume, or *intensity* of the binomial point process, is given by

$$\lambda = \frac{n}{v_d(W)} \qquad (2.3)$$

and

$$\mathbf{E}\big(\Phi_{W^{(n)}}(A)\big) = \lambda v_d(A) \qquad \text{for any Borel set } A \subset W. \qquad (2.4)$$

Numbers of points in different subsets of W are *not* independent even if the subsets are disjoint. This is clear since $\Phi_{W^{(n)}}(A) = m$ necessarily implies $\Phi_{W^{(n)}}(W \setminus A) = n - m$.

The distribution of $\Phi_{W^{(n)}}$ as a point process is completely characterised by the so-called *finite-dimensional distributions*:

$$\mathbf{P}(\Phi_{W^{(n)}}(A_1) = n_1, \ldots, \Phi_{W^{(n)}}(A_k) = n_k) \qquad \text{for } k = 1, 2, \ldots, \qquad (2.5)$$

where $A_1, \ldots, A_k$ are arbitrary Borel sets and $n_1, \ldots, n_k$ are nonnegative integers satisfying $n_1 + \cdots + n_k \leq n$.

Obviously, the binomial process is simple; all points are isolated.

A mathematical theorem says that the distribution of a simple point process is determined by its so-called *void-probabilities* $v_K = \mathbf{P}(\Phi_{W^{(n)}}(K) = 0)$, where K is an arbitrary compact subset of W. Section 6.1 discusses this way of specifying distributions in the much more general context of random sets; see also the remarks on void-probabilities in Section 2.3.1.

The void-probabilities for the binomial point process are given by

$$v_K = \mathbf{P}(\Phi_{W^{(n)}}(K) = 0) = \frac{\left(v_d(W) - v_d(K)\right)^n}{v_d(W)^n}. \tag{2.6}$$

When $A_1, \ldots, A_k$ are disjoint Borel sets with $A_1 \cup \cdots \cup A_k = W$ and $n_1 + \cdots + n_k = n$, the finite-dimensional distributions are given by the multinomial probabilities

$$\mathbf{P}(\Phi_{W^{(n)}}(A_1) = n_1, \ldots, \Phi_{W^{(n)}}(A_k) = n_k)$$
$$= \frac{n!}{n_1! \cdots \cdots n_k!} \cdot \frac{v_d(A_1)^{n_1} \cdots \cdots v_d(A_k)^{n_k}}{v_d(W)^n}. \tag{2.7}$$

As noted above, $\Phi_{W^{(n)}}(A)$ is of binomial distribution with parameters n and $p(A)$. The well-known Poisson limit theorem yields that if the total number of points n tends to infinity and the second parameter $p(A)$ tends to zero in such a way that the product $\lambda \cdot v_d(A)$ remains fixed then $\Phi_{W^{(n)}}(A)$ is asymptotically of Poisson distribution of mean $\lambda \cdot v_d(A)$. This limit can be obtained if the region W is enlarged to fill out all of $\mathbb{R}^d$ while n is allowed to tend to infinity. If the ratio $n/v_d(W)$ remains fixed as n increases and W is enlarged then the Poisson limit will hold for $\Phi_{W^{(n)}}(A)$ for any fixed bounded Borel set A. So if there is a limiting point process Φ then it should possess the property:

$$\Phi(A) \text{ is Poisson of mean } \lambda \cdot v_d(A) \text{ for each bounded Borel set } A,$$

where $\Phi(A)$ is the number of points of Φ in A. It is an implication of (2.7) that such a limiting process should be 'independently scattered':

$$\Phi(A_1), \ldots, \Phi(A_k) \text{ are independent if } A_1, \ldots, A_k \text{ are disjoint bounded Borel sets.}$$

These points are discussed again in Section 2.3.1.

2.2.3 Simulation

The simulation of a binomial point process follows easily from superposition once one knows how to simulate a random point uniformly distributed over the required region; the binomial point process can be obtained by n independent replications of simulating random points.

It is straightforward to simulate a random point uniformly distributed in $[0, 1]^2$. If $\{u_j\}$ is a sequence of independent random numbers uniformly distributed in $[0, 1]$ then the points

$$x_i = (u_{2i-1}, u_{2i}) \qquad \text{for } i = 1, 2, \ldots \tag{2.8}$$

form a sequence of independent random points uniformly distributed in $[0, 1]^2$. Figure 2.1 shows a sample of 100 points obtained in this way. The numbers $\{u_j\}$ were produced by a

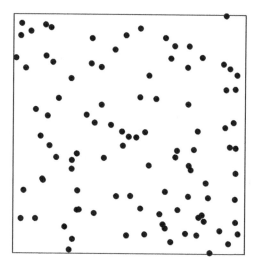

Figure 2.1   A pattern of 100 simulated random points uniformly distributed in a square.

random number generator; see Gentle (2003) and Kroese *et al.* (2011) for more on random number generation.

A sequence of random points uniformly distributed in the hypercube $[0, 1]^d$ is produced by

$$x_i = (u_{(i-1)d+1}, \ldots, u_{id}) \qquad \text{for } i = 1, 2, \ldots. \tag{2.9}$$

Translation and scale changes can be used to produce sequences of points uniformly distributed in any fixed rectangle or hypercube.

Such simulation procedures cover a large number of cases in practice, as the binomial point process to be simulated will frequently take place in a square or cube. However, other regions do arise and then further work must be done. Simulation of a uniform random point in a general bounded region W is tackled using one of three main techniques. (To be definite only the common planar case is discussed.)

(a) *Rejection sampling.* A rectangle R containing W is found and a sequence of independent uniform random points is simulated in R until a point first falls in W. This point will be uniformly distributed in W. To obtain a binomial point process this whole procedure is repeated until n points have fallen in W, and these n points constitute the sample for the binomial point process. To maximise efficiency R should be chosen to be as small as possible. Figure 2.2 illustrates the process.

(b) *Approximation.* The region W is replaced by a disjoint union of k open squares approximating W. A random point distributed uniformly over this union is simulated by first choosing a square with probability proportional to its area and then simulating a random point uniformly distributed over that square.

Exact simulations for complicated regions can be obtained by combining this technique with rejection sampling.

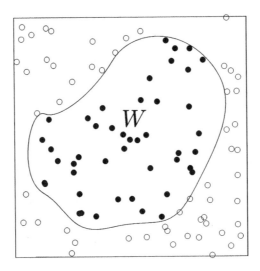

Figure 2.2   A simulation of 45 random points. The points • are uniformly distributed in the set W. The points o are those points of Figure 2.1 that do not lie in W.

(c) *Transformation of coordinates.* If the region W exhibits some symmetry then transformation of coordinates may be useful. For example, if W is the unit disc $B(o, 1)$ then a uniform random point can be described in polar coordinates

$$x = (r, \theta) \qquad \text{for } r \in [0, 1] \text{ and } \theta \in (0, 2\pi].$$

The random variables r and θ are independent, with θ uniform over $(0, 2\pi]$ and r satisfying the law

$$\mathbf{P}(r \leq t) = t^2 \qquad \text{if } 0 \leq t \leq 1.$$

Thus if u_1 and u_2 are independent random numbers uniform over $[0, 1]$ then the formulae

$$r = \sqrt{u_1} \qquad \text{and} \qquad \theta = 2\pi u_2 \tag{2.10}$$

provide a method of simulating $x = (r, \theta)$.

As explained in Section 2.3.1 paragraph (e) on p. 43 the binomial point process arises from the homogeneous Poisson point process by conditioning. Consequently it is not necessary to give statistical methods for testing the hypothesis that a given point pattern is a realisation of a binomial point process; one can use the methods for homogeneous Poisson point processes described in Section 2.6.4. On the other hand, the simulation procedures for homogeneous Poisson point processes and general Poisson point processes depend heavily on the simulation methods for binomial point processes described here.

2.3   The homogeneous Poisson point process

2.3.1   Definition and defining properties

Since a general treatment of point processes appears in Chapter 4, only a brief explanation of terminology is given here. As seen above, a random point pattern or *point process* Φ can be regarded either as a random sequence (more strictly a random set) $\Phi = \{x_1, x_2, \ldots\}$ or else as a random counting measure: for each Borel set B the symbol $\Phi(B)$ denotes the random number of points of Φ which lie in the set B.

As a random set, Φ can be intersected with other sets; if B is a Borel set then $B \cap \Phi$ is the random set of points of Φ that also belong to B. Because the random point patterns considered are all *locally finite* the random set Φ will always be closed and $B \cap \Phi$ will be finite whenever B is bounded.

A homogeneous Poisson point process Φ is characterised by two fundamental properties, which have already appeared as asymptotic properties in Section 2.2.2. They are:

(1) *Poisson distribution of point counts.* The random number of points of Φ in a bounded Borel set B has a Poisson distribution of mean $\lambda v_d(B)$ for some constant λ, that is,

$$\mathbf{P}\big(\Phi(B) = m\big) = \frac{\mu^m}{m!} \exp(-\mu) \qquad \text{for } m = 0, 1, 2, \ldots, \tag{2.11}$$

where

$$\mu = \lambda v_d(B). \tag{2.12}$$

(2) *Independent scattering.* The numbers of points of Φ in k disjoint Borel sets form k independent random variables, for arbitrary k.

Property (2) is also known as the 'completely random' or 'purely random' property.

The number λ occurring in property (1) is the characteristic parameter of the homogeneous Poisson point process. It gives the mean number of points in a set of unit volume, and satisfies

$$\lambda v_d(B) = \mathbf{E}\big(\Phi(B)\big) \qquad \text{for all bounded Borel sets } B. \tag{2.13}$$

It is called the *intensity* or *density* of the homogeneous Poisson point process Φ. The following always assumes that λ is positive and finite. If $\lambda = 0$ then the point pattern contains no points while an infinite λ corresponds to a pathological case.

The intensity can also be interpreted in infinitesimal terms. If B is a set of small Lebesgue measure then by property (1) the following asymptotics hold:

$$\mathbf{P}\big(\Phi(B) = 0\big) = 1 - \lambda v_d(B) + o\big(v_d(B)\big),$$

$$\mathbf{P}\big(\Phi(B) = 1\big) = \lambda v_d(B) + o\big(v_d(B)\big),$$

$$\mathbf{P}\big(\Phi(B) > 1\big) = o\big(v_d(B)\big).$$

Let Φ be a homogeneous Poisson point process of intensity λ. From properties (1) and (2) the whole distribution of the homogeneous Poisson point process can be determined once the intensity λ is known. The following summarises some basic properties of Φ.

(a) *Finite-dimensional distributions.* It can be shown directly from properties (1) and (2) that if $B_1, \ldots, B_k$ are disjoint bounded Borel sets then $\Phi(B_1), \ldots, \Phi(B_k)$ are independent Poisson random variables with means $\lambda v_d(B_1), \ldots, \lambda v_d(B_k)$. Thus

$$
\begin{aligned}
&\mathbf{P}(\Phi(B_1) = n_1, \ldots, \Phi(B_k) = n_k) \\
&= \frac{\lambda^{n_1 + \cdots + n_k} \left(v_d(B_1)\right)^{n_1} \cdot \ldots \cdot \left(v_d(B_k)\right)^{n_k}}{n_1! \cdot \ldots \cdot n_k!} \exp\left(-\sum_{i=1}^{k} \lambda v_d(B_i)\right).
\end{aligned} \tag{2.14}
$$

From this formula the joint probabilities $\mathbf{P}(\Phi(B_1) = n_1, \ldots, \Phi(B_k) = n_k)$ can be evaluated for general (possibly overlapping) $B_1, \ldots, B_k$.

(b) *Stationarity and isotropy.* For a point process $\Phi = \{x_n\}$ to be stationary the translated process $\Phi_x = \{x_n + x\}$ must have the same distribution for all x in $\mathbb{R}^d$. A point process is isotropic if the same is true for all rotated processes $r\Phi = \{rx_n\}$ where r is a rotation about the origin. A process is motion-invariant if it possesses both of these properties. The homogeneous Poisson point process Φ is defined by properties (1) and (2) above and the specification of the intensity λ. These properties and the characteristic λ are clearly invariant under rotation and translation. Therefore the homogeneous Poisson point process Φ must be stationary and isotropic; that is to say, motion-invariant.

This may be verified directly by noting that the finite-dimensional distributions above remain the same whether one uses a homogeneous Poisson point process Φ or its translation Φ_x or its rotation $r\Phi$.

More generally, the Poisson process has the following conservation property. Let $\mathbf{A}$ be a nonsingular linear mapping from $\mathbb{R}^d$ to $\mathbb{R}^d$. If Φ is a homogeneous Poisson process of intensity λ then $\mathbf{A}\Phi = \{\mathbf{A}x : x \in \Phi\}$ is also a homogeneous Poisson process and its intensity is $\lambda |\det(\mathbf{A}^{-1})|$, where $\det(\mathbf{A}^{-1})$ is the determinant of the inverse of $\mathbf{A}$.

This book uses the term 'homogeneous' Poisson process instead of 'stationary' Poisson process, in contrast to the usage in SKM95 and in many mathematical texts on stochastic geometry. The use of the term 'homogeneous' Poisson process follows long traditions and may make understanding for non-mathematicians easier. See also the discussion of the terms 'stationary' and 'homogeneous' in Section 4.1.1.

(c) *Void-probabilities.* The *void-probabilities* of a point process are the probabilities of there being no point of the process in given test sets B:

$$
v_B = \mathbf{P}\big(\Phi(B) = 0\big).
$$

In the case of Φ being a homogeneous Poisson point process,

$$
v_B = \exp\big(-\lambda v_d(B)\big). \tag{2.15}
$$

(d) *Contact distribution functions.* The *contact distribution functions* are closely associated with the void-probabilities. If B is a Borel set with $v_d(B) > 0$ having the property that $r_1 \leq r_2$ implies $r_1 B \subset r_2 B$ then the contact distribution function $H_B(r)$ (with respect to B) is given by

$$
H_B(r) = 1 - v_{rB} = 1 - \mathbf{P}(\Phi(rB) = 0) \qquad \text{for } r \geq 0. \tag{2.16}
$$

It is easy to see that $H_B(r)$ is really a distribution function: in particular $0 \cdot B = \{o\}$ and $\Phi(\{o\}) = 0$ and thus $H_B(0) = 0$; $r_1 B \subset r_2 B$ and thus $H_B(r_1) \leq H_B(r_2)$ holds if $r_1 \leq r_2$; '$\Phi(\infty B) = \infty$' and thus '$P(\Phi(\infty B) = 0) = 0$' and $H_B(\infty) = 1$.

A particularly important case uses $B = B(o, 1)$ the unit ball, yielding the *spherical contact distribution* or *first contact distribution function*

$$H_s(r) = 1 - P(\Phi(B(o, r)) = 0) = 1 - \exp(-\lambda b_d r^d) \qquad \text{for } r \geq 0. \qquad (2.17)$$

This can be reinterpreted as the distribution function of the distance from o to the nearest point of Φ.

(e) *Conditioning and binomial point processes.* If Φ is a homogeneous Poisson point process then one can consider the restriction of Φ to a compact set W under the condition that $\Phi(W) = n$, that is, that in W there are exactly n points. This conditioning yields a finite point process, which is the binomial point process in W with n points.

This assertion can be proved by showing that the finite-dimensional distributions of the two processes coincide. In fact, it is sufficient to consider the void-probabilities. If K is a compact subset of W then the void-probability for K of the conditioned homogeneous Poisson point process is given by

$$P(\Phi(K) = 0 | \Phi(W) = n) = P(\Phi(K) = 0, \Phi(W) = n)/P(\Phi(W) = n)$$

$$= P(\Phi(K) = 0)P(\Phi(W \setminus K) = n)/P(\Phi(W) = n)$$

$$= \frac{(\nu_d(W) - \nu_d(K))^n}{\nu_d(W)^n}$$

after substitution and cancellation. This formula coincides with Formula (2.6) for the void-probability of the binomial point process.

The general theory of simple point processes contains a theorem asserting equality of the distributions of point processes if their systems of void-probabilities are equal. However, in the case just considered it is straightforward to directly compute the finite-dimensional distributions of the conditioned process (in a manner similar to the above) and to check that these are equal to the finite-dimensional distributions of the binomial point process as given by Formula (2.7).

2.3.2 Characterisation of the homogeneous Poisson point process

The properties (1) and (2) on p. 41 of the homogeneous Poisson point process are not logically independent; see Kingman (1993) and Daley and Vere-Jones (2003). Rényi (1967) shows that (1) implies (2); the Poisson distribution property forces the point process to have the independently scattered property. Indeed this follows from the remark after paragraph (e) above. However, property (2) does *not* follow if property (1) holds only for the class of all convex subsets of $\mathbb{R}^d$. Moreover, Moran (1976) shows that it is not sufficient to assume the independence and Poisson distribution of counts of points in k disjoint convex sets for some fixed k. Thus if property (1) is weakened then property (2) may be necessary for characterisation of the homogeneous Poisson point process.

Another set of properties also characterises the Poisson process, and should be borne in mind as it frequently provides a *prima facie* case for assuming that an empirical point pattern is a realisation of a homogeneous Poisson point process. This characterisation asserts that a process must be a homogeneous Poisson point process if the following three properties are satisfied:

(I) *Simplicity.* No two points coincide, so that the process is a simple point process. That is to say, there are no multiple points.

(II) *Stationarity.* This is as defined above.

(III) *Independent scattering.* This is property (2).

In stochastic geometry the property of simplicity generally holds by definition. Usually it is evident from the nature of the considered point pattern whether or not it is simple.

The stationarity assumption has greater significance. It implies that the statistical properties of the point process do not depend on the location of the observer. In some cases, some aspects of stationarity or homogeneity can be tested statistically but often it must be assumed provisionally in order to make progress in the analysis of the point pattern.

The most important assumption is that of independent scattering. In effect it asserts that there is no interaction between the points of the pattern. It can be tested statistically. In some cases, it is suggested by underlying biological or physical theories. Plausible or not, it frequently provides the starting point for statistical analysis, even if only as a null hypothesis.

2.3.3   Moments and moment measures

Just as moments such as mean and variance are important characteristics for random variables, so are the corresponding entities for the Poisson process and for general point processes. However, there is increasing complexity in moving from a random variable to a random point process. While random variables have moments that are real numbers, a point process has moments that are measures. The following describes these measures for the specific case of the homogeneous Poisson point process. In this case there are explicit formulae of relatively simple forms. Section 4.3 discusses moment measures for general point processes.

Consider the homogeneous Poisson point process Φ, regarded as a random measure on the Borel sets in $\mathbb{R}^d$. If B is a Borel set then $\Phi(B)$ is a random variable with first moment or mean

$$\Lambda(B) = \mathbf{E}\big(\Phi(B)\big). \tag{2.18}$$

If B_1 and B_2 are two Borel sets then $\Phi(B_1)$ and $\Phi(B_2)$ are two random variables with a noncentred covariance

$$\mu^{(2)}(B_1 \times B_2) = \mathbf{E}\big(\Phi(B_1)\Phi(B_2)\big). \tag{2.19}$$

Both Λ and $\mu^{(2)}$ are moments of random variables but depend on Borel sets (B) or products of Borel sets ($B_1 \times B_2$). The nature of this dependence is σ-additive (this follows from Φ being a random measure) and so they can both be represented as measures.

The first-order quantity Λ has a straightforward form. By property (1) the random variable $\Phi(B)$ is Poisson of mean $\lambda \cdot v_d(B)$ and so

$$\Lambda(B) = \mathbf{E}\big(\Phi(B)\big) = \lambda v_d(B) \qquad \text{for all Borel sets } B. \tag{2.20}$$

Thus in the case of the homogeneous Poisson point process the *first moment measure* or *intensity measure* Λ is a constant multiple of Lebesgue measure and the multiple is given by the intensity of the Poisson process.

The second-order quantity $\mu^{(2)}(B_1 \times B_2)$ given by (2.19) is evaluated by using both properties (1) and (2). To evaluate it one notes that both B_1 and B_2 can be decomposed into disjoint unions

$$B_1 = (B_1 \cap B_2) \cup (B_1 \setminus B_2), \qquad B_2 = (B_1 \cap B_2) \cup (B_2 \setminus B_1).$$

Using property (2), and the fact that Φ is a random measure, one can establish

$$\mu^{(2)}(B_1 \times B_2) = \mathbf{E}\big(\Phi(B_1)\Phi(B_2)\big)$$
$$= \mathbf{E}\big(\Phi(B_1 \setminus B_2)\big)\mathbf{E}\big(\Phi(B_2 \setminus B_1)\big) + \mathbf{E}\big(\Phi(B_1 \cap B_2)\big)\mathbf{E}\big(\Phi(B_2 \setminus B_1)\big)$$
$$+ \mathbf{E}\big(\Phi(B_1 \setminus B_2)\big)\mathbf{E}\big(\Phi(B_1 \cap B_2)\big) + \mathbf{E}\big((\Phi(B_1 \cap B_2))^2\big)$$
$$= \mathbf{E}\big(\Phi(B_1)\big)\mathbf{E}\big(\Phi(B_2)\big) + \mathbf{E}\big((\Phi(B_1 \cap B_2))^2\big) - \big(\mathbf{E}\big(\Phi(B_1 \cap B_2)\big)\big)^2.$$

Property (1) shows that $\Phi(B_1 \cap B_2)$ is Poisson of mean and variance $\Lambda(B_1 \cap B_2)$. So with the aid of (2.20) the final formula can be derived:

$$\mu^{(2)}(B_1 \times B_2) = \Lambda(B_1)\Lambda(B_2) + \Lambda(B_1 \cap B_2)$$
$$= \lambda^2 v_d(B_1)v_d(B_2) + \lambda v_d(B_1 \cap B_2). \tag{2.21}$$

Thus the *second moment measure* $\mu^{(2)}$, which is a measure on $\mathbb{R}^d \times \mathbb{R}^d$, can be expressed in terms of λ and Lebesgue measure.

Variances and covariances can be calculated directly from the second moment measure. The relevant formulae are

$$\mathbf{var}\big(\Phi(B)\big) = \mu^{(2)}(B \times B) - \big(\Lambda(B)\big)^2 \tag{2.22}$$

and

$$\mathbf{cov}\big(\Phi(B_1), \Phi(B_2)\big) = \mu^{(2)}(B_1 \times B_2) - \Lambda(B_1) \cdot \Lambda(B_2) \tag{2.23}$$

for all Borel sets B, B_1 and B_2. These formulae follow immediately from the definitions of variance and covariance. By using (2.21) these simplify to

$$\mathbf{var}\big(\Phi(B)\big) = \lambda v_d(B) \tag{2.24}$$

and

$$\mathbf{cov}\big(\Phi(B_1), \Phi(B_2)\big) = \lambda v_d(B_1 \cap B_2). \tag{2.25}$$

As a random measure, Φ counts the number of points falling in the set which is the argument. Therefore $\mu^{(2)}(B_1 \times B_2)$ can also be expressed as the expectation of a sum:

$$\mu^{(2)}(B_1 \times B_2) = \mathbf{E}\big(\#\{(x_1, x_2) : x_1 \in \Phi \cap B_1, x_2 \in \Phi \cap B_2\}\big)$$

$$= \mathbf{E}\left(\sum_{x_1, x_2 \in \Phi} \mathbf{1}_{B_1}(x_1)\mathbf{1}_{B_2}(x_2)\right), \tag{2.26}$$

where Φ, as noted before, is also used to denote the random set of the point process. The two terms in (2.21) correspond to the dissection of this sum into the sum over distinct pairs of points $x_1, x_2 \in \Phi$ and the sum over equal points $x_1 = x_2 \in \Phi$.

For some purposes it is convenient to subtract out the second of these terms. The result is the *second-order factorial moment measure* $\alpha^{(2)}$:

$$\alpha^{(2)}(B_1 \times B_2) = \mathbf{E}\big(\#\{(x_1, x_2) : x_1 \in \Phi \cap B_1, x_2 \in \Phi \cap B_2, x_1 \neq x_2\}\big)$$

$$= \mathbf{E}\left(\sum_{x_1, x_2 \in \Phi}^{\neq} \mathbf{1}_{B_1}(x_1)\mathbf{1}_{B_2}(x_2)\right). \tag{2.27}$$

Here $\sum^{\neq}$ stands for summation over all pairs (x_1, x_2) such that $x_1 \neq x_2$.

The difference between $\mu^{(2)}(B_1 \times B_2)$ and $\alpha^{(2)}(B_1 \times B_2)$ lies in the expectation of the sum

$$\sum_{\substack{x_1, x_2 \in \Phi \\ x_1 = x_2}} \mathbf{1}_{B_1}(x_1)\mathbf{1}_{B_2}(x_2) = \sum_{x \in \Phi} \mathbf{1}_{B_1}(x)\mathbf{1}_{B_2}(x) = \sum_{x \in \Phi} \mathbf{1}_{B_1 \cap B_2}(x).$$

Hence

$$\mu^{(2)}(B_1 \times B_2) = \alpha^{(2)}(B_1 \times B_2) + \Lambda(B_1 \cap B_2). \tag{2.28}$$

In the case of the homogeneous Poisson point process, the following elegant formula is obtained:

$$\alpha^{(2)}(B_1 \times B_2) = \lambda^2 v_d(B_1)v_d(B_2) = \Lambda(B_1)\Lambda(B_2). \tag{2.29}$$

So in this case the second-order factorial moment measure $\alpha^{(2)}$ is simply a constant multiple of Lebesgue measure on $\mathbb{R}^d \times \mathbb{R}^d$.

The density of $\alpha^{(2)}$ with respect to Lebesgue measure is known as the *second-order product density* $\varrho^{(2)}$ and the above shows that for a homogeneous Poisson process

$$\varrho^{(2)}(x_1, x_2) \equiv \lambda^2 \qquad \text{for } x_1, x_2 \in \mathbb{R}^d, \tag{2.30}$$

and normalising $\varrho^{(2)}(x_1, x_2)$ by λ^2 gives the *pair correlation function* $g(x_1, x_2)$:

$$g(x_1, x_2) = \varrho^{(2)}(x_1, x_2)/\lambda^2. \tag{2.31}$$

For a stationary and isotropic process both $\varrho^{(2)}(x_1, x_2)$ and $g(x_1, x_2)$ depend solely on the distance $\|x_1 - x_2\| = r$ and in this case it is more convenient to write

$$\varrho^{(2)}(r) = \varrho^{(2)}(x_1, x_2) \quad \text{and} \quad g(r) = g(x_1, x_2).$$

For a homogeneous Poisson process,

$$g(r) \equiv 1. \tag{2.32}$$

Note that $\mu^{(2)}$ does not have a density with respect to Lebesgue measure. This is the principal motivation for using $\alpha^{(2)}$. The following holds

$$E\big(\Phi(B_1)\Phi(B_2)\big) = \int_{B_2} \int_{B_1} \varrho^{(2)}(x_1, x_2)\mathrm{d}x_1\mathrm{d}x_2.$$

The second-order product density has an infinitesimal interpretation. If B_1 and B_2 are two infinitesimally small disjoint Borel sets of volumes $\mathrm{d}x_1$ and $\mathrm{d}x_2$, respectively, and if $x_1 \in B_1$ and $x_2 \in B_2$ then

$$\varrho^{(2)}(x_1, x_2)\mathrm{d}x_1\mathrm{d}x_2 = \lambda^2 \mathrm{d}x_1\mathrm{d}x_2 \tag{2.33}$$

is the probability that Φ places a point in each of B_1 and B_2.

Note that the 'first-order product density' has already been defined: by analogy with the above it is simply the intensity λ.

Just as higher moments can be defined for random variables so higher moment measures can be defined for point processes. The nth *moment measure* $\mu^{(n)}$ is a measure on $\mathbb{R}^{nd}$ defined by

$$\mu^{(n)}(B_1 \times \cdots \times B_n) = E\big(\Phi(B_1) \cdot \ldots \cdot \Phi(B_n)\big)$$

$$= E\left(\sum_{x_1, \ldots, x_n \in \Phi} 1_{B_1}(x_1) \cdot \ldots \cdot 1_{B_n}(x_n)\right) \tag{2.34}$$

for Borel sets $B_1, \ldots, B_n$. The nth-*order factorial moment measure* $\alpha^{(n)}$ is a measure on $\mathbb{R}^{nd}$ given by

$$\alpha^{(n)}(B_1 \times \cdots \times B_n) = E\left(\sum_{x_1, \ldots, x_n \in \Phi}^{\neq} 1_{B_1}(x_1) \cdot \ldots \cdot 1_{B_n}(x_n)\right) \tag{2.35}$$

where $\sum^{\neq}$ stands for summation over n-tuples $(x_1, \ldots, x_n)$ of *distinct* points. The nth-*order product density* $\varrho^{(n)}$ is the density of $\alpha^{(n)}$ with respect to Lebesgue measure on $\mathbb{R}^{nd}$. In the case of homogeneous Poisson point process these product densities are given by the simple formula

$$\varrho^{(n)}(x_1, \ldots, x_n) \equiv \lambda^n \tag{2.36}$$

and $\alpha^{(n)}$ has the form

$$\alpha^{(n)}(B_1 \times \cdots \times B_n) = \lambda^n \nu_d(B_1) \cdot \ldots \cdot \nu_d(B_n). \tag{2.37}$$

For $n = 3$ the third moment measure is given by

$$\mu^{(3)}(B_1 \times B_2 \times B_3) = \lambda^3 v_d(B_1)v_d(B_2)v_d(B_3)$$
$$+ \lambda^2 v_d(B_1)v_d(B_2 \cap B_3) + \lambda^2 v_d(B_2)v_d(B_1 \cap B_3)$$
$$+ \lambda^2 v_d(B_3)v_d(B_1 \cap B_2) + \lambda v_d(B_1 \cap B_2 \cap B_3)$$

for Borel sets B_1, B_2 and B_3. For larger n the relationships between $\alpha^{(n)}$ and $\mu^{(n)}$ are quite complicated.

There is an infinitesimal interpretation for $\varrho^{(n)}(x_1, \ldots, x_n)$ which follows that of $\varrho^{(2)}(x_1, x_2)$.

2.3.4   The Palm distribution of a homogeneous Poisson point process

Many problems in point process theory require the consideration of an arbitrary 'typical' point of a point process Φ. It can be viewed informally as the result of a random selection procedure in which every point of the process has the same chance to be selected. For example, the nearest-neighbour distance distribution function $D(r)$ describes the distribution of the random distance from a typical point x of Φ to the nearest other point in the process, that is, in $\Phi \setminus \{x\}$.

The idea of a typical point is heuristically clear but needs to be made mathematically precise. Points sampled by some systematic method (such as by choosing the nearest to a given origin) are *not* typical just because they *have* been so sampled. Palm distribution theory makes precise the notion of a typical point. In intuitive terms the Palm distribution probabilities are the conditional probabilities of point process events given that a point (the typical point) is observed at a specific location.

Two heuristic approaches to the Palm distribution are discussed in this section, and calculations are made for the case of a homogeneous Poisson point process. One approach is essentially local and involves conditioning the distribution of the point process on there being a point at a given position x. The other approach, which is described on p. 50, is more global, and applies only to stationary point processes. It is related to statistical methods for estimation of Palm probabilities.

(a) Local approach

To discuss the local approach some notation is introduced here. It is necessary to discuss probabilities which involve conditioning of Φ on the event that Φ contains a point at x. If Y is some configuration set, describing some point process property (for example, the property that $\Phi(B) = n$ for some fixed Borel set B) then let

$$\mathbf{P}(\Phi \text{ has property } Y \parallel x) = \mathbf{P}(\Phi \text{ has property } Y \mid x \in \Phi).$$

This conditional probability will be understood as the *Palm distribution of* Φ and the point at x is the *typical point*. In accordance with notation introduced in Chapter 4, the event 'Φ has property Y' is written as '$\Phi \in Y$', interpreting Y as a configuration set, a member of a σ-algebra of sets of point-process realisations. By stationarity

$$\mathbf{P}(\Phi \in Y \parallel x) = \mathbf{P}(\Phi_{-x} \in Y \parallel o),$$

in which Φ_{-x} denotes the shifted point process $\{x_1 - x, x_2 - x, \ldots\}$, where Φ is the set $\{x_1, x_2, \ldots\}$.

In this notation the nearest-neighbour distance distribution function $D(r)$ for a homogeneous point process is given by

$$D(r) = \mathbf{P}\big(\Phi(B(o, r)) > 1 \parallel o\big) = 1 - \mathbf{P}\big(\Phi(B(o, r)) = 1 \parallel o\big),$$

since the distance from o (at which the typical point stands) to its nearest neighbour is not greater than r if the ball of radius r centred at o contains points of Φ.

The conditioning event $\{o \in \Phi\}$ always has probability zero if Φ is a homogeneous point process. This means that the conditioning must be carefully defined. Chapter 4 uses the Radon–Nikodym theorem, which frequently illuminates the meaning of conditional probability constructions in general probability theory. However, in the case of homogeneous Poisson point process the Palm probabilities can be calculated by a limit procedure, which is demonstrated below.

Calculation of the nearest-neighbour distance distribution function for the homogeneous Poisson point process

First consider a similar conditional distribution function, in which the conditioning is on a point being *near* to o. The conditional probability

$$D_\varepsilon(r) = 1 - \mathbf{P}\Big(\Phi\big(B(o, r) \setminus B(o, \varepsilon)\big) = 0 \mid \Phi\big(B(o, \varepsilon)\big) = 1\Big)$$

is well-defined for positive ε smaller than r, since the probability that the small ball $B(o, \varepsilon)$ contains a point, $\mathbf{P}\big(\Phi(B(o, \varepsilon)) = 1\big) = \lambda b_d \varepsilon^d \exp(-\lambda b_d \varepsilon^d)$ [where (2.11) is used], is positive. Using the definition of conditional probability

$$\mathbf{P}(A \mid B) = \mathbf{P}(A \cap B)/\mathbf{P}(B)$$

and property (2) of the homogeneous Poisson point process, one obtains

$$D_\varepsilon(r) = 1 - \frac{\mathbf{P}\big(\Phi(B(o, r) \setminus B(o, \varepsilon)) = 0\big)\mathbf{P}\big(\Phi(B(o, \varepsilon)) = 1\big)}{\mathbf{P}\big(\Phi(B(o, \varepsilon)) = 1\big)}$$

$$= 1 - \mathbf{P}\big(\Phi(B(o, r) \setminus B(o, \varepsilon)) = 0\big)$$

$$= 1 - \exp\big(-\lambda\big(v_d(B(o, r)) - v_d(B(o, \varepsilon))\big)\big).$$

It is reasonable to think of the nearest-neighbour distance distribution function as the limit of the above as $\varepsilon \to 0$. Setting $D(r) = \lim_{\varepsilon \to 0} D_\varepsilon(r)$ yields the result

$$D(r) = 1 - \exp\big(-\lambda v_d(B(o, r))\big) = 1 - \exp(-\lambda b_d r^d) \qquad \text{for } r \geq 0. \qquad (2.38)$$

This result can be established rigorously with the ideas of Section 4.4.4 using Palm distributions.

The right-hand sides of (2.38) and (2.17) are equal. This shows that for a homogeneous Poisson process the spherical contact distribution function and nearest-neighbour distance distribution are equal,

$$D(r) = H_s(r) \qquad \text{for } r \geq 0. \qquad (2.39)$$

The mean and variance of the random variable corresponding to the distribution function $D(r)$ are given by

$$
\mu_D = \begin{cases} \dfrac{1}{2\sqrt{\lambda}} & \text{for } d = 2, \\[2ex] \left(\dfrac{3}{4\pi\lambda}\right)^{1/3} \Gamma\!\left(\dfrac{4}{3}\right) \approx \dfrac{0.554}{\lambda^{1/3}} & \text{for } d = 3, \end{cases} \tag{2.40}
$$

and

$$
\sigma_D^2 = \begin{cases} \dfrac{1}{\lambda}\left(\dfrac{1}{\pi} - \dfrac{1}{4}\right) \approx \dfrac{0.0683}{\lambda} & \text{for } d = 2, \\[2ex] \left(\dfrac{3}{4\pi\lambda}\right)^{2/3} \left(\Gamma\!\left(\dfrac{5}{3}\right) - \Gamma^2\!\left(\dfrac{4}{3}\right)\right) \approx \dfrac{0.0405}{\lambda^{2/3}} & \text{for } d = 3. \end{cases} \tag{2.41}
$$

Illian *et al.* (2008, p. 76) presented formulae for the distributions of the distances to the second-nearest, third-nearest, ... neighbours.

Arguing as for (2.38) leads to the result

$$
\mathbf{P}\big(\Phi(B) = n \,\|\, o\big) = \mathbf{P}\big(\Phi(B) = n\big)
$$

for $n = 0, 1, 2, \ldots$ and for any compact set B not containing o. This suggests that the Palm distribution for the homogeneous Poisson point process is essentially the distribution of the original process together with a point adjoined at o. This is the content of the *Slivnyak–Mecke theorem*:

$$
\mathbf{P}(\Phi \text{ has property } Y \,\|\, o) = \mathbf{P}(\Phi \cup \{o\} \text{ has property } Y). \tag{2.42}
$$

In the σ-algebra language of Chapter 4,

$$
\mathbf{P}(\Phi \in Y \,\|\, o) = \mathbf{P}(\Phi \cup \{o\} \in Y). \tag{2.43}
$$

A proof of the Slivnyak–Mecke theorem is given in Section 4.4.4.

(b) Global approach

The second, more global, approach to Palm probabilities examines all points of Φ that fall within some fixed bounded Borel test set B with $v_d(B) > 0$. Denote by $N(\Phi, B, Y)$ the number of those points x in B such that Φ_{-x} has property Y. Then one can put

$$
\mathbf{P}(\Phi \text{ has property } Y \,\|\, o) = \mathbf{P}(\Phi \in Y \,\|\, o)
$$

$$
= \mathbf{E}\big(N(\Phi, B, Y)\big)/\lambda v_d(B). \tag{2.44}
$$

Since $\lambda v_d(B)$ is the expected number of points falling in B, this equation exhibits the Palm probability for the property Y as the fraction of points x expected to fall in B such that Φ_{-x} has the property Y. By the stationarity of Φ this definition does not depend on B.

It can be shown that the two approaches yield the same Palm distribution.

So as $D(r)$, another important distributional point process characteristic, the *reduced second moment function* $K(r)$, is also related to Palm distribution theory. This will be examined at greater length in Section 4.5. A brief indication of the nature of the relationship follows from the definition

$$\lambda K(r) = \mathbf{E}\big(\Phi(B(o,r)) \parallel o\big) - 1, \tag{2.45}$$

which means that $\lambda K(r)$ is the mean number of further points in a ball of radius r and centred at the typical point, which itself is not counted in the mean. The Slivnyak–Mecke theorem (2.42) yields

$$\lambda K(r) = \mathbf{E}\big(\Phi(B(o,r)) + 1\big) - 1 = \mathbf{E}\big(\Phi(B(o,r))\big). \tag{2.46}$$

Thus for the homogeneous Poisson process

$$K(r) = b_d r^d \qquad \text{for } r \geq 0. \tag{2.47}$$

2.4 The inhomogeneous and general Poisson point process

The homogeneous Poisson point process has a constant point density λ and an intensity measure which is proportional to Lebesgue measure. The mean number of its points per unit area does not vary over space. However, many point patterns arising in applications exhibit fluctuations that make such a lack of spatial variation implausible. For example, they may display an apparent trend of increase in a certain direction, as can be observed in the pattern of locations of trees in some forests. Suppose that the forest is in an area for which height above sea level increases, or soil becomes progressively drier, in a given direction. Then the mean number of trees per unit area may correspondingly decrease. Figures 2.3 and 2.4 on p. 52 show point patterns which clearly exhibit such directional dependence.

In such situations, it makes sense to consider a point process model with an *intensity function* $\lambda(x)$. It has an appealing and intuitive infinitesimal interpretation; $\lambda(x)dx$ is the infinitesimal probability that there is a point of Φ in a region of infinitesimal volume dx situated at x. The corresponding intensity measure Λ is given by

$$\Lambda(B) = \int_B \lambda(x)dx \qquad \text{for Borel sets } B. \tag{2.48}$$

Still a further generalisation is possible. Assume that there is a diffuse (without atoms) Radon measure Λ on $\mathbb{R}^d$. It can be used to construct a *general Poisson point process* Φ with *intensity measure* Λ as a point process possessing the two following properties:

(1') *Poisson distribution of point counts*: the number of points in a bounded Borel set B has a Poisson distribution with mean $\Lambda(B)$

$$\mathbf{P}(\Phi(B) = n) = \frac{\Lambda(B)^n}{n!} \exp\big(-\Lambda(B)\big) \qquad \text{for } n = 0, 1, 2, \ldots. \tag{2.49}$$

(2') *Independent scattering*: the numbers of points in k disjoint Borel sets form k independent random variables.

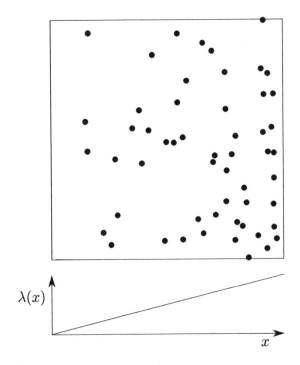

Figure 2.3 A simulated sample of a Poisson process which has a linearly increasing intensity function $\lambda(x)$. The pattern has been generated from that of Figure 2.1 by the thinning procedure described in Section 2.5.2.

It is clear from property (1′) that such a process Φ is *not* stationary in general. Without the assumption that Λ is diffuse, the process could have multiple points at the positions of the atoms.

The measure Λ is called the *intensity measure* of Φ and corresponds to the intensity measure previously introduced for the homogeneous Poisson point process. A point process

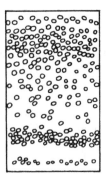

Figure 2.4 Cell nuclei developing in the pyramidal layer of the cerebral cortex of an embryo. The diagram is taken from von Economo and Koskinas (1925), see also Stephan (1975).

possessing an intensity function is called *inhomogeneous Poisson process*. For example, such a function does not exist when the points are randomly scattered on a deterministic system of lines. (If the lines are random, one is confronted with a Cox process; see Section 5.2.)

As for the homogeneous Poisson point process, here one can also define *moment measures*

$$\mu^{(n)}(B_1 \times \cdots \times B_n) = E\big(\Phi(B_1) \cdot \ldots \cdot \Phi(B_n)\big)$$

$$= E\left(\sum_{x_1, \ldots, x_k \in \Phi} \mathbf{1}_{B_1}(x_1) \cdot \ldots \cdot \mathbf{1}_{B_k}(x_k) \right)$$

and *factorial moment measures*

$$\alpha^{(n)}(B_1 \times \cdots \times B_n) = E\left(\sum_{x_1, \ldots, x_k \in \Phi}^{\neq} \mathbf{1}_{B_1}(x_1) \cdot \ldots \cdot \mathbf{1}_{B_k}(x_k) \right).$$

The factorial moment measures are given by

$$\alpha^{(n)}(B_1 \times \cdots \times B_n) = \Lambda(B_1) \cdot \ldots \cdot \Lambda(B_n). \tag{2.50}$$

Consequently in the case of an inhomogeneous Poisson process with intensity function $\lambda(x)$ the *product densities* $\varrho^{(n)}$ are given by

$$\varrho^{(n)}(x_1, \ldots, x_n) = \lambda(x_1) \cdot \ldots \cdot \lambda(x_n). \tag{2.51}$$

Further generalisations of the Poisson process are considered in Sections 5.2 and 5.3.

2.5   Simulation of Poisson point processes

2.5.1   Simulation of a homogeneous Poisson point process

The starting point for simulating a homogeneous Poisson point process is property (e) on p. 43: conditioning on the total number of points in a compact set produces a binomial point process. Thus the simulation of a homogeneous Poisson point process in a compact region W falls naturally into two stages. First the number of points in W is determined by simulating a Poisson random variable, and then the positions of the points in W are determined by simulating a binomial point process in W with that number of points. The second stage has already been covered in Section 2.2.3. There are various possibilities for the first stage, simulation of a Poisson random variable. Two of them are described now. Which of the two methods here is appropriate for simulating a Poisson random variable depends on its mean $\mu = \lambda v_d(W)$.

The usual method for small μ is to simulate a linear Poisson process, exploiting the fact that its inter-point distances are independent exponential random variables; see for example Kingman (1993, p. 39). Random variables e_i are generated to be independent and exponential of mean 1. This can be achieved by the transform method; if u_i is uniform on $[0, 1]$ then $e_i = -\ln(u_i)$ is as required. The desired Poisson random variable is the smallest n for which

$$\sum_{i=1}^{n+1} e_i > \mu. \tag{2.52}$$

Since addition of logarithms is equivalent to multiplication, the Poisson variable can be determined also as the smallest n satisfying

$$\prod_{i=1}^{n+1} u_i < \exp(-\mu). \tag{2.53}$$

An advantage of this method is its flexibility. It is an appropriate method to meet successive requirements to generate Poisson random variables of different (small) means.

In the case when μ is large then some form of acceptance/rejection technique should be used; see Gentle (2003, p. 188) and Kroese *et al.* (2011, pp. 100–1). Alternatively, the central limit theorem may simply be exploited. It states that for large μ, a Poisson random variable with mean μ approximately follows a Gaussian distribution with mean μ and variance μ. (This equality is because mean and variance coincide for a Poisson distribution.) Thus a Gaussian random number may be generated by well-known methods and then rounded to an integer.

In some cases, for example for a simulation of the typical Poisson-Voronoi polyhedron, it is useful to generate samples of a homogeneous Poisson process of intensity λ in a radial way, that is, as a sequence of points with increasing distance from the origin o. This method, called 'radial generation', is described in Quine and Watson (1984) for the d-dimensional case; here the formulae for $d = 2$ and $d = 3$ are given.

A point $x = (x_1, x_2)$ or $x = (x_1, x_2, x_3)$ is represented in polar coordinates by (r, θ) or (r, θ, θ') as

$$x_1 = r \cos \theta, \qquad x_2 = r \sin \theta,$$

or

$$x_1 = r \cos \theta \cos \theta', \qquad x_2 = r \sin \theta \cos \theta', \qquad x_3 = r \sin \theta'.$$

The simulation yields a sequence of points $\{(r_i, \theta_i)\}$ or $\{(r_i, \theta_i, \theta_i')\}$ with

$$r_i = \left((-\lambda b_d)^{-1} \sum_{j=1}^{i} \ln u_{ij} \right)^{1/d}$$

and

$$\theta_i = 2\pi u_i', \qquad \theta_i' = \sin^{-1}(1 - u_i''),$$

where u_{ij}, u_i' and u_i'' are independent uniform random numbers in $[0, 1]$. The sequence $\{r_i^d / b_d\}$ is a sample of a linear Poisson process with intensity λ.

2.5.2 Simulation of an inhomogeneous Poisson point process

For simulating a sample of an inhomogeneous Poisson point process with intensity function $\lambda(x)$ a thinning procedure can be used, as suggested by Lewis and Shedler (1979); see also Ogata (1981). Section 5.1 discusses thinning operations for point processes in a general setting.

The procedure, which assumes that the intensity function is bounded above by a number λ^*, falls into two stages. Firstly, a homogeneous Poisson point process of intensity λ^* is simulated as above. Secondly, the resulting point pattern is thinned by deleting each point x

independently of the others with probability $1 - \lambda(x)/\lambda^*$. If the points of the homogeneous Poisson point process pattern are $\{x_1, x_2, \ldots\}$ then this thinning can be performed with the aid of an independent sequence $u_1, u_2, \ldots$ of random numbers uniformly distributed over $[0, 1]$. The point x_k is deleted if $u_k > \lambda(x_k)/\lambda^*$. In the terminology of Section 5.1 the location-dependent thinning probability is $p(x) = \lambda(x)/\lambda^*$. Formula (5.2) ensures that the point pattern of undeleted points is indeed a sample of an inhomogeneous Poisson point process of intensity function $\lambda(x)$. The pattern in Figure 2.3 was generated in this manner.

2.6 Statistics for the homogeneous Poisson point process

2.6.1 Introduction

The discussion of the central rôle of the homogeneous Poisson point process (in Section 2.1) establishes the importance of statistical methods for this process, including the important problem of deciding whether or not a given point pattern is Poisson. This section is a brief survey of such methods. It considers mainly the case of *planar* point patterns, which is most studied in the literature: The methods presented carry over to point patterns on the line, in space, and on a sphere. In particular, there is special literature on the case of Poisson processes on the line; see, for example, Cox and Lewis (1966), Snyder (1975), Brillinger (1978) and Karr (1991). References for the planar case are Diggle (1983, 2003), Kutoyants (1998) and Illian *et al.* (2008).

2.6.2 Estimating the intensity

A fundamental statistical question for the homogeneous Poisson point process concerns the estimation of the intensity λ when the process is observed through a window W. A general unbiased estimator for the intensity λ of a homogeneous point process is given by

$$\hat{\lambda} = \frac{\Phi(W)}{v_d(W)}, \tag{2.54}$$

which quite naturally means that the observed point number in the window W is divided by the window's volume. This estimator is the most important also in the non-Poisson process case.

Distance methods, that is, methods employing inter-point distances or distances between points and chosen locations, are discussed in for example Diggle (1983, 2003) and Illian *et al.* (2008).

Confidence intervals for λ can be based on (2.54) since $\hat{\lambda} \cdot v_d(W) = \Phi(W)$ has a Poisson distribution. In the case of large $\Phi(W)$ Krebs (1999) and Armitage *et al.* (2002) give simple approximate $100(1 - \alpha)$ % confidence intervals for λ, employing the normal approximation and a continuity correction. For example,

$$\left(\frac{z_{\alpha/2}}{2} - \sqrt{\Phi(W)}\right)^2 \leq \lambda v_d(W) \leq \left(\frac{z_{\alpha/2}}{2} + \sqrt{\Phi(W) + 1}\right)^2. \tag{2.55}$$

Here the $z_{\alpha/2}$ are quantiles of the standard normal distribution:

$$z_{\alpha/2} = 1.645, \ 1.96, \ 2.576 \qquad \text{for } \alpha = 0.10, 0.05, 0.01, \text{ respectively.}$$

This confidence interval is useful in simple planning of the size of window required for a given accuracy of estimation. If δ is the desired width of the confidence interval and α is the required confidence level then

$$\delta \cdot \nu_d(W) \approx \left(\frac{z_{\alpha/2}}{2} + \sqrt{\lambda \nu_d(W) + 1}\right)^2 - \left(\frac{z_{\alpha/2}}{2} - \sqrt{\lambda \nu_d(W)}\right)^2,$$

from which follows the specification

$$\nu_d(W) \approx \frac{4z_{\alpha/2}^2 \lambda}{\delta^2},$$

where λ must itself be estimated by a preliminary analysis or by using *a priori* information.

The next statistical questions to be considered concern the model assumptions – those of homogeneity and of the Poisson process hypothesis.

2.6.3   Testing the hypothesis of homogeneity

If the point pattern is a homogeneous Poisson point process then the quantity F given below has a probability distribution that is approximately an F-distribution of $2n_1 + 1$ and $2n_2 + 1$ degrees of freedom:

$$F = \frac{\nu_d(W_1)(2n_2 + 1)}{\nu_d(W_2)(2n_1 + 1)}. \tag{2.56}$$

Here n_1 and n_2 are the point numbers in two disjoint subregions W_1 and W_2 of the window W. It is assumed that the order of the indices '$_1$' and '$_2$' is chosen to make F greater than one. The homogeneity hypothesis is rejected at the significance level α if

$$F > F_{2n_1+1,\, 2n_2+1;\, \alpha/2}. \tag{2.57}$$

This test is strictly appropriate only if the subregions W_1 and W_2 are chosen *a priori* to investigate specific possibilities of trend. For example, such a trend might relate the intensity of a Poisson process of tree locations to some measure of soil dryness, which is *a priori* known. It should also be noted that the significance calculations above depend on the Poisson nature of the point process and that homogeneity has many aspects, which are difficult to test.

There are other statistics for comparing the point counts in two subregions; see Chiu (2010) for a comparison. Chiu and Wang (2009) discussed the statistics for the case of more than two subregions. For model-free statistics for testing the hypothesis of a constant intensity, see Guan (2008) and Chiu and Liu (2013).

2.6.4   Testing the Poisson process hypothesis

There is a vast range of tests to which one can subject the hypothesis that a given point pattern is Poisson. In the statistical literature one speaks about the CSR hypothesis, with CSR stands for complete spatial randomness. Naturally, there can be no single criterion to give the 'best' test to apply; see the discussion in Diggle (1983) and Illian *et al.* (2008). Which test is most appropriate depends on the nature of the alternative hypothesis envisaged and on limitations imposed by measurement methods. Here only two cases are considered.

The L-test

When the point pattern has been mapped exhaustively, so that measurements of the locations of all points in the pattern are available, then it becomes worthwhile to use Ripley's K-function for a goodness-of-fit test. Formula (2.47) gives this function

$$K(r) = b_d r^d \qquad \text{for } r \geq 0.$$

If an estimate of $K(r)$ deviates considerably from this simple form then the Poisson hypothesis is thrown into doubt. Statistical experience since Ripley (1977) suggests that tests using this idea have good statistical properties (are rather powerful), are the 'best' tests available; see also Myles *et al.* (1995) and Illian *et al.* (2008). Moreover, comparison of the theoretical form with an estimate can do much to reveal the structure of the point pattern, and the comparison can be refined and improved by using estimates of the pair correlation function. Section 4.7.4 contains a brief discussion of estimators for $\lambda^2 K(r)$ for general point processes.

In practice, usually the L-function is used rather than $K(r)$, where

$$L(r) = \sqrt[d]{\frac{K(r)}{b_d}} \qquad \text{for } r \geq 0. \tag{2.58}$$

The root transform stabilises estimation variance, and $L(r)$ has the simple theoretical linear form

$$L(r) = r \qquad \text{for } r \geq 0. \tag{2.59}$$

It is then natural to consider the *deviation test* statistics

$$\tau = \max_{r \leq r_{\max}} |\hat{L}(r) - r| \tag{2.60}$$

with

$$\hat{L}(r) = \sqrt[d]{\frac{\hat{K}(r)}{b_d}}. \tag{2.61}$$

If τ is large, then the Poisson hypothesis has to be rejected. (The alternative hypothesis is 'not Poisson point process' without further specification.) If $\hat{K}(r)$ is obtained via Ripley's estimator,

$$\hat{K}(r) = \frac{\hat{k}_{\text{iso}}(r)}{\hat{\lambda}^2}, \tag{2.62}$$

where $\hat{k}_{\text{iso}}(r)$ is defined in Formula (4.111) in Section 4.7.4, then in the planar case a critical value of τ for the significance level $\alpha = 0.05$ is

$$\tau_{0.05} = \frac{1.45\sqrt{a}}{n}, \tag{2.63}$$

where a is the window area and n the number of points observed (Ripley, 1988, p. 46). This value was obtained by simulations and can be used if $nlr_{\max}^3/a^2$ (where l is the boundary length of W) is small for a 'wide range of' $r_{\max}$ (and $r_{\max} = 1.25/\sqrt{n}$ was recommended for W being a unit square). Many statisticians believe that this $\tau_{0.05}$ can be used also in the case of other estimators of the K-function.

To obtain the value of $\tau_{0.01}$, the factor 1.45 has to be replaced by 1.75; see Chiu (2007).

If Ripley's approximation appears to be inappropriate then simulation can be employed to provide Monte Carlo deviation tests. In such tests a suitable value of r_{max} is chosen, for example the half length of the diagonal for a rectangular window. Then k independent binomial process samples with n points are simulated in the window. For each of the samples the L-function is estimated and the value

$$\tau_i = \max_{r \leq r_{max}} |\hat{L}_i(r) - r| \qquad \text{for } i = 1, 2, \ldots, k,$$

is determined. These values plus the corresponding value τ_{emp} for the empirical data are sorted in ascending order. If the rank of τ_{emp} in the series is too large, the CSR hypothesis is rejected. If, for example, $k = 999$, then ranks larger than 950 (respectively 990) lead to rejection for $\alpha = 0.05$ (respectively $\alpha = 0.01$). The p-value of the test is (approximately) given by

$$\hat{p} = \frac{s + 1}{k + 1}, \tag{2.64}$$

where s is the number of τ_i larger than or equal to τ_{emp}.

Instead of using the L^∞-norm to define τ_{emp} and τ_i, also the L^2-norm can be used, leading to

$$\tau = \int_0^{r_{max}} \left(\hat{L}(r) - r\right)^2 dr \tag{2.65}$$

and the corresponding τ_i from simulated patterns. It depends on *a priori* knowledge which τ-definition is used; see Example 2.3.

In applications, the choice of r_{max} may be crucial. For a pattern of n points observed in a unit square, Ripley (1979) suggest $r_{max} = 1.25/\sqrt{n}$ and Diggle (2003, p. 87) recommend that r_{max} should not be bigger than 0.25. Ho and Chiu (2006) show empirically that if $\hat{\lambda}^2$ in Formula (2.62) is some distance-adapted intensity estimator (Stoyan and Stoyan, 2000), the power of the L-test will be robust against different choices of r_{max}. See also Ho and Chiu (2009) for another suggestion of modification.

Quadrat count methods

The case when the point pattern is sampled by means of counts of points falling in several subregions of the window W is intermediate between the case where one can usefully estimate the reduced second moment function and the case where one is restricted to distance methods. The sampling window W is divided into several subregions (often *quadrats*, squares or cubes) of equal area or volume $v_d(Q)$. Under the hypothesis of a homogeneous Poisson point process the number of points in each quadrat is Poisson of mean $\lambda \cdot v_d(Q)$ and counts in disjoint quadrats are independent. Statistical tests can be based on these distributional properties.

The simplest test is the index-of-dispersion test; the Greig-Smith method below is a refined variant which investigates the extent to which the independence properties are valid.

The *index-of-dispersion* I is defined by

$$I = \frac{(k-1)s^2}{\bar{x}},$$

where k is the number of quadrats, $\bar{x}$ is the mean number of points per quadrat, and s^2 is the sample variance of the number of points per quadrat. This test statistic is exactly that of a χ^2 goodness-of-fit test of the hypothesis that the n points are independent and uniformly distributed in W. Consequently, the index I follows approximately a χ^2-distribution of $k - 1$ degrees of freedom provided that $k > 6$ and $\lambda v_d(Q) > 1$ (Diggle, 1983, 2003). So if I exceeds $\chi^2_{k-1;\alpha}$ or is smaller than $\chi^2_{k-1;1-\alpha}$ then the test rejects the homogeneous Poisson point process hypothesis at the significance level α. In the first case the alternative hypothesis is that the variability in the process is greater than that for the Poisson process, typically clustering. Analogously, in the second case the alternative hypothesis is that there is some regularity in the point pattern as the variability is smaller.

Of course, quadrat count tests have the disadvantage that quite different processes may yet have quadrat counts of similar joint distribution.

The *Greig-Smith method* (see Greig-Smith, 1952, 1983) uses not only the quadrat counts on their own but also counts grouped by neighbouring quadrats. This enables detection of clustering at different scales. Example 2.2 below illustrates this.

It is assumed that the number of quadrats is $n = 2^q$ for some integer q. The quantities $s_1, s_2, \ldots$ are calculated:

$$s_1^2 = \sum_j \left(\begin{array}{c} \text{count in the} \\ j^{\text{th}} \text{ quadrat} \end{array} \right)^2 - \frac{1}{2} \sum_k \left(\begin{array}{c} \text{count in the } k^{\text{th}} \\ \text{pair of quadrats} \end{array} \right)^2 ,$$

$$s_2^2 = \sum_j \left(\begin{array}{c} \text{count in the } j^{\text{th}} \\ \text{pair of quadrats} \end{array} \right)^2 - \frac{1}{2} \sum_k \left(\begin{array}{c} \text{count in the } k^{\text{th}} \\ \text{foursome of quadrats} \end{array} \right)^2 ,$$

and so on.

In the case $q = 6$ the quadrats can be numbered in chessboard notation, since $2^q = 64$. The pairs of quadrats are then the pairs $(a1, a2), (a3, a4), \ldots, (b1, b2), (b3, b4), \ldots$. The foursomes of quadrats are then $(a1, a2, b1, b2), (a3, a4, b3, b4), \ldots$.

The quantities $s_j^2/2^{q-j}$ are unbiased estimators for the variance of the random number of points in a quadrat. Thus the index-of-dispersion formula can be generalised, calculation made of the indices

$$I_j = s_j^2/\bar{x}$$

and the χ^2-test used on each index. Here $\bar{x}$ is as in the index-of-dispersion test. The degrees of freedom for the χ^2-statistics $I_1, I_2, \ldots$ are $2^{q-1}, 2^{q-2}, \ldots$.

The results of the tests should be interpreted as follows. The hypothesis of a homogeneous Poisson point process can be accepted if all the indices I_j lie between two-sided critical values of the appropriate χ^2-statistics. Otherwise, the values of the indices indicate the nature of the deviation from the hypothesis. Thus if $I_3/2^{q-3}$ is significantly greater than one then there is strong evidence of clustering on the scale of groups of four quadrats, etc.

Example 2.1. *Longleaf pine data*

Figure 2.5 on p. 60 shows the locations of 584 longleaf pine trees in a 200 m × 200 m window of an old-growth forest in Thomas County, Georgia in 1979.

The empirical L-function, given in Figure 2.6(a), shows clearly that the mean number of pines in a disc centred at an arbitrarily chosen one is larger than expected in a homogeneous Poisson process. Such an empirical L-function is typical for a clustered pattern.

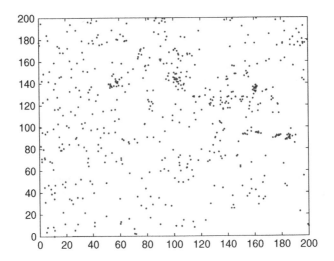

Figure 2.5   The longleaf pine data in a 200 m × 200 m window. (Data source: Cressie, 1993, Table 8.1.)

Using $r_{max} = 10$ m in (2.60) leads to $\tau = 3.48$, which is much larger than $\tau_{0.01} = 0.60$ obtained by (2.63) but replacing the factor 1.45 by 1.75 there, and so the CSR hypothesis is rejected at the 1% level.

The empirical pair correlation function shown in Figure 2.6(b), which has the typical form of that of a cluster process, reveals that points with short inter-point distance are strongly positive correlated, that is, pines of short inter-point distance are more likely than in the Poisson case, indicating clustering in the pattern. It is plausible that $g(r)$ has a pole at $r = 0$, and Stoyan and Stoyan (1996) and Ghorbani (2012) showed that some generalisations of

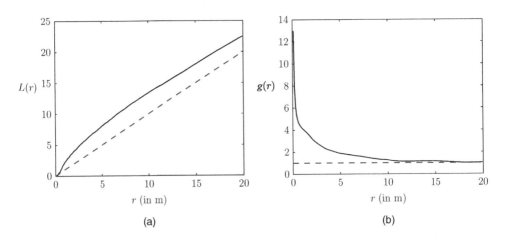

Figure 2.6   Estimate (solid line) for (a) $L(r)$ and (b) the pair correlation function $g(r)$ for the pattern of longleaf pine illustrated in Figure 2.5 in comparison with the theoretical line (dashed line) in the case of the homogeneous Poisson process, for which $L(r) = r$ and $g(r) \equiv 1$.

the Thomas process (see Section 5.3) give better fits to the pattern than the Thomas model suggested by Cressie (1993, p. 667).

Example 2.2. *Midpoints of large Martian craters*

Midpoints of large craters on Mars (where here 'large' is interpreted as having diameter greater than 200 km) form a point pattern on the sphere which is the surface of Mars. Lipskij *et al.* (1977) presented these data. Because this is a point pattern on the sphere rather than the plane, the theory above does not strictly apply; one must speak of the *homogeneous* Poisson point process hypothesis in the sense of homogeneity under rotations of the sphere. However, the modifications to the theory are entirely routine.

The surface of Mars being divided into 32 regions of equal area, the point counts of the regions yielded the numbers in Table 2.1. In this case, the Greig-Smith method can be applied with $q = 5$ and $\bar{x} = 0.97$. Table 2.1 also indicates the means by which the quadrats were grouped for the application of the Greig-Smith method. The dispersion indices are presented in Table 2.2 and compared with the χ^2-values at $\alpha = 0.05$. For $i = 1$, 3 and 5 the index values exceed the critical values. This implies that the differences between North and South hemispheres ($i = 5$), between quarter-spheres ($i = 3$), and between $\frac{1}{32}$-spheres ($i = 1$) are greater than is acceptable for a homogeneous Poisson point process, and thus clustering at these scales can be assumed.

Clearly the results depend on the choice of the division of the planet's surface.

The homogeneous Poisson point process hypothesis is also rejected by the index-of-dispersion test. The index-of-dispersion $I = 31 \times 2.10/0.97 = 67.06$, while the corresponding χ^2-value is 43.77 at $\alpha = 0.05$. When used to compare North and South hemispheres the

Table 2.1 Counts of large Martian craters in 32 segments of the planet's surface.

0°–90°	90°–180°	180°–270°	270°–360°	Latitude	Hemisphere
0	0	0	1	48.5°	
0	1	0	0	30°	Northern
0	1	0	2	14.5°	hemisphere
1	0	0	2	0°	
0	0	0	4	14.5°	
0	0	1	0	30°	Southern
0	2	1	2	48.5°	hemisphere
6	4	2	1		

Table 2.2 Test statistics referred to the homogeneous Poisson point process hypothesis for the Martian crater data.

i	Degrees of freedom	$\chi^2_{0.95}$	I_i	$\chi^2_{0.05}$
1	16	8.0	32.5	26.3
2	8	2.7	6.9	15.5
3	4	0.7	20.0	9.5
4	2	0.1	0.3	6.0
5	1	0.0	7.2	3.8

homogeneity test yields $n_1 = 8, n_2 = 23$ and $F = 2.76$. Thus the hypothesis of equal intensity on both hemispheres must be rejected at the 5% level, as $F_{17,47;0.025} \approx 2.1$.

All these deviations from the homogeneous Poisson point process hypothesis presumably result either from the fact that the crater pattern originated by cause of showers of meteors or else from modifications of the original pattern by areological processes. Marcus (1972) gives a detailed study of point processes on planetary and lunar surfaces; see also Greeley (1987) and Malin *et al.* (2006).

Example 2.3. *Nodes of a network adapted to the pores of Fontainebleau sandstone*

Figure 2.7 shows the three-dimensional pattern of nodes of a network that is adapted to a sample of Fontainebleau sandstone; see for more details Tscheschel and Stoyan (2003).

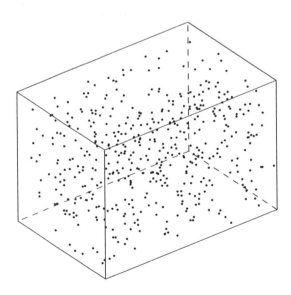

Figure 2.7 Nodes of a network adapted to the pores of a sample of Fontainebleau sandstone. The cuboid has side lengths 2 mm × 2 mm × 3 mm, the number of nodes is 470.

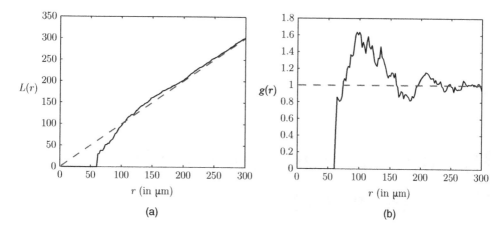

Figure 2.8   Estimate (solid line) for (a) $L(r)$ and (b) the pair correlation function $g(r)$ for the pattern of nodes illustrated in Figure 2.7 in comparison with the theoretical line (dashed line) in the case of the homogeneous Poisson process, for which $L(r) = r$ and $g(r) \equiv 1$.

There are pros and cons for the hypothesis that the pattern is of CSR type:

Pro: The pattern looks as a rather random pattern, and networks with nodes that form a Poisson process are indeed possible; see Section 9.11.5.

Con: The way in which the network was constructed by Sok *et al.* (2002) is perhaps not compatible with the idea of CSR, because pores close together may be considered as one large pore.

Thus it is of some interest to test the CSR hypothesis for the point pattern in Figure 2.7.

Figure 2.8(a) shows the empirical L-function of the pattern. For small r it deviates clearly from the theoretical form of a Poisson process, where $L(r) = r$. There is a hard-core distance r_0 of about 60 μm, that is, $L(r) = 0$ for $r \le 60$ μm. However, for larger r, $L(r)$ comes quite close to r; thus, globally, the Poisson process seems to be a good model.

Under these conditions a Monte Carlo test was chosen with defining τ by (2.60). With $r_{\max} = 300$ μm as suggested by visual inspection of Figure 2.8(a), the value $\tau_{\mathrm{emp}} = 61.644$ μm was obtained (which is just the hard-core distance) and the p-value obtained by (2.64) with $k = 9999$ was 0.0002. Thus the CSR hypothesis was rejected.

Some explanation for the rejection is given in Figure 2.8(b), which shows the empirical pair correlation function of the point pattern. There is a hard-core radius r_0 and inter-point distances around 80 μm are very frequent, much more frequent than in a Poisson process pattern. These two effects are canceled out in the cumulative function $L(r)$, resulting in $L(r) \approx r$ for $r \ge 250$ μm.

By the way, the hard-core distance in the point pattern is so large that a test suggested in Ripley and Silverman (1978), which is based only on the minimum inter-point distance, rejects the CSR hypothesis quite clearly.

3

Random closed sets I – The Boolean model

3.1 Introduction and basic properties

3.1.1 Model description

The *Boolean model* (also known as *Boolean scheme, Poisson germ–grain model, Poissonian penetrable grain model, fully penetrable grain system*, or *homogeneous system of overlapping particles*) is an important and relatively simple example of a random closed set. It is both flexible and amenable to calculation.

Before a general definition is given it may be helpful to consider one of the simplest possible examples of a planar Boolean model. Suppose points are scattered in the plane according to a homogeneous Poisson process of intensity λ. On each of these points a disc of fixed radius r is placed. The union of all of these discs is an example of a Boolean model; a particular realisation is displayed in Figure 3.1 on p. 65. In the terminology to be introduced the points of the Poisson process are the *germs* of the model while the discs are the *(primary) grains*. (Objects that could be called 'secondary grains' are below called 'clumps'.)

This construction can be generalised to produce the general stationary Boolean model. The discs are replaced by independent realisations of a random compact set. In this introduction the notions of random compact sets and random closed sets have been used without a careful treatment merited by the concept. Chapter 6 gives a more detailed discussion. For this chapter an intuitive understanding will suffice.

Suppose $\Phi = \{x_1, x_2, \ldots\}$ is a homogeneous Poisson process in $\mathbb{R}^d$ of intensity λ. Note that the x_i just list the points of Φ in some (arbitrary) manner. Let $\Xi_1, \Xi_2, \ldots$ be a sequence of independent identically distributed random compact sets in $\mathbb{R}^d$ that are independent of

Stochastic Geometry and its Applications, Third Edition.
Sung Nok Chiu, Dietrich Stoyan, Wilfrid S. Kendall and Joseph Mecke.
© 2013 John Wiley & Sons, Ltd. Published 2013 by John Wiley & Sons, Ltd.

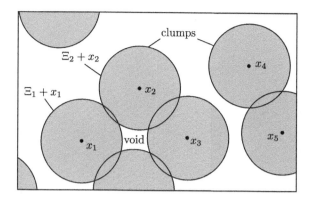

Figure 3.1 Diagrammatic image of a planar Boolean model. The empty space enclosed by four of the grains is a 'void'. Key: germ x_n, •; grain, Ξ_n.

the Poisson process Φ and satisfy a technical condition given below. The *Boolean model* Ξ is constructed by using the *germs* x_n and the *grains* Ξ_n as follows:

$$\Xi = \bigcup_{n=1}^{\infty} (\Xi_n + x_n) = (\Xi_1 + x_1) \cup (\Xi_2 + x_2) \cup \cdots . \qquad (3.1)$$

The technical condition mentioned above is that

$$\mathbf{E}\big(\nu_d(\Xi_0 \oplus K)\big) < \infty \qquad \text{for all compact } K. \qquad (3.2)$$

Here, and in the following, Ξ_0, called the *typical grain*, denotes a further random compact set of the same distribution as the grains Ξ_n but independent of both the grains and the germ process Φ. The union Ξ is a random closed set called *Boolean model with typical grain Ξ_0*.

The technical condition (3.2) ensures that only finitely many of the grains $\Xi_n + x_n$ meet any given compact set. Thus, in particular, it ensures that the property of being a closed set is inherited by Ξ from the grains. The condition is further motivated by Formula (3.7) below for the probability of Ξ hitting a compact set.

A simpler but more restrictive condition than (3.2) is as follows: if R is the radius of the ball circumscribing Ξ_n then

$$\mathbf{E}(R^d) < \infty, \qquad (3.3)$$

see Heinrich (1992b).

The construction of Ξ is illustrated in Figure 3.1, while Figure 3.2 gives an example of a simulated realisation of a Boolean model.

A more theoretical and less constructive point of view defines the Boolean model as the union of sets arising from a Poisson process in the space $\mathbb{K}'$ of all non-empty compact subsets of $\mathbb{R}^d$. Treatments using this approach can be found in Matheron (1975) and Schneider and Weil (2008). The technical condition mentioned above is then naturally expressed as requiring the governing measure of the Poisson process to be a σ-finite Borel measure on $\mathbb{K}'$, using a suitable topology on $\mathbb{K}'$.

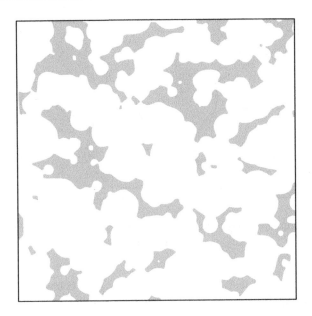

Figure 3.2   Computer simulation of a Boolean model with white discoidal grains. The parameters of the simulation are those fitted to the pattern of a section of a potassium deposit. The white areas correspond to potassium-bearing regions; see Ohser and Stoyan (1980).

Typical possibilities for the grains in the planar case include: discs of random radius (generalising the simple example above), various random polygons, segments of random length and orientation, and random finite clusters of points. The grains are characterised by their distribution M. As they are random sets this distribution is a probability measure on $\mathbb{K}'$. It is the *mark distribution* of the marked point process $\{[x_n; \Xi_n]\}$ as in Section 4.2 and Chapter 6. In the case of a Boolean model the marks are independent.

If the grains are convex then statistical averages of various numerical measures of convex sets play an important rôle in the theory. Their notation and explanation are given in Table 3.1 on p. 67. In the case of nonconvex grains several of the quantities still have meaning and in such cases the same symbols will be used.

The Boolean model has been thoroughly studied by Matheron, Serra and their colleagues; see Matheron (1967, 1975) and Serra (1982) and references therein. More recent references are Hall (1988), Cressie (1993), Stoyan and Mecke (2005), Weil (2007), Schneider and Weil (2008) and Baccelli and Błaszczyszyn (2009a).

3.1.2   Applications

The Boolean model is a basic model in stochastic geometry. As Hadwiger and Giger (1968) show, it also arises from classical assumptions of independence and uniformity in a manner similar to some characterisations of the Poisson process.

There are two general trends of application for the Boolean model. In the first place, it is a natural model for sparse systems of particles distributed at random. Here the sparse nature of the system is modelled by a low value for the intensity λ of the Poisson process of germs.

Table 3.1 Statistical averages of numerical measures of convex sets.

In general Euclidean space $\mathbb{R}^d$:

$\overline{V}_k$ mean of kth intrinsic volume

For spatial sets, $d = 3$:

$\overline{V}$ mean volume, $\overline{V} = \overline{V}_3$

$\overline{S}$ mean surface area, $\overline{S} = 2\overline{V}_2$

$\overline{M}$ mean of integral of mean curvature, $\overline{M} = \pi\overline{V}_1$

$\overline{\overline{b}}$ mean average breadth, $\overline{M} = 2\pi\overline{\overline{b}}$, and $\overline{\overline{b}} = \frac{1}{2}\overline{V}_1$

For planar sets, $d = 2$:

$\overline{A}$ mean area, $\overline{A} = \overline{V}_2$

$\overline{L}$ mean boundary length, $\overline{L} = 2\overline{V}_1$

Note: $\overline{\overline{b}}$ is the statistical mean of a geometric average taken over all possible directions.

If the part of the space covered by Ξ is small, then grains will not often overlap and so Ξ will consist in the main of separated particles. With increasing λ the number of overlappings increases. Of course such overlappings do occur in nature; consider for example pores in cheese or areas of weeds in agricultural fields.

In the second place, the Boolean model may provide a good description for an irregular pattern observed in nature; see for example Figure 3.2. As can be seen from the figure (and in contrast to what is often the case for sparse systems), the grains and their locations at germs generally fail to have operational interpretations. The rôle of the Boolean model in such a case is to provide irregular random sets of a suitable form, and its method of construction needs not correspond to any physical reality. This point is returned to in Section 3.6.

The Boolean model has been used, at least for the special case of spherical grains, since the middle of 19th century, for example by the physicist Clausius (1858); see Guttorp (2007) and Guttorp and Thorarinsdottir (2012). The following list gives some examples (in alphabetical order of the authors of the earliest papers) drawn mainly from the natural and engineering sciences:

- Armitage (1949): random clumping of dust or powder particles;

- Cahn (1956): kinetics of nucleation at grain boundaries;

- Kallmes and Corte (1960), Corte and Kallmes (1962), Räisänen *et al.* (1997), Niskanen *et al.* (1998) and Åström *et al.* (2000): microstructure of paper; the grains are paper fibres, modelled as long thin rectangles or line segments;

- Diggle (1981): distribution of heather in a forest; see also Møller and Helisová (2010);

- Jacod and Joathon (1971): form of geological structures generated by sedimentation; see also Ohser and Stoyan (1980): form of geological deposits of potassium, Figure 3.2 shows a simulated realisation of a Boolean model using as model parameters those

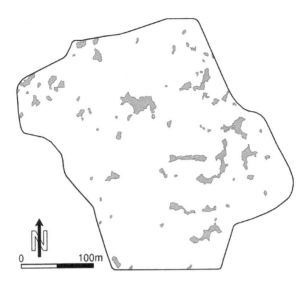

Figure 3.3   Canopy gaps (in grey) in a forest mapped from aerial photographs. Here the tree crowns (not shown) may be considered as standing for the grains, while the gaps stand for the complement of the random set. Reprinted from Nuske *et al.* (©2009) with permission of Elsevier.

obtained by fitting to the geological data, and the white zones in the figure correspond to areas that are rich in potassium; by the way, a figure in Nuske *et al.* (2009), reproduced in Figure 3.3, showing the canopy gaps in a forest, is quite similar to Figure 3.2;

- Hermann (1991): various examples of the application of the Boolean model in materials science;

- Jeulin *et al.* (2001): microstructure of plaster made of needle-shaped gypsum crystals; the grains are modelled as Poisson polyhedra or random parallelepipeds;

- Kolmogorov (1937), Hermann (1998) and Capasso (2003): modelling the crystallisation in metals and polymers;

- Kopp-Schneider *et al.* (1998) and Groos and Kopp-Schneider (2010): modelling the progression of altered hepatocytes in cancer research; the spherical grains stand for cells forming focal lesions;

- Mack (1954, 1956): corrections for clumping in estimation of number of particles; see also Kellerer (1983);

- Nutting (1913) and Marchant and Dillon (1961): patterns in photographic emulsion; see also Frieden (1983);

- Serra (2009): modelling the spread of forest fire by a Markov chain of Boolean models;

- Thovert *et al.* (2001) and Arns *et al.* (2002): Fontainebleau sandstone, a porous medium; in the former paper the grains are balls, in the latter ellipsoids;

- Thovert and Adler (2004) and Adler *et al.* (2005): fracture networks, where the grains are portions of plane surfaces;

- Widom and Rowlinson (1970): systems of water droplets in the study of liquid-vapour phase transitions.

The grains of a Boolean model are not required to be connected sets. For example, they may be sets of discrete points, in which case the Boolean model is a point process, in fact, a Neyman–Scott point process; see Section 5.3.

The so-called Boolean functions can be viewed as generalisations of Boolean models. Here a number of random functions are randomly translated and combined using the supremum-operation; see Serra (1982, 1988, 1989). This model is applied to modelling random surfaces; see Jeulin (1987, 2000, 2002), Serra (1987), Goulard *et al.* (1994), Heinrich and Molchanov (1994), Chadœuf *et al.* (1996) and Dominguez and Torres (1997), Another model closely related to the Boolean model is the shot-noise process; see Section 5.6.

Bombing problems form a very natural application context for the Boolean model. Robbins (1944, 1945) and Neyman and Scott (1972) studied random closed sets with such applications in view. Here the germs are the points of impact of the bombs, and the grains are the areas covered by the splinters from the corresponding bomb explosions. Robbins' account was confined to theoretical issues; Neyman and Scott briefly discussed details of one particular application concerning the clearance of landmines from the landing beaches in Normandy. Generally in such problems the Boolean model seems plausible for representing the total area covered by splinters, while the shot-noise model is more suitable for considerations of intensity of covering. Hall (1988, Section 1.6) gives a short review of further modelling of the effect of salvoes of weapons fired at a target. Typically, the Boolean model is inhomogeneous in such applications, that is, the intensity of the Poisson process of germs is location dependent, given by an intensity function $\lambda(x)$, which should have its maximum in the target's centre; see Figure 3.4 on p. 70.

In the one-dimensional case, when the grains are intervals, the Boolean model is sometimes described as an $M/G/\infty$ queue. This is the D. G. Kendall notation for a queueing model in which the customer arrivals ('germs') are according to a Poisson process, the service times (intervals of those lengths are the 'grains') are independent and identically distributed, and there are infinitely many servers. In passing, note that Baccelli and Błaszczyszyn (2009a) use Kendall's notation also for shot-noise processes and related models.

The Boolean model has a three-fold value in the description of given samples and situations in nature. First, a relatively parsimonious description is available for the random set in question, by means of the intensity λ of the germ process and the various characteristics of the grains. Second, it is at least possible that the model assumptions may be suggestive of the process of formation of the structure, though it will be emphasised later that this is by no means invariably the case; see Section 3.6. Third, the formulae to be derived for the Boolean model may be used for estimation of quantities not available for direct measurements.

3.1.3   Stationarity and isotropy

From the stationarity of the Poisson process Φ of germs and the identical distribution of the grains, it follows that the Boolean model as defined above is stationary. That is to say, its distribution is translation-invariant; see Section 6.1. As a consequence of more general

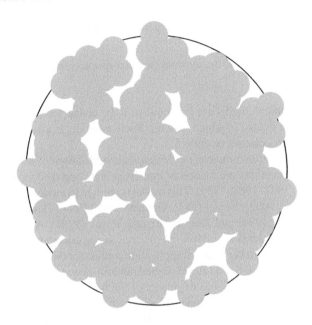

Figure 3.4   A simulated sample of an inhomogeneous Boolean model with discoidal grains and intensity function having a constant value within a disc and vanishing outside.

considerations Nguyen and Zessin (1979a) showed that the Boolean model is *ergodic* and even *mixing*; see Section 6.1 for a discussion of these concepts and Schneider and Weil (2008) for more details. A particular consequence of ergodicity is that

$$\lim_{r \to \infty} \frac{v_d\big(\Xi \cap B(o, r)\big)}{v_d\big(B(o, r)\big)} = p, \tag{3.4}$$

where $p = \mathbf{P}(o \in \Xi)$ is the volume fraction of the Boolean model, that is, that fraction of the $\mathbb{R}^d$ that is covered by Ξ. Note that (3.4) is not the definition of p — this is given by Formula (3.13). Similar relations hold for many other functionals replacing v_d in the numerator of (3.4), for example the intrinsic volumes.

If the distribution of the grains is isotropic (distribution invariant under rotations about the origin o), then the Boolean model is also isotropic in distribution. However, a Boolean model can be isotropic even if the grains are not isotropic; see Molchanov and Stoyan (1994). For example, consider the spatial case, with Ξ_0 a ball with south pole at the origin; see also Remark (3) on p. 73. Isotropy and stationarity together imply that the Boolean model has a distribution invariant under rigid motions.

Some applications consider *inhomogeneous* Boolean models. An important particular case is where the grains are i.i.d. as in the stationary case but the Poisson process of germs is inhomogeneous with intensity function $\lambda(x)$. This model is used in the military applications mentioned above; it is also related to Cressie's tumor model (see Cressie and Hulting, 1992). Figure 3.4 shows a simulated set of this kind. In the case of the figure

$$\lambda(x) = \lambda \mathbf{1}_A(x)$$

with $A = B(o, R)$ and $\Xi_0 = B(o, r)$. Another application is discussed in Hahn *et al.* (1999), where $\lambda(x)$ follows a vertical gradient. The theory of inhomogeneous Boolean models is presented in Schneider and Weil (2008, Chapter 11, especially Theorem 11.1.3).

3.1.4 Simulation

The simulation of Boolean models is not difficult for readers who know how to simulate Poisson processes, homogeneous as well as inhomogeneous. (See Lantuéjoul, 2002, for a very brief sketch.) The second step, the simulation of grains, is usually not very difficult. Of course, for particular grains, for example Poisson polyhedra, particular simulation methods are necessary. There is only one nontrivial problem, that of *edge-effects*.

If the grains are bounded, that is, there is a radius R such that $\Xi_0 \subset B(o, R)$ with probability one, then it suffices to simulate the Boolean model in an enlarged window: instead of the original window W the larger window $W_{\oplus R}$ is used to obtain a sample which is correct also at the boundary of W.

For the case of grains for which such a bound does not exist, methods of 'exact simulation' are useful. In principle the ideas in Brix and Kendall (2002) for the simulation of Poisson cluster processes can be applied also for Boolean models. Another, simpler, possibility is to generalise the method described above, using enlargement on window based on the enclosing radius R: the grains are binned by size and for each bin the R-method is used separately and the patterns obtained are superposed.

Exact simulation can be also used for conditional simulation of Boolean models: simulation of Boolean models such that the given locations $y_1, \ldots, y_n$ are covered by grains; see Thönnes (2001), Cai and Kendall (2002) and Lantuéjoul (2002).

The simulation can be organised so that the data generated are either vector data, that is, coordinates of germs and parameters of grains, as needed when the aim is to determine model parameters numerically, or pixel data, which are needed in the context of statistics of pixel data. Then one starts with an empty image, where all pixels are set to 0. In turn for all pixels it is checked whether these are overlapped by at least one grain. If so, the pixel value is set to 1.

3.1.5 The capacity functional

The distribution of a Boolean model is uniquely determined by its *capacity functional* or hitting distribution T_Ξ:

$$T_\Xi(K) = \mathbf{P}(\Xi \cap K \text{ is not empty})$$
$$= \mathbf{P}(\Xi \cap K \neq \emptyset) \qquad \text{for all compact sets } K. \tag{3.5}$$

Note that in the same form as (3.5), the capacity functional is also defined for any random closed set Ξ. The theory of random closed sets asserts that the functional uniquely determines the distribution of a random closed set; see Section 6.1.

In the case of the Boolean model, $T_\Xi(K)$ can be expressed in a relatively simple form, in pleasant contrast to the complexity of expressions for many other summary characteristics connected with this model. By application of the Poisson assumption, $T_\Xi(K)$ can be shown to take the form

$$T_\Xi(K) = 1 - \exp\left(-\lambda \mathbf{E}\left(\nu_d(\check{\Xi}_0 \oplus K)\right)\right) \tag{3.6}$$

or

$$T_\Xi(K) = 1 - \exp\left(-\lambda \mathbf{E}\left(v_d(\Xi_0 \oplus \check{K})\right)\right).$$ (3.7)

The proof of Formula (3.7) is of methodological interest.

Proof. Let K be an arbitrary fixed compact set. From the original germ process Φ a *thinned* process Φ_K can be produced. The thinning takes place by deleting x_n from Φ if $(\Xi_n + x_n) \cap K$ is empty. Thus

$$\Phi_K = \{x_n \in \Phi : (\Xi_n + x_n) \cap K \neq \emptyset\}.$$

Whether or not a germ x_n is deleted by this thinning procedure is independent of thinning happening to other germs. This follows from the independence properties of the Boolean model: deletion depends only on the location of the germ and on the corresponding grain. This independence implies that Φ_K is also a Poisson process (see Chapter 2), but now inhomogeneous with intensity function $\lambda_K(x)$ and only finitely many points. Here

$$\lambda_K(x) = \lambda p(x)$$ (3.8)

and $p(x) = \mathbf{P}\left((\Xi_0 + x) \cap K \neq \emptyset\right)$. Note that $p(x)$ is the probability that a germ placed at x is not deleted (the so-called retention probability). See Section 5.1 for a discussion of thinnings of point processes.

The total number of points of Φ_K has a Poisson distribution with mean μ_K,

$$\mu_K = \lambda \int_{\mathbb{R}^d} p(x)\, dx.$$ (3.9)

Thus the probability that Φ_K is empty is

$$\mathbf{P}(\Phi_K \text{ has no points}) = \exp(-\mu_K).$$

Consequently the value $T_\Xi(K)$ is obtained as

$$T_\Xi(K) = 1 - \exp(-\mu_K).$$ (3.10)

So it suffices to evaluate μ_K.

The probability $p(x)$ is given by

$$p(x) = \mathbf{P}(x \in \check{\Xi}_0 \oplus K)$$

since $(\Xi_0 + x)$ meets K precisely when $x \in \check{\Xi}_0 \oplus K$. Thus

$$\mu_K = \lambda \int_{\mathbb{R}^d} \mathbf{P}(x \in \check{\Xi}_0 \oplus K)\, dx$$

$$= \lambda \int_{\mathbb{R}^d} \mathbf{E}\left(1_{\check{\Xi}_0 \oplus K}(x)\right) dx$$

$$= \lambda \mathbf{E}\left(v_d(\check{\Xi}_0 \oplus K)\right),$$

where in the last step the expectation is interchanged with the integral. This last expectation is finite, by the imposition of the technical condition required in the definition of the Boolean model.

Thus Formula (3.6) is established:

$$T_\Xi(K) = 1 - \exp(-\mu_K)$$
$$= 1 - \exp\big(-\lambda \mathbf{E}\big(\nu_d(\check{\Xi}_0 \oplus K)\big)\big),$$

and Formula (3.7) follows by noting that

$$(\check{\Xi}_0 \oplus K) = -(\Xi_0 \oplus \check{K}).$$ □

Remarks (1) It is useful to introduce the notation

$$\psi(K) = -\ln\big(T_\Xi(K)\big). \tag{3.11}$$

Then

$$\psi(K) = \lambda \mathbf{E}\big(\nu_d(\Xi_0 \oplus \check{K})\big). \tag{3.12}$$

(2) The proof suggests that similar formulae hold for inhomogeneous Boolean models, such as the one that produces Figure 3.4.

(3) From Formula (3.7), one can see that there is an arbitrary element in the choice of location for each grain Ξ_n. Suppose that $z_1, z_2, \ldots$, are points produced by a construction depending on the corresponding grain, so that z_n depends on Ξ_n alone and not on any other grains nor on any other germs. Then a new Boolean model Ξ^* can be constructed using the original Φ for the germs and $\Xi_n^* = \Xi_n + z_n$ for the grains. For example, take $\Xi_n = B(o, R_n)$ where R_n is a random radius, and then pick

$$z_n = (0, 0, \ldots, 0, R_n).$$

In any case, by the translation-invariance of Lebesgue measure

$$\psi^*(K) = \lambda \mathbf{E}\big(\nu_d(\Xi_0^* \oplus \check{K})\big)$$
$$= \lambda \mathbf{E}\big(\nu_d((\Xi_0 \oplus \check{K}) + z_0)\big)$$
$$= \lambda \mathbf{E}\big(\nu_d(\Xi_0 \oplus \check{K})\big)$$
$$= \psi(K).$$

Hence, the two Boolean models Ξ and Ξ^* have the same capacity functional and so must have the same distribution.

3.1.6 Basic characteristics

Formulae (3.6) and (3.7) and the Poisson process property related to μ_K provide means of calculating some important characteristics of the Boolean model.

The volume fraction

The volume fraction p is the mean fraction of volume occupied by Ξ in a region B of unit volume,

$$p = \mathbf{E}\big(v_d(\Xi \cap B)\big) \qquad \text{for } v_d(B) = 1. \tag{3.13}$$

This value does not depend on the choice of the region B, by virtue of stationarity of Ξ. Note that p is defined without a limit operation, Formula (3.4) is a consequence of (3.13) and ergodicity. A further consequence of stationarity is that

$$p = \mathbf{P}(o \in \Xi), \tag{3.14}$$

a result that holds for general stationary random sets; it is proved in Section 6.3.1.
      The formula for p is obtained by putting $K = \{o\}$ in Formula (3.7). Then

$$\mathbf{P}(o \in \Xi) = 1 - \exp\big(-\lambda \mathbf{E}(v_d(\Xi_0))\big), \tag{3.15}$$

where Ξ_0 is as before the typical grain of Ξ. Hence

$$p = 1 - \exp(-\lambda \overline{V}), \tag{3.16}$$

in which $\overline{V} = \mathbf{E}\big(v_d(\Xi_0)\big)$ denotes the mean volume of the typical grain. Of course, if the grains of the Boolean model all have zero volume (for example, if they are points or line segments in the plane, or points, segments or flat objects in space), then the volume fraction will be zero.
      Note that even if the volume fraction is unity it need not follow that $\Xi = \mathbb{R}^d$; see Hall (1988, Theorem 3.3) and Chiu (1995c). If $\Xi \neq \mathbb{R}^d$, then there can be a relation to fractals.

The covariance or two-point probability function

If a random set has volume fraction $p > 0$, then important second-order properties are summarised in its covariance or *two-point probability function*, as discussed in Chapter 6.
      The (noncentred) *covariance* is defined as

$$C(\mathbf{r}) = \mathbf{P}(o \in \Xi \text{ and } \mathbf{r} \in \Xi) = \mathbf{P}(\{o, \mathbf{r}\} \subset \Xi) \qquad \text{for } \mathbf{r} \in \mathbb{R}^d. \tag{3.17}$$

A simple manipulation reveals that $C(\mathbf{r})$ is the volume fraction of $\Xi \cap (\Xi - \mathbf{r})$. Though this intersection is not in general a Boolean model, this interpretation is important for statistical determination of $C(\mathbf{r})$.
      If Ξ is a Boolean model with typical grain Ξ_0, then

$$C(\mathbf{r}) = 2p - 1 + (1 - p)^2 \exp\big(\lambda \mathbf{E}(\gamma_{\Xi_0}(\mathbf{r}))\big), \tag{3.18}$$

where $\gamma_{\Xi_0}(\mathbf{r}) = v_d\big(\Xi_0 \cap (\Xi_0 - \mathbf{r})\big)$ is the set covariance as in Section 1.7.2.

Proof. Formula (3.18) can be proved by considering the following evaluation of $C(\mathbf{r})$:

$$C(\mathbf{r}) = \mathbf{P}\big(o \in \Xi \cap (\Xi - \mathbf{r})\big)$$
$$= 1 - \big(\mathbf{P}(o \notin \Xi) + \mathbf{P}(o \notin \Xi - \mathbf{r}) - \mathbf{P}(o \notin \Xi \cup (\Xi - \mathbf{r}))\big)$$
$$= 2p - 1 + \mathbf{P}\big(o \notin \Xi \cup (\Xi - \mathbf{r})\big).$$

An application of (3.7) shows that

$$\mathbf{P}(\Xi \cap \{o, \mathbf{r}\} \neq \emptyset) = 1 - \exp(-\psi(\{o, \mathbf{r}\})), \tag{3.19}$$

where ψ is defined by (3.11). On evaluation of the Minkowski sum,

$$\psi(\{o, \mathbf{r}\}) = \lambda \mathbf{E}(v_d(\check{\Xi}_0 \cup (\check{\Xi}_0 + \mathbf{r})))$$
$$= \lambda \left(\mathbf{E}(v_d(\check{\Xi}_0)) + \mathbf{E}(v_d(\check{\Xi}_0 + \mathbf{r})) - \mathbf{E}(v_d(\check{\Xi}_0 \cap (\check{\Xi}_0 + \mathbf{r}))) \right)$$
$$= \lambda \left(2\mathbf{E}(v_d(\check{\Xi}_0)) - \mathbf{E}(v_d(\Xi_0 \cap (\Xi_0 - \mathbf{r}))) \right),$$

and the proof of (3.18) is concluded by noting that

$$1 - p = \exp(-\lambda \mathbf{E}(v_d(\Xi_0))). \qquad \square$$

As discussed in Section 6.3.2, the covariance of a random set is closely related to the concepts of covariance function and correlation function for random fields. Note, however, that the covariance defined above differs from the conventional covariance function in that the covariance is not centred about the mean.

If the random set Ξ is also isotropic, then $C(\mathbf{r})$ depends only on the length $r = \|\mathbf{r}\|$ of the vector $\mathbf{r}$. In such a case a convenient abuse of notation is to consider the covariance as a function of length r alone and to use the symbol $C(r)$.

If the typical grain Ξ_0 is isotropic, then $\gamma_{\Xi_0}(r\mathbf{u})$ can be replaced by the isotropised set covariance $\overline{\gamma}_{\Xi_0}(r)$ introduced in Section 1.7.2. If the mean $\mathbf{E}(\overline{\gamma}_{\Xi_0}(r))$ is denoted as $\overline{\overline{\gamma}}_{\Xi_0}(r)$, then Formula (3.18) takes the form

$$C(r) = 2p - 1 + (1 - p)^2 \exp\left(\lambda \overline{\overline{\gamma}}_{\Xi_0}(r)\right). \tag{3.20}$$

For some random convex bodies Ξ_0 the mean isotropised set covariance $\overline{\overline{\gamma}}_{\Xi_0}(r)$ is known. For a three-dimensional ball of random radius, with radius probability density function $f_R(r)$, it is given by

$$\overline{\overline{\gamma}}_{\Xi_0}(r) = \int_{r/2}^{\infty} \frac{4}{3}\pi x^3 \left(1 - \frac{3r}{4x} + \frac{r^3}{16x^3}\right) f_R(x)\,dx \qquad \text{for } r \geq 0, \tag{3.21}$$

and for a disc in the plane, it is

$$\overline{\overline{\gamma}}_{\Xi_0}(r) = \int_{r/2}^{\infty} \left(2x^2 \cos^{-1}\frac{r}{2x} - \frac{r}{2}\sqrt{4x^2 - r^2}\right) f_R(x)\,dx \qquad \text{for } r \geq 0. \tag{3.22}$$

In the literature further formulae can be found for the typical cells of the Voronoi tessellation (see Gilbert, 1962; Brumberger and Goodisman, 1983), the Poisson polyhedron (the typical cell of the Poisson plane tessellation, see Section 9.5.2) and the typical cell of the dead leaves model (see Gille, 2002).

Typically there exists an r_0 that $\overline{\overline{\gamma}}_{\Xi_0}(r) = 0$ for $r > r_0$. If it is the case, then $C(r) = p^2$ for $r > r_0$, and the Boolean model is said to have the finite range of correlation r_0.

Sometimes only the correlation function $\kappa(r)$, given by

$$\kappa(r) = \frac{C(r) - p^2}{p(1 - p)} \qquad \text{for } r \geq 0, \tag{3.23}$$

is known, resulting, for example, from small-angle scattering analysis. If so, then it may be of value to use the following formula, which yields p:

$$p = \lim_{r \to 0} \frac{\kappa''(r)}{\left(\kappa'(r)\right)^2}. \tag{3.24}$$

A useful asymptotic form relating $\kappa(r)$ to $\overline{\overline{\gamma}}_{\Xi_0}$ is given by Sonntag *et al.* (1981). As the intensity λ of the process Φ tends to zero, the grains remaining fixed, so p tends to zero and

$$\lim_{\lambda \to 0} \kappa(r) = \frac{\overline{\overline{\gamma}}_{\Xi_0}(r)}{\overline{\overline{\gamma}}_{\Xi_0}(0)}.$$

Thus for a 'dilute' Boolean model the correlation function $\kappa(r)$ is asymptotically the same as the normalised isotropised set covariance of Ξ_0.

Complicated formulae exist for the n-point probability functions for the case of Boolean models with identical balls, see Torquato (2002, pp. 122–4).

Coverage characteristics

The Boolean model is a union of grains and hence the following characteristics may be of interest:

- the mean number $\overline{c}$ of grains that contain a fixed point (it suffices to consider the origin o), and

- the volume fraction p_k of the random set of all points that are contained in more than k grains.

The proof of Formula (3.7) has shown that the random number of grains hitting a fixed compact set K follows a Poisson distribution with mean μ_K given by Formula (3.9). For the case $K = \{o\}$ this simplifies to

$$\mu_0 = \lambda \mathbf{E} \nu_d(\Xi_0), \tag{3.25}$$

which yields

$$\overline{c} = \mu_0. \tag{3.26}$$

Similar to Formula (3.14), due to stationarity the volume fraction p_k is equal to the probability that the origin o has been covered by more than k grains, that is,

$$p_k = 1 - \sum_{i=0}^{k} \frac{\overline{c}^i}{i!} e^{-\overline{c}} \qquad \text{for } k = 1, 2, \ldots. \tag{3.27}$$

3.1.7   Contact distribution functions

In its full generality as a function of compact sets K, the capacity functional $T_\Xi(K)$ is not suitable for practical work. Matheron had the idea to use families of standard sets K, which led to the notion of contact distribution functions.

Let B be a convex body of $\mathbb{R}^d$ containing the origin o, which is called in this context *test set*, *gauge set* or *structuring element*. Then

$$T_\Xi(rB) = \mathbf{P}(\Xi \cap rB \neq \emptyset)$$

is a function of the scaling factor r, which could be graphically presented and yield structural-distributional information on Ξ. However, more elegant is the use of the *contact distribution function* $H_B(r)$ given by

$$H_B(r) = 1 - \frac{\mathbf{P}(\Xi \cap rB = \emptyset)}{1 - p} \qquad \text{for } r \geq 0, \tag{3.28}$$

where p is the volume fraction of Ξ. This function has all properties of a distribution function. Clearly, it is

$$H_B(r) = 1 - \frac{1 - T_\Xi(rB)}{1 - p},$$

which shows the close relationship between the contact distribution function and the capacity functional.

The contact distribution function with a suitably chosen structuring element is valuable in particular in situations where identification of particles or of single vacant regions is impossible or inappropriate. Delfiner (1972) gave an excellent discussion of this using the closely related concept of size distribution functions.

In passing let it be noted that the relationship between contact distribution and size distribution, while close, is not explicit save in one special case. Matheron (1975, p. 53) discussed the essential ideas here and established an explicit connection for the special case when the structuring element is a line segment. His discussion concerns granulometries, from which the notion of size distribution is derived.

Since B contains the origin, $\Xi \cap rB$ being empty implies that Ξ does not contain the origin. So the value of $H_B(r)$ is the conditional probability that the scaled set rB hits (or makes contact with) Ξ, conditional on Ξ not containing o.

As another interpretation, $H_B(r)$ is the conditional distribution function of R:

$$H_B(r) = \mathbf{P}(R \leq r | R > 0), \tag{3.29}$$

where

$$R = \inf\{s : \Xi \cap sB \neq \emptyset\},$$

that is, R is the distance from o to Ξ measured with respect to B.

The shape of the structuring element B plays an important rôle. Two particularly important cases are the extremes where on the one hand, B is a line segment of unit length leading to the *linear contact distribution function* $H_l(r)$, and on the other hand B is the unit ball $B(o, 1)$ leading to the *spherical contact distribution function* $H_s(r)$. Two other cases appear on p. 83. Of course in anisotropic cases the orientation of the line segment is crucial for the values of $H_l(r)$. Section 6.3.5 discusses briefly the use of the linear contact distribution function in orientation analysis.

Note that $H_B(r)$ is produced in the first instance from the *complement* of Ξ rather than Ξ. While $H_l(r)$ is related to the distribution function of chord lengths *outside* of Ξ by (6.66), the chord lengths *in* Ξ are difficult to analyse, see p. 86.

The spherical contact distribution function $H_s(r)$ may be interpreted as the distribution function of the distance from a point chosen randomly outside Ξ, measured to the nearest point of Ξ. Hence, $H_s(r)$ is referred to as 'the law of first contact'.

In general a formula for the contact distribution function follows directly from (3.7) and (3.15):

$$H_B(r) = 1 - \exp\left(-\lambda\left(\mathbf{E}(v_d(\Xi_0 \oplus rB)) - \mathbf{E}(v_d(\Xi_0))\right)\right) \qquad \text{for } r \geq 0. \qquad (3.30)$$

3.2 The Boolean model with convex grains

In practice a frequent assumption is that the Boolean model has isotropic and convex grains. This has the advantage of considerable simplification of various formulae. Moreover, practical experience shows that the resulting structures are still sufficiently flexible to be useful in many statistical problems; see also pp. 106–7.

3.2.1 The simplified formula for the capacity functional

If Ξ_0 is isotropic and convex, then $\psi(K)$, which gives the capacity functional by (3.11), can be calculated for convex K by means of the generalised Steiner formula (6.29) on p. 216:

$$\psi(K) = \lambda \mathbf{E}\left(v_d(\Xi_0 \oplus \check{K})\right) = \frac{\lambda}{b_d} \sum_{k=0}^{d} \frac{b_k b_{d-k}}{\binom{d}{k}} \overline{V}_k V_{d-k}(K), \qquad (3.31)$$

where $\overline{V}_k$ denotes the mean of the kth intrinsic volume of Ξ_0, that is,

$$\overline{V}_k = \mathbf{E}\left(V_k(\Xi_0)\right) \qquad \text{for } k = 0, 1, \ldots, d,$$

and b_k the volume of the k-dimensional unit ball.

Thus an important simplification is obtained:

Theorem 3.1. *For convex bodies K, the value of the capacity functional $T_\Xi(K)$ of a Boolean model Ξ with isotropic convex grains depends only on λ and the means of the intrinsic volumes of the grains.*

As an example, let Ξ_0 be a ball $B(o, R)$ of random radius R with $\mathbf{E}(R^d) < \infty$. Then, for convex bodies K, by the Steiner formula (1.23) in the planar case,

$$\psi(K) = \lambda \left(A(K) + L(K)\mathbf{E}(R) + \pi\mathbf{E}(R^2) \right), \qquad (3.32)$$

and by Formula (1.24) in the spatial case,

$$\psi(K) = \lambda \left(V(K) + S(K)\mathbf{E}(R) + 2\pi\overline{b}(K)\mathbf{E}(R^2) + \frac{4}{3}\pi\mathbf{E}(R^3) \right). \qquad (3.33)$$

By means of the generalised Steiner formula (6.29) both formulae can be extended to the case of isotropic grains.

Note that the distribution of Ξ is of course not determined by the values of $\psi(K)$ for convex bodies K alone, as the distribution of radii is not determined by the first two or three moments.

3.2.2 Intensities or densities of intrinsic volumes

From a Boolean model various random measures can be derived. For example a three-dimensional Boolean model Ξ yields the *surface measure* S_Ξ, defined by

$$S_\Xi(B) = h_2(B \cap \partial\Xi) \qquad \text{for all Borel sets } B, \tag{3.34}$$

where the right hand side denotes the surface area of the boundary $\partial\Xi$ of the Boolean model Ξ in the set B; h_2 is the two-dimensional Hausdorff measure. The random measure S_Ξ inherits stationarity from the Boolean model Ξ. Therefore the mean measure $\mathbf{E}(S_\Xi(B))$ is invariant under translations and so is a multiple of Lebesgue measure:

$$\mathbf{E}(S_\Xi(B)) = S_V v_3(B). \tag{3.35}$$

The constant S_V is the *intensity* of the random measure S_Ξ. It is called the *specific surface (area)* because S_V can be interpreted as the *mean surface area of Ξ per unit volume*.

An analogous quantity for a planar Boolean model is L_A, the *mean boundary length per unit area*. Similar intensities can be defined also with respect to the other intrinsic volumes; see Section 7.3.4.

Some readers may wonder whether this approach is over-complicated. However it generalises well, for example to the nonstationary case, where Radon–Nikodym densities are used. The alternative density approach provides an easier route:

$$S_V = \lim_{r\to\infty} \frac{\mathbf{E}(S_\Xi(B(o,r)))}{v_3(B(o,r))} = \lim_{r\to\infty} \frac{2\mathbf{E}(V_2(\Xi \cap B(o,r)))}{v_3(B(o,r))}. \tag{3.36}$$

For the kth intrinsic volume in $\mathbb{R}^d$ the *density* v_k is

$$v_k = \lim_{r\to\infty} \frac{\mathbf{E}(V_k(\Xi \cap B(o,r)))}{v_d(B(o,r))} \qquad \text{for } k = 0, 1, \ldots, d. \tag{3.37}$$

Instead of $B(o, r)$, the set rW can be used, where W is a convex body with positive volume. Since the Boolean model is ergodic, the 'E' can be omitted in the equations above (see Theorem 6.2 on p. 210). More specific notations for the v_k are, in the planar case

$$A_A = v_2, \tag{3.38}$$
$$L_A = 2v_1, \tag{3.39}$$
$$N_A = v_0, \tag{3.40}$$

and in the spatial case

$$V_V = v_3, \tag{3.41}$$
$$S_V = 2v_2 \tag{3.42}$$
$$M_V = \pi v_1, \tag{3.43}$$
$$N_V = v_0. \tag{3.44}$$

The quantities N_A and N_V are called *specific connectivity numbers* and M_V is the *specific integral of mean curvature*. The intensities N_A, N_V and M_V belong to signed measures and can be negative.

There are 'positive' counterparts N_A^+, N_V^+, called *specific convexity numbers*, and M_V^+. The quantity N_A^+ is the intensity of lower convex tangent points of Ξ in $\mathbb{R}^2$. Each grain Ξ_n has a lower convex tangent point l_n, which is the point with minimum second coordinate in Ξ_n. (If this minimum is attained by more than one point, then the point having the minimum first coordinate is chosen, i.e. the lexicographic minimum point.) The points $x_n + l_n$ that are not covered by other grains $\Xi_m + x_m$ ($m \neq n$) are the lower convex tangent points of Ξ. The analogous intensity in the spatial case is N_V^+.

The densities or intensities satisfy the following equations, often called *Miles' formulae*:

$$V_V = p = 1 - \exp(-\lambda \overline{V}), \qquad (3.45)$$

$$S_V = \lambda(1-p)\overline{S} = \lambda \exp(-\lambda \overline{V})\overline{S}, \qquad (3.46)$$

$$M_V = \lambda(1-p)\left(\overline{M} - \frac{\pi^2 \lambda \overline{S}^2}{32}\right), \qquad (3.47)$$

$$N_V = \lambda(1-p)\left(1 - \frac{\lambda \overline{M}\,\overline{S}}{4\pi} + \frac{\pi \lambda^2 \overline{S}^3}{384}\right), \qquad (3.48)$$

$$N_V^+ = \lambda(1-p) = \lambda \exp(-\lambda \overline{V}), \qquad (3.49)$$

and

$$A_A = p = 1 - \exp(-\lambda \overline{A}), \qquad (3.50)$$

$$L_A = \lambda(1-p)\overline{L} = \lambda \exp(-\lambda \overline{A})\overline{L}, \qquad (3.51)$$

$$N_A = \lambda(1-p)\left(1 - \frac{\lambda \overline{L}^2}{4\pi}\right), \qquad (3.52)$$

$$N_A^+ = \lambda(1-p) = \lambda \exp(-\lambda \overline{A}), \qquad (3.53)$$

see Miles (1976), Weil (2007, p. 211) and Schneider and Weil (2008, p. 389).

Remarks (1) In the case $p = 0$ (for example in the case of a segment process) Formulae (3.51) and (3.46) should be replaced by

$$2L_A = \lambda \overline{L} \qquad \text{and} \qquad 2S_V = \lambda \overline{S}.$$

This change is necessary since in this case overlappings do not reduce the boundary lengths or surface areas of grains, and for a convex body K of volume zero, the perimeter $L(K)$ and the surface area $S(K)$ should be twice the length and the area of K, respectively; see p. 12.

(2) Many of the formulae from (3.45) to (3.53) above are true also in the anisotropic case. Isotropy is needed only for the formulae which contain a 'π'.

Sketch of the proof of (3.46) and (3.53). Formula (3.46) is a direct consequence of (3.103), which implies that for every grain the ratio of 'mean surface area not covered by other grains' to 'total mean surface area of grain' is $1 - p$. Summing this relation over all grains yields (3.46).

In the case of (3.53) the starting point is the definition of N_A^+ above. Without loss of generality it can be assumed that $l_n = o$ for all n; the grains $\Xi_n^* = \Xi_n - l_n$ have this property, and by Remark (3) on p. 73 the Boolean model Ξ^* formed by the system of the grains $\{\Xi_n^*\}$ and the original germs $\{x_1, x_2, \ldots\}$ is still a Boolean model, with the same distribution as Ξ.

The intensity of the point process of lower convex tangent points of Ξ^* is $p^*\lambda$, where p^* is the probability that the typical point of the germ process is not covered by grains belonging to other germs. By the Slivnyak–Mecke theorem the union of the other grains has the distribution of the original Boolean model and p^* is the probability that the origin o is not covered. Thus $p^* = 1 - p$. □

By the way, the lower convex tangent points of the Boolean model form a stationary point process, which can be interpreted as resulting from a (dependent) thinning of the original germ process. Because tangents parallel to the x-axis play the main rôle in the process construction, the thinned process is in general not isotropic, not even in the case of spherical grains. Its second-order characteristics can be computed; see Molchanov and Stoyan (1994).

The proof of the formulae for v_0 in the cases $d = 2$ and $d = 3$ (which up to a factor are the specific connectivity numbers N_A and N_V), and v_1 in the case $d = 3$ (which up to a factor is the specific integral of mean curvature M_V) are much more complicated.

For the particular case that the grains are random segments of constant length ℓ, in the planar case it holds

$$N_A = \lambda - N_3 \tag{3.54}$$

with

$$N_3 = \lambda^2 \ell^2 / \pi.$$

The term N_3 is linked to the topological concept of Betti numbers; see Robins (2002).

3.2.3 Contact distribution functions

By means of (3.31) the general Formula (3.30) for contact distribution functions can be simplified for a convex compact structuring element B. It becomes

$$H_B(r) = 1 - \exp\left(-\frac{\lambda}{b_d} \sum_{k=0}^{d-1} \frac{b_k b_{d-k}}{\binom{d}{k}} \overline{V}_k V_{d-k}(B) r^{d-k}\right) \qquad \text{for } r \geq 0. \tag{3.55}$$

Note that the last term, corresponding to $k = d$, of the original summation cancels with the $-\lambda \overline{V}_d$ or $-\lambda \mathbf{E}(v_d(\Xi_0))$ term of the exponent in (3.30).

The important planar and spatial cases are as follows.

Planar case ($d = 2$):

$$H_B(r) = 1 - \exp\left(-\lambda\left(\frac{\overline{L}L(B)}{2\pi}r + A(B)r^2\right)\right)$$

$$= 1 - \exp\left(-\frac{1}{1 - A_A}\left(\frac{L_A L(B)}{2\pi}r + N_A^+ A(B)r^2\right)\right) \qquad \text{for } r \geq 0. \qquad (3.56)$$

Spatial case ($d = 3$):

$$H_B(r) = 1 - \exp\left(-\lambda\left(\frac{\overline{S}M(B)}{4\pi}r + \frac{\overline{M}S(B)}{4\pi}r^2 + V(B)r^3\right)\right)$$

$$= 1 - \exp\left(-\frac{1}{1 - V_V}\left(\frac{S_V M(B)}{4\pi}r + \frac{M_V^+ S(B)}{4\pi}r^2 + N_V^+ V(B)r^3\right)\right) \qquad \text{for } r \geq 0.$$

$$(3.57)$$

In the formulae for $H_B(r)$ the physical dimensionality should be noted. For example, in (3.56) it seems to be natural to give $\overline{L}$ the right physical dimension of length. However, if r is given the dimension length, as it seems to be reasonable, then $L(B)$ and $A(B)$ should be dimension-free. The corresponding numerical values should be chosen such that $rL(B)$ and $r^2 A(B)$ have the right values in comparison to $\overline{L}$ and $\overline{A}$, respectively.

Linear contact distribution function $H_l(r)$

This is given by

$$H_l(r) = 1 - \exp\left(-\lambda\frac{2b_{d-1}\overline{V}_{d-1}}{db_d}r\right) = 1 - \exp(-\lambda_1^{(d)}r) \qquad \text{for } r \geq 0. \qquad (3.58)$$

Note that this is an exponential distribution of parameter $\lambda_1^{(d)}$. The parameter $\lambda_1^{(d)}$ is defined by its occurrence here, and also by Formula (3.68), which comes from (3.66).

Example 3.1.   Consider the *range of vision in a forest* where the set of tree cross sections can be modelled (after Pólya, 1918) as a planar Boolean model realisation, with discoidal grains. If the mean radius of a grain (radius of a typical trunk) is $E(R)$ then

$$H_l(r) = 1 - \exp(-2\lambda r E(R)) \qquad \text{for } r \geq 0.$$

Taking for example the values $E(R) = 0.20\,\text{m}$ and $\lambda = 0.01\,\text{m}^{-2}$, then

$$H_l(r) = 1 - \exp(-0.004r),$$

and hence $H_l(500) = 1 - 0.135$. That is to say that in a forest which can be described by such a Boolean model, with parameters as above, an observer standing at the typical point and looking in an arbitrary direction would have a probability of 13.5% of being able to see more than 500 m.

Further visibility properties are summarised in the *star*, or visible volume outside of Ξ, which is the set

$$\text{star}_x(\Xi) = \{y \in \mathbb{R}^d : [x, y] \cap \Xi = \emptyset\}$$

defined for x not in Ξ; see also Serra (1982), Wieacker (1985), Yadin and Zacks (1985, 1988), Schneider (1987), Molchanov (1994) and Zacks (1994). The mean volume of the star in the planar case is

$$\mathbf{E}\big(\text{star}_o(\Xi)\big) = \frac{2}{\pi \lambda_1^{(2)}}, \tag{3.59}$$

and in the spatial case

$$\mathbf{E}\big(\text{star}_o(\Xi)\big) = \frac{8}{\pi \lambda_1^{(3)}}. \tag{3.60}$$

Spherical contact distribution function $H_s(r)$

This is given by

$$H_s(r) = 1 - \exp\left(-\lambda \sum_{k=1}^{d} b_k \overline{V}_{d-k} r^k\right) \qquad \text{for } r \geq 0. \tag{3.61}$$

Consider, for example, a Boolean model with spherical grains in three-dimensional space, that is, random balls whose centres are distributed by a Poisson process in $\mathbb{R}^3$. The spherical contact distribution function is the distribution function of the distance to the nearest ball from a point chosen at random but lying outside all balls. It is

$$H_s(r) = 1 - \exp\left(-\frac{4}{3}\lambda \pi r (3\mathbf{E}(R^2) + 3r\mathbf{E}(R) + r^2)\right). \tag{3.62}$$

Quadrat contact distribution function $H_q(r)$

In statistical applications with $d = 2$, this function is useful. Here B is a square of unit area and

$$H_q(r) = 1 - \exp\left(-\lambda r\left(\frac{2\overline{U}}{\pi} + r\right)\right) \qquad \text{for } r \geq 0. \tag{3.63}$$

Discoidal contact distribution function $H_d(r)$

Analogously, in statistical applications with $d = 3$ and planar sections, this function is useful. Here B is a disc of radius 1 in $\mathbb{R}^3$ and

$$H_d(r) = 1 - \exp\left(-\lambda \pi r\left(\frac{\overline{\overline{S}}}{4} + \overline{\overline{b}}r\right)\right) \qquad \text{for } r \geq 0. \tag{3.64}$$

This function was introduced in Hug *et al.* (2006) and called there 'disc contact distribution'; instead of a circular disc, general convex flats were considered. That paper shows that the linear or polynomial form of $\ln\big(1 - H_B(r)\big)$ for the linear and discoidal contact distribution functions of stationary Boolean models is valid only for convex grains. However, note that there are non-Boolean random set models for which the logarithm above is also linear or polynomial.

3.2.4 Morphological functions

In 1991 Klaus Mecke suggested that one could transform the intrinsic volume densities to functions, making them powerful tools in random set statistics; see Mecke (1994, 2000) and Stoyan and Mecke (2005). In his papers he speaks about 'Minkowski functions' and denotes these as $W_k(r)$, following the notation for the Minkowski functionals. The exposition here uses instead the intrinsic volumes, and the functions are called *Mecke's morphological functions*. The following explains the idea for S_V; the same can be done for M_V and N_V. The case of V_V can be seen as covered by the spherical contact distribution function.

The function $S_V(r)$ is simply the specific surface of the parallel set $\Xi \oplus B(o, r)$. Since for a Boolean model Ξ the set $\Xi \oplus B(o, r)$ is the Boolean model with the same intensity λ and typical grain $\Xi_0 \oplus B(o, r)$, it is easy to give a formula for $S_V(r)$ by means of the Steiner formula and its counterparts for surface and average breadth:

$$S_V(r) = \lambda(\overline{S} + 4\pi\overline{\overline{b}}r + 4\pi r^2)\exp\left(-\lambda\left(\overline{V} + \overline{S}r + 2\pi\overline{\overline{b}}r^2 + \frac{4}{3}\pi r^3\right)\right) \qquad \text{for } r \geq 0. \quad (3.65)$$

It is derived by means of (1.47), (1.48) and (3.46).

Similarly, the formulae for $M_V(r)$ and $N_V(r)$ can be derived by means of (1.47), (1.48), (3.47) and (3.48). These functions can also be defined for negative values of r, as the specific surface of $\Xi \ominus B(o, r)$, see Stoyan and Mecke (2005).

3.2.5 Intersections with linear subspaces

Let L be an affine linear subspace of $\mathbb{R}^d$ (so not necessarily passing through the origin) of dimension l. For Ξ a Boolean model in $\mathbb{R}^d$ consider the intersection $\Xi_L = \Xi \cap L$. Then Ξ_L is again a Boolean model on L, when L is identified with $\mathbb{R}^l$. That Ξ_L is a Boolean model follows from considering its capacity functional. If the grains of Ξ are convex, then so of course are those of Ξ_L.

In the case of isotropy, when Ξ has convex grains, the induced Boolean model Ξ_L is also isotropic. Moreover, the intensity of its germ process is $\lambda_l^{(d)}$ with

$$\lambda_l^{(d)} = \lambda \frac{b_{d-l}b_l}{b_d} \frac{\overline{V}_{d-l}}{\binom{d}{d-l}}, \qquad (3.66)$$

and the mean value $\overline{V}_k^{(l)}$ of the kth intrinsic volume of the grain in $\mathbb{R}^l$ is

$$\overline{V}_k^{(l)} = \frac{b_{d-l+k}}{b_k b_{d-l}} \binom{d}{d-l} \frac{\overline{V}_{d-l+k}}{\overline{V}_{d-l}} \qquad \text{for } k \leq l. \quad (3.67)$$

These formulae were proved by Matheron (1975, p. 146), where there is also a discussion of the anisotropic case and the second moments of the volume of the induced grain; see also Schneider and Weil (2008). In the particular case of intersection of a planar Boolean model with a line ($l = 1$) the proof of (3.66) and (3.67) is as follows.

Proof. Suppose Ξ is a Boolean model in the plane which is isotropic and has convex grains. Let L be a line and let K be an arbitrary compact subset of L. Clearly $\mathbf{P}(\Xi \cap K = \emptyset) = \mathbf{P}(\Xi_L \cap K = \emptyset)$ thus $\psi(K) = \psi_L(K)$, which shows that indeed Ξ_L is a Boolean model, where $\psi_L(K) = -\ln(T_{\Xi_L}(K))$.

Let K be a bounded closed interval in L of length $l(K)$. Let $\lambda_1^{(2)}$ be the intensity of the germ process of Ξ_L. Then Formula (3.12) yields

$$\psi_L(K) = \lambda_1^{(2)}\big(l(K) + \overline{\ell}\big),$$

where $\overline{\ell}$ is the mean length of the grains, which are segments, of Ξ_L. On the other hand

$$\psi(K) = \lambda \mathbf{E}\big(A(\Xi_0 \oplus \check{K})\big) = \lambda\left(\overline{A} + \frac{\overline{L}\,l(K)}{\pi}\right)$$

from the planar case ($d = 2$) of (3.31). Equating the coefficients of $l(K)$ gives

$$\lambda_1^{(2)} = \lambda\frac{\overline{L}}{\pi} = \lambda\frac{b_1}{b_2}\overline{V}_1 \qquad \text{as } \overline{L} = 2\overline{V}_1$$

and

$$\overline{\ell} = \lambda\frac{\overline{A}}{\lambda_1^{(2)}} = \frac{b_2}{b_1}\frac{\overline{V}_2}{\overline{V}_1} \qquad \text{as } \overline{A} = \overline{V}_2. \qquad \square$$

The case of planar sections of three-dimensional Boolean models is considered in Section 10.3.1, Example 10.1.

Distribution function of chord lengths for Boolean models

Let Ξ be a Boolean model of convex grains and assume that Ξ is intersected with a line L. The intersection $\Xi_L = \Xi \cap L$ yields two alternating sequences of chords or intervals on L; one sequence forms Ξ_L and the other the exterior $L \setminus \Xi_L$. Because of the independence assumptions for the Boolean model and convexity of the grains, these intervals are independent, both kinds of chords follow fixed distributions and the resulting structure is called in the mathematical literature an 'alternating renewal process'; see for example Cox and Isham (1980).

The determination of the two types of chord length distributions differs greatly. It is easy to show that the *exterior* chords in the complement of Ξ follow an exponential distribution. If furthermore the Boolean model Ξ is isotropic, then the parameter of this exponential distribution is $\lambda_1^{(d)}$, where

$$\lambda_1^{(d)} = \lambda\frac{2db_{d-1}}{b_d}\overline{V}_{d-1}, \qquad (3.68)$$

which is the special case of (3.66) for $l = 1$. Then the chord length probability density function $f_{ex}(l)$ is

$$f_{ex}(l) = \lambda_1^{(d)} \exp(-\lambda_1^{(d)}l) \qquad \text{for } l \geq 0. \tag{3.69}$$

The mean chord length is

$$\bar{\ell}_{ex} = 1/\lambda_1^{(d)}. \tag{3.70}$$

Clearly in the anisotropic case the parameter depends on the direction of the line.

The *interior* chord lengths for the grain phase have a complicated distribution depending on the distribution of the grains. Tools such as the Fourier transform have been used to obtain results, see Torquato and Lu (1993) and Gille (2011). The curves in Figure 3.5 for the probability density function $f_{in}(l)$ in the case of identical balls of diameter 1 were obtained numerically by inverting the Fourier transform. Nevertheless, for the first two moments formulae are known. The mean chord length $\bar{\ell}_{in}$ can be obtained from

$$p = \frac{\bar{\ell}_{in}}{\bar{\ell}_{in} + \bar{\ell}_{ex}}, \tag{3.71}$$

and the second moment is

$$\overline{\ell^2}_{in} = \frac{\bar{\ell}_{in}^2 \cdot l_c}{\bar{\ell}_{ex} \cdot p}, \tag{3.72}$$

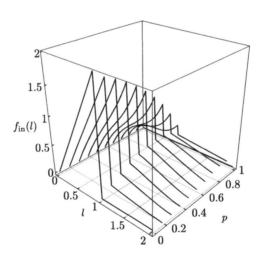

Figure 3.5 Probability density functions $f_{in}(l)$ of chord lengths for a Boolean model with identical spherical grains of diameter 1 in dependence of volume fraction p. For small p isolated balls dominate and so $f_{in}(l)$ looks like the chord length probability density function for a ball as in Section 1.7. For larger values of p clumping plays a rôle and the chords become longer. Courtesy of W. Gille.

with

$$l_c = 2 \int_0^\infty \kappa(r) dr \qquad (3.73)$$

and $\kappa(r)$ is given in Equation (3.23), as shown by Gille (2011).

3.2.6 Formulae for some special Boolean models with isotropic convex grains

Two-dimensional models:

(1) Grains = *random segments* with uniform (or isotropic) orientation and of mean length $\overline{\ell}$. The linear contact distribution is an exponential distribution with parameter

$$\lambda_1^{(2)} = \frac{2\lambda\overline{\ell}}{\pi}. \qquad (3.74)$$

The spherical contact distribution function is

$$H_s(r) = 1 - \exp\big(-\lambda r (2\ell + \pi r)\big) \qquad \text{for } r \geq 0. \qquad (3.75)$$

(2) Grains = *random discs* with random radius R. The area fraction is given by

$$p = 1 - \exp\left(-\lambda\pi\mathbf{E}(R^2)\right). \qquad (3.76)$$

Moreover

$$\lambda_1^{(2)} = 2\pi\lambda\mathbf{E}(R), \qquad (3.77)$$

$$H_s(r) = 1 - \exp\big(-\lambda\pi r\left(2\mathbf{E}(R) + r\right)\big) \qquad \text{for } r \geq 0. \qquad (3.78)$$

(3) Grains = *Poisson polygons* with parameter ϱ (see Section 9.5.1). The expected area of a grain is then

$$\overline{A} = \frac{4}{\pi\varrho^2}. \qquad (3.79)$$

The other formulae are:

$$p = 1 - \exp\left(\frac{-4\lambda}{\pi\varrho^2}\right), \qquad (3.80)$$

$$\lambda_1^{(2)} = \frac{4\lambda}{\pi\varrho}, \qquad (3.81)$$

$$H_s(r) = 1 - \exp\left(-\lambda r \left(\frac{4}{\varrho} + \pi r\right)\right) \qquad \text{for } r \geq 0, \qquad (3.82)$$

$$C(r) = 2p - 1 + (1-p)^2 \exp\left(\frac{4\lambda}{\pi\varrho^2}\exp(-\varrho r)\right) \qquad \text{for } r \geq 0. \qquad (3.83)$$

Three-dimensional models:

(4) Grains = *random segments* with uniform orientations of mean length $\bar{\ell}$. The discoidal and spherical contact distribution functions are

$$H_d(r) = 1 - \exp(-\lambda \pi^2 r^2 \bar{\ell}) \qquad \text{for } r \geq 0, \qquad (3.84)$$

$$H_s(r) = 1 - \exp\left(-\lambda \pi r^2 \left(\bar{\ell} + \frac{4r}{3}\right)\right) \qquad \text{for } r \geq 0. \qquad (3.85)$$

(5) Grains = *random discs* with uniform orientation and random radius R. Then

$$\lambda_1^{(3)} = \frac{\lambda \pi \mathbf{E}(R^2)}{2}, \qquad (3.86)$$

$$H_d(r) = 1 - \exp\left(-\lambda \pi r \left(\frac{\pi \mathbf{E}(R^2)}{2} + \frac{\pi \mathbf{E}(R)r}{2}\right)\right) \qquad \text{for } r \geq 0, \qquad (3.87)$$

$$H_s(r) = 1 - \exp\left(-\lambda \pi r \left(2\mathbf{E}(R^2) + \pi r \mathbf{E}(R) + \frac{4r^2}{3}\right)\right) \qquad \text{for } r \geq 0. \qquad (3.88)$$

(6) Grains = *random balls* with random radius R. Then

$$p = 1 - \exp\left(-\frac{4}{3}\lambda \pi \mathbf{E}\left(R^3\right)\right), \qquad (3.89)$$

$$\lambda_1^{(3)} = \lambda \pi \mathbf{E}(R^2), \qquad (3.90)$$

$$H_d(r) = 1 - \exp\left(-\lambda \pi r \left(\pi \mathbf{E}(R^2) + 2\mathbf{E}(R)r\right)\right) \qquad \text{for } r \geq 0, \qquad (3.91)$$

$$H_s(r) = 1 - \exp\left(-\frac{4}{3}\lambda \pi r \left(3\mathbf{E}(R^2) + 3r\mathbf{E}(R) + r^2\right)\right) \qquad \text{for } r \geq 0. \qquad (3.92)$$

(7) Grains = *Poisson polyhedra* with parameter ϱ; see Section 9.5.2. (The first paragraph of Section 3.3.5 has an explanation of why the Poisson polyhedron grain can serve as a good model in a wide variety of applications.) The expected volume of a grain is then

$$\bar{V} = \frac{6}{\pi \varrho^3}.$$

The other formulae are:

$$p = 1 - \exp\left(\frac{-6\lambda}{\pi \varrho^3}\right), \qquad (3.93)$$

$$\lambda_1^{(3)} = \frac{6\lambda}{\pi \varrho^2}, \qquad (3.94)$$

$$H_d(r) = 1 - \exp\left(-\lambda \pi r \left(\frac{6}{\pi \varrho^2} + \frac{3}{2\varrho}r\right)\right) \qquad \text{for } r \geq 0, \qquad (3.95)$$

$$H_s(r) = 1 - \exp\left(-\lambda r \left(\frac{24}{\pi \varrho^2} + \frac{3\pi}{\varrho}r + \frac{4}{3}\pi r^2\right)\right) \qquad \text{for } r \geq 0, \qquad (3.96)$$

$$C(r) = 2p - 1 + (1-p)^2 \exp\left(\frac{6\lambda \exp(-\varrho r)}{\pi \varrho^3}\right) \qquad \text{for } r \geq 0. \qquad (3.97)$$

The isotropised set covariance of the typical grain is

$$\overline{\gamma}_{\Xi_0}(r) = \frac{6\exp(-\varrho 2r)}{\pi \varrho^3} \qquad \text{for } r \geq 0. \tag{3.98}$$

The case of cylindrical grains is considered in Peyrega *et al.* (2009).

3.3 Coverage and connectivity

3.3.1 Coverage probabilities

Coverage probabilities are quantities of the form

$$\mathbf{P}(K \subset \Xi),$$

where K is any compact subset of $\mathbb{R}^d$. The volume fraction $p = \mathbf{P}(o \in \Xi)$ can be written in the form

$$p = \mathbf{P}(\{o\} \subset \Xi)$$

and is therefore a coverage probability with $K = \{o\}$. Also the covariance $C(r)$ can be interpreted as a coverage probability, that for the two-point set $\{o, \mathbf{r}\}$.

Clearly, it holds for general compact K

$$\mathbf{P}(K \subset \Xi) \leq \mathbf{P}(K \cap \Xi \neq \emptyset); \tag{3.99}$$

that is, typically the coverage probability is smaller than the hitting probability.

As might be expected, the calculation of coverage probabilities is difficult. Hall (1988, p. 180) discussed the coverage problem in detail. For the particular case $d = 2$, $K = [0, 1]^2$ and $\Xi_0 = B(o, r)$ (r deterministic) he showed that

$$1 - 3\min\{1, (1 + \lambda^2 \pi r^2)e^{-\lambda \pi r^2}\} < \mathbf{P}(K \subset \Xi) < 1 - 0.05\min\{1, (1 + \lambda^2 \pi r^2)e^{-\lambda \pi r^2}\}$$
$$\tag{3.100}$$

if $\lambda \geq 1$ and $r \leq 0.5$. This is a rather rough bound for the coverage probability.

Janson (1986) found asymptotic relationships for coverage probabilities for large λ and small grains; see also Baccelli and Błaszczyszyn (2009a, p. 327).

The coverage of the whole space $\mathbb{R}^d$ is also of interest. As shown by Hall (1988, p. 130), the condition given in (3.2) ensures that with probability one the uncovered part of $\mathbb{R}^d$ has infinite Lebesgue measure. If $\mathbf{E}(\nu_d(\Xi_0)) = \infty$, which violates (3.2), then with probability one almost all of $\mathbb{R}^d$ is covered by Ξ. However, $\Xi = \mathbb{R}^d$ is not necessarily true.

Now consider the coverage or hitting of parts of Ξ by the rest of Ξ.

Let x_* be the typical point of the germ point process, and let Ξ_* be the corresponding grain. This grain has the same distribution as the typical grain Ξ_0 and can be interpreted heuristically as a grain chosen at random in such a way that every grain is equally likely to be chosen. Thus, it is also called the typical grain (in the Palm sense). The relationship between Ξ_* and the rest of the Boolean model is of interest. Let

$$\Xi' = \bigcup\{\Xi_i + x_i : x_i \neq x_*\}$$

be the union of the remaining grains. Then the following quantities are of interest:

$$p_0 = \mathbf{P}(x_* \text{ is covered by } \Xi'),$$
$$p_G = \mathbf{P}(\Xi_* + x_* \text{ intersects } \Xi'),$$
$$p_s = \text{fraction of boundary of } \Xi_* + x_* \text{ which is covered by } \Xi'.$$

Of course $p_0 > 0$ only if $\mathbf{E}(v_d(\Xi_0)) > 0$, and p_s only makes sense if the grains of Ξ have finite nonzero surface area. The value $1 - p_G$ is the probability that the typical grain is isolated. If it is positive, then there are even infinitely many isolated grains. Clearly, the mean number of isolated grains with germ points in a given compact set K is $\lambda(1 - p_G)v_d(K)$.

These probabilities satisfy

$$p_0 = p, \tag{3.101}$$

$$p_G = 1 - \mathbf{E}(\exp(-\psi(\Xi_0))), \tag{3.102}$$

$$p_s = p, \tag{3.103}$$

with $\psi(\Xi_0)$ as defined in equation (3.11). In the particular case of the grains being balls of fixed radius r,

$$p_G = 1 - \exp(-\lambda v_d(B(o, 2r))) = 1 - \exp(-2^d \lambda b_d r^d). \tag{3.104}$$

In the two-dimensional case, with grains being discs of fixed radius, it is

$$p_G = 1 - (1 - p)^4, \tag{3.105}$$

independently of the particular values of λ and r.

Formula (3.101) can be established as follows. By an analogue of the theorem of Slivnyak–Mecke (see Sections 2.3.4 and 4.4.4) the grain $\Xi_* + x_*$ may be taken to be independent of the other grains, and $\Xi' \overset{d}{=} \Xi$. Thus, by taking $x_* = o$, $p_0 = \mathbf{P}(o \in \Xi) = p$.

The proofs of Formulae (3.102) and (3.103) are similar. In the same way the mean number n_0 of grains that hit the typical grain can be obtained:

$$n_0 = \lambda \mathbf{E}(v_d(\Xi_0 \oplus \check{\Xi}_0)). \tag{3.106}$$

The corresponding distribution is a Poisson distribution. Here $\check{\Xi}_0$ denotes an independent random compact set with the same distribution as $-\Xi_0$. The quantity $v_d(\Xi_0 \oplus \check{\Xi}_0)$ is called *excluded* volume. Note that in the case of isotropy the generalised Steiner formula (6.29) yields formulae for the mean excluded volume $\overline{V}_{\text{ex}} = \mathbf{E}(v_d(\Xi_0 \oplus \check{\Xi}_0))$:

$$\overline{V}_{\text{ex}} = 2\overline{A} + \frac{\overline{L}^2}{2\pi} \qquad \text{for } d = 2, \tag{3.107}$$

$$\overline{V}_{\text{ex}} = 2\overline{V} + \frac{\overline{M}\,\overline{S}}{2\pi} \qquad \text{for } d = 3. \tag{3.108}$$

In the case of fibrous grains n_0 is called the 'mean number of fibre crossings per fibre' (Sampson, 2004). If Ξ_0 is a randomly oriented flat convex object, Formula (3.108) becomes

$$\overline{V}_{\text{ex}} = \frac{\overline{A}\,\overline{L}}{2}, \tag{3.109}$$

where $\overline{A}$ is the mean area and $\overline{L}$ the mean perimeter of Ξ_0. Such grains play an important rôle in the context of fracture networks, see Thovert and Adler (2004).

3.3.2 Clumps

A clump is a maximal connected cluster of overlapping grains. Its size is defined as the number of grains constituting it; it may be infinite. Hall (1988, Chapter 4) gave properties of clumps for sparse models. They include asymptotics for the number of n-clumps (formed by n overlapping grains) in a bounded region and the expected number of clumps per unit volume. Błaszczyszyn et al. (1999) considered the moments of clump size.

3.3.3 Connectivity

Two germs x_i and x_j are said to connected if $(\Xi_i + x_i) \cap (\Xi_j + x_j) \neq \emptyset$. This leads to the *Boolean connectivity graph* or, in the case of identical spherical grains, *random geometric graph* or *Gilbert graph*; see Figure 3.6. The name 'Gilbert graph' is used to honour E. Gilbert, who studied this graph in his pioneering paper Gilbert (1961).

This graph has been extensively studied in the literature, see Penrose (2003). The probability that the typical vertex of the Gilbert graph is isolated is $1 - p_G$. The degree of the typical vertex has a Poisson distribution with mean n_0 given by (3.106). Furthermore, an asymptotic result (for $\lambda \to \infty$) for *connectivity in a compact set*, that is, for the connectivity of the finite random graph formed by the germs in a given compact set K, is shown in Baccelli and Błaszczyszyn (2009a, Section 13.1.2), using Penrose (1997).

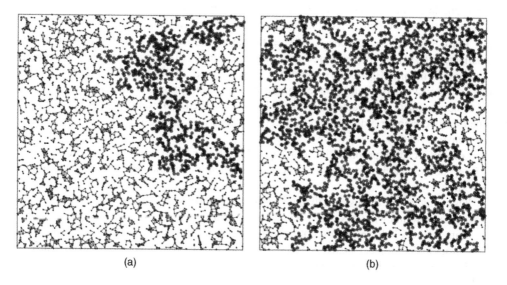

(a)   (b)

Figure 3.6 Planar random geometric graph with (a) $\lambda = 0.98\lambda_c$ and (b) $\lambda = 1.02\lambda_c$ and the radius such that the grain area is 1; the critical intensity λ_c is believed to be 1.1281. The largest component is highlighted in black. Courtesy of B. Błaszczyszyn.

3.3.4   Percolation

The problem of the occurrence of giant clumps or of *continuum percolation* is of particular interest. One says that a Boolean model *percolates* if there exists an infinite connected component. (Such a component consists of infinitely many overlapping grains, and it is almost surely an unbounded set.) Percolation in the Boolean model is a practically important question since it often serves as a model for porous media. Fluid flow is impossible in a porous medium with isolated grains. Furthermore, 'continuum percolation is becoming an important tool in the investigation of various telecommunication networks' (van der Hofstad, 2010a).

Percolation is studied by considering that the distribution of the typical grain is fixed, while the intensity λ acts as a variable model parameter. For $d = 1$, the Boolean model does not percolate if the mean length of the typical grain is finite (Meester and Roy, 1996, Theorem 3.1). Thus, only the case $d \geq 2$ is discussed here.

Two percolation characteristics are considered:

(a) the probability p_∞ that the typical grain $\Xi_* + x_*$ of the Boolean model is part of a clump of infinite order, that is, there are infinitely many $\Xi_k + x_k$ which are connected with $\Xi_* + x_*$ by overlapping;

(b) the expected order o_∞ of the clump (= number of its grains) containing the typical grain $\Xi_* + x_*$.

For the case of spherical grains, Hall (1985b) showed that if and only if $\mathbf{E}\big(\nu_d(\Xi_0)^2\big) < \infty$, there is a positive finite constant λ'_c such that $o_\infty < \infty$ for $\lambda < \lambda'_c$ and $o_\infty = \infty$ for $\lambda > \lambda'_c$; if $\mathbf{E}\big(\nu_d(\Xi_0)^{2-1/d}\big) < \infty$, there exists another positive finite λ_c such that $p_\infty = 0$ for $\lambda < \lambda_c$ and $p_\infty > 0$ for $\lambda > \lambda_c$. (Note that $p_\infty = 1$ does not hold, since isolated grains always exist. Nevertheless, since there are infinitely many grains, whenever $p_\infty > 0$, the Boolean model does percolate. Thus, λ_c and λ'_c are the *critical intensities* for the so-called *phase transition*.) The two different moment conditions lead to the interesting case that when $\mathbf{E}\big(\nu_d(\Xi_0)^2\big) = \infty$ and $\mathbf{E}\big(\nu_d(\Xi_0)^{2-1/d}\big) < \infty$, then $\lambda_c > \lambda'_c = 0$, that is, if a Boolean model has a sufficiently small intensity, it does not percolate but the expected number of grains in the clump containing the origin is infinite. Clearly in general $0 \leq \lambda'_c \leq \lambda_c \leq \infty$. Meister and Roy (1994) showed that the Boolean model with spherical grains can have at most one infinite clump.

Other percolation characteristics have also been studied. An example is the existence of a path of overlapping grains from one facet of a cube to the opposite facet, see Zuyev and Sidorenko (1985a,b).

Local percolation probabilities, used by Hilfer (1991, 2000) to characterise the connectivity in porous media, give the fraction of cells where percolation is from one side to the opposite.

It is characteristic of this field that there is a great gap between what mathematicians could prove until now and what physicists 'know' by simulations.

Many authors believe that except for pathological cases, $\lambda_c = \lambda'_c$. The critical intensity λ'_c is easier to determine, usually as the asymptote of the graph of mean clump size against λ. For $\Xi_0 = B(o, r)$ (r chosen so that disc area and ball volume are 1, respectively) simulations yielded

$$\lambda'_c \approx 1.1281 \quad (d = 2) \qquad \text{and} \qquad \lambda'_c \approx 0.3419 \quad (d = 3), \tag{3.110}$$

see Figure 3.6 on p. 91. The corresponding values of p_∞ are

$$p_\infty \approx 0.6763 \quad (d = 2) \qquad \text{and} \qquad p_\infty \approx 0.2896 \quad (d = 3), \tag{3.111}$$

see Lorenz and Ziff (2001) for the spatial case by developing a cluster growth algorithm, and Quintanilla and Ziff (2007) for the planar case by employing the continuum frontier-walk method, using inhomogeneous Boolean models; the latter approach has been theoretically justified by Zuyev and Quintanilla (2003). Note that

$$p_\infty = 1 - e^{-\lambda_c}. \tag{3.112}$$

Various authors have studied the influence of the radius distribution on λ_c' for the case of random discs or balls; see Torquato (2002, p. 253). Meester *et al.* (1994) conjectures that the monodisperse case minimises λ_c'. Torquato (2012) studies the influence of dimensionality d.

Physicists have suggested various approximations for λ_c' in cases of nonspherical grains; see Bretheau and Jeulin (1989), Mecke and Wagner (1991) and Torquato (2002). The following exposition follows Stoyan and Mecke (2005).

A simple lower bound uses the mean excluded volume $\overline{V}_{ex}$ given by (3.107) and (3.108); see Balberg *et al.* (1984). From (3.106), the equality

$$\lambda_c' \overline{V}_{ex} = 1 \tag{3.113}$$

yields an approximation, based on the idea that on the average in a percolating Boolean model each grain should be connected with at least one other grain.

In the particular case of segments of length l as grains of planar Boolean models Formula (3.113) yields the approximation $\lambda_c' = \alpha/l^2$ with $\alpha = \pi/2$. Simulations showed that probably $\lambda_c' = 5.7/l^2$, that is, the true coefficient α is larger than that predicted by (3.113). Nevertheless, the qualitative behaviour of λ_c' seems to be well predicted.

A more precise approach uses a topological argument. It considers the specific connectivity number N_V. For small λ the grains are mainly isolated and N_V is positive. For large λ the Boolean model becomes sponge-like with many small holes and N_V is negative. Therefore, it is natural to assume that the smallest zero of the function $N_A(\lambda)$ in (3.52) or $N_V(\lambda)$ in (3.48) provides a good approximation for λ_c'. The zeros are

$$\lambda_0 = \frac{4\pi}{\overline{L}^2} \qquad\qquad \text{for } d = 2, \tag{3.114}$$

$$\lambda_0 = \frac{48\overline{M}}{\pi^2 \overline{S}^2}\left(1 - \left(1 - \frac{\pi^3 \overline{S}}{6\overline{M}^2}\right)^{1/2}\right) \qquad \text{for } d = 3. \tag{3.115}$$

Mecke and Wagner (1991), Mecke and Seyfried (2002) and Stoyan and Mecke (2005) demonstrate for various grain shapes the good quality of the approximation

$$\lambda_c' = \lambda_0. \tag{3.116}$$

For the case of discs of area 1 in $\mathbb{R}^2$ Formula (3.114) yields the approximation $\lambda_c' = 1$ and for balls in $\mathbb{R}^3$ (3.115) gives the value 0.38.

Formula (3.116) together with (3.114) and (3.115) shows that the shape of the grains has a great influence on the value of λ_0. For long and thin grains λ_0 is clearly smaller than for spherical grains; thus percolation happens for such grains earlier than for balls.

Oriented grains have also been considered, for example the cases of squares ($d = 2$), cubes ($d = 3$) (see Torquato, 2002, p. 254) and cylinders (see Jeulin and Moreaud, 2007).

The field of percolation is far wider than is represented in the above account. The interested reader may look into Hughes (1996), Meester and Roy (1996), Grimmett (1999), Torquato (2002, Chapters 9 and 10), Penrose (2003), Hunt (2005), Bollobás and Riordan (2006) and van der Hofstad (2010a).

3.3.5 Vacant regions

Sometimes also the complement Ξ^c of a Boolean model Ξ is of interest. Its closure $(\Xi^c)^{cl}$ is also a random closed set. (Since the Boolean model Ξ is a closed set, its complement is not closed – the boundaries belong to Ξ and not to Ξ^c. Therefore, it is necessary to take the closure.) Each connected component of $(\Xi^c)^{cl}$ is called a *vacancy*. In the case of a volume fraction p close to 1 and not very small grains, the vacancies are randomly scattered and each resembles a convex polygon or polyhedron, or a union of such shapes; see Figures 3.2 and 3.3. Thus a Boolean model with grains that are Poisson polygons or polyhedra may be often a good approximation for $(\Xi^c)^{cl}$.

This heuristic idea can be partly made mathematically rigorous; see Hall (1988, Chapter 3), Aldous (1989, Chapter H), Molchanov (1996), Michel and Paroux (2003) and Calka *et al.* (2009).

A particular case is as follows: Consider in $\mathbb{R}^d$ a sequence of Boolean models with the same typical grain Ξ_0, which is isotropic, and with intensities λ_n with $\lambda_n \to \infty$. Denote by V_n the vacancy containing the origin o given that o is not covered. Then it holds

$$\lambda_n V_n \to C_0 \qquad \text{for } n \to \infty, \tag{3.117}$$

where C_0 is the zero cell (the cell containing the origin) of the Poisson hyperplane tessellation of intensity ϱ (which is S_V for $d = 3$ and L_A for $d = 2$) equal to the mean surface area or mean perimeter of Ξ_0; see Molchanov (1996). (In practical application, some length scale is chosen and fixed during all calculations, in which physical dimensions are ignored.) The convergence of the random compact sets is almost-sure convergence with respect to the Hausdorff metric in $\mathbb{K}'$; see Calka *et al.* (2009), who also considered the convergence of the difference between $\lambda_n V_n$ and C_0 in fixed direction. (These types of convergence are explained in Molchanov, 2005.)

The convergence in the 'local' limit theorem (3.117) is still true, at least in distribution, for grains of more general shapes, see Molchanov (1996) and Michel and Paroux (2003).

A 'global' limit theorem says that, for $d = 2$, the mean number of vacancies of a Boolean model with $\Xi_0 = B(o, r)$ per unit area is approximately

$$N_A = \lambda^2 \pi r^2 e^{-\lambda \pi r^2};$$

see Hall (1985a) and Aldous (1989, p. 149).

The problem of connectivity of unbounded components of Ξ^c has been studied by Meester and Roy (1996); see also Torquato (2002, p. 253).

For a Boolean model Ξ with typical grain whose volume is of finite second moment, Hall (1988, pp. 142 and 190) gives asymptotic expressions, as $\lambda \to \infty$, for the variance $\mathbf{var}\big(v_d(\Xi^c \cap W)\big)$ of the volume of the union of vacant regions in a window W and shows that it is minimised when the grains are uniformly rotated. Rau and Chiu (2011) study the continuity properties of this asymptotic variance, when the grains are subject to both nonuniform rotations and random shape distortions, where the latter belong to a certain class of linear transformations that are not necessarily volume-preserving.

3.4 Statistics

3.4.1 General remarks

This section discusses in detail statistical analysis for the Boolean model, both because of its intrinsic interest and as an example of parametric statistics for random closed sets. The general nonparametric statistics for random sets will be presented in Section 6.4. There it will be explained how summary characteristics such as the intrinsic volume densities (e.g. p or S_V) and functions such as $C(r)$ or the various contact distribution functions can be estimated.

The main objective of statistical analysis for the Boolean model is the determination of model parameters such as λ, $\overline{L}$, $\overline{A}$, $\overline{b}$, $\overline{S}$ and $\overline{V}$, and the distribution of the typical grain, in particular the probability density function of the radii in the case of balls. A further and also very important problem concerns the decision as to whether or not the Boolean model is appropriate in describing a particular data set, that is, testing the Boolean model hypothesis.

These problems present less difficulty if it is possible to discriminate clearly between the grains. Such a discrimination can be carried out, albeit with some difficulty, in the case of lower dimensional grains, that is, in the cases of segments ($d = 2$ or 3) or of convex flats ($d = 3$). In this case the germs can be identified and measured individually, leading to separate studies of the grain distributions and of the point pattern of germs. Study of the grains is a problem in random-compact-set statistics while study of the germs involves point-process statistics.

The statistics become more complicated when the grains have full dimension and overlap. Then the individual grains cannot be identified and the statistics must go indirect ways, using the summary characteristics mentioned in the first paragraph of this section. Since in the corresponding formulae parameters, such as λ and the means of the intrinsic volumes (e.g. $\overline{V}$ or $\overline{S}$), appear and they can be estimated – usually not by maximum likelihood method but by method of moments and related procedures.

When an empirical two-phase structure is given, sometimes it is not clear which phase of the image should be interpreted as the realisation of a Boolean model. For example, when a porous medium is modelled by a Boolean model, either the system of pores or its complement may be fitted by a Boolean model. Thus it may make sense to perform some steps of statistical analysis in parallel, comparing results obtained from assuming first that one and then the other component forms a Boolean model. See also the remark in the first paragraph of Section 3.3.5 about the possibility of approximating the complement of a Boolean model by another Boolean model. A final decision in favour of one component may result from considering empirical contact distribution functions.

This section mainly presents the theory and methods for the stationary and isotropic case. The key references for the anisotropic case are Weil (1995) and Schneider and Weil (2008). For nonparametric estimation of the intensity function of a nonstationary or partially stationary Boolean model, see Molchanov and Chiu (2000).

The data are usually two- or three-dimensional pixel images, the planar images may belong to true planar structures or to planar sections through three-dimensional structures. Also data

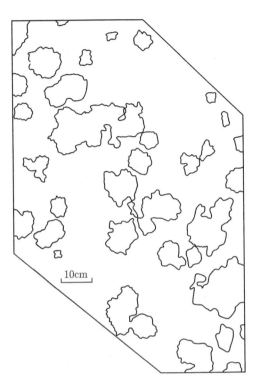

Figure 3.7   A pattern of lichen on a stone. It is in some sense similar to a sample from a Boolean model with convex grains if the boundary roughnesses are ignored.

such as Figure 3.7 are usually digitised for the determination of most summary characteristics. Such situations require the use of numerically stable and robust methods that work well for digital data and automatic analysis, and hence exclude methods that use for example tangent points or curvatures in boundary points. In this context the 'precision hierarchy' of the intrinsic volume density estimators (see p. 99) should be considered. One should try to get along with V_V and S_V and, if possible, to renounce the use of M_V and N_V.

3.4.2   Testing model assumptions

Whether or not the Boolean model assumption is appropriate for given data is usually evaluated by considering functional summary characteristics such as the covariance $C(r)$, contact distribution functions $H_B(r)$ or Mecke's morphological functions. Which function should be used depends mainly on the measurement conditions: if small-angle scattering data are analysed, there is no other choice than to work with $C(r)$, though it is known to be not a powerful test characteristic and the corresponding formulae are rather complicated. The best characteristics for test purposes seem, if applicable, to be the quadrat, discoidal and spherical contact distributions and the morphological functions; see the classical papers by Ohser (1980), Ohser and Stoyan (1980), Diggle (1981), Serra (1982) and Mecke (2000). The linear contact distribution function is not a good mean for such tests since it is also approximately exponential for many random set models that are not Boolean.

Many authors simply show images with empirical summary characteristic diagrams and theoretical counterparts corresponding to the fitted Boolean model. They consider some degree of visual similarity as a proof of a good fit. A correct significance test is a *deviation test* by the parametric bootstrap procedure as already demonstrated in Diggle (1981). It is explained in the following for the case that the summary characteristic used is $H_s(r)$, as also in Diggle (1981).

The fitted Boolean model with the theoretical spherical contact distribution function $H_s(r)$ is simulated k times, and for each pattern the corresponding estimate $\hat{H}_{s,i}(r)$ is determined. Then the deviations

$$\tau_{emp} = \max_{r \leq r_{max}} \left| H_s(r) - \hat{H}_s(r) \right|, \tag{3.118}$$

where $\hat{H}_s(r)$ is the empirical spherical contact distribution function, and

$$\tau_i = \max_{r \leq r_{max}} \left| H_s(r) - \hat{H}_{s,i}(r) \right| \qquad \text{for } i = 1, 2, \ldots, k \tag{3.119}$$

are determined for a suitable r_{max}. Under the Boolean model hypothesis, τ_{emp} has the same distribution as τ_i and so (if there is no tie) has a rank, when pooled together with all τ_i, uniformly distributed on integers between 1 and $k + 1$. A large rank suggests rejection of the Boolean model hypothesis, analogous to the case of testing the Poisson process hypothesis in Section 2.6.4. The (approximate) *p*-value of this test is given by (2.64).

Instead of using the L^∞-norm, the L^2-norm

$$\tau_{emp} = \int_0^{r_{max}} \left(H_s(r) - \hat{H}_s(r) \right)^2 dr$$

for a suitable r_{max} and corresponding τ_i can be used to measure deviation. Sometimes *a priori* knowledge may help decide which τ-definition should be used.

This procedure is well explained in Mrkvička and Mattfeldt (2011), albeit in the slightly different context of the quadrat contact distribution function.

If such a test is applied to an image in the form of pixel data, which is usually the case in practice, it is important to digitise the simulated Boolean models in a similar (ideally in the same way) way as the original pixel image, in particular the proportions of pixel size to grain size should be similar. The estimation of $H_s(r)$ has to be carried out as for pixel images and not for images with vector data (which would be possible for simulated data).

When contact distribution functions are used, it is popular to visualise the differences between empirical and theoretical values in the case of convex and isotropic grains for $d = 2$ and 3 as follows.

As a result of Formula (3.55) the logarithm of $1 - H_B(r)$ must be a polynomial of the form:

$$\ln\left(1 - H_B(r)\right) = -a(B)r - b(B)r^2 - c(B)r^3 \qquad \text{for } r \geq 0, \tag{3.120}$$

with nonnegative coefficients $a(B)$, $b(B)$ and $c(B)$ for each convex B. For the linear contact distribution it is

$$b(B) = c(B) = 0,$$

and for the cases of $d = 2$ and discoidal contact distribution function it is

$$c(B) = 0.$$

Consequently the following informal test suggests itself. For various shapes of B the empirical contact distribution function $\hat{H}_B(r)$ is computed from the given image. If the normalised logarithm

$$\ln\left(1 - \hat{H}_B(r)\right)/r \tag{3.121}$$

is well approximated by polynomials as in Formula (3.120), for each B considered, then the Boolean model can be considered not inappropriate as a model to fit the data.

The case of nonconvex grains is considered in Chadœuf *et al.* (2008).

3.4.3 Estimation of model parameters

Method of densities

The best method for parameter estimation for a Boolean model is probably the method of densities or intensities, which goes back to Santaló (1976) and Weil (1984). It is based on Formulae (3.50) to (3.53) for $d = 2$ and (3.45) to (3.49) for $d = 3$. The idea is to estimate these intensities and then to use formulae which contain them and model parameters, such as λ and parameters of grain size distributions.

For example, the densities $A_A = p$, L_A, N_A^+ and N_A are connected with model parameters by the four equations known already from p. 80:

$$A_A = 1 - \exp(-\lambda\overline{A}), \tag{3.122}$$

$$L_A = \lambda(1 - A_A)\overline{L}, \tag{3.123}$$

$$N_A^+ = \lambda(1 - A_A), \tag{3.124}$$

$$N_A = \lambda(1 - A_A)\left(1 - \frac{\lambda\overline{L}}{4\pi}\right). \tag{3.125}$$

Consequently, when estimates of the densities are given, the parameters λ, $\overline{L}$ and $\overline{A}$ can be estimated.

The statistical properties of this method are not yet thoroughly investigated. The main problem is the accuracy of the density estimators. Several authors have determined estimation variances by simulation. The asymptotical normality of estimators of A_A, L_A and N_A^+ is known: see Section 6.4 for A_A and Molchanov and Stoyan (1994) for N_A^+ and L_A. The estimator

$$\hat{\lambda} = \frac{\hat{N}_A^+}{1 - \hat{A}_A}$$

for the intensity λ has the property that for a large window W the quantity

$$A(W)^{1/2}(\hat{\lambda} - \lambda) \tag{3.126}$$

follows a Gaussian distribution with zero mean and variance $\lambda/(1 - p)$.

The problems that appear in the context of pixel images are discussed in Ohser and Schladitz (2009). An important message of that book is that there is a 'hierarchy in precision': The most precise estimation is possible for A_A and V_V; then comes estimation of L_A and S_V; followed by M_V, N_A^+ and N_V^+; and finally N_A and N_V. Therefore one should try to use in the case of

(i) a *two-parameter model* only Equations (3.122) and (3.123), or (3.45) and (3.46), respectively; important two-parameter models are Boolean models with identical balls (parameters: intensity λ and radius R) and with Poisson polyhedral grains (parameters: λ and ϱ); for example, for the planar model with identical discs, the right-hand terms $\overline{A}$ and $\overline{L}$ can be both expressed by the radius R, and one has two equations for the two unknowns λ and R;

(ii) a *three-parameter model* only Equations (3.122) to (3.125), or (3.45) to (3.49); an important example of a three-parameter Boolean model is a model with balls and a radius distribution with two parameters, for example a gamma (or Schulz), Weibull or lognormal distribution.

In the case of *segments* as grains, using N_A^+ is reasonable. However, unfortunately, the equations (3.123) to (3.125) can yield only the mean segment length. The estimation of the length distribution is nontrivial since usually there are segments that cross the boundary of the window and so their lengths are censored. In this case more sophisticated methods have to be applied; see Chadœuf *et al.* (2000).

The method of densities has a long history. For the case of isotropic grains it first appeared in Santaló (1976) and Kellerer (1983). The general stationary case (with anisotropic grains) was studied in Weil (1988), and Bindrich and Stoyan (1991) used the specific convexity number N_V^+.

Minimum contrast method

The minimum contrast method consists in determining such parameter values that bring a theoretical summary characteristic function as close as possible to its empirical counterpart. In Boolean model statistics the use of the covariance $C(r)$ or correlation function $\kappa(r)$ can be recommended, as it may lead to model parameters different from $\overline{V}$, $\overline{S}$, etc., which could be determined using contact distributions but are better determined by the method of densities. Sometimes minimising contrast is the only method for parameter estimation; see Example 3.3 below.

Let $F(r; \theta_1, \theta_2)$ be a summary characteristic function depending on the unknown parameters θ_1 and θ_2, and let $\hat{F}(r)$ be its empirical counterpart. Then, for example, minimum contrast estimators $\hat{\theta}_1$ and $\hat{\theta}_2$ of θ_1 and θ_2 are those values of ζ_1 and ζ_2 which minimise

$$\int_{r_1}^{r_2} \left(F(r; \zeta_1, \zeta_2) - \hat{F}(r)\right)^2 dr \tag{3.127}$$

for suitable r_1 and r_2.

Often the integral is replaced by a sum over squared differences of $F(r_i; \zeta_1, \zeta_2)$ and $\hat{F}(r_i)$ for a series of r_i-values.

Heinrich (1993) investigated such estimators for the Boolean model and showed their consistency and asymptotic normality.

Example 3.2. *A pattern of lichen (genus Placodium) on a stone*

The pattern is illustrated in Figure 3.7 on p. 96. Is it possible to fit a planar Boolean model with discoidal grains to these data? Of course, in a strict sense such a model is implausible because of the irregularity of the boundaries of the regions occupied by the lichen – the grains are definitely neither convex nor discoidal. However, if only the large-scale features of the pattern are considered, then the model makes some sense: visual inspection suggests the suitability of a model involving grains. And also biologically, a Boolean model, which is based on the ideas of random germ points and independent grain growth, is not misleading.

The set-theoretic summary characteristics of the pattern of lichen were estimated using an image analyser. The area of the window W shown in Figure 3.7 is $7400\,\text{cm}^2$. Area fraction and mean boundary length per unit area were estimated as

$$\hat{p} = 0.35,$$

$$\hat{L}_A = 0.011\,\text{mm}^{-1}.$$

Counting of lower tangent points for a bit smoothed boundaries led to 38, which gives $\hat{N}_A^+ = 0.0051\,\text{cm}^{-2}$.

Now Formula (3.124) yields $\hat{\lambda} = 0.0079\,\text{cm}^{-2}$, which means that on the average in a window like W, 58 grains are expected.

Formula (3.122) then results in $\overline{A} = 54.5\,\text{cm}^2$ and Formula (3.123) $\overline{L} = 21.4\,\text{cm}$.

For demonstration purposes now the case of a gamma distribution for the radii of discoidal grains is considered. (There is no indication that this distribution is the only candidate; visual inspection of Figure 3.7 suggests that the radius distribution is skewed.) The radius probability density $f_R(r)$ is assumed to be

$$f_R(r) = \frac{\mu^\alpha r^{\alpha-1}}{\Gamma(\alpha)} \exp(-\mu r).$$

Since mean and variance are

$$\mathbf{E}(R) = \frac{\alpha}{\mu} \quad \text{and} \quad \mathbf{var}(R) = \frac{\alpha}{\mu^2},$$

the estimates

$$\hat{\alpha} = 2.03 \quad \text{and} \quad \hat{\mu} = 0.60$$

are obtained.

Example 3.3. *Small-angle scattering analysis of materials*

Small-angle scattering of X-rays or neutrons is able to yield information on random three-dimensional structures at the scale of 1 to 100 nm. The small-angle intensity $I(s)$ is connected with the correlation function $\kappa(r)$ of the random set by

$$\kappa(r) = \frac{1}{2\pi^2 r} \int_0^\infty s I(s) \sin sr \, ds, \tag{3.128}$$

where $s = \frac{4\pi}{\omega} \sin\theta$, in which 2θ is the scattering angle and ω the wavelength; see Hosemann and Bagchi (1962).

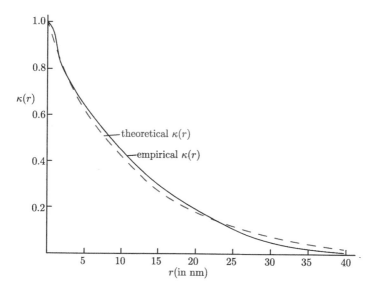

Figure 3.8   The empirical correlation function (solid line) for Co precipitate in a $Co_{75}Fe_5B_{20}$ alloy and the theoretical correlation function (dashed line) of a Boolean model with Poisson polyhedra as grains, in which the parameter of the Poisson polyhedra was fitted by a least squares method.

The first example considers an amorphous $Co_{75}Fe_5B_{20}$ alloy, following Sonntag *et al.* (1981). Figure 3.8 shows the experimental correlation function $\hat{k}(r)$ obtained from small-angle scattering. One possible model for the set Ξ of Co precipitate is a Boolean model of Poisson polyhedra with parameter ϱ as grains. The model correlation function is $\kappa(r)$, where

$$\kappa(r) = \frac{1}{p}\left(\exp\left(6\lambda\frac{\exp(-\varrho r)-1}{\pi \varrho^3}\right) - 1 + p\right) \qquad \text{for } r \geq 0. \qquad (3.129)$$

Fitting this to the experimental correlation function $\hat{k}(r)$, via minimisation of

$$\sum_{i=1}^{N}\left(\hat{k}(r_i) - \kappa(r_i)\right)^2$$

using a Marquardt procedure yields the estimates

$$\hat{p} = 0.1,$$
$$\hat{\varrho} = 0.027\,\text{nm}^{-1}.$$

Here the r_i are equidistant values of r for which the experimental correlation function has been calculated from the small-angle scattering data. The dashed curve on Figure 3.8 is the model correlation function $\kappa(r)$ for these parameters.

In order to show that also non-exponential correlation functions are observed, statistical results for a sample of nanoporous silica is investigated. Figure 3.9 shows a TEM image of a nanoporous silica sample. The pores, shown in black, appear to be spherical.

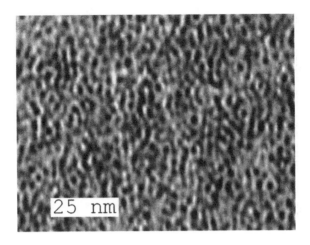

Figure 3.9 TEM image of a nanoporous silica sample. The pores appear in black. Courtesy of H. Hermann.

Methods of image analysis and intuition suggested a truncated power-law distribution for the radii, that is,

$$f_R(r) = \begin{cases} c(r/r_2)^a & \text{for } r_1 \leq r \leq r_2, \\ 0 & \text{otherwise.} \end{cases}$$

In the paper Hermann *et al.* (2005) the following estimates for the parameters are found:

$$\hat{r}_1 = 1 \, \text{nm},$$
$$\hat{r}_2 = 5 \, \text{nm},$$
$$\hat{a} = -3.6,$$

and c is a normalising factor.

In Hermann *et al.* (2005) the structure is analysed using methods for hard ball systems, with the comment that in reality the porous phase looks like a system of overlapping balls. Therefore H. Hermann kindly re-analysed the data using a Boolean model with spherical grains. Using the radius distribution above and Formulae (3.18) and (3.22) for $C(r)$ he found a suitable estimate of λ and calculated the dashed curve for $\kappa(r)$ in Figure 3.10. The estimate $\hat{\lambda} = 0.048 \, \text{nm}^{-3}$ finally led to the estimate of V_V as 0.7, which is considered by Hermann as a bit too large. (However, in the approach of Hermann *et al.*, 2005, no estimate of V_V can be given.)

Estimation of the radius distribution of spherical grains

Various authors have developed statistical methods for the determination of the radius probability density function $f_R(r)$ in the case of Boolean models with *spherical* (or *discoidal*) *grains*. Many of them seem numerically instable, require difficult measurements and yield in the best case only qualitative information. One of these methods is that in Thovert *et al.* (2001). They consider a grid of test points in Ξ and determine for each of them the largest ball entirely lying in Ξ and covering it, the test point. The sample of the corresponding radii

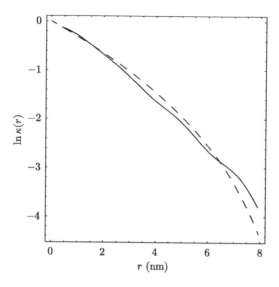

Figure 3.10   The empirical correlation function (solid line) for the silica sample, and the theoretical correlation function (dashed line) of a Boolean model with spherical grains, the radii of which follow a truncated power-law distribution. The ordinate is scaled logarithmically in order to show clearly deviations from exponential curves. Courtesy of H. Hermann.

(which tend to be equal for test points close together) is considered as a sample corresponding to the volume-weighted version of the density $f_R(r)$. The papers Thovert *et al.* (2001) and Thovert and Adler (2011) describe how to determine the radii numerically, based on the co-variance, and discuss the accuracy of the method. The authors confess that some '*a posteriori* verification' is necessary. See also Emery *et al.* (2012).

Molchanov and Stoyan (1994) propose a kernel estimator of the diameter probability density, whose asymptotic normality is proved by Heinrich and Werner (2000).

Others

Further methods of statistics for Boolean models are given in Hall (1988), Schmitt (1991), Weil (1995), Molchanov (1997) and Schneider and Weil (2008).

3.5 Generalisations and variations

Some generalisations of the Boolean model may be of use for description of structures of greater complexity. A straightforward generalisation involves the weakening of model assumptions. It seems especially useful to suspend the assumption that the germs constitute a Poisson process. Random structures that are essentially Boolean models, but that use general point processes to provide the germs, are called germ–grain models. They are studied in Chapter 6. Such models may display a greater or a lesser irregularity of structure than the Boolean model, according to the particular properties of the germ process.

Two independent Boolean models $\Xi^{(1)}$ and $\Xi^{(2)}$ considered together can be used to produce patterns of greater irregularity. Clearly, the union $\Xi^{(1)} \cup \Xi^{(2)}$ is again a Boolean model. Three more complicated variations are considered now.

Variation 1. Suppose that $\Xi^{(1)}$ and $\Xi^{(2)}$ have volume fractions $p^{(1)}$ and $p^{(2)}$ and covariances $C^{(1)}(r)$ and $C^{(2)}(r)$. Then the intersection

$$\Xi^{(i)} = \Xi^{(1)} \cap \Xi^{(2)} \tag{3.130}$$

is a closed random set; see Figure 3.11(a) for an illustration. Its basic characteristics are: volume fraction

$$p^{(i)} = p^{(1)} p^{(2)} \tag{3.131}$$

and covariance

$$C^{(i)}(r) = C^{(1)}(r) \cdot C^{(2)}(r) \qquad \text{for } r \geq 0. \tag{3.132}$$

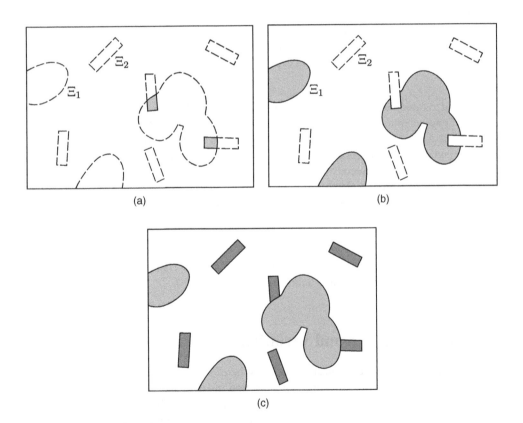

(a)

(b)

(c)

Figure 3.11 Three variations of Boolean models: (a) intersection of two Boolean models; (b) intersection of a Boolean model with the complement of another; (c) a simple three-component model.

The calculation of the capacity functional is complicated. This model appears in Savary *et al.* (1999) and is used in the context of clusters of small grains. Jeulin (2012) speaks about 'multi scale Boolean models'.

Variation 2. Another closed set which can be built from $\Xi^{(1)}$ and $\Xi^{(2)}$ is the intersection of one and the closure of the complement of the other:

$$\Xi^{(ic)} = \Xi^{(1)} \cap \left((\Xi^{(2)})^c \right)^{cl}. \tag{3.133}$$

Here $\left((\Xi^{(2)})^c \right)^{cl}$ is the closure of the complement of $\Xi^{(2)}$. This is illustrated in Figure 3.11(b). Such constructions are used in Molchanov (1992) to describe spatial censoring in Boolean model statistics.

The basic characteristics of $\Xi^{(ic)}$ are: volume fraction

$$p^{(ic)} = p^{(1)}(1 - p^{(2)}) \tag{3.134}$$

and covariance

$$C^{(ic)}(r) = C^{(1)}(r) \left(1 - 2p^{(2)} + C^{(2)}(r) \right) \qquad \text{for } r \geq 0. \tag{3.135}$$

Savary *et al.* (1999) also uses $\Xi^{(1)} \cap \Xi^{(2)} \cap \left((\Xi^{(3)})^c \right)^{cl}$. In a similar way, but extending beyond the province of random closed sets, models can be constructed for multi-component structures. A simple example is the three-component model below, studied by Greco *et al.* (1979) and Serra (1982). Statistical problems related to this are considered in Molchanov (1997).

Variation 3. The sets $\Xi^{(1)}$ and $\Xi^{(2)}$ may be combined to produce a three-phase pattern as in Figure 3.11(c). Its components are:

- component 1: $\Xi^{(1)}$;
- component 2: $\Xi^{(2)} \setminus \Xi^{(1)}$ (the set-difference can be interpreted as a pattern given by $\Xi^{(2)}$ but destroyed in part by $\Xi^{(1)}$);
- component 3: the remainder, $(\Xi^{(1)} \cup \Xi^{(2)})^c$.

This model is used in Greco *et al.* (1979) for the description of sinter metals. In such a case component 1 models the haematites, component 2 the slag, and component 3 the calcium ferrites.

A further model closely related to the Boolean model is Matheron's 'dead-leaves' model; see Serra (1982). Boolean models also appear in the context of birth-and-growth processes; see Section 6.6.4 and Molchanov and Chiu (2000).

Finally, it must be mentioned that in particular physicists also consider discrete Boolean models, defined on lattices. There the germs are random lattice points and the grains are subsets of the lattice; see Hall (1988), Mecke (1994, 1996, 2000) and Stoyan and Mecke (2005).

3.6 Hints for practical applications

Planar case

In the majority of applications a Boolean model with discoidal grains should be a good first choice. The estimators considered above then yield values of λ and of the first two moments of the random radius. If a two-parameter distribution (such as normal, truncated normal, lognormal, gamma or Weibull) is plausible, then the Boolean model is completely specified, and simulations are possible. This may be sufficient if the boundaries are curved. Sometimes this model is also in approximate agreement with physical or biological theories using terms such as 'germs' and 'circular growth'.

   If the boundaries are straight, then the Poisson polygon is indicated for use as a grain. In this case the model has only two parameters, intensity λ and polygon parameter ϱ. This is convenient for estimating, but for fitting the model to empirical data this may turn out to be too special. If the area fraction is small, then the Poisson-polygonal Boolean model may be justified by the arguments of Section 3.3.5.

Spatial case

Analogously to the planar case, spherical and Poisson-polyhedric grains are particularly interesting. Usually the real structures are small and/or opaque, and their direct investigation is not possible. In such cases the corresponding parameters can be estimated using data obtained from computerised tomography, X-ray scattering (see Example 3.3) or from planar sections. In Section 10.3 stereological methods for the Boolean model are presented. Clearly, it is then important that planar sections of a Boolean model are again Boolean models.

Two-step statistical procedure

Most known statistical procedures which yield more than the densities and mean values of intrinsic volumes must be considered as numerically unstable. Therefore their results should be considered as preliminary and as a first step of statistics. A typical preliminary result is qualitative information of grain shape and radius probability density function type. Based on the first step then the second step yields the model parameters by the method of densities. It guarantees that, for example in the spatial case, V_V and S_V of the model coincide with the corresponding data values. Careful statisticians will make a fine-tuning based on simulations: the estimated parameters are *a posteriori* so modified that summary characteristics of simulated samples are as close as possible to the empirical ones.

A critical assessment of Boolean models

Finally, the following remarks about difficulties and intrinsic limitations of the Boolean model may be helpful to the interested reader.

- *Interpretation.* In many cases, the suggestive nature of the Boolean model terminology ('germs' as origins of growth, or locations of physical entities) will not coincide with physical reality. This is not strictly speaking a disadvantage of the Boolean model, but more a tendency to be resisted in its interpretation. The Boolean model may fit a data set to a greater or lesser degree, but that of itself does not justify a physical interpretation of the germ–grain mechanism.

Two kinds of interpretation can be distinguished. In the first, which might be called the *genetic* interpretation, the individual grains of the Boolean model correspond to physically identifiable entities. The second interpretation, the 'representation' interpretation, avoids drawing out such a correspondence, but rather uses the Boolean model as a moderately flexible means of producing patterns to compare with the image and to describe it by few parameters only. Example 3.2 on lichen falls in the second category: it is tempting here to identify grains of the Boolean model with individual lichen entities, but in fact such an identification cannot be substantiated without rigorous evidence derived from the botanical context.

The point here is that conclusions appropriate to a genetical interpretation will be misleading, even nonsensical, in a representation context. In an application to dust-particle counting, clearly one is interested in estimating the number of particles. Conceivably one might attempt to estimate lichen 'growth sites' in a similar way, but without evidence favouring a genetic interpretation such an attempt is ill-judged. However, the representation interpretation still provides a possible description of the data set as an image.

- *Interaction between grains.* The previous section has already introduced the generalisation of replacing the Poisson germ process by a more general point process. In many cases, the assumption of independence between the grains is implausible. Consider for example various packing and coverage problems, where intersection of grains is prohibited. In such a case the Boolean model is clearly inadequate. Then germ–grain models form a possible alternative; see Section 6.5. Particularly important models are the 'morphological Gibbs processes'; see Brodatzki and Mecke (2001), Stoyan and Mecke (2005), Dereudre and Lavancier (2009) and Møller and Helisová (2010).

  When some interaction between the grains is merely a possibility to be considered, the Boolean model can serve as a convenient benchmark, so that the strength of the interaction, if any, can be assessed.

- *Nonconvex grains.* Many of the useful formulae for the Boolean model are for the case of convex grains. However, in many data sets arising in practice, such as the one in Example 3.2, the grains are clearly nonconvex. It is indeed the case that larger-scale features of the data set can still be modelled by a Boolean model with convex grains. However, care must be taken over the interpretation of quantities calculated from the model. As shown above, it is possible to derive formulae for the case of nonconvex grains.

4

Point processes II – General theory

4.1 Basic properties

4.1.1 Introduction

Point processes provide models for irregular patterns of points. The theory was developed in response to various problems of physics, biology, and queueing theory. Important pioneers include Neyman (1939) (as reported by Neyman and Scott, 1972), Palm (1943), Bartlett (1954), Cox (1955), Khintchin (1955) and Matérn (1960); see Guttorp and Thorarinsdottir (2012) for more details of the history of various point process models. The work by Kallenberg, Kerstan, Krickeberg, Matthes, Mecke and Ryll-Nardzewski during 1960 to 1975 laid the foundation of the modern theory of point processes. Books on the subject include: Srinivasan (1974), Kallenberg (1976b, 1983a, 1986, 2002), Brillinger (1978), Matthes *et al.* (1978), Cox and Isham (1980), Brémaud (1981), Karr (1986, 1991), Daley and Vere-Jones (1988, 2003, 2008), Snyder and Miller (1991), Reiss (1993), Last and Brandt (1995), Sigman (1995), van Lieshout (2000), Baccelli and Brémaud (2003), Srinivasanm and Vijayakumar (2003a,b), Møller and Waagepetersen (2004), Jacobsen (2006), Last (2010) and Gelfand *et al.* (2010). This chapter presents the general theory, which is a development of the special case of the Poisson process that is discussed in Chapter 2. Chapter 5 presents some special point process models.

The mathematical definition of a point process on $\mathbb{R}^d$ is as a random variable taking values in a measurable space $[\mathbb{N}, \mathcal{N}]$, where $\mathbb{N}$ is the family of all sequences φ of points of $\mathbb{R}^d$ satisfying two regularity conditions:

(i) the sequence φ is *locally finite*, that is, each bounded subset of $\mathbb{R}^d$ must contain only a finite number of points of φ,

(ii) the sequence is *simple*, that is, $x_i \neq x_j$ if $i \neq j$.

Stochastic Geometry and its Applications, Third Edition.
Sung Nok Chiu, Dietrich Stoyan, Wilfrid S. Kendall and Joseph Mecke.
© 2013 John Wiley & Sons, Ltd. Published 2013 by John Wiley & Sons, Ltd.

The condition (ii) of simplicity is relaxed in some extensions of the theory, but not in this book.

For convenience sometimes the notation $\varphi = \{x_n\}$ is used in order to emphasise the sequential nature of φ. The reader should note that the x_n are dummy variables, and have no particular interpretation. Thus for example x_1 need not be the point closest to the origin.

The σ-algebra $\mathcal{N}$ is defined as the smallest σ-algebra on $\mathbb{N}$ to make all mappings $\varphi \mapsto \varphi(B)$ measurable, for B running through the bounded Borel sets. Here $\varphi(B)$ is simply the number of points in the set B. This $\mathcal{N}$ contains the so-called *configuration sets*, often denoted as Y and Z. An example is $Y = \{\varphi : \varphi(B) = 0\}$, the set of all point sequences that have no point in a given set B.

Thus formally a point process Φ is a measurable mapping of a probability space $[\Omega, \mathcal{A}, \mathbf{P}]$ into $[\mathbb{N}, \mathcal{N}]$. More intuitively it is a random choice of one of the φ in $\mathbb{N}$. It generates a distribution on $[\mathbb{N}, \mathcal{N}]$, the *distribution P* of Φ.

Perhaps the following imaginary scenario may help readers who lack training in measure theory. A simple example of a random variable describes the outcome of a roll of a die. Here the random variable takes on the values 1 to 6, and when the die is fair the probability for each of the six values is $\frac{1}{6}$. Now assume that it is possible to construct a die with infinitely many sides, with a (two-dimensional) point pattern on each of the sides. Every time the die is thrown, a point pattern is generated. This die represents the point process Φ, and a point pattern $\varphi = \Phi(\omega)$ is assigned to each sample point ω represented by a roll of die. Now assume somebody is observing the independent repetitive rolling of the die. This person sees different point patterns and, in particular, fluctuating values of the number of points in some fixed, deterministic subset B of the plane. In other words, every throw of the 'point pattern die' produces a different realisation of the random variable 'number of points in B'.

This way of thinking of point processes (and later of random sets and other structures) is fundamental to the way in which mathematicians approach point process theory. The way in which application-oriented people consider point patterns is sometimes different. They often focus on only one irregular point pattern and study its variability. The mathematical theory can be used in such applied studies using the idea of ergodicity, see p. 114.

The word 'process' in the term 'point process' does not imply a dynamic evolution over time. Historically, the first researchers into the theory had in mind random sequences of temporal events, such as instants of arrivals of calls at telephone exchanges, arrivals of customers at a queue, occurrences of earthquakes or mine disasters, etc. However, a notion of time is often absent in applications of point processes in $\mathbb{R}^2$ or $\mathbb{R}^3$. The phrase 'random point field' would be a more exact term:

$$\boxed{\text{point process} \;=\; \text{random point field}}.$$

It is used in Stoyan and Stoyan (1994) and Ohser and Schladitz (2009). Nevertheless, the abstract-thinking mathematicians retained the term 'point process'. *Spatio-temporal point processes* explicitly involving temporal as well as spatial dispersion of points constitute a separate theory; see Diggle (2007), Daley and Vere-Jones (2008), Illian *et al.* (2008) and Cressie and Wikle (2011).

A similar point can be made about the adjective 'stationary', as in 'stationary point processes', occurring later in the chapter. In this context 'stationary' and 'homogeneous' are equivalent terms and sometimes an adverb such as 'statistically' (as in Torquato, 2002), 'spatially' or 'macroscopically' (as in Ohser and Schladitz, 2009) is added to the latter to distinguish it from other kinds of homogeneity:

stationary = homogeneous = statistically / spatially / macroscopically homogeneous .

4.1.2 The distribution of a point process

The distribution P of a point process Φ is determined by the probabilities

$$P(Y) = \mathbf{P}(\Phi \in Y)$$
$$= \mathbf{P}(\{\omega \in \Omega : \Phi(\omega) \in Y\}) \qquad \text{for } Y \in \mathcal{N}.$$

The term $\Phi \in Y$ means that Φ has some property, for example that it has no point in the set B. Then $\mathbf{P}(\Phi \in Y)$ denotes the probability that Φ has this property.

The *finite-dimensional distributions* are of particular importance. These are probabilities of the form

$$\mathbf{P}\big(\Phi(B_1) = n_1, \ldots, \Phi(B_k) = n_k\big)$$

where $B_1, \ldots, B_k$ are bounded Borel sets and $n_1, \ldots, n_k \geq 0$. The term denotes the probability that Φ has n_1 points in the set B_1, $\ldots$, and n_k points in B_k. The distribution of Φ on $[\mathbb{N}, \mathcal{N}]$ is uniquely determined by the system of all these values for all $k = 1, 2, \ldots$. In fact, it is determined by the subsystem for which the constituent B_i are pairwise disjoint.

An even smaller but still sufficient subsystem is that of the *void-probabilities*

$$v_B = P(\{\varphi \in \mathbb{N} : \varphi(B) = 0\})$$
$$= \mathbf{P}\big(\Phi(B) = 0\big)$$
$$= \mathbf{P}(\Phi \cap B = \emptyset) \qquad \text{for Borel sets } B.$$

The quantity v_B is the probability that B is empty (i.e. does not contain a point of Φ). If the point process is simple, as described above, then P is determined by the system of values of v_K as K ranges through the compact sets. This last and strongest characterisation follows by interpreting Φ as a random closed set with capacity functional $T_\Phi(K) = 1 - v_K$, and then applying the Choquet theorem in Chapter 6. (A simple direct proof is possible; see for example Møller and Waagepetersen, 2004, Theorem B.1.) The system of void-probabilities (or the 'avoidance function') specifies the probabilities that points do *not* occur in a given region.

An example of the use of void-probabilities is the proof of the Slivnyak–Mecke theorem (4.69) on p. 132.

4.1.3 Notation

Point processes can be considered either as random sets of discrete points or as random measures counting the numbers of points lying in spatial regions. Corresponding to these two interpretations is the following notation:

$x \in \Phi$ asserts that the point x belongs to the random sequence Φ;

$\Phi(B) = n$ asserts that the set B contains n points of Φ.

Since Φ is a random closed set, results from the theory of random closed sets can also be applied to point processes.

Correspondingly there is a variety of notation for point process formulae. The process Φ can be written as

$$\Phi = \{x_1, x_2, \ldots\} = \{x_n\}$$

which emphasises the interpretation of Φ as a sequence of points or a random closed set. Let f be a measurable function on $\mathbb{R}^d$. The sum of $f(x)$ over x in Φ can be written variously as

$$f(x_1) + f(x_2) + \cdots, \qquad \sum_{x \in \Phi} f(x),$$

or (as physicists would write)

$$\int_{\mathbb{R}^d} f(x)p(x)\,dx \quad \text{where } p(x) = \sum_{y \in \Phi} \delta(x - y) \text{ and } \delta \text{ is the Dirac delta function,}$$

or

$$\int f(x)\Phi(dx).$$

The mean value of the sum is written variously as

$$\mathbf{E}\left(\sum_{x \in \Phi} f(x)\right), \qquad \int_{\mathbb{N}} \sum_{x \in \varphi} f(x) P(d\varphi) \qquad \text{or} \qquad \int_{\mathbb{N}} \int_{\mathbb{R}^d} f(x)\varphi(dx)P(d\varphi).$$

The number of points of φ in a set B can be written as

$$\varphi(B) = \sum_{x \in \varphi} 1_B(x),$$

where $1_B(x)$ is the indicator function of B. Its mean value is written as

$$\mathbf{E}(\Phi(B)) = \int \varphi(B)P(d\varphi)$$

$$= \mathbf{E}\left(\sum_{x \in \Phi} 1_B(x)\right)$$

$$= \int \sum_{x \in \varphi} 1_B(x)P(d\varphi)$$

$$= \int \int 1_B(x)\varphi(dx)P(d\varphi).$$

Such a plethora of notation is perhaps confusing for the beginner but reflects the variety of approaches to the theory.

4.1.4   Stationarity and isotropy

This book mainly considers point processes with infinitely many points; physicists would perhaps speak about the 'thermodynamic limit'. But no limit procedures are used in introducing the fundamental notions of stationarity and isotropy; limits will only appear later, in the context of ergodicity.

A point process Φ, or its distribution P, is said to be *stationary* if its characteristics are invariant under translation: the processes $\Phi = \{x_n\}$ and $\Phi_x = \{x_n + x\}$ have the same distribution for all x in $\mathbb{R}^d$. So

$$\mathbf{P}(\Phi \in Y) = \mathbf{P}(\Phi_x \in Y) \tag{4.1}$$

for all configuration sets Y and all x in $\mathbb{R}^d$. Equivalent to (4.1) is

$$P(Y) = P(Y_x), \tag{4.2}$$

where $Y_x = \{\varphi_x : \varphi \in Y\}$.

The notion of *isotropy* is entirely analogous: Φ is isotropic if its characteristics are invariant under rotation; that is to say, Φ and $r\Phi$ have the same distribution for every rotation r around the origin. So

$$\mathbf{P}(\Phi \in Y) = \mathbf{P}(r\Phi \in Y) \tag{4.3}$$

and

$$P(Y) = P(rY) \tag{4.4}$$

for all r and all Y, with $rY = \{r\varphi : \varphi \in Y\}$.

Stationarity and isotropy together yield *motion-invariance*. A motion-invariant point process Φ has the same distribution as $m\Phi$ for all rigid motions m of $\mathbb{R}^d$.

Figures 2.3, 2.4 and 4.1 show point patterns which behave like typical realisations of nonstationary point processes, while Figures 2.1, 5.4, 5.6, 5.7 and 5.9 resemble point patterns of stationary processes. Figure 4.1 illustrates that nonstationarity can arise even if the point density is in some sense constant in space. Here the kind of distribution of the points depends on spatial location: lower in the figure there is a stronger degree of clustering.

4.1.5   Intensity measure and intensity

The *intensity measure* Λ of Φ is a characteristic analogous to the mean of a real-valued random variable. Its definition is

$$\Lambda(B) = \mathbf{E}\big(\Phi(B)\big) = \int \varphi(B) P(d\varphi) \qquad \text{for Borel sets } B. \tag{4.5}$$

So $\Lambda(B)$ is the mean number of points in B.

In particular, if Φ is stationary, then the intensity measure simplifies; it must be translation-invariant since

$$\Lambda(B) = \mathbf{E}\big(\Phi(B)\big) = \mathbf{E}\big(\Phi_x(B)\big) = \mathbf{E}\big(\Phi(B_{-x})\big) = \Lambda(B_{-x}) \tag{4.6}$$

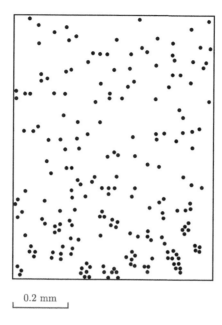

0.2 mm

Figure 4.1   A point pattern which should by no means be modelled by a stationary point process. There is obviously a gradient in the vertical direction. The dots are the midpoints of cell nuclei in a specimen of joint cartilage. The lower part of the figure is the side near the bone tissue, while the upper part near the joint.

for all x. By (1.83) this translation-invariance implies that

$$\Lambda(B) = \lambda v_d(B) \tag{4.7}$$

for some (possibly infinite) nonnegative constant λ, which is called the *intensity* of Φ. Choosing B to have volume 1 shows that λ may be interpreted as the mean number of points of Φ per unit volume or the *point density*. It is assumed that $0 < \lambda < \infty$.

In general, under some continuity conditions, which are usually satisfied in practical applications of point process theory, $\Lambda(x)$ has a density function $\lambda(x)$, which is called the *intensity function* of Φ:

$$\Lambda(B) = \int_B \lambda(x)\,dx. \tag{4.8}$$

The continuity condition is violated if, for example, the points are arranged on a lattice. It is clear that $\lambda(x)$ is proportional to the local point density around a location x. If dx is the volume of an infinitesimal ball centred at x, then $\lambda(x)dx$ is the probability that there is a point in this ball. This shows that the intensity function can be also interpreted in a local sense; see Daley and Vere-Jones (2008, Section 9.3).

Some calculations may be simplified by using the *Campbell theorem*:

Theorem 4.1. *For any nonnegative measurable function* $f(x)$,

$$\mathbf{E}\left(\sum_{x\in\Phi}f(x)\right) = \int\sum_{x\in\varphi}f(x)P(d\varphi) = \int\int f(x)\varphi(dx)P(d\varphi)$$

$$= \int f(x)\Lambda(dx). \tag{4.9}$$

This is essentially an application of Fubini's theorem. (The name Campbell theorem refers to early work by Campbell, 1909.) In the stationary case the Λ-integral in (4.9) becomes a volume integral so

$$\mathbf{E}\left(\sum_{x\in\Phi}f(x)\right) = \lambda\int f(x)\,dx. \tag{4.10}$$

Similarly, when the point process possesses an intensity function then (4.9) can be re-written as:

$$\mathbf{E}\left(\sum_{x\in\Phi}f(x)\right) = \int f(x)\lambda(x)\,dx. \tag{4.11}$$

4.1.6 Ergodicity and central limit theorem

Often one is interested in *spatial averages*

$$\lim_{n\to\infty}\frac{1}{v_d(W_n)}\int_{W_n}f(\Phi_x)\,dx$$

or in *spatial point averages*

$$\lim_{n\to\infty}\frac{1}{v_d(W_n)}\sum_{x_i\in W_n}f(\Phi_{-x_i})$$

for a nonnegative integrable (with respect to **P**) function f and $W_n\uparrow\mathbb{R}^d$. Under ergodicity, these limits exist as deterministic numbers. The following gives some facts of ergodic theory, for details the reader is referred to Kallenberg (2002, Chapter 10), Daley and Vere-Jones (2008, Section 12.2) and Baccelli and Błaszczyszyn (2009a, Section 1.6).

The sequence $\{W_n\}$ is a *convex averaging sequence*, that is, a sequence of subsets of $\mathbb{R}^d$ satisfying:

(1) each W_n is convex and compact;

(2) $W_n\subset W_{n+1}$;

(3) $\sup\{r\geq 0: B(x,r)\subset W_n \text{ for some } x\}\to\infty$ as $n\to\infty$.

An example is $W_n = B(o,n)$.

A stationary point process Φ is said to be *ergodic* if

$$\lim_{t \to \infty} \frac{1}{(2t)^d} \int_{[-t,t]^d} \mathbf{1}(\Phi_x \in Y, \Phi \in Z) \, dx = P(\Phi \in Y) P(\Phi \in Z) \qquad (4.12)$$

for all configuration sets Y and Z.

A property that implies ergodicity is *mixing*; Φ is said to be mixing if

$$P(Y \cap Z_x) \to P(Y) P(Z) \qquad \text{for } \|x\| \to \infty \qquad (4.13)$$

for all configuration sets Y and Z. Various mixing conditions are given in Daley and Vere-Jones (2008, Section 12.2).

The homogeneous Poisson process is mixing, whereas the mixed Poisson process (see p. 166) is not ergodic.

For a stationary point process which is ergodic, it holds

$$\lim_{n \to \infty} \frac{1}{v_d(W_n)} \int_{W_n} f(\Phi_x) \, dx = \mathbf{E}\big(f(\Phi)\big), \qquad (4.14)$$

where the convergence is almost surely. A consequence is

$$\lim_{n \to \infty} \frac{\Phi(W_n)}{v_d(W_n)} = \lambda. \qquad (4.15)$$

Also *central limit theorems*, that is, theorems ensuring the convergence of the distribution of point numbers in large sets towards normal distributions, can be proved; see Schreiber (2010). Formally, for $\Phi(W_n)$ with a convex averaging sequence $\{W_n\}$

$$\frac{\Phi(W_n) - \lambda v_d(W_n)}{\sqrt{v_d(W_n)}} \to N(0, \sigma^2) \qquad \text{for } n \to \infty, \qquad (4.16)$$

where

$$\sigma^2 = \lim_{n \to \infty} \frac{\text{var}\big(\Phi(W_n)\big)}{v_d(W_n)},$$

and the stationary point process Φ has to satisfy strong mixing conditions; see Ivanoff (1982) and Heinrich and Schmidt (1985).

4.1.7 Contact distributions

Let B be a convex compact (or, more generally, star-shaped compact) set in $\mathbb{R}^d$ with $o \in B$ and $v_d(B) > 0$. The *contact distribution function* $H_B(r)$ of the stationary point process Φ with respect to the structuring element B is defined by

$$H_B(r) = 1 - P\big(\Phi(rB) = 0\big) \qquad \text{for } r \geq 0. \qquad (4.17)$$

It is a distribution function. In the special case of $rB = B(o, r)$ the contact distribution function is written as $H_s(r)$ and called the *spherical contact distribution function, empty space distribution function* or *first contact distribution function*. It can be interpreted as the distribution function of the distance from o to the nearest point of Φ. In the language of statisticians it is the distribution of the distance from an arbitrary test point to its nearest neighbour in Φ.

Another interpretation sees $H_s(r)$ as the volume fraction of the random set $\Phi_{\oplus r}$. This aspect is considered in more detail in Sections 3.2.3 and 6.3.3.

For $H_B(r)$ there always exists a probability density function; see Hansen *et al.* (1999). An excellent survey on contact distribution functions is Hug *et al.* (2002b), which contains formulae for many point process models and considers also the nonstationary case.

Closely related are Klaus Mecke's morphological functions, which describe the behaviour of other geometrical characteristics of the random set $\Phi_{\oplus r}$ as a function of r; see pp. 84 and 229.

Note that the notion of a linear contact distribution function as in Section 6.3.3 is ineffective for point processes since for 'sufficiently random' point processes a line starting in a process point will never hit another process point, and $\Phi \oplus s(o, r)$ has null Lebesgue measure, where $s(o, r)$ denotes a segment of length r of a given direction. (It is $v_d(s(o, r)) = 0$.)

4.2 Marked point processes

4.2.1 Fundamentals

A point process is made into a *marked point process* by attaching a characteristic (a *mark*) to each point of the process. Thus a marked point process on $\mathbb{R}^d$ is a random sequence $\Psi = \{[x_n; m_n]\}$ from which the points x_n together constitute a point process (unmarked, called the *ground process*) in $\mathbb{R}^d$ and the m_n are the marks corresponding to the respective points x_n. The marks m_n belong to a given *space of marks* $\mathbb{M}$ which is assumed to be a Polish space. The Borel σ-algebra of $\mathbb{M}$ is denoted by $\mathcal{M}$. Important special cases are discrete marks, where $\mathbb{M}$ is a finite set, or real-valued marks, where $\mathbb{M}$ is $\mathbb{R}$.

Specific examples are:

- for x the centre of a particle, m the volume of the particle;
- for x the position of a tree, m the stem diameter of the tree or its growth function;
- for x the centre of an atom, m the type of the atom;
- for x the location (suitably defined) of a convex compact set, m the centred (shifted to origin) set itself.

The marks can be continuous variables, as in the first two examples, indicators of types as in the third example (in which case the terms 'multivariate point process' or 'multitype point process' are often used; 'bivariate' point processes have only two marks), or rather complicated indeed, as in the last example, which occurs in the marked point process interpretation of a germ–grain model (see Section 6.5). Figures 4.2 and 4.3(a) on p. 117 show two point patterns that can be interpreted as samples of marked point processes.

A formally important case, which often serves as null-model, is that of independent marks, where the m_n form a sequence of i.i.d. random variables.

Matthes (1963) and his co-workers established the theory of marked point processes. Its initial application was in queueing theory; such applications up to early 1980s are discussed in the books of Franken *et al.* (1981) and Baccelli and Brémaud (2003). Today marked point processes are a valuable tool in many branches of applied probability.

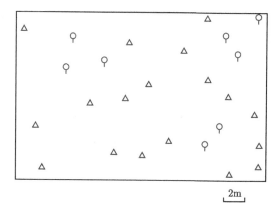

Figure 4.2   Positions of spruces and birches in a forest. The point pattern can be interpreted as a sample of a bivariate point process; that is to say, a marked point process with marks 0 and 1.

It is of course possible to interpret the marked point process Ψ as an ordinary point process in $\mathbb{R}^d \times \mathbb{M}$. The notation of a random counting measure can be used: for Borel $B \in \mathcal{B}^d$ and $L \in \mathcal{M}$ the number of points of Ψ in B with marks in L is denoted by $\Psi(B \times L)$. However, there is a particular feature of marked point processes which makes it worthwhile to consider them separately: usually one defines rigid motions of marked point processes as transforms which move the points but leave the marks unchanged. So Ψ_x, the translate of $\Psi = \{[x_n, m_n]\}$ by x, is given by

$$\Psi_x = \{[x_1 + x; m_1], [x_2 + x; m_2], \ldots\}. \tag{4.18}$$

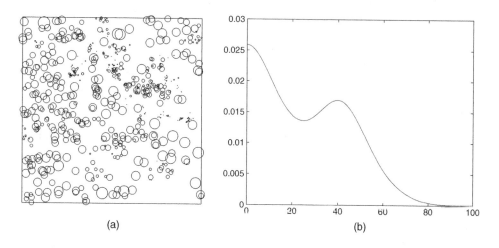

(a)   (b)

Figure 4.3   (a) Locations of longleaf pines in a North American forest. The circles indicate the stem diameters. They can be interpreted as marks for the point process of locations. (b) Statistically determined mark (stem diameter in cm) probability density function for the point pattern of longleaf pine trees.

Rotations act on marked point processes by rotating the points but *not* altering the marks. (However, in the case of set-marks as above this may be not so natural: it may be appropriate to subject the set-marks to rotations as well; see Baddeley and Jensen, 2005, p. 271; in contrast, Rau and Chiu, 2011, considered independent random rotations of the set-marks without moving the points.)

4.2.2 Intensity and mark distribution

The distribution of a marked point process is defined analogously as that of a unmarked point process. A marked point process is said to be *stationary* if for all x the translated process, Ψ_x has the same distribution as Ψ; it is *motion-invariant* if for all rigid motions m, the process $m\Psi$ has the distribution of Ψ.

The definition of the *intensity measure* Λ of a marked point process Ψ is analogous to that of the intensity measure of Ψ when Ψ is interpreted as an unmarked point process:

$$\Lambda(B \times L) = \mathbf{E}\big(\Psi(B \times L)\big). \tag{4.19}$$

$\Lambda(B \times L)$ is the mean number of points in B having mark in L.

A *Campbell theorem* (cf. Theorem 4.1) also holds:

$$\mathbf{E}\left(\sum_{[x;m]\in\Phi} f(x, m) \right) = \int f(x, m)\Lambda\big(\mathrm{d}(x, m)\big) \tag{4.20}$$

for any nonnegative measurable function f on $\mathbb{R}^d \times \mathbb{M}$.

The measure $\Lambda(\cdot \times L)$ for fixed L in $\mathcal{M}$ is absolutely continuous with respect to the intensity measure Λ_g of the ground point process $\{x_n\}$. Consequently, it can be shown that

$$\Lambda\big(\mathrm{d}(x, m)\big) = M_x(\mathrm{d}m)\Lambda_g(\mathrm{d}x), \tag{4.21}$$

where M_x is a probability measure on $[\mathbb{M}, \mathcal{M}]$, which can be interpreted as the *distribution of the mark* of a point at x. Section 4.4 explains the close relationship of M_x to Palm distribution theory. The Polish space property of $\mathbb{M}$ is needed here to establish that M_x is indeed a measure.

When Ψ is stationary

$$\Lambda(B \times L) = \mathbf{E}\big(\Psi(B \times L)\big) = \mathbf{E}\big(\Psi(B_x \times L)\big) = \Lambda(B_x \times L) \qquad \text{for all } x \in \mathbb{R}^d. \tag{4.22}$$

If L is a fixed member of $\mathcal{M}$, then $\Lambda(\cdot \times L)$ is a translation-invariant measure. Hence, by (1.83)

$$\Lambda(B \times L) = \lambda_L \nu_d(B). \tag{4.23}$$

The quantity λ_L is the *intensity of Ψ with respect to L*; it is the mean number of points of Ψ per unit volume with marks in L. Of course,

$$\lambda_{\mathbb{M}} = \lambda, \tag{4.24}$$

where λ is the intensity of the ground point process.

As a function of L the quantity λ_L is a measure on $[\mathbb{M}, \mathcal{M}]$; the quotients λ_L/λ provide the so-called *mark distribution* M of the stationary marked point process Ψ:

$$M(L) = \frac{\lambda_L}{\lambda}. \tag{4.25}$$

In the case of stationarity, the intensity measure satisfies

$$\Lambda(B \times L) = \lambda \cdot \nu_d(B) \cdot M(L) \tag{4.26}$$

and $M = M_x$ for all x, where M_x is given in Formula (4.21).

In the case of independent marks, M is simply the distribution of the m_n.

In the ergodic case the following holds for every nonnegative integrable function $h(m)$ and any convex averaging sequence $\{W_n\}$

$$\frac{1}{\nu_d(W_n)} \sum_{i:\, x_i \in W_n} h(m_i) \to \int h(m)M(dm) \qquad \text{as } n \to \infty. \tag{4.27}$$

Thus M can be interpreted as the distribution of the mark of a 'randomly chosen point' of Ψ:

$$M(L) = \lim_{n \to \infty} \frac{\Psi(W_n \times L)}{\Psi(W_n \times \mathbb{M})} \tag{4.28}$$

for convex averaging sequences $\{W_n\}$. In this book the term *typical point* is often used instead of 'randomly chosen point'.

In the stationary case the Campbell theorem (4.20) takes the form

$$\mathbf{E}\left(\sum_{[x;m] \in \Psi} f(x, m) \right) = \lambda \int \int f(x, m)M(dm)\, dx. \tag{4.29}$$

A typical application of this theorem is to establish Formula (6.124). An important particular case is

$$\mathbf{E}\left(\sum_{[x;m] \in \Psi} \mathbf{1}_B(x)h(m) \right) = \lambda \int h(m)M(dm) \tag{4.30}$$

for any Borel set B with $\nu_d(B) = 1$, and for $h(m)$ a nonnegative measurable function on $\mathbb{M}$.

In the case of real-valued marks the following characteristics are useful. The *mark distribution function* $F_M(m)$ is given by

$$F_M(m) = \int_{\mathbb{R}} \mathbf{1}_{(-\infty, x]}(k)M(dk) \tag{4.31}$$

and the *mean mark* is

$$\overline{m} = \int_{-\infty}^{\infty} m\, dF_M(m). \tag{4.32}$$

If the marks are positive, then the *mark sum measure* S_m can be considered:

$$S_m(B) = \sum_{[x;m] \in \Psi} m\mathbf{1}_B(x) \qquad \text{for Borel sets } B. \tag{4.33}$$

This is a random measure, namely the sum of the random marks of all points within B. In the stationary case its mean satisfies

$$\mathbf{E}\big(S_m(B)\big) = \lambda \overline{m} v_d(B). \tag{4.34}$$

The second-order properties of the mark sum measure are studied in Stoyan (1984c).

Example 4.1. *Longleaf pines in a North American forest (Cressie, 1993, p. 579ff)*

Figure 4.3(a) is the point pattern of the positions of 583 longleaf pines in a 200 m × 200 m region of a forest. The marks shown by circles are the stem diameters. Figure 4.3(b) shows a kernel-smoothed estimate of the mark probability density function (giving the stem diameter distribution). The estimated mean mark is

$$\hat{\overline{m}} = 26.8\,\text{cm},$$

and the intensity of the ground point process is estimated as

$$\hat{\lambda} = 0.0146\,\text{m}^{-2}.$$

In the remainder of this chapter most formulae are given for unmarked point processes. Corresponding formulae for a marked point process Ψ can be deduced either by interpreting Ψ as a point process in $\mathbb{R}^d \times \mathbb{M}$ or by noting that for each fixed L in $\mathcal{M}$ the ground process $\{x_n : [x_n; m_n] \in \Psi, m_n \in L\}$ is an unmarked point process.

4.3 Moment measures and related quantities

4.3.1 Moment measures

In the classical theory of random variables the moments (particularly mean and variance), generating functions, and Laplace–Stieltjes transforms are important tools and provide useful means of describing distributions. Point process theory has analogues to these. For example, numerical means and variances of random variables are replaced by moment *measures*, as described below and in more detail in Daley and Vere-Jones (2008, Section 9.5).

Throughout the section (except when noted otherwise), $B, B_1, \ldots, B_n$ stand for Borel sets in $\mathbb{R}^d$.

General definition

The nth moment measure of the point process Φ is the measure $\mu^{(n)}$ defined on $\mathcal{B}^{nd}$ by

$$\int_{\mathbb{R}^{nd}} f(x_1, \ldots, x_n) \mu^{(n)}\big(\mathrm{d}(x_1, \ldots, x_n)\big) = \int_{\mathbb{N}} \sum_{x_1, \ldots, x_n \in \varphi} f(x_1, \ldots, x_n) P(\mathrm{d}\varphi)$$

$$= \mathbf{E}\left(\sum_{x_1, \ldots, x_n \in \Phi} f(x_1, \ldots, x_n) \right), \tag{4.35}$$

where f is any nonnegative measurable function on $\mathbb{R}^{nd}$. In particular, if $f(x_1, \ldots, x_n) = \mathbf{1}_{B_1}(x_1) \cdot \ldots \cdot \mathbf{1}_{B_n}(x_n)$, then

$$\mu^{(n)}(B_1 \times \cdots \times B_n) = \mathbf{E}\big(\Phi(B_1) \cdot \ldots \cdot \Phi(B_n)\big), \tag{4.36}$$

and if $B_1 = \cdots = B_n = B$, then

$$\mu^{(n)}(B^n) = \mathbf{E}\big(\Phi(B)^n\big). \tag{4.37}$$

Thus $\mu^{(n)}$ yields the nth moment of the real-valued random variable $\Phi(B)$, which is the number of points in B.

Particular cases

$$n = 1: \qquad \mu^{(1)}(B) = \mathbf{E}\big(\Phi(B)\big) = \Lambda(B); \tag{4.38}$$

$$n = 2: \qquad \mu^{(2)}(B_1 \times B_2) = \mathbf{E}\big(\Phi(B_1)\Phi(B_2)\big), \tag{4.39}$$

$$\mathbf{var}\big(\Phi(B)\big) = \mu^{(2)}(B \times B) - \big(\Lambda(B)\big)^2. \tag{4.40}$$

The covariance of the random variables $\Phi(B_1)$ and $\Phi(B_2)$, that is, of the point numbers in two different sets, satisfies

$$\mathbf{cov}\big(\Phi(B_1), \Phi(B_2)\big) = \mathbf{E}\big(\Phi(B_1)\Phi(B_2)\big) - \mathbf{E}\big(\Phi(B_1)\big)\mathbf{E}\big(\Phi(B_2)\big)$$
$$= \mu^{(2)}(B_1 \times B_2) - \Lambda(B_1)\Lambda(B_2). \tag{4.41}$$

If Φ is stationary, then the $\mu^{(n)}$ are translation-invariant in an extended sense:

$$\mu^{(n)}(B_1 \times \cdots \times B_n) = \mu^{(n)}\big((B_1 + x) \times \cdots \times (B_n + x)\big) \qquad \text{for all } x \text{ in } \mathbb{R}^d. \tag{4.42}$$

Section 4.5 gives a simplified description of the second moment measure in the stationary case.

4.3.2 Factorial moment measures

The nth-order factorial moment measure $\alpha^{(n)}$ of the point process Φ is defined on $\mathcal{B}^{nd}$ by

$$\int f(x_1, \ldots, x_n)\alpha^{(n)}\big(\mathrm{d}(x_1, \ldots, x_n)\big) = \int \sum_{x_1, \ldots, x_n \in \varphi}^{\neq} f(x_1, \ldots, x_n)P(\mathrm{d}\varphi). \tag{4.43}$$

where, again, f is any nonnegative measurable function on $\mathbb{R}^{nd}$. Now the summation is over all n-tuples of *distinct* points in φ including all permutations of given points, as indicated by notation $\sum^{\neq}$. This is the difference between $\alpha^{(n)}$ and $\mu^{(n)}$. In (4.43) the sum omits all n-tuples with two or more equal members. Thus if $B_1, \ldots, B_n$ are pairwise disjoint, then

$$\mu^{(n)}(B_1 \times \cdots \times B_n) = \alpha^{(n)}(B_1 \times \cdots \times B_n). \tag{4.44}$$

For $n = 2$,

$$\mu^{(2)}(B_1 \times B_2) = \Lambda(B_1 \cap B_2) + \alpha^{(2)}(B_1 \times B_2) \tag{4.45}$$

as can be seen from the definitions of $\mu^{(2)}$, Λ and $\alpha^{(2)}$, and the relation

$$\sum_{x_1, x_2 \in \varphi} \mathbf{1}_{B_1}(x_1)\mathbf{1}_{B_2}(x_2) = \sum_{x_1 \in \varphi} \mathbf{1}_{B_1 \cap B_2}(x_1) + \sum_{x_1, x_2 \in \varphi}^{\neq} \mathbf{1}_{B_1}(x_1)\mathbf{1}_{B_2}(x_2).$$

Since

$$\alpha^{(n)}(B^n) = \mathbf{E}\big(\Phi(B)(\Phi(B) - 1) \cdot \ldots \cdot (\Phi(B) - n + 1)\big), \qquad (4.46)$$

the quantity $\alpha^{(n)}(B^n)$ is the nth-order factorial moment of the random variable $\Phi(B)$: this motivates the term 'factorial moment measure'.

The factorial moment measures for a Poisson process are given in Section 2.3.3.

4.3.3 Product densities

Suppose that $\alpha^{(n)}$ is locally finite and absolutely continuous with respect to Lebesgue measure ν_{nd}. Then $\alpha^{(n)}$ has a density $\varrho^{(n)}$, the nth-*order product density*:

$$\alpha^{(n)}(B_1 \times \cdots \times B_n) = \int_{B_n} \cdots \int_{B_1} \varrho^{(n)}(x_1, \ldots, x_n) \, dx_1 \ldots dx_n. \qquad (4.47)$$

Moreover, for any nonnegative bounded measurable function f

$$\mathbf{E}\left(\sum_{x_1, \ldots, x_n \in \Phi}^{\neq} f(x_1, \ldots, x_n)\right) = \int \cdots \int f(x_1, \ldots, x_n)\varrho^{(n)}(x_1, \ldots, x_n) \, dx_1 \ldots dx_n.$$

The product densities have an intuitive infinitesimal interpretation, which probably accounts for their historical precedence over product measures. Suppose that $C_1, \ldots, C_n$ are pairwise disjoint balls with centres $x_1, \ldots, x_n$ and infinitesimal volumes $dV_1, \ldots, dV_n$, respectively. Then $\varrho^{(n)}(x_1, \ldots, x_n) \, dV_1 \ldots dV_n$ is the probability that there is a point of Φ in each of $C_1, \ldots, C_n$.

Clearly, $\varrho^{(1)}(x)$ is the same as the intensity function $\lambda(x)$.

For pairwise disjoint $B_1, \ldots, B_n$, the Formulae (4.44) and (4.47) yield

$$\mu^{(n)}(B_1 \times \cdots \times B_n) = \int_{B_1} \cdots \int_{B_n} \varrho^{(n)}(x_1, \ldots, x_n) \, dx_1 \ldots dx_n.$$

If Φ is stationary, then $\varrho^{(1)} \equiv \lambda$, and $\varrho^{(2)}$ depends only on the difference of its arguments:

$$\varrho^{(2)}(x_1, x_2) = f_{\text{st}}(x_1 - x_2) \qquad \text{for all } x_1, x_2 \in \mathbb{R}^d.$$

If furthermore Φ is motion-invariant, then this f_s depends only on the distance r between x_1 and x_2:

$$\varrho^{(2)}(x_1, x_2) = f_{\text{inv}}(\|x_1 - x_2\|) = f_{\text{inv}}(r).$$

The function $f_{\text{inv}}(r)$ is often still denoted by $\varrho^{(2)}(r)$ and is also called the second-order product density.

Product densities for the Poisson process can be found in Sections 2.3.3 and 2.4, and for other point processes models in Chapter 5.

4.3.4 The Campbell measure

The Campbell measure $\mathscr{C}$ is defined as a measure on $[\mathbb{R}^d \times \mathbb{N}, \mathcal{B}^d \times \mathcal{N}]$ by

$$\int \sum_{x \in \varphi} f(x, \varphi) P(\mathrm{d}\varphi) = \int f(x, \varphi) \mathscr{C}(\mathrm{d}(x, \varphi)), \tag{4.48}$$

where f is any nonnegative measurable function on $\mathbb{R}^d \times \mathbb{N}$. Since

$$\mathscr{C}(B \times Y) = \int \varphi(B) \mathbf{1}_Y(\varphi) P(\mathrm{d}\varphi) \tag{4.49}$$

for Borel $B \in \mathcal{B}^d$ and $Y \in \mathcal{N}$, one can also write

$$\mathscr{C}(B \times Y) = \mathbf{E}\big(\Phi(B)\mathbf{1}_Y(\Phi)\big) = \mathbf{E}\big(\Phi(B)\mathbf{1}(\Phi \in Y)\big). \tag{4.50}$$

It is sometimes useful to consider the *reduced Campbell measure* $\mathscr{C}^!$ defined by

$$\int \sum_{x \in \varphi} f(x, \varphi \setminus \{x\}) P(\mathrm{d}\varphi) = \int \sum_{x \in \varphi} f(x, \varphi - \delta_x) P(\mathrm{d}\varphi)$$

$$= \int f(x, \varphi) \mathscr{C}^!\big(\mathrm{d}(x, \varphi)\big). \tag{4.51}$$

(Note that Daley and Vere-Jones, 2008, used the term *modified Campbell measure* for $\mathscr{C}^!$.) Here $\varphi \setminus \{x\}$ and $\varphi - \delta_x$ are alternative notations, in set-theoretic and measure-theoretic language respectively, for the point pattern φ with the point $x \in \varphi$ deleted.

Campbell measures and reduced Campbell measures of higher order can also be defined; see Daley and Vere-Jones (2008, Section 13.1). For example, the second-order Campbell measure is

$$\mathscr{C}^{(2)}(B_1 \times B_2 \times Y) = \mathbf{E}\big(\Phi(B_1)\Phi(B_2)\mathbf{1}(\Phi \in Y)\big).$$

4.3.5 The mark correlation function

Moment measures for marked point processes can be defined in a fashion similar to the unmarked case. Details are given in Illian *et al.* (2008) both for discrete and continuous (i.e. real-valued) marks.

Here only one example of a useful second-order characteristic for marked point processes is considered, to give the reader some idea: the *mark correlation function* $k_{mm}(r)$ for a motion-invariant marked point process $\Psi = \{[x_n; m_n]\}$ with continuous marks. Heuristically $k_{mm}(r)$ is a normalised mean of the product of marks of points at two positions separated by a distance r, under the condition that there are indeed points of Ψ in these two positions. By motion-invariance it suffices to consider the positions o and $\mathbf{r}$, with $r = \|\mathbf{r}\|$. So

$$k_{mm}(r) = \frac{\mathbf{E}_{o,\mathbf{r}}\big(m(o)m(\mathbf{r})\big)}{\overline{m}^2} \qquad \text{for } r > 0, \tag{4.52}$$

where $\mathbf{E}_{o,\mathbf{r}}$ is expectation subject to the conditioning that Ψ has points at positions o and $\mathbf{r}$ with marks $m(o)$ and $m(\mathbf{r})$; $\overline{m}$ is the mean mark.

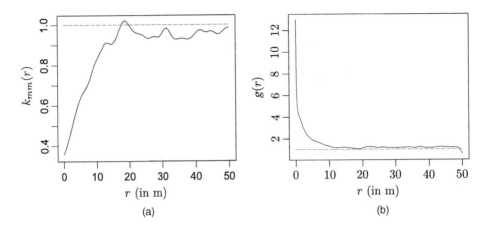

Figure 4.4 (a) Statistically estimated mark correlation function $k_{mm}(r)$ for the longleaf pines stem diameters. This function tends to increase with r up to $r = 15$ m, and the values smaller than 1 indicate that trees close together tend to have diameters smaller than the mean diameter. The distance $r = 15$ m may be seen as an estimate of the range of correlation. (b) Statistically estimated pair correlation function for the longleaf pines. Comparing it with Figure 4.7 suggests that it is similar to the pair correlation function of a Poisson cluster process.

Example 4.1 (Continued). *Longleaf pines in a North American forest*

The mark correlation function $k_{mm}(r)$ was calculated using as marks the stem diameters of the 583 pine trees. The result shown in Figure 4.4(a) is as perhaps expected: for small r (≤ 10 m) $k_{mm}(r)$ is smaller than one. This indicates that trees close together tend to have smaller diameters than the typical tree; this is the price trees have to pay for being close together. That this correlation goes only up to 15 m shows that the forest considered has only a weak short-range order; it is similar in most forests. For completeness Figure 4.4(b) shows also the estimated pair correlation function. It looks like the pair correlation function of a Poisson cluster process; see also the discussion in Cressie (1993) and Stoyan and Stoyan (1996).

Similar studies for other forests and for other tree characteristics such as damage characteristics can be found in Penttinen *et al.* (1992), Stoyan and Penttinen (2000) and Illian *et al.* (2008); growth functions as marks appear in Comas *et al.* (2013). Stoyan (1993) analysed a collage of Hans Arp by this method.

Stoyan (1984c) and Penttinen and Stoyan (1989) gave a mathematical description of $k_{mm}(r)$. The main idea is as follows: consider the second-order factorial moment measure of the marked point process in question

$$\alpha_m^{(2)}(B_1 \times B_2 \times L_1 \times L_2) = \mathbf{E} \left(\sum_{\substack{[x_1;m_1] \in \Psi \\ [x_2;m_2] \in \Psi}}^{\neq} \mathbf{1}_{B_1}(x_1) \mathbf{1}_{B_2}(x_2) \mathbf{1}_{L_1}(m_1) \mathbf{1}_{L_2}(m_2) \right),$$

where the sum is taken over all pairs $[x_1; m_1]$ and $[x_2; m_2]$ in Ψ but excluding the cases when $x_1 = x_2$, as indicated by the symbol $\Sigma^{\neq}$. Let $\alpha^{(2)}$ be the second-order factorial moment measure of the unmarked ground point process $\Phi = \{x_n\}$. Then for fixed Borel subsets L_1 and L_2 of $[0, \infty)$, the measure $\alpha_m^{(2)}(\cdot \times \cdot \times L_1 \times L_2)$ is absolutely continuous with respect to $\alpha^{(2)}$. If $\alpha_m^{(2)}$ is σ-finite, then by the Radon–Nikodym theorem there is a 'density' M_{x_1,x_2} such that

$$\alpha_m^{(2)}(B_1 \times B_2 \times L_1 \times L_2) = \int_{B_1 \times B_2} M_{x_1,x_2}(L_1 \times L_2)\alpha^{(2)}\big(\mathrm{d}(x_1, x_2)\big).$$

For fixed x_1 and x_2 the term M_{x_1,x_2} can be interpreted as a measure on $\mathbb{M} \times \mathbb{M}$, the 'two-point mark distribution'. It gives the joint distribution of the marks of two points at the locations x_1 and x_2, under the condition that there are indeed points of Ψ at x_1 and x_2.

Under the assumption of motion-invariance, M_{x_1,x_2} depends only on $r = \|x_1 - x_2\|$ and the simpler notation M_r is used. In that case

$$k_{mm}(r) = \int m_1 m_2 M_r\big(\mathrm{d}(m_1, m_2)\big)\Big/ \overline{m}^2 \qquad \text{for } r \geq 0. \tag{4.53}$$

For more details see Illian *et al.* (2008), as well as Schlather (2001a), who also discussed the case of $r = 0$.

4.3.6 The probability generating functional

Recall from probability theory that the probability generating function G_X of a nonnegative integer-valued random variable X is given by

$$G_X(t) = \mathbf{E}(t^X) = \sum_{n=0}^{\infty} t^n \mathbf{P}(X = n) \qquad \text{for } t \in [0, 1].$$

The mean and variance of X can be derived from G_X:

$$\mathbf{E}(X) = G'(1) \qquad \text{and} \qquad \mathbf{var}(X) = G''(1) + G'(1) \cdot \big(1 - G'(1)\big)$$

under suitable regularity conditions.

The probability generating functional G of a point process $\Phi = \{x_n\}$ is defined by analogy with G_X. Let $\mathbf{U}$ be the family of all nonnegative bounded measurable functions u on $\mathbb{R}^d$ whose support $\{x \in \mathbb{R}^d : u(x) > 0\}$ is bounded. Furthermore let $\mathbf{V}$ be the family of all functions $v = 1 - u$ for u in $\mathbf{U}$, $0 \leq u \leq 1$. Then G of Φ is defined by

$$G(v) = \mathbf{E}(v(x_1)v(x_2) \cdot \ldots) = \mathbf{E}\left(\prod_{x \in \Phi} v(x)\right)$$

$$= \int \prod_{x \in \varphi} v(x) P(\mathrm{d}\varphi) \qquad \text{for } v \in \mathbf{V}. \tag{4.54}$$

Since Φ is locally finite, the functional G is well-defined for any measurable function $v = 1 - u$ on $\mathbb{R}^d$ such that $0 \leq u \leq 1$ and $\int \big(1 - u(x)\big)\Lambda(\mathrm{d}x) < \infty$; see Daley and Vere-Jones (2008, p. 59).

The distribution P of Φ is determined uniquely by G. The joint probability generating function $G_{\Phi(B_1), \ldots, \Phi(B_n)}$ of the random vector $(\Phi(B_1), \ldots, \Phi(B_n))$ for bounded Borel sets $B_1, \ldots, B_n$ is given by

$$G_{\Phi(B_1), \ldots, \Phi(B_n)}(t_1, \ldots, t_n) = \mathbf{E}(t_1^{\Phi(B_1)} \cdot \ldots \cdot t_n^{\Phi(B_n)}) = G(v),$$

where $v(x) = v_1(x) \cdot \ldots \cdot v_n(x)$ in which $v_i(x) = 1 + (t_i - 1)\mathbf{1}_{B_i}(x)$ and $t_1, \ldots, t_n$ all belong to $[0, 1]$. Thus in particular for $n = 1$

$$G_{\Phi(B)}(t) = G(v) \qquad \text{with } v(x) = 1 + (t - 1)\mathbf{1}_B(x).$$

The factorial moment measures of Φ can be derived from G

$$\alpha^{(n)}(B_1 \times \cdots \times B_n) = (-1)^n \left(\left(\frac{\partial}{\partial s_1} \cdots \frac{\partial}{\partial s_n} \right) G(1 - s_1 \mathbf{1}_{B_1} - \cdots - s_n \mathbf{1}_{B_n}) \right) \bigg|_{s_1 = \cdots = s_n = 0}.$$

Conversely, under suitable regularity conditions the probability generating functional is determined by the factorial moment measures; for u in $\mathbf{U}$ with $0 \le u \le 1$

$$G(1 - u) = 1 + \sum_{n=1}^{\infty} \frac{(-1)^n}{n!} \int u(x_1) \cdot \ldots \cdot u(x_n) \alpha^{(n)}(\mathrm{d}(x_1, \ldots, x_n))$$

if the $\alpha^{(n)}$ are locally finite measures for all n and if the right-hand side converges.

Example 4.2. *The probability generating functional of the Poisson process*
Let Φ be a Poisson process of intensity measure Λ. Then

$$G(v) = \exp\left(- \int (1 - v(x))\Lambda(\mathrm{d}x) \right) \qquad \text{for } v \in \mathbf{V}. \tag{4.55}$$

Proof. Consider $v(x) = 1 - \sum_{i=1}^{n}(1 - t_i)\mathbf{1}_{C_i}(x)$ for t_i all in $[0, 1]$ and $C_1, \ldots, C_n$ pairwise disjoint compact sets. Then

$$G(v) = \int t_1^{\varphi(C_1)} \cdot \ldots \cdot t_n^{\varphi(C_n)} P(\mathrm{d}\varphi)$$

$$= \mathbf{E}(t_1^{\Phi(C_1)} \cdot \ldots \cdot t_n^{\Phi(C_n)})$$

$$= \mathbf{E}(t_1^{\Phi(C_1)}) \cdot \ldots \cdot \mathbf{E}(t_n^{\Phi(C_n)}),$$

where the last step uses the independence of $\Phi(C_1), \ldots, \Phi(C_n)$ for disjoint $C_1, \ldots, C_n$. Then each $\mathbf{E}(t_i^{\Phi(C_i)})$ can be determined, since $\Phi(C_i)$ has a Poisson distribution of mean $\Lambda(C_i)$. Therefore

$$G(v) = \exp\big(-(1 - t_1)\Lambda(C_1)\big) \cdot \ldots \cdot \exp\big(-(1 - t_n)\Lambda(C_n)\big)$$

$$= \exp\left(-\sum_{i=1}^{n} \int_{C_i} (1 - t_i)\Lambda(dx)\right)$$

$$= \exp\left(-\int \big(1 - v(x)\big)\Lambda(dx)\right).$$

Thus (4.55) holds for piecewise constant v. The general result follows by standard arguments from measure theory.   □

A point process can be considered also as a random counting measure. Thus ideas of the theory of random measures can be transferred to point processes. For example, the Laplace functional of a point process can be defined analogously to Section 7.2.1.

4.4 Palm distributions

4.4.1 Heuristic introduction

In the study of point processes it is often natural to ask questions concerning the conditional expectations or distributions of some random quantities, given that there is a point of the process at the fixed location x. Section 2.3.4 gives an elementary example, which is related to the nearest-neighbour distance distribution and which is reconsidered in this section. Suppose $\mathbf{P}\big(\Phi(B(x, r)) = 1 \parallel x\big)$ represents the conditional probability, given that Φ has a point at x, that there is no further point within the ball of radius r centred at x. This conditional probability is ambiguous without further explanation since the conditioning event has probability zero. It must be defined by requiring it to satisfy some extra properties to be described in the following.

What properties might one reasonably require of such conditional distributions $\mathbf{P}(\cdot \parallel x)$? To begin with one would require in the stationary case that the relation

$$\mathbf{P}(\Phi \in Z_x \parallel x) = \mathbf{P}(\Phi \in Z \parallel o)$$

holds for $Z \in \mathcal{N}$ and $Z_x = \{\varphi_x : \varphi \in Z\}$. Substituting $Z = Y_{-x}$ into it yields

$$\mathbf{P}(\Phi \in Y \parallel x) = \mathbf{P}(\Phi_x \in Y \parallel o) = \mathbf{P}(\Phi \in Y_{-x} \parallel o) \qquad (4.56)$$

for $x \in \mathbb{R}^d$ and $Y \in \mathcal{N}$, where Y_{-x} is defined in the same manner as Y_x after Formula (4.1). Given (4.56) one could define a function

$$D(r) = 1 - \mathbf{P}\big(\Phi(B(o, r)) = 1 \parallel o\big),$$

which could be interpreted as the distribution function of the distance of the typical point to its nearest neighbour. (If it is sure that the ball $B(o, r)$ contains o as a point of Φ, then the event $\Phi \in Z$ with $Z = \{\varphi \in \mathbb{N} : \varphi(B(o, r)) = 1\}$ is the same as that the distance from o to its nearest neighbour is larger than r.)

In order to introduce further properties required, consider again the example related to $D(r)$. Suppose that Φ is a point process, not necessarily stationary, of distribution P and locally finite intensity measure Λ. For the set Y in N given by

$$Y = \{\varphi \in N : \varphi(B(o, r)) = 1\},$$

consider the function h on $\mathbb{R}^d \times N$ defined by

$$h(x, \varphi) = 1_B(x)1_Y(\varphi - x)$$

for some bounded Borel set B; as usual if $\varphi = \{x_n\}$, then

$$\varphi_{-x} = \varphi - x = \{x_n - x\}.$$

With the distribution function $D(r)$ given above in mind, one might wish to evaluate the mean number of points of Φ in B whose neighbours are all at distance at least r:

$$\mathbf{E}\left(\sum_{x \in \Phi} h(x, \Phi)\right) = \int \sum_{x \in \varphi} h(x, \varphi) P(\mathrm{d}\varphi).$$

More generally one might wish to evaluate this for arbitrary nonnegative measurable functions h on $\mathbb{R}^d \times N$.

If $\mathbb{R}^d$ is partitioned into domains $D_1, D_2, \ldots$ each of positive volume, then the expectation above can be written as

$$\mathbf{E}\left(\sum_{x \in \Phi} h(x, \Phi)\right) = \sum_k \mathbf{E}\left(\sum_{x \in \Phi \cap D_k} h(x, \Phi) \;\middle|\; \Phi(D_k) > 0\right) \cdot \mathbf{P}\big(\Phi(D_k) > 0\big).$$

Suppose that each D_k tends to an infinitesimal volume element $\mathrm{d}x$. Then $\mathbf{P}(\Phi(D_k) > 0)$ should converge to $\Lambda(\mathrm{d}x)$, while the conditional mean should converge to the mean $\mathbf{E}(h(x, \Phi) \| x)$ of $h(x, \varphi)$ with respect to $\mathbf{P}(\cdot \| x)$. Hence one is led to the relation

$$\mathbf{E}\left(\sum_{x \in \Phi} h(x, \Phi)\right) = \int \mathbf{E}\big(h(x, \Phi) \| x\big) \Lambda(\mathrm{d}x). \tag{4.57}$$

Under the assumption that Φ is stationary, it is $\Lambda(\cdot) = \lambda v_d(\cdot)$ with $0 < \lambda < \infty$. The relations (4.56) and (4.57) then yield

$$\mathbf{E}\left(\sum_{x \in \Phi} h(x, \Phi)\right) = \lambda \int \mathbf{E}\big(h(x, \Phi_x) \| o\big) \mathrm{d}x. \tag{4.58}$$

If the notation $P_o(Y) = \mathbf{P}(\Phi \in Y \parallel o)$ is used for $Y \in \mathcal{N}$, then (4.58) takes the form

$$\mathbf{E}\left(\sum_{x\in\Phi} h(x, \Phi)\right) = \lambda \int \int h(x, \varphi_x) P_o(\mathrm{d}\varphi)\,\mathrm{d}x \tag{4.59}$$

and so for $h(x, \varphi) = \mathbf{1}_B(x) \cdot \mathbf{1}_Y(\varphi_{-x})$ as above

$$\int \sum_{x\in\varphi\cap B} \mathbf{1}_Y(\varphi - x) P(\mathrm{d}\varphi) = \lambda v_d(B) P_o(Y) \tag{4.60}$$

for Borel $B \in \mathcal{B}^d$ and $Y \in \mathcal{N}$.

In the following, a distribution on $[\mathbb{N}, \mathcal{N}]$ will be constructed which will have the behaviour of the hypothetical P_o above: it will satisfy (4.59) and its definition will be inspired by (4.60).

4.4.2   The Palm distribution: First definition

Suppose that Φ is a stationary point process, with finite nonzero intensity λ. The *Palm distribution* (at o) of P is a distribution defined on $[\mathbb{N}, \mathcal{N}]$ by

$$P_o(Y) = \frac{1}{\lambda v_d(B)} \int \sum_{x\in\varphi\cap B} \mathbf{1}_Y(\varphi_{-x}) P(\mathrm{d}\varphi) \qquad \text{for } Y \in \mathcal{N}, \tag{4.61}$$

where B is an arbitrary Borel set of positive volume; the definition does not depend on the choice of the set B.

The definition can be understood intuitively using an approach of Matthes, based on ideas from the theory of marked point processes as follows. To each point x of Φ a mark of 1 or 0 is given according as to whether the shifted process Φ_{-x} belongs to the configuration set Y or not. For example, consider $Y = \{\varphi \in \mathbb{N} : \varphi \cap B(o, r) = \{o\}\}$. Then x has mark 1 precisely when its nearest neighbour is further away than r. The result of this is a stationary marked point process of marks 0 and 1. Its mark distribution M, defined in (4.25), is a 'ratio of rates', given by

$$M(\{1\}) = \frac{\lambda_{\{1\}}}{\lambda},$$

where $\lambda_{\{1\}}$ is the intensity of the subpoint process of points with mark 1. (It is analogous for the points with mark 0.) Thus for any fixed Borel set B the number of points of Φ in B of mark 1 has the mean $\lambda_{\{1\}} \cdot v_d(B)$. The same value is yielded by the integral in (4.61) and so

$$P_o(Y) = M(\{1\}), \tag{4.62}$$

and the choice of B does not play a rôle.

Note in passing that this approach is closely related to methods of statistical estimation of Palm probabilities, see around Formula (4.128).

If Φ is ergodic, then the Palm probability $P_o(Y)$ may be interpreted as the probability that the typical point x of Φ has the property that Φ_{-x}, the point process shifted with x into the origin o, belongs to Y. This is comparable with the ergodic interpretation of a mark distribution as on p. 119. The following ergodic theorem holds:

$$\lim_{n\to\infty} \frac{1}{\Phi(W_n)} \sum_{x_i \in W_n} f(\Phi_{-x_i}) = \int f(\varphi)P_o(d\varphi). \tag{4.63}$$

It is plausible when one considers the definition of the Palm distribution in (4.61) and Formula (4.15).

Because of (4.61) the Palm distribution can be shown to satisfy the so-called *Campbell–Mecke theorem*:

Theorem 4.2. *(Mecke, 1967) For any nonnegative measurable function h on $\mathbb{R}^d \times \mathbb{N}$,*

$$\mathbf{E}\left(\sum_{x \in \Phi} h(x, \Phi)\right) = \int \sum_{x \in \varphi} h(x, \varphi)\, P(d\varphi)$$

$$= \lambda \int \int h(x, \varphi_x)P_o(d\varphi)\, dx. \tag{4.64}$$

Substituting $h(x, \varphi) = f(x)$ into (4.64) gives (4.10), the simple Campbell theorem.

Many problems in point process theory can be resolved by calculations of quantities in the form of the left-hand side of (4.64). The Campbell–Mecke theorem helps reduce these calculations to relatively simple integrations.

4.4.3 The Palm distribution: Second definition

The approach in this section uses more measure-theoretic tools, namely the Radon–Nikodym theorem and the Campbell measure $\mathscr{C}$. Suppose that the intensity measure Λ is σ-finite: this implies that $\mathscr{C}$ is likewise σ-finite, and furthermore, for any configuration set Y the measure $\mathscr{C}(\cdot \times Y)$ is absolutely continuous with respect to Λ. So there is a density function $d(x)$ with

$$\mathscr{C}(B \times Y) = \int_B d(x)\Lambda(dx) \qquad \text{for Borel sets } B.$$

Since this density function depends on Y, it is denoted by $P_x(Y)$. Soon it will turn out that it can be interpreted as a distribution. Thus

$$\mathscr{C}(B \times Y) = \int_B P_x(Y)\Lambda(dx). \tag{4.65}$$

Indeed, for fixed x, $P_x(\cdot)$ can be taken to be a distribution on $[\mathbb{N}, \mathcal{N}]$: the *Palm distribution of P with respect to x*. In the stationary case, with $0 < \lambda < \infty$, it *can be taken to satisfy* relation (4.56) and so

$$P_o(Y) = P_x(Y_x) \qquad \text{for } Y \in \mathcal{N}, \tag{4.66}$$

as

$$\lambda \int_B P_z(Y)\mathrm{d}z = \mathscr{C}(B \times Y) = \mathscr{C}(B_x \times Y_x)$$

$$= \lambda \int_{B_x} P_z(Y_x)\,\mathrm{d}z = \lambda \int_B P_{x+z}(Y_x)\,\mathrm{d}z$$

for all $B \in \mathcal{B}^d$, $x \in \mathbb{R}^d$ and $Y \in \mathcal{N}$. This implies $P_z(Y) = P_{x+z}(Y_x)$. Putting $z = 0$ yields (4.66).

The elliptical language here ('can be taken to satisfy ...') is used because relations such as (4.65) define $P_x(Y)$ for all x off a null-set that may *a priori* depend on Y. The measure-theoretic machinery of *regular conditional probabilities* (see Kallenberg, 2002) can be used to resolve this problem. Nevertheless, it can be ignored for the practical purposes of Palm distribution theory.

The Campbell–Mecke theorem given in Theorem 4.2 for stationary point processes is a consequence of this second definition:

$$\mathbf{E}\left(\sum_{x \in \Phi} h(x, \Phi)\right) = \int h(x, \varphi)\mathscr{C}(d(x, \varphi))$$

$$= \lambda \int \int h(x, \varphi)P_x(\mathrm{d}\varphi)\,\mathrm{d}x = \lambda \int \int h(x, \varphi_x)P_o(\mathrm{d}\varphi)\,\mathrm{d}x. \qquad (4.67)$$

The two definitions agree in the stationary case (the first definition is not otherwise applicable).

A further possible way to define the Palm distribution locally is sketched in Section 2.3.4: it is to consider the conditional distribution of Φ, given that $\Phi\big(B(o, \varepsilon)\big) \geq 1$, and to take the limit as $\varepsilon \to 0$. Mathematical details can be found in Daley and Vere-Jones (2008, pp. 306–8).

4.4.4 Reduced Palm distributions

As the ordinary Palm distributions are related to $\mathscr{C}$, so the reduced Palm distributions are related to the reduced Campbell measure $\mathscr{C}^!$. If the appropriate replacement is made in (4.65), then the *reduced Palm distribution* $P_x^!$ is defined:

$$P_x^!(Y) = P(\Phi \setminus \{x\} \in Y \parallel x) \qquad \text{for } Y \in \mathcal{N}, \qquad (4.68)$$

where $P_x^!(Y)$ denotes the probability that Φ without point x belongs to the configuration set Y, under the condition that Φ has a point at x. In the stationary case, Formula (4.61) can be employed to define $P_o^!(Y)$ if $\mathbf{1}_Y(\varphi - x)$ is replaced by $\mathbf{1}_Y\big((\varphi - x) \setminus \{o\}\big)$.

Analogously as the usual Palm distribution, the reduced Palm distribution characterises the distribution of the point process under the condition that at o there is a point. But now the distribution of the point process is given with the point o removed after conditioning. This simplifies notation, as the following two examples show.

The *nearest-neighbour distribution function* $D(r)$ can be expressed via P_o and $P_o^!$, by

$$D(r) = 1 - P_o\big(\varphi \in \mathbb{N} : \varphi(B(o, r)) = 1\big)$$

$$= 1 - P_o^!\big(\varphi \in \mathbb{N} : \varphi(B(o, r)) = 0\big).$$

The latter formula shows more clearly that the ball $B(o, r)$ is empty as it is expected for the complement of the nearest-neighbour distance distribution function.

Example 4.3. *The Palm distribution of a Poisson process*

The *theorem of Slivnyak–Mecke* gives the Palm distribution P_x of a Poisson process of intensity measure Λ and distribution P:

$$P_x = P * \delta_{\delta_x} \qquad \text{for all } x \in \mathbb{R}^d. \tag{4.69}$$

Here '$*$' denotes convolution of distributions, which corresponds to the superposition of point processes; see Section 5.1. The symbol δ_{δ_x} denotes the distribution of the degenerate point process that consists solely of the (nonrandom) point x. Formula (4.69) can be interpreted as

$$P_x(Y) = \mathbf{P}(\Phi \in Y \parallel x) = \mathbf{P}(\Phi \cup \{x\} \in Y) \qquad \text{for } Y \in \mathcal{N},$$

or

$$\int f(\varphi) P_x(\mathrm{d}\varphi) = \int f(\varphi \cup \{x\}) P(\mathrm{d}\varphi)$$

for all measurable nonnegative functions f. In words: the Palm distribution with respect to x is simply the distribution of the Poisson process plus an added point at x.

If the reduced Palm distribution is used, then (4.69) takes on a more elegant form, called *Mecke's formula*:

$$P_x^! = P \qquad \text{for all } x \in \mathbb{R}^d. \tag{4.70}$$

That is to say, the Palm conditioning does not change the distribution of the Poisson process, apart from a given point at x. An equivalent equation is

$$\mathbf{E}\left(\sum_{x \in \Phi} f(x, \Phi \setminus \{x\})\right) = \lambda \mathbf{E}\left(\int f(x, \Phi) \mathrm{d}x\right). \tag{4.71}$$

The Poisson process is characterised by (4.69) or (4.70); see Mecke (1968) and Daley and Vere-Jones (2008, p. 281).

Proof that a Poisson process satisfies Formula (4.69). Both $P * \delta_{\delta_x}$ and P_x are distributions of simple point processes. So by the discussion in Section 4.1 their equality is established if the corresponding systems of void-probabilities are equal, that is,

$$P * \delta_{\delta_x}(V_K) = P_x(V_K) \qquad \text{for all compact } K \text{ in } \mathbb{R}^d,$$

where $V_K = \{\varphi \in \mathbb{N} : \varphi(K) = 0\}$ is the set of all point sequences without a point in K.

Suppose A is any bounded Borel set. Then

$$\int_A (P * \delta_{\delta_x})(V_K)\Lambda(\mathrm{d}x) = \int_{A \setminus K} P(V_K)\Lambda(\mathrm{d}x) = P(V_K) \cdot \Lambda(A \setminus K)$$

$$= \mathbf{E}\big(\Phi(A \setminus K) \cdot \mathbf{1}(\Phi(K) = 0)\big)$$

$$= \mathscr{C}\big((A \setminus K) \times V_K\big),$$

where the second line results from the independence property of the Poisson process. Clearly $\mathscr{C}\big((A \cap K) \times V_K\big) = \mathbf{E}\big(\Phi(A \cap K)\mathbf{1}(\Phi(K) = 0)\big) = 0$. Hence

$$\int_A P * \delta_{\delta_x}(V_K)\Lambda(\mathrm{d}x) = \mathscr{C}(A \times V_K) = \int_A P_x(V_K)\Lambda(\mathrm{d}x)$$

by using (4.65). Since A was arbitrary, a standard argument of measure theory establishes (4.69). □

4.4.5 Isotropy of Palm distribution

Clearly the Palm distribution P_o is never a stationary distribution since under P_o a point process must always contain o; but Example 4.3 shows that $P_o^!$ can be stationary. However, a Palm distribution can be isotropic, that is, invariant with respect to rotations about the origin o: if the point process Φ is motion-invariant, then its Palm distribution is isotropic.

Proof. Rotation invariance means

$$P_o(Y) = P_o(rY) \qquad \text{for } Y \in \mathcal{N} \tag{4.72}$$

for every rotation r about o. To establish this, consider

$$\lambda P_o(Y)v_d\big(B(o, r)\big) = \int\int_{B(o,r)} \mathbf{1}_Y(\varphi - x)\varphi(\mathrm{d}x)P(\mathrm{d}\varphi)$$

$$= \int\int_{B(o,r)} \mathbf{1}_Y(r^{-1}\varphi - x)\,r^{-1}\varphi(\mathrm{d}x)P(\mathrm{d}\varphi) \qquad \text{(by isotropy of } \Phi)$$

$$= \int\int_{B(o,r)} \mathbf{1}_Y\big(r^{-1}(\varphi - y)\big)\varphi(\mathrm{d}y)P(\mathrm{d}\varphi) \qquad \text{(by letting } x = r^{-1}y)$$

$$= \int\int_{B(o,r)} \mathbf{1}_{rY}(\varphi - y)\varphi(\mathrm{d}y)P(\mathrm{d}\varphi)$$

$$= \lambda P_o(rY) \cdot v_d\big(B(o, r)\big).$$

□

Suppose a system of polar coordinates is based on the typical point of a motion-invariant point process. Then as a direct consequence of (4.72) the random angle α of the coordinates of the nearest neighbour is uniformly distributed over $(0, 2\pi]$; see Figure 4.5.

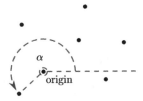

Figure 4.5   The angle α between a point x and its nearest neighbour $n(x)$.

4.4.6 Inversion formulae

It is theoretically possible to determine the distribution of Φ from λ and P_o if the point process Φ is stationary. Here, for example, is a simple relationship between Φ and P_o:

$$k\,\mathbf{P}\big(\Phi(B) = k\big) = \lambda \int_B P_o\big(\{\varphi \in \mathbb{N} : \varphi(B - x) = k\}\big)\,\mathrm{d}x \qquad (4.73)$$

for $k = 1, 2, \ldots$ and all bounded Borel sets B. The stationary distribution can be retrieved by integration over the typical Voronoi cell (to be discussed in Chapter 9); see Daley and Vere-Jones (2008, p. 306):

Theorem 4.3. *For every integrable nonnegative function f, it holds*

$$\mathbf{E}\big(f(\Phi)\big) = \lambda \int \int_{\mathbb{R}^d} f(\varphi_{-x})\mathbf{1}\big(x \in C_0(\varphi)\big)\mathrm{d}x\,P_o(\mathrm{d}\varphi), \qquad (4.74)$$

where $C_0(\varphi)$ is the zero cell of the Voronoi tessellation with respect to the point pattern φ.

4.4.7 *n*-fold Palm distributions

The n-fold Palm distributions of Φ are derived from its higher-order Campbell measures; see Kallenberg (1983a, 1986) and Hanisch (1982). An n-fold Palm distribution is a conditional distribution for a point process given that points of the process occur at n fixed positions. (The mark correlation function $k_{mm}(r)$ described on p. 123 is closely related to a two-fold Palm distribution.)

The n-fold Palm distribution of a Poisson process with distribution P is, analogously to (4.69), equal to

$$P * \delta_{\delta_{x_1}} * \cdots * \delta_{\delta_{x_n}} \qquad \text{for } x_1, \ldots, x_n \in \mathbb{R}^d. \qquad (4.75)$$

This means, it is the distribution of a Poisson process plus n points at the (deterministic) locations $x_1, \ldots, x_n$.

4.4.8 Palm distributions for marked point processes

The definitions of these distributions follow the case of an unmarked point process, using either the Campbell measure or, in the stationary case, a generalised version of Formula (4.61). For a marked point process Ψ with distribution P the Campbell measure is defined by

$$\mathcal{C}(B \times L \times Y) = \mathbf{E}\big(\Psi(B \times L)\mathbf{1}(\Psi \in Y)\big) \tag{4.76}$$

for Borel $B \in \mathcal{B}^d$, $L \in \mathcal{M}$ and $Y \in \mathcal{N}_M$, where $\mathcal{N}_M$ is the smallest σ-algebra on $\mathbb{N}_M$ (the family of all marked point sequences) that makes every mapping

$$\psi \mapsto \psi(B \times L)$$

measurable for all Borel $B \in \mathcal{B}^d$ and $L \in \mathcal{M}$.

Again, assume that Λ is σ-finite. Since for all Y the measure $\mathcal{C}(\cdot \times \cdot \times Y)$ is absolutely continuous with respect to Λ, the Radon–Nikodym derivative $P_x^m(Y)$ exists such that

$$\mathcal{C}(B \times L \times Y) = \int_{B \times L} P_x^m(Y) \Lambda\big(\mathrm{d}(x, m)\big). \tag{4.77}$$

Here, for fixed x and m, P_x^m is a distribution on $\mathcal{N}_M$. It can be interpreted as the probability of Y given that Ψ has a point at x of mark m.

From (4.26) and (4.66) in the stationary case, Formula (4.77) takes the form

$$\mathcal{C}(B \times L \times Y) = \lambda \int_B \int_L P_o^m(Y_{-x})M(\mathrm{d}m)\,\mathrm{d}x, \tag{4.78}$$

where M is the mark distribution of Ψ.

The *Campbell–Mecke theorem* for a stationary marked point process is

$$\int \sum_{[x;m]\in\psi} h(x, m, \psi)P(\mathrm{d}\psi) = \lambda \int \int \int h(x, m, \psi_x)P_o^m(\mathrm{d}\psi)M(\mathrm{d}m)\,\mathrm{d}x \tag{4.79}$$

for any nonnegative measurable function h on $\mathbb{R}^d \times \mathbb{M} \times \mathbb{N}_M$.

Again in the stationary case, suppose that $L \in \mathcal{M}$ with $M(L) > 0$. Then one can define the *Palm distribution of Ψ with respect to the mark set L*, denoted by P_L, as

$$P_L(Y) = \int_L P_o^m(Y)M(\mathrm{d}m)/M(L) \qquad \text{for } Y \in \mathcal{N}_M.$$

It has an interpretation as the conditional distribution of Ψ given that Ψ has a point at o with mark in L.

The mark distribution M itself can be thought of as a Palm distribution. Set $Y_L = \{\psi \in \mathbb{N}_M : [o; m] \in \psi, m \in L\}$. Then

$$P_{\mathbb{M}}(Y_L) = \int_{\mathbb{M}} P_o^m(Y_L)M(\mathrm{d}m) = M(L),$$

since $P_o^m(Y_L)$ is 1 if $m \in L$, and 0 otherwise.

For brevity and convenience, throughout this book and also in the literature the phrase *typical point* is used. This is to be interpreted using Palm distributions and understood intuitively as a 'randomly chosen point'. For example, if under the Palm distribution the distance from o to the nearest point has an exponential distribution, then it is said that the distance of the typical point to its nearest neighbour is exponentially distributed. Similarly the statement that the mark of the typical point has probability p of lying in L means that $M(L) = p$.

4.4.9 Point-stationarity

Let Φ be a stationary point process with Palm distribution P_o. It has been noticed in Section 4.4.5 that P_o is not stationary. Still P_o has important invariance properties. To discuss these it is convenient to introduce the notion of a *Palm version* of Φ, that is a point process Φ^o with distribution P_o.

In one dimension it is possible to write Φ^o as a point sequence, $\Phi^o = \{T_n\}$, where n is running through all integers, with $T_0 := 0$ and $T_n < T_{n+1}$ for all n. It can be shown that

$$\mathbf{P}(\Phi^o_{-T_n} \in Y) = \mathbf{P}(\Phi^o \in Y) \tag{4.80}$$

for all integers n and all configuration sets Y. This means, moving the origin $o \in \Phi^o$ to the nth point of Φ^o does not change the distribution of Φ^o, and the sequence $\{T_n - T_{n-1}\}$ of inter-point distances is stationary. Moreover, there is an essentially unique correspondence between such sequences and stationary point processes on the line; see for example Kallenberg (2002, Theorem 11.4) or Daley and Vere-Jones (2008, Theorem 13.3.I). These facts are of fundamental importance in one-dimensional point process theory and its applications.

In higher dimensions there is no obvious analogue of (4.80). Assume for instance that Φ is a homogeneous Poisson process. According to Example 4.3 one can then take $\Phi^o = \Phi \cup \{o\}$. Let $T \in \Phi$ be the nearest neighbour of o in Φ. Then Φ^o_{-T} has not the same distribution as Φ^o. This is because $-T$ is a point of Φ^o_{-T} that is closer to the origin than to any other point of Φ^o_{-T}. It is not hard to see that a Poisson process cannot have this property almost surely.

A generalisation of (4.80) can be based on the concept of the *bijective point map*, introduced in Thorisson (2000). A point map is a mapping $\tau : \mathbb{N} \times \mathbb{R}^d \to \mathbb{R}^d$ satisfying

$$\tau(\varphi, x) \in \varphi \qquad \text{whenever } x \in \varphi,$$

and the covariance property

$$\tau(\varphi_x, y - x) = \tau(\varphi, y) - x \qquad \text{for all } \varphi \in \mathbb{N}, \text{ and } x, y \in \mathbb{R}^d. \tag{4.81}$$

A point map τ is called bijective if $\tau(\varphi, \cdot)$ is a bijection on φ for all $\varphi \in \mathbb{N}$. In this case

$$\mathbf{P}(\Phi^o_{-T} \in Y) = \mathbf{P}(\Phi^o \in Y) \qquad \text{for all } Y \in \mathcal{N}, \tag{4.82}$$

where $T = \tau(\Phi, o)$. This was proved in Thorisson (2000) and Heveling and Last (2005), but can also be derived from results in Mecke (1975). Note that (4.82) contains (4.80) as a special case.

An example of bijective point map τ is as follows: For $\varphi \in \mathbb{N}$ and $x \in \varphi$, let $\tau(\varphi, x) = y$ if $y \in \varphi$ is the unique nearest neighbour of $x \in \varphi$ and if x is the unique nearest neighbour of y. In all other cases let $\tau(\varphi, x) := x$. Hence (4.82) holds if $T = \tau(\Phi, o)$ and o are (unique) mutually nearest neighbours (even though $\mathbf{P}(T = o) > 0$ for most Φ). Other examples of bijective point maps can be found in Heveling and Last (2005), which also contains a construction of a bijective point map with $\mathbf{P}(T = o) = 0$.

Heveling and Last (2005) also proved that (4.82) is even characterising Palm versions. The attractive feature of this result is that (4.82) is an *intrinsic* invariance property, which does not refer to a stationary distribution. The first intrinsic characterisation of Palm measures was discovered in Mecke (1967).

Proof of (4.82). Let B be a Borel set of $\mathbb{R}^d$ with $v_d(B) = 1$. The definition (4.61) implies for all nonnegative measurable functions g on $\mathbb{N}$ that

$$\lambda \mathbf{E}\big(g(\Phi^o_{-T})\big) = \mathbf{E}\left(\sum_{x \in \Phi \cap B} g\big((\Phi_{-x})_{-\tau(\Phi-x,o)}\big) \right) = \mathbf{E}\left(\sum_{x \in \Phi \cap B} g(\Phi_{-\tau(\Phi,x)}) \right),$$

where (4.81) has been used with $x = y$. For any $\varphi \in \mathbb{N}$ let $\tau^{-1}(\varphi, \cdot)$ denote the inverse of the bijection $\tau(\varphi, \cdot)$. It is easy to check that τ^{-1} has the covariance property (4.81). Now one can replace $\tau(\Phi, x)$ in the above summation by y and x by $\tau^{-1}(\Phi, y)$ to obtain

$$\lambda \mathbf{E}\big(g(\Phi^o_{-T})\big) = \mathbf{E}\left(\sum_{y \in \Phi} \mathbf{1}\big(\tau^{-1}(\Phi, y) \in B\big) g(\Phi_{-y}) \right) = \mathbf{E}\left(\sum_{y \in \Phi} \mathbf{1}\big(\tau^{-1}(\Phi_y, o) + y \in B\big) g(\Phi_{-y}) \right)$$

$$= \lambda \iint \mathbf{1}\big(\tau^{-1}(\varphi, o) + y \in B\big) g(\varphi) \, dy \, P_o(d\varphi),$$

in which the Campbell–Mecke theorem (4.64) and the Fubini theorem have been used. Since the inner integration yields $g(\varphi)v_d(B) = g(\varphi)$, this implies $\mathbf{E}\big(g(\Phi^o_{-T})\big) = \mathbf{E}\big(g(\Phi^o)\big)$ and therefore (4.82). $\qquad \square$

4.4.10   Stationary and balanced partitions

Given a homogeneous Poisson process Φ one may ask for the existence of a random variable T such that $T \in \Phi$ and Φ_{-T} have the same distribution as $\Phi \cup \{o\}$. This section will show that not only does such a T exist but also it may even be chosen as a function $T \equiv T(\Phi)$ of Φ. In fact the result can be formulated for any ergodic point process.

Let Φ be a stationary point process on $\mathbb{R}^d$ with finite intensity λ. A *stationary partition* (based on Φ) is a measurable mapping $\pi : \mathbb{N} \times \mathbb{R}^d \to \mathbb{R}^d$ satisfying the covariance property (4.81) and such that almost surely $\pi(\Phi, x) \in \Phi$ for all $x \in \mathbb{R}^d$; see Last (2006, 2010). For $x \in \varphi \in \mathbb{N}$, the set

$$C_x(\varphi) = \{y \in \mathbb{R}^d : \pi(\varphi, y) = x\}$$

is referred to as *cell* with *centre* x. It is not required that $x \in C_x(\varphi)$ and some of the cells might be empty. The cells $C_x(\Phi)$, $x \in \Phi$, form a random partition of $\mathbb{R}^d$ into Borel sets. By stationarity and (4.81) the statistical properties of $\{C_x(\Phi) + z : x \in \Phi\}$ are the same for all $z \in \mathbb{R}^d$. For simplicity (and in contrast to Last, 2006, 2010), the attention here is restricted to partitions that do only depend on Φ but not on any additional source of randomness. An example is the Voronoi tessellation generated by Φ; see Section 9.2.

Let π be a stationary partition based on Φ and let Φ^o be a Palm version of Φ. It is easy to see that again almost surely $\pi(\Phi^o, x) \in \Phi^o$ for all $x \in \mathbb{R}^d$. By Last (2006, Theorem 4.1),

$$\mathbf{E}\big(f(\Phi)g(\Phi_{-\pi(\Phi,o)})\big) = \lambda \mathbf{E}\left(g(\Phi^o) \int_{C_0(\Phi^o)} f(\Phi^o_{-x}) \, dx \right) \qquad (4.83)$$

for all measurable nonnegative functions f, g on $\mathbb{N}$. This result is a simple consequence of (4.64). A special case will be proved in Proposition 9.1.

Letting $g \equiv 1$ in (4.83) generalises the inversion formula (4.74), and letting $f \equiv 1$ yields

$$E\big(g(\Phi_{-\pi(\Phi,o)})\big) = \lambda E\big(g(\Phi^o)\nu_d(C_0(\Phi^o))\big). \qquad (4.84)$$

Applying this with $g(\varphi) = \nu_d\big(C_0(\varphi)\big)^\alpha$ for some $\alpha \geq 0$ gives

$$E\big(\nu_d(V)^\alpha\big) = \lambda E\big(\nu_d(C_0(\Phi^o))^{\alpha+1}\big), \qquad (4.85)$$

where $V = C_{\pi(\Phi,o)}(\Phi) = \{x \in \mathbb{R}^d : \pi(\Phi,x) = \pi(\Phi,o)\}$ is the *zero cell* of π. In particular,

$$E\big(\nu_d(C_0(\Phi^o))\big) = \lambda^{-1}, \qquad (4.86)$$

cf. (9.15). In the notation of Chapter 9 it is $C_0(\Phi^o) = C_0$.

If $P\big(0 < \nu_d(C_0(\Phi^o)) < \infty\big) = 1$, then (4.85) extends to all real α, cf. Last (2006).

A stationary partition π is called *balanced* if

$$P\big(\nu_d(C_x(\Phi)) = \lambda^{-1} \text{ for all } x \in \Phi\big) = 1. \qquad (4.87)$$

In the ergodic case such a partition can be used to construct the Palm version Φ^o from Φ by a random shift. By stationarity, the origin $o \in \mathbb{R}^d$ might be interpreted as a uniformly distributed random test point in $\mathbb{R}^d$. This point picks one of the cells $C_x(\Phi)$, $x \in \Phi$, namely $C_T(\Phi)$, where $T := \pi(\Phi,o)$. If π is balanced, then all cells have the same chance of being picked by the origin. Therefore T is the typical point of Φ. Ergodicity suggests that Φ_{-T} and Φ^o should have the same distribution. The following result of Holroyd and Peres (2005) shows in particular that this is indeed true.

Theorem 4.4. *A stationary partition π is balanced if and only if*

$$P(\Phi_{-\pi(\Phi,o)} \in Y) = P_o(Y) \qquad \text{for all } Y \in \mathcal{N}. \qquad (4.88)$$

Proof. If π is balanced, then (4.88) is a direct consequence of (4.84). Assume, conversely, that (4.88) holds. Taking $g(\varphi) := \nu_d\big(C_0(\varphi)\big)$ in (4.84) yields

$$E\big(\nu_d(C_0(\Phi^o))\big) = \lambda E\big(\nu_d(C_0(\Phi^o))^2\big).$$

In view of (4.86) this means that $E\big(\nu_d(C_0(\Phi^o))^2\big) = \lambda^{-2}$. Therefore the variance of $\nu_d\big(C_0(\Phi^o)\big)$ vanishes, implying that $P\big(\nu_d(C_0(\Phi^o)) = \lambda^{-1}\big) = 1$. This is equivalent to (4.87). $\square$

By the ergodic theorem for stationary point processes (see e.g. in Kallenberg, 2002, Corollary 10.19) a balanced partition can only exist if the *sample intensity* $\lim_{n\to\infty} \Phi(W_n)/\nu_d(W_n)$ equals λ almost surely, where $\{W_n\}$ is a convex averaging sequence as in Section 4.1.6. This is the case, for instance, if Φ is ergodic. (In the general case the definition of a balanced partition has to be modified.) It is a quite remarkable fact that this assumption is already enough to guarantee the existence of a balanced partition.

The following construction is due to Holroyd and Peres (2005). Place a small ball around each point of the point process Φ and expand the balls simultaneously at the same rate in all directions. Each expanding ball occupies space that has not yet been previously occupied by other balls; the expansion of a ball stops when the size of its occupied space is λ^{-1}. In this way $\mathbb{R}^d$ is partitioned into regions (cells) of size λ^{-1}, each containing one point of Φ. Now

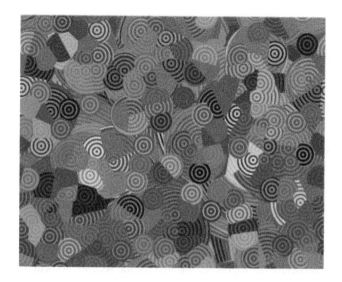

Figure 4.6 A balanced partition for a sample of a homogeneous Poisson process (with periodic boundary conditions). All cells have the same area, but not all cells are connected. The figure shows the generating points with the occupied areas. Concentric circles around the points are used to aid the identification of the cells. Courtesy of A. E. Holroyd.

map each location to the point of its cell. It should be noted that the cells of this partition need not be spherical or convex, they can be even disconnected. Figure 4.6 shows a balanced partition for a sample of a homogeneous Poisson process.

4.5 The second moment measure

Just as the variance is a fundamental parameter of the distribution of a random variable, the second moment measure is an important characteristic of a point process Φ. A simple description is provided in the stationary case by the Palm distribution. The second-order factorial moment measure $\alpha^{(2)}$ can be written as

$$\alpha^{(2)}(B_1 \times B_2) = \lambda^2 \int_{B_1} \mathcal{K}(B_2 - x)\, dx$$

$$= \lambda^2 \int_{\mathbb{R}^d} \int_{\mathbb{R}^d} \mathbf{1}_{B_1}(x)\mathbf{1}_{B_2}(x+h)\, \mathcal{K}(dh)\, dx \qquad \text{for Borel sets } B_1 \text{ and } B_2,$$

$$(4.89)$$

where $\mathcal{K}$ is a measure defined by

$$\lambda \mathcal{K}(B) = \int_{\mathbb{N}} \varphi(B \setminus \{o\})\, P_o(d\varphi)$$

$$= \int \varphi(B)\, P_o^!(d\varphi) \qquad \text{for Borel sets } B. \qquad (4.90)$$

So $\lambda\mathcal{K}(B)$ can be interpreted as the mean number of points in $B \setminus \{o\}$ under the condition that at o there is a point of Φ; $\mathcal{K}$ is called the *reduced second moment measure*. It possesses a symmetry property under reflection:

$$\mathcal{K}(B) = \mathcal{K}(\check{B}) \qquad \text{for Borel sets } B. \qquad (4.91)$$

Formula (4.89) follows from the Campbell–Mecke theorem (4.64):

$$\alpha^{(2)}(B_1 \times B_2) = \mathbf{E}\left(\sum_{x_1,x_2 \in \Phi}^{\neq} \mathbf{1}_{B_1}(x_1)\mathbf{1}_{B_2}(x_2) \right)$$

$$= \int_{\mathsf{N}} \sum_{x \in \varphi} \mathbf{1}_{B_1}(x)\varphi(B_2 \setminus \{x\}) P(\mathrm{d}\varphi)$$

$$= \lambda \int_{\mathbb{R}^d} \int_{\mathsf{N}} \mathbf{1}_{B_1}(x)\varphi\big((B_2 - x) \setminus \{o\}\big) P_o(\mathrm{d}\varphi)\,\mathrm{d}x.$$

From (4.45) one can derive the following formula for the second moment measure:

$$\mu^{(2)}(B_1 \times B_2) = \lambda v_d(B_1 \cap B_2) + \lambda^2 \int \int \mathbf{1}_{B_1}(x)\mathbf{1}_{B_2}(x+h)\,\mathcal{K}(\mathrm{d}h)\,\mathrm{d}x. \qquad (4.92)$$

It shows that the second moments given on $\mathcal{B}^d \times \mathcal{B}^d$ can be described in terms of intensity λ and measure $\mathcal{K}$.

Moreover, it holds the general formula for nonnegative measurable function f:

$$\mathbf{E}\left(\sum_{x_1,x_2 \in \Phi}^{\neq} f(x_1, x_2) \right) = \lambda^2 \int_{\mathbb{R}^d} \int_{\mathbb{R}^d} f(x_1, x_2)\,\mathcal{K}(\mathrm{d}h)\,\mathrm{d}x. \qquad (4.93)$$

According to Section 4.3 the second-order product density $\varrho^{(2)}$, if it exists, also describes $\mu^{(2)}$. So there is also a relationship between $\varrho^{(2)}$ and $\mathcal{K}$, namely

$$\lambda^2 \mathcal{K}(B) = \int_B \varrho^{(2)}(x)\,\mathrm{d}x \qquad \text{for Borel sets } B. \qquad (4.94)$$

In the planar case one can introduce polar coordinates and describe the reduced second moment measure by the function $K(r, \alpha)$,

$$K(r, \alpha) = \mathcal{K}\big(S(r, \alpha)\big) \qquad \text{for } r \geq 0 \text{ and } 0 \leq \alpha \leq \pi,$$

where $S(r, \alpha)$ is the sector of radius r, centred at the origin and given by the angle α.

For fixed r, the ratio

$$K(r, \alpha)/K(r, \pi)$$

is a distribution function for directions. This idea is useful for directional analysis of point processes; see Illian *et al.* (2008, Section 4.5). The present book limits itself to the case of isotropic point processes. Note that it makes sense to consider also 'local anisotropies', for example local tendencies to parallelism.

The description of the second moment measure simplifies in the motion-invariant case (when isotropy is added to stationarity). Then the reduced second moment measure can be replaced by the *reduced second moment function* $K(r)$ which is a function that can be plotted. This function, often called *Ripley's K-function*, is defined by

$$K(r) = \mathcal{K}\big(B(o, r)\big) \qquad \text{for } r \geq 0. \tag{4.95}$$

The quantity $\lambda K(r)$ is the mean number of points of Φ within a ball of radius r centred at the typical point, which is not itself counted.

In the case of a homogeneous Poisson process

$$K(r) = b_d r^d \qquad \text{for } r \geq 0, \tag{4.96}$$

which follows from the Slivnyak–Mecke theorem as

$$\lambda K(r) = \int \varphi\big(B(o, r)\big) \, P_o^!(\mathrm{d}\varphi)$$

$$= \int \varphi\big(B(o, r)\big) \, P(\mathrm{d}\varphi)$$

$$= \mathbf{E}\big(\Phi(B(o, r))\big) = \lambda v_d\big(B(o, r)\big) = \lambda b_d r^d.$$

Asymptotically, for all stationary and isotropic point processes $K(r)$ behaves as $b_d r^d$.

Other functions than $K(r)$ are often used to describe the second-order behaviour of a point process. Which function is to be preferred depends mainly on convenience, but also on statistical considerations and traditions. Some functions originate from the physical literature in which they have been used for a long time, much longer than by statisticians. An example is the pair correlation function, for which an early reference is Ornstein and Zernike (1914).

Four examples of such functions are

product density $\varrho^{(2)}$:

$$\varrho^{(2)}(r) = \lambda^2 \, \frac{\mathrm{d}K(r)}{\mathrm{d}r} \Big/ (d b_d r^{d-1}) \qquad \text{for } r \geq 0, \tag{4.97}$$

pair correlation function g:

$$g(r) = \frac{\varrho^{(2)}(r)}{\lambda^2} \qquad \text{for } r \geq 0, \tag{4.98}$$

radial distribution function RDF:

$$RDF(r) = \lambda \frac{\mathrm{d}K(r)}{\mathrm{d}r} = \frac{d b_d r^{d-1} \varrho^{(2)}(r)}{\lambda} \qquad \text{for } r \geq 0, \tag{4.99}$$

L-function:

$$L(r) = \left(\frac{K(r)}{b_d}\right)^{1/d} \qquad \text{for } r \geq 0. \tag{4.100}$$

In the Poisson process case all these functions satisfy simple formulae:

$$\varrho^{(2)}(r) = \lambda^2, \tag{4.101}$$

$$g(r) \equiv 1, \tag{4.102}$$

$$RDF(r) = db_d r^{d-1} \lambda, \tag{4.103}$$

$$L(r) = r. \tag{4.104}$$

For other models the forms of these functions correspond to various properties of the underlying point process. Maxima of $g(r)$, or values of $K(r)$ larger than $b_d r^d$ for r in specific intervals, indicate frequent occurrences of inter-point distances at such values; likewise, minima of $g(r)$ or low values of $K(r)$ indicate inhibition at these distances. Either way, some form of inner order or clustering in the point pattern may be responsible. (For this purpose, $g(r)$ is the more suitable tool, while the cumulative nature of $K(r)$ or $L(r)$ makes the interpretation sometimes difficult.) Model identification may be suggested by comparison of empirical pair correlation functions, or reduced second moment functions, with their theoretical versions. Figure 4.7 exhibits pair correlation functions for four different spatial point process models.

The possible forms of pair correlation functions and their interpretation are discussed in of Illian *et al.* (2008, Section 4.3.4).

The K-function (as well as $g(r)$) does *not* represent all distributional information about a stationary isotropic point process. Baddeley and Silverman (1984) gave an example of a planar point process which is quite different from a Poisson process and yet has the same K-function:

$$K(r) = \pi r^2.$$

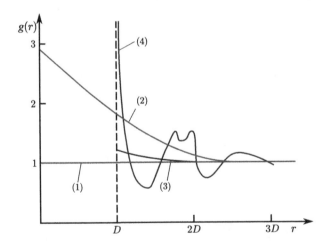

Figure 4.7 Pair correlation functions g for various point process models: (1) Poisson process; (2) Matérn cluster process with parameters $R = 2.5$ and $\lambda = 0.027, \mu = 5$ (the values of $g(r)$ greater than 1 for small r result from clustering); (3) Matérn hard-core process with parameters $D = 2$ and $p_h \simeq 0.03$ (the hard-core distance is 2); (4) random close packing of balls of diameter D. The peaks occur in a typical pattern at D, at around $\sqrt{3}D$ and below $2D$.

It is simple to give also examples of pairs of other processes with equal $K(r)$. For example, Stoyan (1991) gave an example, where one of the processes is even anisotropic; there $K(r)$ is defined as $\mathscr{K}(B(o, r))$.

If $\mathscr{K}$, K or g is given, then (at least in principle) all variances and covariances of point-counts can be calculated. In particular, if B is a bounded Borel set, then

$$\mathbf{var}\big(\Phi(B)\big) = \lambda^2 \int \gamma_B(h)\, \mathscr{K}(\mathrm{d}h) + \lambda \nu_d(B) - \big(\lambda \nu_d(B)\big)^2, \qquad (4.105)$$

and in the isotropic case

$$\mathbf{var}\big(\Phi(B)\big) = \lambda^2 \int_0^\infty \overline{\gamma}_B(r)\, \mathrm{d}K(r) + \lambda \nu_d(B) - \big(\lambda \nu_d(B)\big)^2, \qquad (4.106)$$

where $\gamma_B(h)$ is the set covariance of B.

If B is large ('spherically infinite' in the terminology of Girling, 1982), then

$$\frac{\mathbf{var}\big(\Phi(B)\big)}{\nu_2(B)} \simeq \lambda + 2\pi\lambda^2 \int_0^\infty \big(g(r) - 1\big) r\, \mathrm{d}r \qquad (d = 2),$$

and

$$\frac{\mathbf{var}\big(\Phi(B)\big)}{\nu_3(B)} \simeq \lambda + 4\pi\lambda^2 \int_0^\infty \big(g(r) - 1\big) r^2\, \mathrm{d}r \qquad (d = 3).$$

For point processes with a tendency to regularity these formulae tend to produce approximations somewhat too large, while approximations for cluster processes tend to be too small.

In the case of marked point processes there exist similar characteristics; see Illian *et al.* (2008, Section 5.3). They are of a different nature in the cases of discrete and continuous marks.

4.6 Summary characteristics

It would not be surprising if one feels that the distribution P of a point process is a rather complex mathematical thing and much too complicated for intuitive understanding or for visualisation. Therefore various summary characteristics have been proposed which describe certain aspects of P. These are typically real numbers or functions, usually of distance r as they are based on inter-point distances.

The following tries to give the reader some orientation for the case of stationary point processes. For the case of nonstationary and of marked point processes the reader is referred to Illian *et al.* (2008).

It is recommended that one uses some summary characteristics in parallel since each of them gives only specific information. A single summary characteristic can hardly completely describe all interesting aspects of a point process distribution.

The correct interpretation of summary characteristics is a question of experience and somewhat of an art.

Intensity λ

The intensity gives very important global information about the point process. It is another term for 'point density'.

Of course, λ alone is of little value, similarly as the mean for a random variable in classical statistics. And λ influences the other summary characteristics, though some of them are constructed with the aim of eliminating the influence of point density.

Second-order characteristics

Second-order characteristics are for many scientists the favourite summary characteristics, in particular for physicists, who have used the pair correlation function since the beginning of the 20th century and very rarely consider other second-order characteristics and non-second-order summary characteristics.

Second-order characteristics give information on many scales of distances, describe so-to-say the average behaviour of the point process of interest.

The K-function is only rarely used in practical statistics and model visualisation. The L-function is more popular because of its simpler graphical form and the better properties in the statistical context. It is useful in statistical tests. However, on the other hand its interpretation is a bit difficult because of its cumulative character. Frequent inter-point distances in a distance interval (r_1, r_2) lead to large values of $L(r)$ not only in the interval (r_1, r_2), but also for r-values larger than r_2.

The pair correlation function $g(r)$ does not have the 'cumulative' disadvantage just mentioned. Hence it is ideal for the interpretation and understanding of point process distribution; but it would be wrong to assume that it contains more information than $K(r)$ or $L(r)$. However, its statistical estimation is a bit difficult since a bandwidth must be carefully chosen. (This is similar to the case of density function estimation in classical statistics.) Moreover, the interpretation needs experience or some support; see Illian *et al.* (2008).

All three characteristics, $K(r)$, $L(r)$ and $g(r)$, are normalised by division of λ^2. Nevertheless, it should not be said that they are independent of intensity λ. Still intensity influences them, for example via the hard-core distance, which may be closely related to λ.

Nearest-neighbour distance distribution function $D(r)$

The nearest-neighbour distance distribution function $D(r)$ is a rather natural characteristic. Frequently it is used spontaneously as the first functional summary characteristic. It is easy to understand, and therefore is used in Section 4.4 to explain Palm characteristics.

For lattice-like patterns $D(r)$ yields the lattice-distance distribution. For processes with clusters it gives information on the distances of the points within clusters, but hardly on distances between clusters. The distribution function is said to be 'short-sighted', since it gives information only about the nearest neighbour; what happens further away does not play a rôle. This is different for second-order characteristics.

Spherical contact distribution function $H_s(r)$

Also the spherical contact distribution function is quite natural. It goes also under the name 'empty space function', with the colourless symbol $F(r)$. Now the distance to the nearest neighbour from a random test point is considered. So it is clear that $H_s(r)$ is also short-sighted. In contrast to $D(r)$ it is valuable for describing the extent of empty space between clusters, but of little value for describing the situation within clusters.

4.7 Introduction to statistics for stationary spatial point processes

4.7.1 General remarks

The theory of statistics for point processes comprises a part of spatial statistics as described by Ripley (1981), Cressie (1993) and Gelfand *et al.* (2010). This section presents some ideas of exploratory analysis for stationary spatial point processes; model-based statistics will play some rôle in Chapter 5. All is limited to the case of stationary point processes. The basic ideas of point process statistics apply also to other structures of stochastic geometry, in particular for line and fibre processes. The general aim of this section is to describe methods to obtain unbiased estimators or, at least, *ratio-unbiased* estimators (quotients where numerator and denominator are unbiased) of important summary characteristics. It is always assumed the Cartesian coordinates of all points are given. For other data situations see Illian *et al.* (2008).

Point process statistics in this book are considered for the planar $d = 2$ and spatial $d = 3$ case. The statistics of point processes on the real line ($d = 1$, often described as series of events) have a relatively large literature; see Cox and Lewis (1966), Brillinger (1975, 1978), Snyder (1975), Karr (1991) and Andersen *et al.* (1993). The special nature of the real line makes this case more amenable.

Books providing more details for $d \geq 2$ are Diggle (1983, 2003), Ripley (1981, 1988), Stoyan and Stoyan (1994), Martínez and Saar (2002), Møller and Waagepetersen (2004), Illian *et al.* (2008) and Gelfand *et al.* (2010).

Often statistical analysis of a stationary point process depends on observation of *one sample* only, and that via a bounded sampling window W. Patterns arising in astronomy, ecology, geography and geology are often truly unique samples of a stochastic phenomenon. In other cases data collection is so complicated that only one sample is collected. Typically, in such cases it is assumed that the observed patterns are samples of *stationary ergodic* point processes, an assumption not susceptible to statistical analysis if there is only one sample, but one that is necessary if any statistical analysis is sensible. Therefore ergodicity is also assumed throughout this section. In practice, either it is plausible from the very nature of the data, or else one must proceed on an *ad hoc* basis as if the assumption were true, and subject one's conclusions to the proviso that while they may be of possible value in the nonstationary case, they will not then have the same interpretation. For example, an empirical point density (the mean number of points per unit area) can still be calculated even if the observed point pattern is nonstationary. It has value as a description of the spatial average behaviour of the pattern but does not possess all the properties of an estimator of a stationary point process intensity. See the interesting discussion in Matheron (1989), who justified this approach.

4.7.2 Edge-corrections

An ever present problem of spatial statistics is that of *edge-effects*; Ripley (1982) and Illian *et al.* (2008) discuss these in detail. The problem intervenes in point process statistics in the estimation of g, $\mathcal{K}$, K, H_s and D. As an illustration, consider the problem of estimating the nearest-neighbour distribution function $D(r)$. Two naïve methods suggest themselves: plus-sampling and minus-sampling named according as to whether they require more, or less, information than is given through the window W. For rectangular W the method of periodic boundary conditions is also feasible.

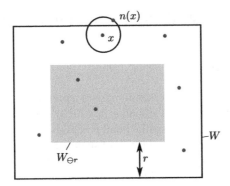

Figure 4.8 The window W and the eroded window $W_{\ominus r}$ for minus-sampling of $D(r)$. The shaded area is the eroded window. The nearest neighbour $n(x)$ of the point x is not in the window and so cannot be used for estimating of $D(r)$ by minus-sampling. Ignoring this edge problem will introduce bias.

Plus-sampling

In this method $D(r)$ is estimated by the empirical distribution function of the sample of all distances $\|x - n(x)\|$, where x runs through the points of the pattern in the whole window W, and $n(x)$ is the nearest neighbour of x, *no matter whether this neighbour is in the sampling window W or not*. If x is near the boundary of W, then its true nearest neighbour may lie outside W. Therefore, plus-sampling may require more information than is contained in the window W. When the point pattern can only be observed within W, this means that for some x it may not be possible to determine $n(x)$. Figure 4.8 gives an illustration. If one simply ignores this edge problem and uses the nearest neighbour to x *lying within* W, then bias is introduced; plus-sampling itself is unbiased.

Tscheschel and Chiu (2008) introduced the so-called *quasi-plus-sampling*. The idea is to reconstruct the point pattern (see Section 6.7) in a larger region containing W and then apply plus-sampling as if the reconstructed pattern were observed.

Minus-sampling

In estimating $D(r)$, one might use only the points x in $W_{\ominus r}$ in creating the sample of distances $\|x - n(x)\|$; see Figure 4.8. Clearly, for these points the nearest neighbours, if within distance r, lie in W. Thus, for each x in the eroded window, one can correctly determine whether its nearest-neighbour distance is larger than r or not. This 'border method' will avoid bias but loses much information for large r.

Periodic edge-correction

If the window W is a rectangle or a parallelepiped, the given point patterns can be continued outside of W, as shown in Figure 4.9 for the planar case. The resulting point pattern can then be analyzed by means of plus-sampling. To do this, the distances have to be redefined. In the planar case this is termed as 'torus metric': if W is the rectangle with side lengths a and b with

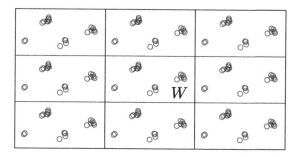

Figure 4.9   A point pattern in a rectangular window W and its periodic continuation.

left lower vertex at the origin, the distance between two points x and $y \in W$ with $x = (\xi_1, \xi_2)$ and $y = (\eta_1, \eta_2)$ is

$$\|x - y\| = \sqrt{(\min\{|\xi_1 - \eta_1|, a - |\xi_1 - \eta_1|\})^2 + (\min\{|\xi_2 - \eta_2|, b - |\xi_2 - \eta_2|\})^2}. \quad (4.107)$$

By this method every point in W gets a natural nearest neighbour, but in some cases they are incorrect. It is difficult to determine the errors of this method; strange and spurious point configurations may appear along the borders, and the method is merely a cheap trick to provide more points.

Clearly, this method is absolutely suitable when the pattern in W is itself a result of a simulation under 'periodic boundary conditions'.

These three edge-correction methods are applied in many cases and also outside of point process statistics. However, for particular summary characteristics there exist adapted, better edge-correction methods, which will be demonstrated in the following.

4.7.3   Estimation of the intensity λ

The natural and best estimator of the intensity λ is $\hat{\lambda}$ defined by

$$\hat{\lambda} = \frac{\Phi(W)}{v_d(W)}. \quad (4.108)$$

It uses the idea that the intensity is the mean number of points per volume unit. Straightforward calculation shows that it is unbiased.

If Φ is ergodic, then $\hat{\lambda}$ is strongly consistent, in the sense that $\hat{\lambda} \to \lambda$ almost surely as the window size increases as a convex averaging sequence.

The variance of $\hat{\lambda}$ depends both on the window's shape and size and on the second moment measure of the point process. In the isotropic case (4.106) implies

$$\mathbf{var}(\hat{\lambda}) = \left(d b_d \lambda^2 \int_0^\infty r^{d-1} \overline{\gamma}_W(r) g(r) \, dr + \lambda v_d(W) - \left(\lambda v_d(W)\right)^2 \right) \Big/ \left(v_d(W)\right)^2, \quad (4.109)$$

where $\overline{\gamma}_W(r)$ is the isotropised set covariance of the convex compact window W.

4.7.4 Estimation of the reduced second moment measure

Estimation of $\mathcal{K}(B)$ — anisotropic case

In the general anisotropic case $\lambda^2\mathcal{K}(B)$ can be estimated by $\kappa_{\mathrm{st}}(B)$, where

$$\kappa_{\mathrm{st}}(B) = \sum_{x,\,y\in\Phi\cap W}^{\neq} \frac{\mathbf{1}_B(y-x)}{v_d(W_x\cap W_y)} \tag{4.110}$$

defined for bounded Borel sets B such that $v_d(W\cap W_z)$ is positive for all z in B.

Proof of the unbiasedness of $\kappa_{\mathrm{st}}(B)$ (Ohser and Stoyan, 1981). Set $f(x_1,x_2)=\mathbf{1}_B(x_2-x_1)\mathbf{1}_W(x_1)\mathbf{1}_W(x_2)/v_2(W_{x_1}\cap W_{x_2})$ and apply (4.43) and (4.89):

$$\mathbf{E}\big(\kappa_{\mathrm{st}}(B)\big) = \int f(x_1,x_2)\,\alpha^{(2)}\big(\mathrm{d}(x_1,x_2)\big)$$

$$= \lambda^2 \int \int f(x,x+h)\,\mathrm{d}x\,\mathcal{K}(\mathrm{d}h)$$

$$= \lambda^2 \int \mathbf{1}_B(h)\,\mathcal{K}(\mathrm{d}h) = \lambda^2\mathcal{K}(B). \qquad \square$$

Estimation of $\mathcal{K}(B)$ and $K(r)$ — isotropic case

An unbiased estimator of $\lambda^2 K(r)$ is $\kappa_{\mathrm{iso}}(r)$, defined by

$$\kappa_{\mathrm{iso}}(r) = \sum_{x,\,y\in\Phi\cap W} \frac{\mathbf{1}(0<\|x-y\|\le r)k(x,y)}{v_d(W^{(\|x-y\|)})} \qquad \text{for } 0\le r<r^*, \tag{4.111}$$

where

$$r^* = \sup\{r : v_d(W^{(r)})>0\} \qquad \text{and} \qquad W^{(r)} = \{x\in W : \partial(B(x,r))\cap W\neq\varnothing\},$$

and $k(x,y)$ is the inverse of the fraction of surface area in W of the sphere of radius $\|x-y\|$ centred at x. In the planar case it is $k(x,y)=2\pi/\alpha_{xy}$, where α_{xy} is the sum of all angles of the arcs in W of a circle centred at x with radius $\|x-y\|$. Figure 4.10 illustrates the definition of α_{xy}. If $\alpha_{xy}=0$, then $k(x,y)=0$.

The estimator $\kappa_{\mathrm{iso}}(r)$ was originally suggested by Ripley (1976); Formula (4.111) uses the modification due to Ohser (1983). The modification extends estimation to the case of large r,

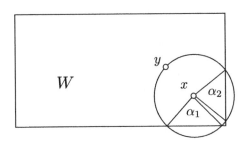

Figure 4.10 Definition of α_{xy} for estimation of $\kappa_{\mathrm{iso}}(r)$. In the example $\alpha_{xy}=2\pi-\alpha_1-\alpha_2$.

while Ripley's original form was unbiased only for

$$r \leq r_0 = \inf\{t : v_2(W^{(t)}) < v_2(W)\}.$$

These and other estimators are discussed in Illian *et al.* (2008). As it seems it makes sense to use always $\kappa_{st}(B)$ with $B = B(o, r)$, since in cases where the isotropy assumption is invalid, $\kappa_{iso}(r)$ may behave badly.

Both estimators have edge-correction built in. For example, the naïve estimator corresponding to (4.111) is

$$\hat{k}(r) = \frac{\text{number of point pairs in } W \text{ separated by distance less than } r}{v_d(W)} \qquad (4.112)$$

and for small r (such that $W = W^{(r)}$) the edge-bias is corrected by a weighting factor $k(x, y)$,

$$\kappa_{iso}(r) = \frac{1}{v_2(W)} \sum_{x,y \in \Phi \cap W} \mathbf{1}(0 < \|x - y\| \leq r)k(x, y).$$

This is the form originally suggested by Ripley (1976). The principle is to correct the bias, by setting the weight $k(x, y)$ greater than 1 whenever one or both of the points x and y are near the boundary of W. Deep in W (when $B(x, \|x - y\|)$ and $B(y, \|x - y\|)$ lie completely in W) the weight is 1. The isotropy assumption is used to estimate the mean number of pairs missed out by the window at a particular location, and thus to calculate the weight. The full form of (4.111) makes a further correction which is necessary for large r.

Stoyan and Stoyan (2000) found that for estimating $K(r)$ the naïve approach of dividing estimators for $\lambda^2 K(r)$ by the squared $\hat{\lambda}$ of (4.108) can be improved by adapted intensity estimators which depend on r; see also Illian *et al.* (2008, p. 231).

Estimation of other second-order characteristics

For statistical purposes it is useful to stabilise variances of $\hat{K}(r)$. This can be done by using the L-function, estimating $L(r)$ by

$$\hat{L}(r) = \left(\frac{\hat{K}(r)}{b_d}\right)^{1/d}, \qquad (4.113)$$

where $\hat{K}(r)$ is either $\kappa_{st}(r)/\hat{\lambda}^2$ or $\kappa_{iso}(r)/\hat{\lambda}^2$. Examples 2.1 and 2.3 illustrate the use of $L(r)$.

The product density $\varrho^{(2)}(r)$ and the pair correlation function $g(r)$ can be estimated by using an edge-corrected kernel estimator such as

$$\hat{\varrho}^{(2)}(r) = \sum_{x,y \in \Phi \cap W} \frac{\mathbf{k}(\|x - y\| - r)}{db_d r^{d-1} v_d(W_x \cap W_y)}, \qquad (4.114)$$

where $\mathbf{k}$ is a kernel function; see Illian *et al.* (2008, pp. 230–1). An estimator of $g(r)$ is then

$$\hat{g}(r) = \hat{\varrho}^{(2)}(r)/\hat{\lambda}^2. \qquad (4.115)$$

Edge-corrections for the estimation of mark correlation functions are analogous and discussed in Illian *et al.* (2008, Section 5.3.4).

4.7.5 Estimation of the spherical contact distribution and of the probability generating functional

The methods of estimating $H_s(r)$ for random closed sets (see Section 6.4.5) can be also used in the particular case of point processes. Since

$$H_s(r) = \mathbf{P}(o \in \Phi_{\oplus r}) \qquad \text{for } r \geq 0, \tag{4.116}$$

$H_s(r)$ is equal to the area fraction of the random closed set

$$\Xi = \Phi_{\oplus r} = \bigcup_{x \in \Phi} B(x, r). \tag{4.117}$$

An unbiased estimator is given by

$$\hat{H}_s(r) = \frac{v_d\left(W_{\ominus r} \cap \bigcup_{x \in \Phi} B(x, r)\right)}{v_d(W_{\ominus r})} \qquad \text{for } 0 \leq r \leq \frac{\text{diam}(W)}{2}. \tag{4.118}$$

This is edge-correction by minus-sampling; see Figure 4.11.

Note that $\hat{H}_s(r)$ may be neither continuous nor monotonic, while the distribution function $H_s(r)$ always has a density function. Section 6.4.5 will present more sophisticated estimators which are free of these disadvantages. Their construction follows an idea of Hanisch, to be explained in the next section.

Estimation of the probability generating functional G follows a similar route. For example, take v in $\mathbf{V}$ with $v(x) = 1$ if x is not in $B(o, r)$. (In this case $G(v)$ is the same $1 - H_s(r)$, only in another expression.) Then

$$\hat{G}(v) = \frac{1}{n} \sum_{i=1}^{n} \prod_{x \in \Phi \cap W} v(x - z_i) \tag{4.119}$$

is an unbiased estimator of $G(v)$ if the points $z_1, \ldots, z_n$ form a grid in $W_{\ominus r}$. Again this estimator is of the minus-sampling type.

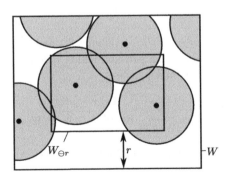

Figure 4.11 Edge-effects on the estimation of $H_s(r)$. Points outside of W contribute to $v_2(W \cap \Phi_{\oplus r})$. If their positions are not known, then only points falling in the eroded window $W_{\ominus r}$ should be used for unbiased estimation.

4.7.6 Estimation of the nearest-neighbour distance distribution function

Recall that the nearest-neighbour distribution function is formally defined for a stationary point process as

$$D(r) = 1 - P_o^! (\{\varphi \in \mathbb{N} : \varphi(B(o,r)) = 0\}) \qquad \text{for } r \geq 0. \tag{4.120}$$

An equivalent formulation uses the marked point process obtained by marking each point x of the process Φ with the distance $d(x)$ from x to its nearest neighbour $n(x)$. The resulting marked point process Ψ inherits the stationarity property from Φ. As can be seen from the mark distribution approach to Palm distributions on p. 129, the mark distribution function for Ψ is precisely $D(r)$. This formulation via Ψ clarifies the logic behind the estimators of $D(r)$. The following demonstrates two ways to handle the problems of edge-effects mentioned in Section 4.7.2.

The first estimator is in the spirit of minus-sampling. The so-called border estimator $\hat{D}_b(r)$ was introduced by Ripley (1977). It is defined as

$$\hat{D}_b(r) = \sum_{[x;d(x)]} \frac{\mathbf{1}_{W_{\ominus r}}(x)\mathbf{1}(0 < d(x) \leq r)}{\Phi(W_{\ominus r})} \qquad \text{for } r \geq 0. \tag{4.121}$$

The idea behind this estimator is quite simple: when estimating $D(r)$, only those points x in the window W are considered which have a distance larger than r from the window's boundary. For these points it is thus clear that there are no points closer than r outside the window and whether or not the nearest-neighbour distance $d(x)$ is not larger than r can be determined within W. These points lie in $W_{\ominus r}$. Since $\Phi(W_{\ominus r})$ is the number of those points, $\hat{D}_b(r)$ is a natural ratio estimator for $D(r)$. Because numerator and denominator are random, the estimator is not unbiased, only ratio-unbiased. However, if Φ is ergodic, $\hat{D}_b(r)$ is asymptotically unbiased.

Unfortunately, the estimator $\hat{D}_b(r)$ has some disadvantages. It is not necessarily monotonically increasing in r, it can exceed 1 in value and it excludes many points from the estimation procedure.

For better understanding of the next estimator, rewrite $\hat{D}_b(r)$ as follows. The quantity

$$\hat{\mathcal{D}}_b(r) = \sum_{[x;d(x)]} \frac{\mathbf{1}_{W_{\ominus r}}(x)\mathbf{1}(0 < d(x) \leq r)}{v_d(W_{\ominus r})} \qquad \text{for } r \geq 0 \tag{4.122}$$

is clearly an unbiased estimator of $\lambda D(r)$, which can easily be shown by the Campbell theorem. Division by the adapted intensity estimator

$$\hat{\lambda}(r) = \Phi(W_{\ominus r})/v_d(W_{\ominus r}), \tag{4.123}$$

yields $\hat{D}_b(r)$. The use of $\hat{\lambda}(r)$ reduces the estimation variance (while the use of the classical $\hat{\lambda}$ results in intolerable biases and large mean squared errors), but it leads to the non-monotonicity mentioned above.

Hanisch (1984c) suggested an unbiased estimator which outperforms the border estimator. It uses the so-called nearest-neighbour edge-correction (see Illian et al., 2008, p. 187), which leads to the estimator

$$\hat{D}_{nn}(r) = \hat{\mathcal{D}}_{nn}(r) / \hat{\lambda}_{nn}(r) \qquad \text{for } 0 \leq r \leq R, \tag{4.124}$$

where

$$\hat{D}_{nn}(r) = \sum_{[x;d(x)]} \frac{\mathbf{1}_{W_{\ominus d(x)}}(x)\mathbf{1}(0 < d(x) \le r)}{\nu_d(W_{\ominus d(x)})} \tag{4.125}$$

with

$$\hat{\lambda}_{nn} = \sum_{[x;d(x)]} \frac{\mathbf{1}_{W_{\ominus d(x)}}(x)}{\nu_d(W_{\ominus d(x)})} \tag{4.126}$$

and

$$R = \sup\{r > 0 : \nu_d(W_{\ominus r}) > 0\}. \tag{4.127}$$

The indicator $\mathbf{1}_{W_{\ominus d(x)}}(x)$ can be rewritten as $\mathbf{1}(d(x) < e(x))$, where $e(x)$ is the distance of x to the boundary of W.

The principle underlying $\hat{D}_{nn}(r)$ is simple: to use precisely those points x for which it is known that their nearest neighbours $n(x)$ are both within W and closer than r to x.

4.7.7 Estimation of Palm characteristics and mark distributions

The estimation procedures used for $K(r)$ and $D(r)$ above can be considered as prototypes for estimation of quantities related to Palm or mark distributions. The estimators given for these cases are complicated because of necessary edge-correction. If no edge-correction is necessary, all is much simpler.

Assume that the marks m of the points are given together with the points, as for example in the case where the points are tree positions and the marks their stem diameters. Then the mark distribution M can be estimated by

$$\widehat{\lambda M}(L) = \sum_{[x;m]\in\Psi} \frac{\mathbf{1}_W(x)\mathbf{1}_L(m)}{\nu_d(W)} \qquad \text{for } L \in \mathcal{M}, \tag{4.128}$$

and

$$\hat{M}(L) = \frac{\widehat{\lambda M}(L)}{\hat{\lambda}}. \tag{4.129}$$

For nonnegative marks, the mark distribution function $F_M(t)$ can be estimated by

$$\hat{F}_M(t) = \sum_{[x;m]\in\Psi} \frac{\mathbf{1}_W(x)\mathbf{1}(0 \le m \le t)}{\Phi(W)} \qquad \text{for } t \ge 0. \tag{4.130}$$

If the marks are 'constructed' marks (such as $d(x)$) that have to be determined from the given configuration of points, then edge-corrections are often necessary, namely when information from outside the window is required to construct the marks of some points, as in the cases of $D(r)$ and $K(r)$ considered above. If such information is indeed accessible, then possibly the *plus-sampling method* will be useful. Consider for example estimation of

$$\lambda P_f = \lambda E_o^!\big(f(\Psi)\big), \tag{4.131}$$

where f is a nonnegative measurable function on $\mathbb{N}$,

$$P_f = \int f(\varphi) P_o^!(\mathrm{d}\varphi).$$ (4.132)

Then

$$\widehat{\lambda P_f} = \sum_{x \in \Phi} \frac{\mathbf{1}_W(x) f(\Phi_{-x} \setminus \{o\})}{v_d(W)}$$ (4.133)

is unbiased for λP_f. It is natural to estimate P_f by $\widehat{\lambda P_f}/\hat{\lambda}$, but this estimator is only ratio-unbiased.

On the other hand, if only the information in W is available, then a universal method is *minus-sampling*. For example, suppose f in (4.131) is such that

$$f(\varphi) = f(\varphi \cap B)$$

for some set B (often but not necessarily $B = B(o, r)$ for some r), so that one only needs the information local to B. Then an unbiased estimator of λP_f is

$$\widehat{\lambda P_f} = \sum_{[x; m] \in \Psi} \frac{\mathbf{1}_{W \ominus \check{B}}(x) f(\Phi_{-x} \setminus \{o\})}{v_d(W \ominus \check{B})}.$$ (4.134)

Calculation of $v_2(W \ominus \check{B})$ is described in Matheron (1978) and Weil (1982b); see also Saxl (1989, p. 157).

An analogous procedure for estimation of $M(L)$ can be carried out if the mark m of point x can be determined when all points of Φ in B_x are known.

4.7.8   Parameter estimation

Maximum likelihood method

Maximum likelihood methods are widely used in classical statistics, and many statisticians believe that they should also be preferred in point process statistics. Readers familiar with estimation methods in classical statistics know that maximum likelihood estimation techniques can only be applied if the likelihood function – describing the probability (or probability density) of observing the data given the model – is known. This likelihood is maximised (with fixed data and variable parameters), yielding parameter estimators that best fit the data. However, often and particularly for stationary point processes, it is extremely difficult to find the likelihood function. Therefore, the maximum likelihood method can only be applied to specific classes of models. These are Poisson processes, Cox processes and finite Gibbs processes; see Diggle (2003), Møller and Waagepetersen (2004) and Gelfand *et al.* (2010).

It is possible to (heuristically) approximate likelihood functions when they are not known explicitly. One of these methods is the pseudo-likelihood method introduced by Besag (1975, 1978); see Møller and Waagepetersen (2004) and Gelfand *et al.* (2010). Another method, introduced by Tanaka *et al.* (2008), uses the pair correlation function (for which a formula must be known) and the constructed point process of difference points

$$\delta = x - y \qquad \text{for } x \neq y,$$ (4.135)

where x and y are points of the observed pattern (see also Illian *et al.*, 2008, pp. 449–50).

Method of moments

The method of moments has many applications in the context of spatial point processes, especially when the likelihood function is not available.

Note that the term 'method of moments' is used here somewhat loosely. The approaches described below are all based on the same general idea, which is applied to moments or moment-measure-related characteristics as well as to other summary characteristics that are not moment-related. The general idea is to find parameters that minimise the difference between a 'suitable' summary characteristic S that is known analytically (or from simulations) and its unbiased estimator $\hat{S}$ obtained from the data. It is important that S depends on the unknown parameter θ; to emphasise this dependence, the characteristic is denoted by S_θ. The value θ for which S_θ and $\hat{S}$ are 'as similar as possible' is used as an estimator. The term 'as similar as possible' means here 'similar in the sense of a specific approximation method', such as the least-squares approach.

As indicated by the vague expression 'suitable', different summary characteristics S may be used. Which of these is deemed suitable depends on the context. A first criterion for the choice of summary characteristics is often whether a formula for S_θ is known. However, if this is not the case, then simulation approaches may be used instead. Another criterion should be that S_θ is sensitive to variation in θ.

In many applications the method of moments is based on a functional summary characteristic, that is, the S_θ above is a function $S_\theta(r)$. Sometimes it is sufficient to simply plot $\hat{S}(r)$ and to identify specific points, for example cusp points or points where $S_\theta(r)$ becomes constant. This approach may be used to find estimators of particular distances such as the range of correlation. In the context of Neyman–Scott processes the radius of the clusters may be found in this way.

Typically, however, the method of moments for functional summary characteristics applies a least squares approach, which is a special case of the *minimum contrast method*. This is based on the simple idea of minimising

$$\Delta(\theta) = \int_{s_1}^{s_2} |\hat{S}(r) - S_\theta(r)|^\beta dr \qquad (4.136)$$

with respect to θ. The value of θ that minimises $\Delta(\theta)$ is the minimum contrast estimator $\hat{\theta}$.

Here, the values of s_1, s_2 and β as well as the summary characteristic $S_\theta(r)$ itself can in principle be chosen arbitrarily; often $\beta = 2$ (i.e., least squares) is used. In the literature most applications use either

$$S_\theta(r) = g(r), \qquad S_\theta(r) = H_s(r), \qquad \text{or} \qquad S_\theta(r) = D(r).$$

For more details see Illian *et al.* (2008) and Gelfand *et al.* (2010). Jolivet (1986) and Heinrich (1992a, 1993) investigated the statistical properties of $\hat{\theta}$ for special cases. In reasonable cases the minimum contrast estimator is strongly consistent, that is, it converges almost surely to the true value as the size of the window W increases, as a convex averaging sequence, to the whole space.

4.7.9 Representative windows, representative volume elements

An important question in spatial statistics in general is the issue of choosing an appropriate window size that ensures pre-defined statistical precision requirements. In this context the term *representative volume element* (RVE) is used.

Of course the RVE depends both on the variability of the point process of interest and on the summary characteristic considered. If the same summary statistic is considered, larger RVEs are necessary when cluster point processes are analysed as opposed to regular processes, where smaller RVEs are sufficient. On the other hand, for a precise estimation of the pair correlation function a larger RVE is necessary than for intensity estimation.

In classical statistics, sample size calculations require some prior knowledge on the nature of the data that will be analysed, such as their variation. In the context of spatial point processes this is similar. It is impossible to determine the RVE without any *a priori* knowledge of the distribution of the point process investigated. A straightforward approach to acquiring *a prior* knowledge is a *pilot study* consisting of a preliminary statistical analysis of a small window or a small number of windows if a series of windows has to be analysed. The expectation is that the pilot study yields a useful yet rough estimate of the intensity λ and fundamental information on the point processes type, that is, whether it is a regular or a clustered pattern. Based on this, estimates of RVEs can be obtained. This is sketched in Illian *et al.* (2008, Section 4.8.2) for estimation of intensity λ and pair correlation function $g(r)$. For the other summary characteristics similar methods can be used; see also Section 6.4.6.

4.7.10   Hypotheses testing

Statistical tests are considered as an important part of point processes statistics. In most cases these are goodness-of-fit tests, typically for proving that a model developed fits the data considered. As Baddeley (2010, p. 361) says: 'A general weakness of goodness-of-fit tests is that the alternative hypothesis is very broad (embracing all point processes other than the model specified in the null hypothesis), so that rejection of the null hypothesis is rather uninformative, and acceptance of the null hypothesis is unconvincing because of weak power against specific alternatives.'

Nevertheless, such tests are standard. It is typical for point process statistics that the distributions of the test characteristics are too complicated to derive analogues for the classical *t*-, χ^2- and *F*-tests. Therefore it is expedient to use *parametric bootstrap tests*, which are applied in point process statistics since Ripley (1977) and Besag and Diggle (1977). For an exposition of the theory of parametric bootstrap tests, see for example Davison and Hinkley (1997).

The following discusses a simple variant of goodness-of fit tests, so-called *deviation tests*, which have nature similar to that of the Kolmogorov–Smirnov and Cramér–von Mises tests.

A cursory description of a typical test runs as follows. Suppose the hypothesis to be considered is that a given point pattern φ, observed through a window W, is a sample of a point process Φ. One chooses a test statistic τ, such as

$$\tau = \max_{r_1 \leq r \leq r_2} |D(r) - \hat{D}(r)| \tag{4.137}$$

or

$$\tau = \int_{r_1}^{r_2} \left(D(r) - \hat{D}(r)\right)^2 dr. \tag{4.138}$$

Here $D(r)$ is the theoretical model distribution function of nearest-neighbour distance and $\hat{D}(r)$ the corresponding empirical function. The deviation measure with L^∞-norm is of Kolmogorov–Smirnov type and that with the L^2-norm is of Cramér–von Mises type. The choice of which one to use may be a matter of taste, but *a priori* knowledge may help; see Section 3.4.2.

One calculates the value τ of the deviation measure for the given sample $\varphi \in W$, denoted by τ_{emp}. This value is compared with analogous τ-values obtained by simulating the specified model k times, always observing it through the window W.

If τ_{emp} takes an extreme position in the series of ordered τ-values, then the hypothesis may be cast in doubt. It is then possible to perform a test of the hypothesis at a significance level α. If no estimated parameter is used in the simulation of the specified model, the significance level is *exact*. One rejects the null hypothesis if τ_{emp}, when pooled together with all τ-values, has a rank larger than or equal to $(1 - \alpha)(k + 1)$. The p-value of the deviation test can be approximated by Formula (2.64). Loosmore and Ford (2006) discussed the variation of $\hat{p}$ as a function of k and suggested $k = 999$ for a good approximation of the p-value. However, for $\alpha = 0.05$, even using only $k = 99$ will not cause a serious loss of power (Davison and Hinkley, 1997, p. 156). For $\alpha = 0.05$ and $k = 999$ the critical ranks are 950 or above, for $\alpha = 0.01$ and $k = 999$ they are 990 or above.

Summary characteristics commonly chosen for constructing deviation measures are $L(r)$, $D(r)$ and $H_s(r)$, playing the rôle of $D(r)$ above. Often the theoretical functions for the specified model must be determined by simulation. Diggle (2003, p. 89) recommended using the k simulated functions and then averaging. In accordance with common practice in classical statistics the density functions $g(r)$, $d(r)$ and $h_s(r)$ are usually not used for testing, because the estimation of $L(r)$, $D(r)$ and $H_s(r)$ is more standardised than that of the pair correlation function $g(r)$ and of the density functions $d(r)$ and $h_s(r)$. A goodness-of-fit test should be based on a different summary characteristic from the one used in estimating the model parameters, for example, by the minimum contrast method.

Deviation tests for goodness-of-fit hypotheses also provide an *ultima ratio* in parameter estimation; see Ripley (1977) or Diggle (1979, 1983). (The minimum χ^2 estimation method of classical statistics parallels this.) A trial-and-error method chooses a set of parameters for a fixed point process model, and then conducts a goodness-of-fit test to see if the parameters are compatible with the empirical data. If they are, then the parameter set can be used as an estimate of the model parameters. If not, then the procedure is repeated for a new set of parameters, and so forth until a satisfactory set is found. This is a rather *ad hoc* procedure, but it can be refined using techniques involving iterative maximisation of likelihood; cf. Diggle and Gratton (1984).

4.8 General point processes

In previous sections, point processes are models for random point patterns in $\mathbb{R}^d$. This consideration is sufficient for most applications in stochastic geometry, but not for all. This chapter concludes with a briefly introduction to the general theory, which is presented in full detail in Daley and Vere-Jones (2008) and Schneider and Weil (2008, 2010).

Now the 'points' are elements of a general space E. To be mathematically precise, E is assumed to be a locally compact space with a countable base or a complete separable metric space. An example, the most important one for this book, is the set $\mathbb{F}$ of all closed subsets of $\mathbb{R}^d$ with the topology sketched in Section 6.1. Then a 'point' is a closed set, for example a line or a ball $B(x, r)$. In this case Schneider and Weil speak about *geometric processes*.

The needed σ-algebra is the Borel σ-algebra $\mathcal{E}$ of E, that is, the σ-algebra generated by the open subsets of E. This σ-algebra contains, for example, sets such as $\mathbb{F}_{B(x,r)}$, which denotes the set of all closed subsets of $\mathbb{R}^d$ which have a non-empty intersection with the ball $B(x, r)$.

A (simple) point process with points in E is then a random locally finite countable set of distinct points in E, or more mathematically, a measurable mapping from some probability space $[\Omega, \mathcal{A}, \mathbf{P}]$ to the set of all locally finite sequences of points in E. As above, the point process Φ denotes the random set of points as well as the corresponding counting measure, that is, $\Phi(A)$ for $A \in \mathcal{E}$ is the random number of points in the set $A \cap \Phi$.

Many ideas of the theory in previous sections can also be used in the general case. It is possible, for example, to define an intensity measure Λ, by the same equation as in the case of $\mathbb{R}^d$:

$$\Lambda(A) = \mathbf{E}\big(\Phi(A)\big) \qquad \text{for } A \in \mathcal{E},$$

and there is a corresponding Campbell theorem.

Poisson processes can also be defined, simply by requiring the two fundamental properties on p. 41 to be valid for Borel sets in E, instead of in $\mathbb{R}^d$.

With stationarity and isotropy it is a bit more complicated. These notions only make sense if for E there exist motions of the types 'translation' and 'rotation'. This is the case for geometric processes. Nevertheless, it is not a straightforward matter to define an intensity or point density, since in these spaces there is not necessarily a counterpart to the Lebesgue measure.

Finally, two important geometric processes should be mentioned here:

(a) Points = compact sets.

In this case Schneider and Weil speak about *particle processes*. A thorough study of these processes shows that these can be well described by marked point processes of the type $\{[x_n; K_n]\}$, where the x_n are points in $\mathbb{R}^d$ and the K_n are compact sets. This leads to the theory of germ–grain processes; see Section 6.5, and the theory of general point processes is not needed there.

(b) Points = flats.

This corresponds to what Schneider and Weil call *flat processes*. The points are here elements in the set $A(d, k)$ of all k-dimensional planes in $\mathbb{R}^d$. For $d = 2$ and $k = 1$, flats are lines in the plane, and for $d = 3$ and $k = 2$ planes in three-dimensional space.

Interpreting a flat process as a marked point process with points x_n in $\mathbb{R}^d$ is not suitable. If one would try to construct such a model, then the marks should be flats. The following scenario illustrates that such a construction would have too many flats. Consider the two-dimensional case and a stationary point process of points in the plane. Assume first that all 'mark' lines are parallel and orthogonal to the x_1-axis. Since in the strip $0 \leq x_1 \leq 1$ there are infinitely many points of the point process, the number of lines intersecting the interval $[0, 1]$ on the x_1-axis is infinite. The same can be expected for the case where the line directions are independent and uniform random. Thus, this approach does not work in general.

In contrast, Figure 8.7 shows a reasonable sample of a line process, constructed as a point process in $A(2, 1)$.

It can be said, extending the wording in Daley and Vere-Jones (2008, p. 484), that the structure of stationary isotropic line processes as well as of the more general stationary flat processes in $\mathbb{R}^d$ differs radically from that of stationary point processes in $\mathbb{R}^d$.

Line and flat processes are explained and discussed in Section 8.2.

5

Point processes III – Models

5.1 Operations on point processes

Model construction is a fundamental step in practical applications of stochastic geometry. It is explained here for the case of point processes; however, many of the modelling and simulation ideas discussed here can be used for other random structures.

This section describes three fundamental operations frequently used in modelling, which produce new point processes from old. Several important models can be derived from simpler ones (such as the Poisson process) by means of these operations. The resulting models, as well as some others, are described in the remainder of this chapter.

The three fundamental operations to be described in this section are:

- thinning,
- clustering,
- superposition.

Matthes *et al.* (1978), Kallenberg (2002) and Daley and Vere-Jones (2008) discuss limit theorems for the results of repeated use of these operations both separately and in combination. In this section only the operations themselves are described.

Thinning

A thinning operation uses some definite rule to delete points of a basic point process Φ_b, thus yielding the *thinned point process* Φ. Considered as a random closed set, Φ is a subset of Φ_b:

$$\Phi \subset \Phi_b. \tag{5.1}$$

A simple form of thinning is *p-thinning*: each point of Φ_b has probability $1 - p$ of suffering deletion, and its deletion is independent of locations and possible deletions of any other points of Φ_b. Thus p is the probability that a point is retained.

Stochastic Geometry and its Applications, Third Edition.
Sung Nok Chiu, Dietrich Stoyan, Wilfrid S. Kendall and Joseph Mecke.
© 2013 John Wiley & Sons, Ltd. Published 2013 by John Wiley & Sons, Ltd.

A natural generalisation allows the retention probability p to depend on the location x of the point. A deterministic function $p(x)$ is given on $\mathbb{R}^d$, with $0 \le p(x) \le 1$. If the point x belongs to Φ_b, it is deleted with probability $1 - p(x)$ and again its deletion is independent of locations and possible deletions of any other points. To emphasise the spatial dependence here, the generalised operation is called $p(x)$-*thinning*.

In a further generalisation the function $p(x)$ is itself random. Formally, a random field $\pi = \{\pi(x) : x \in \mathbb{R}^d\}$ is given, which is independent of Φ_b, where $0 \le \pi(x) \le 1$ for all $x \in \mathbb{R}^d$. A realisation φ of the thinned process Φ is constructed by taking a realisation φ_b of Φ_b and applying $p(x)$-thinning to φ_b, where $\{p(x) : x \in \mathbb{R}^d\}$ is a realisation of the random field π. Given $\pi(x) = p(x)$ and given $\Phi_b = \varphi_b$, the probability of x in Φ_b also belonging to Φ is $p(x)$. This further generalisation is known as $\pi(x)$-*thinning*.

All these thinnings are *independent thinnings*. This means there is no interaction between the points, so that the thinning functions (which are independent of Φ_b) determine the operation completely.

Yet another generalisation allows dependence on the configuration of Φ_b, giving the class of *dependent thinnings*. A practical context for a dependent thinning would be a point pattern representing plant locations. Suppose there is mutual inhibition between plants leading to the death of a fraction of them. This suggests a point process constructed by a thinning with deletion probability which is higher in regions of higher point density. This could be obtained, for example, by making deletion probability depend on the distance to the nearest point in the initial pattern.

Figure 5.1 demonstrates the result of independent and dependent thinnings applied to the same initial point pattern. An important dependent thinning leads to the Matérn hard-core processes; an example is considered in Section 5.4.

If the summary characteristics of the basic process Φ_b are known, then it is straightforward to calculate the characteristics of point processes produced by independent thinning. Indeed, if Φ is the result of $p(x)$-thinning of Φ_b, then its intensity measure Λ is given by

$$\Lambda(B) = \int_B p(x)\Lambda_b(dx), \tag{5.2}$$

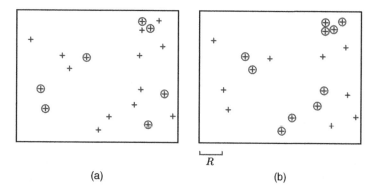

(a)   (b)

Figure 5.1   Thinnings of a point pattern. Points of the origin pattern are marked by $+$, deleted points by $\oplus$. (a) Independent thinning carried out by a p-thinning; (b) dependent thinning deleting all points with nearest neighbour within distance R.

where Λ_b is the intensity measure of Φ_b. If G_b is the generating functional of Φ_b, then that of Φ is

$$G(v) = G_b(v_p) \qquad \text{for } v \in V, \tag{5.3}$$

where

$$v_p(x) = v(x)p(x) + 1 - p(x); \tag{5.4}$$

V is the set of functions defined in Section 4.3.6.

Analogous formulae for $\pi(x)$-thinnings follow by averaging with respect to the distribution of the random field π.

If Φ_b is stationary, then the p-thinned process Φ is also stationary; in general this is false for $p(x)$-thinning. From Formula (5.2) the intensity is given by

$$\lambda = p\lambda_b. \tag{5.5}$$

A $\pi(x)$-thinned point process can be stationary: if Φ is a stationary point process and $\pi = \{\pi(x)\}$ a stationary random field. Formula (5.5) still applies with $p = E(\pi(x))$.

When both Φ_b and π are motion-invariant so is Φ. The existence of a second-order product density $\varrho_b^{(2)}(r)$ for Φ_b implies the same for Φ. The corresponding second-order characteristics are as follows: in the case of a p-thinning the product density is given simply by

$$\varrho^{(2)}(r) = p^2 \varrho_b^{(2)}(r), \tag{5.6}$$

and the reduced second moment functions as well as the pair correlation functions of Φ and Φ_b coincide.

In the case of a $\pi(x)$-thinning let $k(r)$ denote the covariance function of the random field π by

$$k(r) = E(\pi(o)\pi(\mathbf{r})) - p^2 \qquad \text{where } \|\mathbf{r}\| = r. \tag{5.7}$$

Then

$$\varrho^{(2)}(r) = (k(r) + p^2)\, \varrho_b^{(2)}(r), \tag{5.8}$$

and the reduced second moment function $K(r)$ of Φ is given by

$$K(r) = \left. \int_0^r \left((k(x) + p^2)\, dK_b(x) \right) \middle/ p^2. \right. \tag{5.9}$$

If $\varrho_b^{(2)}(r)$ is interpreted intuitively as in Section 4.3.3, then Formulae (5.6) and (5.8) are plausible: if a pair of points are both to contribute to the product density, then both of them must survive.

It is easy to describe Φ if Φ_b is a *Poisson process*. Let Φ_b be Poisson with intensity measure Λ_b. Under a $p(x)$-thinning Φ is again Poisson with intensity measure given by (5.2). This result (sometimes called *Prekopa's Theorem*) follows easily from (5.3) and the formula for the generating functional of a Poisson process. Alternatively, one can compute the void-probabilities of a p-thinned Poisson process:

$$\mathbf{P}(\Phi(K) = 0) = \sum_{k=0}^{\infty} \mathbf{P}(\Phi_{\mathrm{b}}(K) = k) \cdot \mathbf{P}(\text{all these } k \text{ points are deleted})$$

$$= \sum_{k=0}^{\infty} e^{-\Lambda_{\mathrm{b}}(K)} \frac{\Lambda_{\mathrm{b}}(K)^k}{k!} (1 - p)^k = \exp\big(-p\Lambda_{\mathrm{b}}(K)\big).$$

In particular if Φ_{b} is a homogeneous Poisson process of intensity λ_{b}, then its p-thinning is a homogeneous Poisson process of intensity $p\lambda_{\mathrm{b}}$. Note that the points which are *removed* by this thinning also form a Poisson process, and the two processes are independent.

The $\pi(x)$-thinnings of Poisson processes are doubly stochastic Poisson processes (i.e. Cox processes) as described in Section 5.2. Multiple thinnings combined with rescalings lead to Cox processes; see Daley and Vere-Jones (2008, Section 11.3).

Example 5.1.   *An interrupted point process: Lightning gap locations (Kautz et al., 2011)*

Figure 5.2 shows the locations of 78 gaps in the canopy of a Vietnamese mangrove forest resulting from lighting strikes, together with the distribution of the forest matrix, interrupted

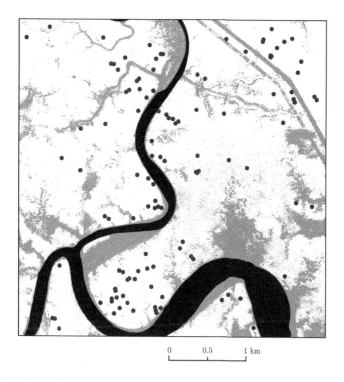

0   0.5   1 km

Figure 5.2   Positions of 78 lightning gaps (marked by •) in a 4 km × 4 km sampling window of a Vietnamese mangrove forest (white regions) with rivers (black) and open (grey) areas such as water channels and mud flats, generated in the years 2003–2007. Reprinted from Kautz *et al.* (©2011) with permission of Elsevier.

by water channels and mud flats. Lightning strikes leave traces only in woodland. Thus here a quite natural process of thinning is acting: the observed gap locations are taken to be result of thinning a motion-invariant point process Φ_b (representing 'all lightning strikes') by using the indicator of a motion-invariant random closed set Ξ (representing the woodland) as the thinning random field π. In the notation of Stoyan (1979a) the resulting point process is an *interrupted point process*; it is a $\pi(x)$-thinning with $\pi(x) = \mathbf{1}_\Xi(x)$. It can also be written as

$$\Phi = \Phi_b \cap \Xi.$$

Let Φ_b have intensity λ_b and pair correlation function $g_b(r)$ and let Ξ have area fraction p and covariance $C(r)$. Note that the covariance function $k(r)$ of the random field $\{\mathbf{1}_\Xi\}$ is given by

$$k(r) = C(r) - p^2.$$

From the formulae above the intensity of Φ is

$$\lambda = p\lambda_b, \tag{5.10}$$

and its pair correlation function is

$$g(r) = \frac{C(r)}{p^2} g_b(r) \qquad \text{for } r \geq 0. \tag{5.11}$$

Naturally this model is only an approximation, both by assuming motion-invariance and independence of lightning and woodland. Kautz *et al.* (2011) confirmed the latter assumption by a test and analysed statistically the data further. Figure 5.3 displays estimates of the pair

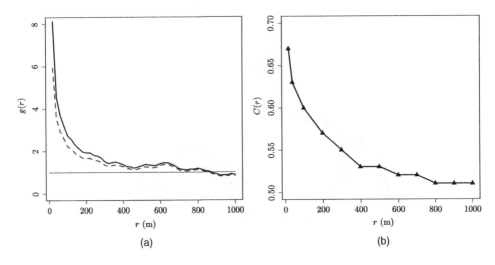

(a)   (b)

Figure 5.3 (a) The empirical pair correlation function $\hat{g}(r)$ (solid line) for the point pattern of Figure 5.2 and the pair correlation function (dashed line) of the full lightning strike process Φ_b obtained by Formula (5.11). (b) The estimated covariance $\hat{C}(r)$ of the forest matrix of Figure 5.2. Reprinted from Kautz *et al.* (©2011) with permission of Elsevier.

correlation function $g(r)$ and covariance $C(r)$. The estimated first-order statistics are:

$$\hat{p} = 0.7,$$
$$\hat{\lambda} = 4.9\,\text{km}^{-2},$$
$$\hat{\lambda}_b = 7.0\,\text{km}^{-2}.$$

Formula (5.11) yields the graph of $\hat{g}_b(r)$ shown in Figure 5.3, which is similar to pair correlation functions of cluster processes. Thus, it is reasonable to assume that the lightning strikes are spatially clustered, and Kautz *et al.* (2011) proposed a suitable process with clustering both in time and space to model the given pattern.

Clustering

In a clustering operation each point x of a given point process Φ_p (the subscript 'p' stands for 'parent') is replaced by a cluster N^x of daughter points. The replacement clusters N^x are themselves point processes, which have only a finite number of points each. The union of all these clusters is the *cluster point process* Φ,

$$\Phi = \bigcup_{x \in \Phi_p} N^x. \tag{5.12}$$

It is tacitly assumed that almost certainly points of different clusters do not coincide, so that $N^x \cap N^y$ is empty whenever $x \neq y$. Furthermore, it is assumed that Φ is locally finite. Depending on the particular model, x may or may not itself belong to N^x. Figure 5.4 shows a sample of a cluster point process.

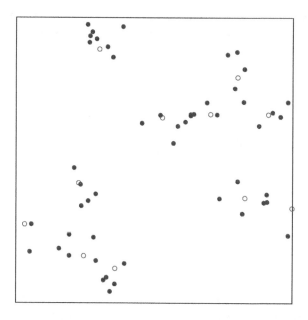

Figure 5.4   A sample of a cluster point process. Parent points are marked by ○, daughter points by ●.

Note in passing that thinning operations can formally be interpreted as special cases of cluster operations: a thinning can be obtained by supposing N^x to be either the singleton set $\{x\}$ or the empty set, depending on the thinning criterion evaluated at x. Random shifts can be interpreted similarly: i.i.d. shifted points are daughter points of a given pattern of parent points; in this case all N^x are i.i.d. singletons.

Cluster processes have been used as models for many natural phenomena, as documented by the references in Section 5.3. For example, parents are locations of plants and daughters are the seedfall locations; the cluster pattern is generated as the locations where new plants grow. Often cluster processes serve as point process models of a higher degree of variability than Poisson processes.

An important special case is the Neyman–Scott process, which is described in detail in Section 5.3. The formulae in the rest of this section are given for a somewhat more general model. It is supposed that the parent point process $\Phi_p = \{x_1, x_2, \ldots\}$ is stationary of intensity λ_p and that the clusters N^x are of the form

$$N^{x_i} = N_i + x_i \tag{5.13}$$

for each x_i in Φ_p. The N_i are i.i.d. finite point sets with the common distribution c, independent of the parent point process Φ_p, and the $+x_i$ terms shift the cluster centres to x_i. This is called *homogeneous independent clustering*. If Φ_p is a Poisson process, then the resultant process Φ is called a *Poisson cluster process*.

Whatever the form of Φ_p in this model is, the intensity λ of Φ is given by

$$\lambda = \lambda_p \bar{c} \tag{5.14}$$

where $\bar{c}$ is the mean number of points in the cluster N_0, which has the same distribution as the N_i and is called the *representative cluster* of the cluster process. Clearly, $\bar{c}$ and λ_p of Φ_p are assumed to be finite.

A general formula for the second-order product density $\varrho^{(2)}(x_1, x_2)$ can be found in Illian *et al.* (2008, p. 370).

If G, G_c and G_p, denote the generating functionals of Φ, N_0 and Φ_p respectively, then

$$G(v) = G_p\left(G_{(\cdot)}(v)\right) = \mathbf{E}\left(\prod_{x\in\Phi_p} G_{(x)}(v)\right) \qquad \text{for } v \in \mathbf{V}, \tag{5.15}$$

where

$$G_{(x)}(v) = G_c\left(v(x+\cdot)\right) = \mathbf{E}\left(\prod_{y\in N_0} v(x+y)\right). \tag{5.16}$$

Formula (5.15) expresses a close similarity to the theory of branching processes, and indeed the clustering operation is precisely what is used to create new generations of a *branching random walk* or *spatial branching process*; see Biggins (1977), Mollison (1977), Matthes *et al.* (1978) and Liemant *et al.* (1988). Biological applications can be found in Shimatani (2002, 2010).

Superposition

Let Φ_1 and Φ_2 be two point processes with distributions P_1 and P_2, intensities λ_1 and λ_2, etc. Consider the union

$$\Phi = \Phi_1 \cup \Phi_2. \tag{5.17}$$

Suppose that with probability one the point sets Φ_1 and Φ_2 do not overlap (in the case of independence this is equivalent to the corresponding intensity measures having no common atoms; see Daley and Vere-Jones, 2008.) The set-theoretic union then coincides with the superposition operation of general point process theory. In the general theory superpositions may have multiple points, a complication not considered here.

It is clear that the intensity measure Λ and intensity λ (in case of stationarity) of Φ are given by

$$\Lambda = \Lambda_1 + \Lambda_2, \tag{5.18}$$

$$\lambda = \lambda_1 + \lambda_2. \tag{5.19}$$

If Φ_1 and Φ_2 are independent, then the distribution P of Φ is written by means of the convolution symbol $*$ as

$$P = P_1 * P_2. \tag{5.20}$$

The generating functional G of Φ satisfies

$$G(v) = G_1(v) \cdot G_2(v) \qquad \text{for } v \in \mathbf{V}, \tag{5.21}$$

and the second-order product density $\varrho^{(2)}$ is

$$\varrho^{(2)}(x_1, x_2) = \varrho_1^{(2)}(x_1, x_2) + \varrho_2^{(2)}(x_1, x_2) + 2\lambda_1\lambda_2 \qquad \text{for } x_1 \text{ and } x_2 \in \mathbb{R}^d, \tag{5.22}$$

which leads easily to formulae for the pair correlation function $g(r)$ and the reduced second moment function $K(r)$; see Illian *et al.* (2008, p. 371).

The nearest-neighbour distance distribution function of Φ is given by

$$1 - D(r) = \frac{\lambda_1}{\lambda}\big(1 - D_1(r)\big)\big(1 - H_{s,2}(r)\big) + \frac{\lambda_2}{\lambda}\big(1 - D_2(r)\big)\big(1 - H_{s,1}(r)\big) \qquad \text{for } r \geq 0, \tag{5.23}$$

see van Lieshout and Baddeley (1996), where $H_{s,i}(r)$ is the spherical contact distribution function of Φ_i. The spherical contact distribution function of Φ is

$$H_s(r) = 1 - \big(1 - H_{s,1}(r)\big)\big(1 - H_{s,2}(r)\big) \qquad \text{for } r \geq 0. \tag{5.24}$$

If Φ_1 and Φ_2 are independent Poisson processes, then so is Φ. This follows directly from the fact that the sum of two independent Poisson variables is itself Poisson.

Multiple superpositions combined with rescalings lead to Poisson processes; see Daley and Vere-Jones (2008, Section 11.2).

5.2 Doubly stochastic Poisson processes (Cox processes)

5.2.1 Introduction

An obvious generalisation of a Poisson process is made by supposing that the intensity measure is itself random, with the point process being Poisson *conditional* on the realisation of the intensity measure. Such a process is called a *doubly stochastic Poisson process* or *Cox process*.

Formally a Cox process is defined by choosing a distribution Q on $[\mathbb{M}, \mathcal{M}]$, the space of all nonnegative locally finite measures on $\mathbb{R}^d$. Suppose P_Λ is the distribution of the Poisson process of intensity measure Λ, and Ψ is a random measure with distribution Q. (See Chapter 7 for elements of the theory of random measures.) Then the *Cox process* Φ *with driving random measure* Ψ has distribution

$$P_\Phi(Y) = \int P_\Lambda(Y) Q(\mathrm{d}\Lambda) \qquad \text{for } Y \in \mathcal{N}. \tag{5.25}$$

To ensure that Φ is simple, that is, that Φ has no multiple points, it is assumed that Q is concentrated on the subset of diffuse measures.

A Cox process can be thought of as arising from a two-step random mechanism, hence the term 'doubly stochastic'. The first step generates a measure Λ on $\mathbb{R}^d$ according to the driving random measure distribution Q and the second step generates a Poisson process of intensity measure Λ.

Cox processes are both general and amenable to calculation; therefore they find important applications as stochastic models. They occur frequently in physics, for example, in optics. Systematic study began with Cox (1955), after earlier examples. Mecke (1968) and Krickeberg (1972) made important early contributions, Grandell (1976) is an excellent early monograph on the subject; further material is to be found in all books on point processes, for example Daley and Vere-Jones (2003, 2008), Møller and Waagepetersen (2004) and Illian *et al.* (2008).

Early applications of Cox processes in spatial statistics are discussed by Matérn (1971), Bartlett (1975) and Grandell (1981). Diggle (1983), Diggle and Milne (1983) and Lotwick (1984) considered a marked Cox process as a model for a pattern of points of two types, such as locations of two different species of trees. Positive and negative dependence are combined in this model. Kingman (1977) showed that Cox processes are more natural candidates than Poisson processes for the distribution of reproducing populations in space.

An interesting classical example is given by Stapper *et al.* (1980), who used a mixed Poisson process (see the next section) to model the pattern of point defects on the surfaces of silicon wafers arising in semiconductor manufacture. (The randomised intensity is given by a gamma distribution.) Modern developments are presented in Kuo *et al.* (2006).

Statistical problems for Cox processes have been studied in many papers. Krickeberg (1982) and Karr (1984, 1986) made early major contributions. The books by Møller and Waagepetersen (2004) and Illian *et al.* (2008) describe the modern approach.

5.2.2 Examples of Cox processes

The mixed Poisson process

The *mixed Poisson process* is a simple instance of a Cox process. It can be thought of as a homogeneous Poisson process with randomised intensity parameter. The driving random measure Ψ is the random measure

$$\Psi = X\nu_d, \tag{5.26}$$

where the randomised intensity X is a nonnegative random variable. Every sample of such a process looks like a sample of some homogeneous Poisson process. This is a typical example of a non-ergodic Cox process.

$\pi(x)$-thinning of a Poisson process

In this case the driving random measure Ψ is given by

$$\Psi(B) = \int_B \pi(x)\Lambda_b(dx) \qquad \text{for Borel } B, \tag{5.27}$$

where Λ_b is the intensity measure of the original Poisson process Φ_b.

The Log-Gaussian Cox process

A random field $\{Z(x)\}$ is Gaussian if, for any finite collection of locations $x_1, \ldots, x_k$, any linear combination $b_1 Z(x_1) + \cdots + b_k Z(x_k)$ with real $b_1, \ldots, b_k$ has a one-dimensional normal distribution. If $\{Z(x)\}$ is stationary and isotropic, its distribution is determined completely by the mean μ_Z and the covariance function $k_Z(r)$. However, such a field cannot be used as the intensity field of a Cox process since it can take negative values. Thus, a suitable transformation has to be applied to the field $\{Z(x)\}$ to yield a Cox process. An elegant transformation, resulting in a mathematically tractable model, is

$$\Lambda(x) = \exp\big(Z(x)\big) \qquad \text{for } x \in \mathbb{R}^d. \tag{5.28}$$

The corresponding process is termed a *log-Gaussian Cox process*. It was independently introduced by Coles and Jones (1991), Rathbun (1996) and Møller *et al.* (1998) and is now a very popular model, which has a simple formula for all product densities; see Example 5.3. It can be seen as a link between point process statistics and geostatistics.

Random-set-generated Cox process

A regular-closed stationary random closed set Ξ divides $\mathbb{R}^d$ into two parts or phases: the set Ξ and its complement Ξ^c. In both phases, a Poisson process is generated with intensities λ_1 and λ_2, respectively.

In so-called 'ideal cluster materials' there exist deterministic or random ellipsoidal domains ('particle clouds') with high concentration of nuclei/reinforcements within and low without; see Buryachenko (2007, pp. 3 and 151). In another application the random-set-generated Cox process was used for a data set of a pattern of seedling locations in a commercial tree plantation, where the soil had been treated in two different ways. The random set reflects these two different soil treatments (Penttinen and Niemi, 2007). The interrupted process of Example 5.1 is a special case with $\lambda_1 = \lambda$ and $\lambda_2 = 0$. The random intensity of this Cox process is simply

$$\Lambda(x) = \lambda_1 \mathbf{1}_\Xi(x) + \lambda_2\big(1 - \mathbf{1}_\Xi(x)\big), \tag{5.29}$$

where $\mathbf{1}_\Xi(x)$ denotes the indicator function of the set Ξ.

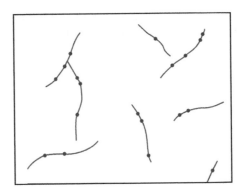

Figure 5.5 A sample of a particular Cox process. The points are randomly scattered on a system of random fibres. Compare this with Hodder and Orton (1976, Figure 7.1), which shows the distribution of finds of late Iron Age inscribed coins and Roman roads in central and southern England. Many coin finds lie on or very close to these roads. In another interpretation the points here can be considered as centres of nucleation.

A closely related class of Cox processes are *shot-noise Cox processes* as discussed in Møller and Waagepetersen (2004). Neyman–Scott processes belong to this class if the number of points per cluster follows a Poisson distribution; an example is the Matérn cluster process on p. 173.

Points scattered on random fibres and networks

Let χ be a random fibre process (see Section 8.3). The random measure Ψ defined by

$$\Psi(B) = N_L h_1(\chi \cap B) \qquad \text{for Borel } B \tag{5.30}$$

yields a driving measure for a Cox process. Here N_L is a positive constant and $h_1(\chi \cap B)$ is the total length of all fibre pieces of χ in B. The resultant Cox process is formed of points scattered uniformly with intensity N_L on the fibres of χ, as illustrated in Figure 5.5. The parameter N_L is the mean number of points per unit length of fibres.

Such Cox processes appear in materials science in the context of nucleation on fibres (or, similarly, on surfaces), see Section 6.6.4, and in telecommunication networks, see Voss *et al.* (2010). There χ is the edge system of a random tessellation, representing the infrastructure (roads, railways), while the points are for example mobile users in wireless networks. In another approach Ang *et al.* (2012) consider random point patterns on deterministic/fixed networks, for example crime locations on the streets of Chicago. They make second-order statistics based on travel distances in the network. This falls outside of the usual domain of Cox point process theory.

5.2.3   Formulae for characteristics of Cox processes

The generating functional G of the Cox process Φ_C is given by

$$G(v) = L_Q(1 - v) \qquad \text{for } v \in \mathbf{V}, \tag{5.31}$$

where L_Q is the Laplace functional of the driving random measure Ψ of Φ. This follows from

$$G(v) = \int_{\mathbb{M}} G_\Lambda(v) Q(d\Lambda)$$

$$= \int_{\mathbb{M}} \exp\left(-\int_{\mathbb{R}^d}(1-v(x))\Lambda(dx)\right) Q(d\Lambda)$$

$$= L_Q(1-v),$$

where G_Λ is the generating functional of the Poisson process of intensity measure Λ.

The void-probabilities of a Cox process are given by

$$v_K = \int_{\mathbb{M}} \exp(-\Lambda(K)) Q(d\Lambda)$$

$$= L_Q(1_K) \qquad \text{for compact } K, \tag{5.32}$$

since

$$v_K = \mathbf{P}(\Phi \cap K = \emptyset) = \mathbf{P}(\Phi(K) = 0)$$

$$= \mathbf{E}(\exp(-\Psi(K))) = \int_{\mathbb{M}} \exp(-\Lambda(K)) Q(d\Lambda).$$

The factorial moment measures $\alpha^{(n)}$ of Φ coincide with the moment measures $\mu_\Psi^{(n)}$ of Ψ,

$$\alpha^{(n)} = \mu_\Psi^{(n)} \qquad \text{for } n = 1, 2, \ldots, \tag{5.33}$$

and, in particular, the intensity measures are equal,

$$\Lambda = \Lambda_\Psi. \tag{5.34}$$

If Ψ is stationary, then so is Φ, and both have the same intensity parameter. In particular, every mixed Poisson process is stationary.

The Palm distributions P_x for Φ are given by

$$P_x = \delta_{\delta_x} * P_{Q_x}, \tag{5.35}$$

where P_{Q_x} is the distribution of a Cox process with driving random measure Q_x, and Q_x is the Palm distribution of Ψ at x. Note that Palm distributions for random measures are defined in Section 7.1.3.

Formula (5.35) is a good starting point for calculating the reduced second moment function $K(r)$ and the nearest-neighbour distance distribution function $D(r)$.

For all Cox processes the number of counts in bounded sets is *super-Poissonian*. That is to say, their variances exceed those for homogeneous Poisson processes of the same intensity. This is a consequence of the mixed Poisson distribution of the number of counts in a bounded set.

Example 5.2.   *Mixed Poisson processes using two intensities λ_1 and λ_2*

Consider a mixed Poisson process Φ as above, with the randomised intensity X having a two-point distribution

$$
X = \begin{cases} \lambda_1, & \text{with probability } 1 - p, \\ \lambda_2, & \text{with probability } p, \end{cases}
$$

where $0 < p < 1$. Its intensity is given by

$$
\lambda = (1 - p)\lambda_1 + p\lambda_2.
$$

Its Palm distribution P_o is derived from the Palm distributions $P_o^{(1)}$ and $P_o^{(2)}$ of Poisson processes with intensities λ_1 and λ_2 respectively:

$$
P_o(Y) = \frac{1}{\lambda}\left(\lambda_1(1 - p)P_o^{(1)}(Y) + \lambda_2 p P_o^{(2)}(Y)\right) \qquad \text{for } Y \in \mathcal{N}.
$$

Consequently the nearest-neighbour distance distribution function is given by

$$
D(r) = \frac{1}{\lambda}\left(\lambda_1(1 - p)\left(1 - \exp(-\lambda_1 b_d r^d)\right) + \lambda_2 p\left(1 - \exp(-\lambda_2 b_d r^d)\right)\right).
$$

The reduced second moment function K of Φ is

$$
K(r) = \frac{\lambda_1^2(1 - p) + \lambda_2^2 p}{\lambda^2} b_d r^d \qquad \text{for } r \geq 0.
$$

Example 5.3.   *Log-Gaussian Cox processes*

Recall that for a log-Gaussian Cox processes the random intensity is $\Lambda(x) = \exp\big(Z(x)\big)$, where $\{Z(x)\}$ is a stationary and isotropic Gaussian random field with mean μ_Z, variance σ_Z^2 and covariance function $k_Z(r)$. Then the first- and second-order point process characteristics are

$$
\lambda = \exp\left(\mu_Z + \frac{\sigma_Z^2}{2}\right) \tag{5.36}
$$

and

$$
g(r) = \exp\big(k_Z(r)\big) \qquad \text{for } r \geq 0. \tag{5.37}
$$

Moreover, the nth-order product density can be expressed in terms of λ and $g(r)$ as

$$
\varrho^{(n)}(x_1, \ldots, x_n) = \lambda^n \prod_{1 \leq i < j \leq n} g(r_{ij}), \tag{5.38}
$$

where $r_{ij} = \|x_i - x_j\|$; see Møller *et al.* (1998).

Example 5.4.   *Random-set-generated Cox processes*

The intensity and pair correlation function of a random-set-generated Cox process are obtained from the first- and second-order characteristics of a stationary and isotropic random set Ξ, namely, the area fraction (volume fraction) p and covariance $C(r)$; see Section 6.3. These two characteristics, together with the Poisson process intensities λ_1 and λ_2 of the two phases, lead to

$$\lambda = p\lambda_1 + (1 - p)\lambda_2 \qquad (5.39)$$

and

$$g(r) = \frac{1}{\lambda^2}\left(C(r)(\lambda_1 - \lambda_2)^2 - 2p(\lambda_1 - \lambda_2)\lambda_2 + \lambda_2^2\right) \qquad \text{for } r \geq 0. \qquad (5.40)$$

5.3 Neyman–Scott processes

Neyman–Scott processes (also called *centre–satellite processes*) are examples of Poisson cluster processes, which are often used in spatial statistics. They result from homogeneous independent clustering applied to a homogeneous Poisson process. The parent points form a homogeneous Poisson process of intensity λ_p while the daughter points of a representative cluster N_0 are random in number and are scattered independently with identical spatial probability density $f(y)$ around the origin. The parent points do not occur in the observed point pattern; only daughter points are included.

Under these assumptions the resultant cluster process Φ is stationary. If the scattering probability density function $f(y)$ is radially symmetric, then Φ is isotropic.

The Neyman–Scott process can be interpreted as a particular kind of Boolean model, where the clusters play the rôle of the 'grains'. This is useful for the derivation of formulae for void-probabilities.

The general Formula (5.14) for the intensity λ of cluster processes applies here:

$$\lambda = \lambda_p \bar{c}, \qquad (5.41)$$

where $\bar{c}$ is the mean number of daughter points per parent. The generating functional in Formula (5.15) specialises to

$$G(v) = \exp\left(-\lambda_p \int_{\mathbb{R}^d} \left(1 - G_n\left(\int_{\mathbb{R}^d} v(x + y)f(y)\,\mathrm{d}y\right)\right)\mathrm{d}x\right), \qquad (5.42)$$

where G_n is the generating function for the random number of points of N_0.

Further progress can be made by considering a fundamental formula for the Palm distribution P_o of a Poisson cluster process. Let P be the distribution of Φ and c_o be the 'Palm' distribution of the cluster N_0. Then

$$P_o = P * c_o \qquad (5.43)$$

as given by Ambartzumian (1966) and Mecke (1967). The Palm distribution c_o is given by

$$c_o(Y) = \frac{1}{\bar{c}}\,\mathbf{E}\left(\sum_{x \in N_0} \mathbf{1}_Y(N_0 - x)\right) \qquad \text{for } Y \in \mathcal{N}, \qquad (5.44)$$

which is the formula for the 'Palm' distribution of a point process with finite mean total number of points.

Formula (5.43) can be seen as plausible if it is considered as a generalisation of the Slivnyak–Mecke theorem for the Palm distribution of a homogeneous Poisson process. The resulting Palm distribution P_o for Φ may be interpreted as the superposition of a finite point process following the Palm distribution c_o for N_0 and an independent point process with the distribution P of the complete Φ.

The distribution c_o can be understood as follows. Let $\{p_n\}$ be the distribution of the number of points in the representative cluster N_0 so that

$$p_n = \mathbf{P}\big(N_0(\mathbb{R}^d) = n\big).$$

Then

$$\bar{c} = \sum_{n=0}^{\infty} n p_n.$$

Simple calculations with conditional probabilities yield $c_o(Y_k)$ for $Y_k = \{\varphi \in \mathbb{N} : \varphi(\mathbb{R}^d) = k\}$, that is, the 'Palm' probability of the event that there are precisely k points in the cluster N_0. It is

$$\pi_0 = c_o(Y_0) = 0, \tag{5.45}$$

$$\pi_k = c_o(Y_k) = \frac{1}{\bar{c}} \, \mathbf{E}\left(\sum_{x \in N_0} \mathbf{1}_{Y_k}(N_0 - x)\right) = \frac{1}{\bar{c}} k p_k \qquad \text{for } k = 1, 2, \ldots. \tag{5.46}$$

The reason that $\pi_0 = 0$ is clear: a cluster which contains the typical point cannot be empty. Thus the distribution c_o is *number-weighted* and in general different from the distribution of the representative cluster N_0.

Formula (5.43) allows other characteristics to be expressed in terms of quantities that are related to Palm distributions of Neyman–Scott processes. In the isotropic case the reduced second moment function $K(r)$ satisfies

$$K(r) = b_d r^d + \frac{1}{\lambda \bar{c}} \sum_{n=2}^{\infty} p_n n(n-1) F(r) \qquad \text{for } r \geq 0, \tag{5.47}$$

where $F(r)$ is the distribution function of the distance between two independent random points of the same cluster.

From Formula (5.47) the pair correlation function $g(r)$ is given by

$$g(r) = 1 + \frac{1}{\lambda \bar{c}} \sum_{n=2}^{\infty} p_n n(n-1) \frac{f(r)}{d b_d r^{d-1}} \qquad \text{for } r \geq 0, \tag{5.48}$$

where $f(r)$ denotes the probability density function of $F(r)$. It is obtained from the spatial probability density $f(y)$ of the cluster points as follows: the probability density function of the vector difference of two independent points of the same cluster is

$$f(\mathbf{r}) = \int f(y) f(y + \mathbf{r}) dy,$$

and changing to polar coordinates $(\mathbf{r} \to r)$ yields $f(r)$.

Formula (5.43) also leads to the nearest-neighbour distance distribution function $D(r)$:

$$1 - D(r) = \mathbf{P}\big(\Phi(B(o, r)) = 0\big) \cdot c_o\big(\{\varphi \in \mathbb{N} : \varphi(B(o, r)) = 1\}\big) \qquad \text{for } r \geq 0, \qquad (5.49)$$

where the first factor on the right-hand side is equal to $1 - H_s(r)$, that is, given by the spherical contact distribution function; see Stoyan and Stoyan (1994, p. 312). The spherical contact distribution function $H_s(r)$ of the cluster process can be obtained by means of Formula (3.30) using the fact that Φ can be interpreted as a Boolean model with typical grain N_0. Unfortunately, the use of these formulae for concrete models is not easy.

Clearly the number of points in a compact convex set is super-Poissonian for a Neyman–Scott process; see Stoyan (1983). That is, its variance is greater than that for a Poisson process of the same intensity. A similar inequality holds for the corresponding K-functions.

Example 5.5. *Matérn cluster process (Matérn, 1960, 1986)*

In this example, the representative cluster N_0 has a distribution as follows. The number of points in N_0 has a Poisson distribution with the positive parameter $\mu = \bar{c}$. The sum in (5.47) and (5.48) is then $\bar{c}^2$. The points of N_0 are independently uniformly scattered in the ball $B(o, R)$, where R is a further model parameter. Hence

$$\lambda = \lambda_{\mathrm{p}}\mu, \qquad (5.50)$$

and the probability density function $f(r)$ occurring in Formula (5.48) for $g(r)$ is of the form

$$f(r) = \begin{cases} \dfrac{1}{R}\left(1 - \dfrac{r}{2R}\right), & \text{if } d = 1, \\[3mm] \dfrac{4r}{\pi R^2}\left(\cos^{-1}\dfrac{r}{2R} - \dfrac{r}{2R}\sqrt{1 - \dfrac{r^2}{4R^2}}\right), & \text{if } d = 2, \\[3mm] \dfrac{3}{2}\dfrac{r^2}{R^6}\left(R - \dfrac{r}{2}\right)^2\left(2R + \dfrac{r}{2}\right), & \text{if } d = 3, \end{cases} \qquad (5.51)$$

for $0 < r < 2R$; otherwise $f(r) = 0$.

These results can be found already in Santaló (1976, p. 212, Note 6).

A sample of a simulated Matérn cluster process is shown in Figure 5.6; Figure 4.7 shows the pair correlation function for particular parameters.

Variations on the scatter distribution produce other models. If in the construction above the uniform distribution in the disc is replaced by the isotropic d-dimensional normal distribution with variance parameter σ, then the result is the *(modified) Thomas process*. For this model the pair correlation function is

$$g(r) = 1 + \frac{\bar{c}}{\lambda}(4\pi\sigma^2)^{-d/2}\exp\left(-\frac{r^2}{4\sigma^2}\right) \qquad \text{for } r \geq 0 \qquad (5.52)$$

(see Møller and Waagepetersen, 2004). Here $\mu = \bar{c}$ is the mean of the Poisson distribution of cluster size and σ^2 is the variance of the symmetric normal distribution. Both the Matérn and Thomas process can be generalised by replacing the Poisson distribution of the random number of cluster points by any other discrete distribution; the formulae for $f(r)$ can be

Figure 5.6   A simulated realisation of a Matérn cluster process in a unit square with $\lambda = 25$, $R = 0.025$ and $\mu = 2$.

plugged into the general formula (5.48). Tanaka *et al.* (2008) generalise the Thomas process model by introducing two types of clusters with σ_1^2 and σ_2^2.

Buryachenko (2007, p. 151) discusses the application of cluster processes in materials science and writes that processes with convex clusters are useful in the context of recrystallisation and residual stresses. Illian *et al.* (2008) give examples for the application of cluster processes in forestry, where sometimes it is necessary to use more general models than the Thomas process; see also Example 2.1.

A simple subclass of Poisson cluster processes is formed by the Gauss-Poisson processes. Milne and Westcott (1972) discuss this subclass, in which each cluster consists of zero, one, or two points only.

Example 5.6.   *Calculation of D(r) for a planar Gauss-Poisson process*

In this example the typical cluster N_0 has an isotropic distribution, being composed of zero, one or two points with probability p_0, p_1 or p_2 respectively. If N_0 consists of only one point, then that point is the origin o. If N_0 is composed of two points, then these are separated by a unit distance and have midpoint o.

Formula (5.49) yields $D(r)$ by means of

$$\mathbf{E}\big(v_2(N_0 \oplus B(o, r))\big) = p_1 \pi r^2 + p_2\big(2\pi r^2 - \gamma_r(1)\big),$$

where

$$\gamma_r(1) = \begin{cases} 2r^2 \cos^{-1} \dfrac{1}{2r} - \dfrac{\sqrt{4r^2 - 1}}{2} & \text{for } 2r \geq 1, \\ 0 & \text{otherwise.} \end{cases}$$

This quantity $\gamma_r(1)$ is the area of intersection of two circles of radius r and centres separated by unit distance. Then (3.30) gives $H_s(r)$. Moreover,

$$\bar{c} = p_1 + 2p_2$$

and

$$c_o(\{\varphi \in \mathbb{N} : \varphi(B(o, r)) = 1\}) = \frac{1}{\bar{c}} \cdot \begin{cases} (p_1 + 2p_2) & \text{for } 0 \leq r < 1, \\ p_1 & \text{for } r \geq 1. \end{cases}$$

Therefore

$$1 - D(r) = \frac{1}{\bar{c}} \exp\left(-\lambda_p \left(p_1 \pi r^2 + p_2(2\pi r^2 - \gamma_r(1))\right)\right) \cdot \begin{cases} p_1 + 2p_2 & \text{for } 0 \leq r < 1, \\ p_1 & \text{for } r \geq 1. \end{cases}$$

$$(5.53)$$

Some processes are simultaneously Neyman–Scott *and* Cox processes. A Neyman–Scott process is Cox if the number of points per cluster has a Poisson distribution. This category includes the Matérn cluster process and the modified Thomas process. For example, the driving measure Ψ of the Matérn cluster process is

$$\Psi(B) = \mu \sum_{x \in \Phi_p} v_d\left(B \cap B(x, R)\right) \qquad \text{for Borel } B,$$

where Φ_p denotes the Poisson process of parent points. Thall (1983) shows that for a Poisson cluster process that is also a Cox process, the number of points per cluster either takes on each positive integer value with positive probability or is identically equal to one. Thus, a Gauss-Poisson process cannot be Cox.

Van Lieshout (1995) and Baddeley *et al.* (1996) show that certain nonstationary Neyman–Scott processes can also be interpreted as Gibbs processes, at least if the Baddeley–Møller 'nearest neighbour' generalisation is used.

The Neyman–Scott process has found many applications as a model for spatial phenomena. Neyman and Scott (1958) introduce it as a model for the locations of galaxies in space; in this context the cluster centres are unobservable. (This model is now superseded; see Martínez and Saar, 2002.) The survey paper by Neyman and Scott (1972) contains further examples: the distribution of larvae in fields and the geometry of bombing. In the case of larvae in fields the parent points are positions of egg masses and the daughter points are the positions of larvae. In the case of the geometry of bombing, the parent points are the aiming points of bomb release and the daughter points are the points of impact of individual bombs. There are many applications of cluster processes in ecology, beginning with Warren (1971); see Stoyan and Penttinen (2000) and Illian *et al.* (2008).

The simulation of Neyman–Scott processes in a window W is easy: first the parent Poisson process is simulated in $W_{\oplus R}$, where R is a radius such that $\mathbf{P}(N_0 \subset B(o, R))$ is very small or zero. Then around each parent point x a cluster distributed as $N_0 + x$ is generated. The sample is then the union of all daughter points in W. Brix and Kendall (2002) propose a procedure that simulates only those parents giving rise to daughter points in W. See Illian *et al.* (2008) for more details.

Statistics for Neyman–Scott processes is surprisingly complicated. Various methods have been tried, in particular the minimum contrast method as in Møller and Waagepetersen (2004) with $D(r)$, $H_s(r)$, $K(r)$ and $g(r)$. Experience gained by Stoyan and Stoyan (1996) and Brix (1999) indicates that it is preferable to work with $g(r)$. Tanaka *et al.* (2008) suggest an approximative maximum likelihood method.

5.4 Hard-core point processes

A hard-core point process is a point process in which the constituent points are forbidden to lie closer together than a certain positive minimum distance. In this section the minimum distance is denoted by D. These hard-core models describe patterns produced by the locations of centres of non-overlapping discs or balls of diameter D. There are many models of this type, which are often rather complicated. In this book some of them are presented in Section 6.5.3 the context of random sets. The reason for this is that these point processes belong to systems of hard balls, which are considered as random sets. The models considered in Section 6.5.3 are:

- the Stienen model,
- the lilypond model,
- packings of hard balls, and
- the random sequential adsorption (RSA) model.

Furthermore, Section 5.5 discusses the Gibbs hard-core process, which assigns an infinite *energy* to point pairs having a distance less than D so that the likelihood that a realisation containing such point pairs is zero.

This section considers the popular *(second) Matérn hard-core process*, for which exact mathematical calculations are possible and the corresponding dependent thinning procedure can be easily described. In his famous booklet, Matérn (1960, 1986) suggests two such models, but here only the model yielding a higher eventual intensity of points is considered. A similar hard-core process has also been proposed by Porod (1952). Its intensity is, however, not higher than that yielded by the model described here.

The model results from a dependent thinning applied to a homogeneous Poisson process Φ_b of intensity λ_b. The points of Φ_b are marked independently by random numbers uniformly distributed over $(0, 1)$. The dependent thinning retains the point x of Φ_b with mark $m(x)$ if the ball $B(x, D)$ contains no points of Φ_b with marks smaller than $m(x)$. Formally the thinned process Φ is given by

$$\Phi = \left\{ x \in \Phi_b : m(x) < m(y) \text{ for all } y \in \Phi_b \cap B(x, D) \setminus \{x\} \right\}. \tag{5.54}$$

It is stationary and isotropic.

The process Φ may be an appropriate model in certain ecological contexts. The points of Φ_b are for example the locations of seedlings while the marks are the instants at which the seedlings commence growth. So Φ consists of the positions x of those plants which are the eldest seedlings in their required growing area $B(x, D)$. A simulation is shown in Figure 5.7.

The intensity λ of Φ is given by

$$\lambda = p\lambda_b, \tag{5.55}$$

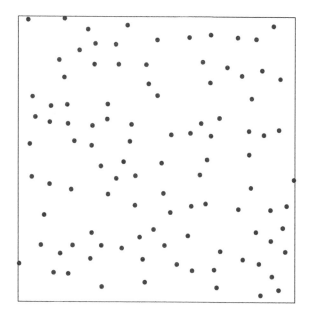

Figure 5.7   A simulated realisation of a planar second Matérn hard-core process in a unit square. The parameters are $\lambda_b = 200$ and $D = 0.05$.

where p is the *Palm retaining probability* of the typical point of Φ_b. It is given by

$$p = \int_0^1 r(t)\,dt = \frac{1 - \exp(-\lambda_b c)}{\lambda_b c} \qquad \text{with } c = b_d D^d, \tag{5.56}$$

where $r(t) = \exp(-\lambda_b c t)$ is the retaining probability of a point of mark t. This formula follows from the observation that the point process

$$\{x \in \Phi_b : m(x) < t\}$$

is simply a t-thinning (the usual p is here t) of a Poisson process, hence itself a Poisson process of intensity $\lambda_b t$. So $r(t)$ is the probability that a ball of radius D contains no points of the t-thinned process. Consequently,

$$\lambda = \frac{1 - \exp(-\lambda_b c)}{c}. \tag{5.57}$$

If $D = 1$, then the maximum intensity λ is approximately 0.32 if $d = 2$, and 0.24 if $d = 3$. These maxima result from letting λ_b tend to infinity.

The second order product density $\varrho^{(2)}(r)$ is given by

$$\varrho^{(2)}(r) = \lambda^2 \begin{cases} 0, & \text{if } r \leq D, \\[2ex] \dfrac{2\Gamma_D(r)\big(1 - \exp(-\lambda_b c)\big) - 2c\big(1 - \exp(-\lambda_b \Gamma_D(r))\big)}{c\Gamma_D(r)\big(\Gamma_D(r) - c\big)}, & \text{if } D < r \leq 2D, \end{cases}$$

$$\tag{5.58}$$

with

$$\Gamma_D(r) = v_d\big(B(o, D) \cup B(\mathbf{r}, D)\big) = 2c - \gamma_D(r) \qquad \text{if } r > D, \tag{5.59}$$

where $r = \|\mathbf{r}\|$. In the practically important cases for $r \le 2D$

$$\gamma_D(r) = \begin{cases} 2D^2 \cos^{-1}\left(\dfrac{r}{2D}\right) - \dfrac{r}{2}\sqrt{4D^2 - r^2}, & \text{if } d = 2, \\[2ex] \dfrac{4\pi}{3}D^3\left(1 - \dfrac{3r}{4D} + \dfrac{r^3}{16D^3}\right), & \text{if } d = 3. \end{cases} \tag{5.60}$$

Note that $\gamma_D(r)$ vanishes for $r \ge 2D$.

In the planar case the approximation $\gamma_D(r) = \pi D^2 - 2Dr$ may be useful. The pair correlation function of this process (with particular values chosen for the parameters) is graphed in Figure 4.7.

These formulae can be obtained by using $\varrho^{(2)}(r) = \lambda_b^2 k(r)$, where $k(r)$ is the two-point Palm probability that two points of Φ_b separated by distance r are both retained. Details of calculation of the above using the second-order factorial moment measure are to be found in Stoyan and Stoyan (1985). In this paper, as well as in Stoyan (1987, 1988), the Matérn model is generalised to have a variable hard-core, leading then to a soft-core model. So it is possible to obtain more realistic models for point patterns for which the original Matérn model is too simple. Statistically analysed examples include systems of locations of trees and villages.

Further generalisations that lead to germ–grain models with variable balls as grains are considered in Månsson and Rudemo (2002) and Andersson et al. (2006). Teichmann et al. (2013) study an elegant and general model where in a point pair of distance r in Φ_b the point with the larger mark is eliminated with probability $p(r)$. The classical second Matérn process is obtained with $p(r) = \mathbf{1}_{[0,D]}(r)$.

Closely related to this class of processes is Matheron's dead leaves model; see Stoyan and Schlather (2000). All these models, if considered as germ–grain models, have only a very small volume fraction $p = \lambda b_d D^d$, usually smaller than $1/2^d$.

The simulation of a Matérn hard-core process is not difficult. First a Poisson process of intensity λ_b is simulated in $W_{\oplus D}$. These points are marked independently with uniform marks in $(0, 1)$. Then the specific thinning rule explained on p. 176 is applied. Note that points deemed to be thinned out still have their job to do in the thinning rule and hence should be removed only after all points are classified as retained or thinned out; the retained points in W then form the required sample.

The thinning leading to the Matérn hard-core process can be refined such that higher intensities λ are possible, by the price of more complicated algorithms; see Møller et al. (2010) and Hörig and Redenbach (2012).

5.5 Gibbs point processes

5.5.1 Introduction

Gibbs point processes (*Gibbs processes* for short) first arose in the theories of statistical physics. They are related to the so-called Boltzmann–Gibbs distributions, which describe equilibrium states of subsystems of very large closed physical systems. In this book the physical background will be largely ignored; the treatment will concentrate on explaining the mathematical reasoning and statistical aspects. Gibbs processes will be treated as a class of point processes and described in terms of the characteristics used to describe the classes of point processes discussed above. Detailed discussion of the mathematical–physical aspects may be found in Minlos (1968), Ruelle (1969, 1970) and Preston (1974, 1976). Historical aspects of the mathematical theory are covered briefly in Kallenberg (1983a, p. 177).

Note that Gibbs processes are *not* a universal class of models which can be used in all situations. They are good models for patterns with some degree of regularity (more regular than Matérn hard-core processes) or for moderate clustering, but can be deficient in cases of strong clustering (but see also the remark by Baddeley *et al.*, 1996, p. 347).

To commence the discussion in the spirit of this chapter, note that new point processes can be produced from old not only by thinning, clustering and superposition, but also by trans-forming their distributions by means of probability densities. The theory of Gibbs processes is a development of this idea.

Example 5.7.   *Altering the distribution of a Poisson process of finite total intensity measure*
Suppose that Ψ is a Poisson process on $\mathbb{R}^d$ with intensity measure Λ of total mass $\Lambda(\mathbb{R}^d) = 1$. Let the distribution of Ψ be Q. A new point process distribution P_λ is defined for positive λ by

$$P_\lambda(Y) = \int_Y f_\lambda(\varphi)Q(d\varphi) \qquad \text{for } Y \in \mathcal{N}, \tag{5.61}$$

where

$$f_\lambda(\varphi) = \lambda^{\varphi(\mathbb{R}^d)} \exp(1 - \lambda). \tag{5.62}$$

The point process under the new probability distribution P_λ is a Poisson process on $\mathbb{R}^d$ of intensity measure $\lambda\Lambda$.

Proof. Let Φ be a Poisson process of intensity measure $\lambda\Lambda$ and with distribution P_Φ. Then calculation shows that P_λ and P_Φ give the same measure to the set

$$Y_{K,n} = \{\varphi \in \mathbb{N} : \varphi(\mathbb{R}^d) = n, \varphi(K) = 0\}$$

for each nonnegative integer n and each compact set K. The void-probabilities of both pro-cesses agree since

$$P_\lambda(Y_{K,n}) = \lambda^n \exp(1 - \lambda)\exp\bigl(-\Lambda(K)\bigr)\exp\bigl(-(1 - \Lambda(K))\bigr)\frac{\bigl(1 - \Lambda(K)\bigr)^n}{n!} = Q(Y_{K,n}),$$

and therefore their distributions are equal.   □

Fundamental to the theory of Gibbs processes is the idea of a general *basic* or *weight process* Ψ of distribution Q. A new probability distribution P can be defined on $[\mathbb{N}, \mathcal{N}]$ by means of a probability density $f(\varphi)$ as

$$P(Y) = \int_Y f(\varphi)Q(d\varphi) \qquad \text{for } Y \in \mathcal{N}. \tag{5.63}$$

If Φ is a point process of distribution P, then $f(\varphi)$ is the *likelihood* that Φ takes the realisation φ compared with Ψ taking on the same realisation. Obviously this construction is very general. In particular whole classes of realisations of Φ can be prohibited by setting f to be zero on the corresponding subset of $\mathbb{N}$.

The form of the density $f(\varphi)$ is often suggested by the field of application. It can be chosen conveniently to model interactions between the points of the process. The Gibbs process approach is straightforward if the process to be modelled contains only a finite number of points confined to a bounded region B. Problems arise when considering more general point processes. However, the construction via (5.63) may fail for point processes in the whole $\mathbb{R}^d$, as the following example shows.

Example 5.8.   *Counter-example*

The homogeneous Poisson process $\Phi^{(1)}$ of intensity λ_1 cannot be obtained in the simple manner described above from another homogeneous Poisson process $\Phi^{(2)}$ of intensity λ_2 if $\lambda_1 \neq \lambda_2$.

Proof. Consider a family $\{W_n\}$ of compact sets with $W_n \uparrow \mathbb{R}^d$. Set, for $\lambda > 0$,

$$Y_\lambda = \{\varphi \in \mathbb{N} : \varphi(W_n)/v_d(W_n) \to \lambda\}.$$

Then clearly $Y_{\lambda_1} \cap Y_{\lambda_2}$ is empty but $\mathbf{P}(\Phi^{(i)} \in Y_{\lambda_i}) = 1, i = 1, 2$. So the distributions of the two processes are mutually singular, and thus a representation of the form (5.63) cannot be obtained. $\square$

To obtain Gibbs processes in the situation of point processes on the whole $\mathbb{R}^d$, the density idea described above must be applied to conditional distributions confined to bounded regions. Section 5.5.3 describes the case of stationary point processes on $\mathbb{R}^d$. Before this, Section 5.5.2 deals with the simpler case of finite point processes.

5.5.2   Gibbs point processes in bounded regions

To facilitate understanding, the following considers a Gibbs point process Φ of distribution P and with *exactly n points in a specific bounded set B* and no points outside B ('absence of an outer field'). Physicists term this a *canonical ensemble*. This case of fixed n is of great practical importance for statistical purposes, where one often works under the condition that the number of points observed in the window of observation is fixed. It is assumed that the distribution of the point process is given by a probability density function $f : \mathbb{R}^{nd} \to [0, \infty)$ so that

$$\mathbf{P}(\Phi \in Y) = \int \ldots \int_{\{x_1,\ldots,x_n\}\in Y} f(x_1, \ldots, x_n)\, dx_1 \ldots dx_n \qquad \text{for } Y \in \mathcal{N}_{B,n}, \qquad (5.64)$$

where $\mathcal{N}_{B,n}$ denotes the trace of $\mathcal{N}$ on the set of all point sequences with n points in B. Because point processes are unordered sets of points, $f(x_1, \ldots, x_n)$ is taken not to depend on the order of the arguments. The form used to describe f is

$$f(x_1, \ldots, x_n) = \exp\big(-E(x_1, \ldots, x_n)\big)/Z. \qquad (5.65)$$

Here Z is a normalising constant called the *(configurational) partition function*, and the function $E : \mathbb{R}^{nd} \to \mathbb{R} \cup \{\infty\}$ is called the *energy function* or *multiparticle functional*; also it does not depend on the order of the arguments. These terms derive from statistical mechanics. The convention is adopted that

$$\exp(-\infty) = 0.$$

Frequently E is chosen to be of a specialised form, as a sum of *pair potentials*

$$E(x_1, \ldots, x_n) = \sum \sum_{1 \leq i < j \leq n} \theta(\|x_i - x_j\|). \qquad (5.66)$$

The name 'pair potential' for the function $\theta : [0, \infty) \to (-\infty, \infty]$ also comes from statistical physics.

This leads to the formula

$$f(x_1, \ldots, x_n) = \exp\left(-\sum \sum_{1 \leq i < j \leq n} \theta(\|x_i - x_j\|)\right) \Big/ Z. \qquad (5.67)$$

Often the function θ is rewritten so that (5.67) takes the form

$$f(x_1, \ldots, x_n) = \exp\left(-\beta \sum \sum_{1 \leq i < j \leq n} \theta(\|x_i - x_j\|)\right) \Big/ Z, \qquad (5.68)$$

where β is called the *inverse temperature*.

A typical example of a pair potential is graphed in Figure 5.8. Its features correspond to features of the point process of density f constructed by using Formulae (5.63), (5.65) and (5.66):

(a) Because $\theta(r)$ is infinite for $r < D$ the inter-point distance can never be less than D. So the point process is in fact a hard-core model.

(b) Because $\theta(r)$ is large when r is only a little larger than D such distances r are possible but are unlikely to occur as inter-point distances.

(c) In contrast, inter-point distances close to r_1, where θ takes its minimum, should occur relatively frequently.

From the theory of statistical mechanics, distributions of the form given in (5.65) arise when the point process Φ is constrained to have fixed mean energy

$$\int_{B^n} E(x_1, \ldots, x_n) f(x_1, \ldots, x_n) \, dx_1 \ldots dx_n$$

and maximum entropy

$$H = -\int_{B^n} f(x_1, \ldots, x_n) \ln\big(f(x_1, \ldots, x_n)\big) \, dx_1 \ldots dx_n$$

among such fixed-energy systems.

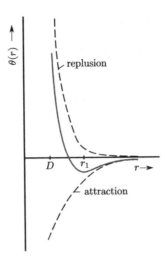

Figure 5.8 A typical pair potential, the result of superposition of attractive and repulsive forces.

If the distribution of Φ is given by Equation (5.64), as an integral of a probability density, then the underlying weight process Ψ has a density proportional to the uniform density on B^n. Clearly the form of B is crucial to the distribution of the resulting Gibbs process Φ.

Often, Formula (5.65) with E as in (5.66) is re-cast in terms of an *interaction function* $e(t) = \exp(-\theta(t))$. This motivates the frequent description of such point processes Φ as *pairwise interaction processes* or (special cases of) *Markov point processes*; see van Lieshout (2000) and Møller and Waagepetersen (2004). Then the exponential of the sum (5.66) is replaced by a product of individual exponential terms.

The normalising constant Z in Formula (5.65) is given by

$$Z = \int_{B^n} \exp(-E(x_1, \ldots, x_n)) \, dx_1 \ldots dx_n. \tag{5.69}$$

Computation of the partition function Z is notoriously difficult. Ogata and Tanemura (1981) propose an approximation method for the case of pair potentials, the second cluster approximation:

$$Z \approx v_d(B)^n \left(1 - \frac{a}{v_d(B)}\right)^{n(n-1)/2}, \tag{5.70}$$

where

$$a = \left(\int_0^\infty (1 - \exp(-\theta(r))) r^{d-1} \, dr\right) db_d.$$

(Recall that db_d is the surface area of the unit ball in $\mathbb{R}^d$.) Ogata and Tanemura (1984) and Gates and Westcott (1986) discuss the quality of this approximation. It is poor if the pair potential $\theta(r)$ is negative for small r or has large deviations from zero. Ripley (1988) and Diggle *et al.* (1994) describe other more precise approximations. The partition function is often determined by MCMC simulation; see Ogata and Tanemura (1984), Geyer (1992), Huang and Ogata (2001) and Møller and Waagepetersen (2004).

Example 5.9.   *The square-well pair potential*

For $0 < D < \rho$ and $-\infty < b < \infty$ let

$$
\theta_{D,b,\rho}(r) = \begin{cases} \infty, & \text{if } r \leq D, \\ -b, & \text{if } D < r \leq \rho, \\ 0, & \text{if } r > \rho. \end{cases}
$$

Then the density $f(\varphi)$ has the form

$$
f(\varphi) = \begin{cases} \exp\big(bs_\varphi(\rho)\big)\big/ Z, & \text{if } \varphi \subset B \text{ and } s_\varphi(D) = 0, \\ 0, & \text{otherwise,} \end{cases}
$$

where for $\varphi = \{x_1, \ldots, x_n\}$,

$$
s_\varphi(\rho) = \sum\sum_{1 \leq i < j \leq n} \alpha_\rho(x_i, x_j)
$$

with

$$
\alpha_\rho(x_i, x_j) = \begin{cases} 1, & \text{if } \|x_i - x_j\| \leq \rho, \\ 0, & \text{otherwise.} \end{cases}
$$

So $s_\varphi(\rho)$ is simply the number of inter-point distances in φ that are less than or equal to ρ.

In Example 5.9 the parameter D is a hard-core distance while ρ is a radius of interaction. The cases $b > 0$ and $b < 0$ correspond respectively to clustering and regularity. Particular cases are:

- the uniform distribution of n points in B (when $b = 0$, $D = 0$);
- the *Strauss model* with fixed number of points (when $D = 0$).

The Strauss model is discussed in many papers on Gibbs processes; see the books by van Lieshout (2000), Møller and Waagepetersen (2004), Daley and Vere-Jones (2008) and Illian *et al.* (2008), often for random n. The model need not then be well-defined for positive b; see Møller (1999). Even for fixed n, problems may appear in simulations for large positive b: the algorithm may get stuck in the same point configuration for many iterations.

Ogata and Tanemura (1981) study other parametric families of pair potentials θ, namely

$$
\theta_1(r) = -\ln\big(1 + (\alpha r - 1)e^{-\beta r^2}\big),
$$
$$
\theta_2(r) = -\ln\big(1 + (\alpha - 1)e^{-\beta r^2}\big),
$$
$$
\theta_3(r) = \beta \left(\frac{\sigma}{r}\right)^n - \alpha \left(\frac{\sigma}{r}\right)^m,
$$

for $\alpha \geq 0$, $\beta, \sigma > 0$ and $n > m$.

The last potential $\theta_3(r)$ is a common model in statistical mechanics, called the *Lennard-Jones potential*. The case $n = 12$, $m = 6$ is of particular interest. It provides both types of interaction, attraction and repulsion, at different scales.

Ogata and Tanemura also describe *maximum likelihood estimation* of parameters for these models in case of sparseness, using approximations to the likelihood via (5.65) and (5.70). A general approximation method by using MCMC is proposed in Gu and Zhu (2001). Asymptotic optimality properties of maximum likelihood estimators are studied by Mase (1992).

Penttinen (1984) proposed an alternative approach. In the planar case, his method gives for the parameter b of $\theta_{D,b,\rho}$ (if D and ρ are known) the estimator

$$\hat{b} = -\ln \frac{2v_2(B)s_\Phi(\rho)}{\pi n^2(\rho^2 - D^2)}.$$

For the Strauss model this simplifies to

$$\hat{b} = -\ln \frac{2v_2(B)s_\Phi(\rho)}{\pi n^2 \rho^2}.$$

Here and below B and W, the window of observation, coincide.

In contrast, the maximum likelihood method leads for the Strauss process to the estimator

$$\hat{b} = -\ln \frac{2v_2(B)s_\Phi(\rho)}{\pi n(n-1)\rho^2}.$$

More general statistical methods for Gibbs processes with finite point number are described in the books by Møller and Waagepetersen (2004), Illian *et al.* (2008) and Gelfand *et al.* (2010). These books also describe a number of varied examples.

Gibbs processes with a *random* total number of points are also studied; these go in physics under the name *grand canonical ensemble*. The reader is referred to the literature, for example van Lieshout (2000), Møller and Waagepetersen (2004), Illian *et al.* (2008) and Gelfand *et al.* (2010).

Simulation of Gibbs processes with fixed point number

For the simulation of Gibbs processes usually an iterative procedure called *Markov chain Monte Carlo* (MCMC) method is applied. The idea is to construct a Markov chain, the states of which are point configurations in B and the stationary distribution of which is that of the Gibbs process. The algorithm simulates the Markov chain for a long time. Once the chain is in its stationary regime, every state of it is a sample from the stationary distribution of the chain; of course, states close together in time are spatially correlated. The following sketches two particular algorithms. Van Lieshout (2000), Møller and Waagepetersen (2004), Gelfand *et al.* (2010) and Huber (2011) provide a detailed overview of MCMC algorithms for point pattern simulation and give the theoretical justifications.

The first method is the classical Metropolis algorithm as in Metropolis *et al.* (1953); see also Torquato (2002) and Billera and Diaconis (2001), who explain the 'magic trick' of this famous algorithm, why it works.

The steps of the *Metropolis algorithm* in the Gibbs process case are as follows.

1. Generate an initial configuration φ of n points in B and compute its energy $E(\varphi)$.

2. Move a point by displacing it along each coordinate axis by amounts randomly and uniformly in the interval $[-\delta, \delta]$, where δ is a positive parameter. This leads to a new configuration φ'. Compute the new energy $E(\varphi')$ and the energy difference $\Delta E = E(\varphi') - E(\varphi)$.

3. Accept the new configuration φ' with probability $\min\{1, \exp(-\Delta E)\}$. This means: if the energy of the new configuration is smaller than that of the old, the move $\varphi \to \varphi'$ is accepted. Otherwise it is accepted with probability $\exp(-\Delta E)$; non-acceptance of φ' means that φ is retained.

4. Continue by going to step 2 or, after sufficiently many moves, goto step 5.

5. Accept the current configuration as a sample of the Gibbs process.

The order in which the points are moved is fixed. Moves that lead to positions outside of B are corrected in suitable form. Torquato (2002) recommended choosing the parameter δ so that about 50% of the proposed moves would be accepted.

Metropolis *et al.* (1953) considered the planar case with B a square under periodic boundary conditions, which means that their intention is to simulate a stationary Gibbs process. As an illustration, they simulated the hard disc case with $n = 224$.

Instead of the random shifts, other 'moves' may also be used, for example jumps; see the literature on MCMC, for example Frenkel and Smit (2002) and Brooks *et al.* (2011).

The second method uses a *birth-and-death algorithm*. As in Illian *et al.* (2008), in the following a simple but nontrivial example is discussed followed by a description of the general algorithm.

Example 5.10. *Simulation of the Gibbs hard-core process with n points*

The starting configuration can be any n-point pattern in B in which the smallest inter-point distance is not less than D, for example a lattice of n points.

Assume that the current state of the Markov chain after l steps is $\{x_1, \ldots, x_n\}$. The $(l+1)^{\text{st}}$ step is:

1. A point in the set $\{x_1, \ldots, x_n\}$ is deleted at random, where each point has the same probability $1/n$ to be deleted. The deleted point is, say, x_k.

2. A new point x is drawn in B with uniform distribution. If

$$\min\{\|x - x_i\| : i = 1, 2, \ldots, n, \ i \neq k\} \geq D,$$

then x replaces x_k to form a new set of n points $\{x_1, \ldots, x_{k-1}, x, x_{k+1}, \ldots, x_n\}$. Otherwise, if

$$\min\{\|x - x_i\| : i = 1, 2, \ldots, n, \ i \neq k\} < D,$$

the point x is rejected and a new proposal point is drawn until the proposed x can be accepted.

When n and D are large, it often takes a long time to have a new point accepted.

Inspection of patterns simulated with this algorithm shows that there is some boundary effect called 'drift towards the boundary': the point density along the edge of B is slightly higher than in its interior. This is a correct result, which is in full agreement with the Gibbs distribution in a bounded set B with no influence from outside. If a pattern looking like a 'stationary sample' is required, periodic boundary conditions can be used.

The structure of the general birth-and-death algorithm closely resembles that of the algorithm for the Gibbs hard-core process above. A randomly chosen point in the pattern is deleted and a new point is generated following the conditional probability density function as above. The simulation starts with an initial point configuration of positive density, which does not 'contradict' the pair potential.

Suppose that after the first l steps the point pattern is $\{x_1, \ldots, x_n\}$. The $(l + 1)^{\text{st}}$ step is:

1. A point in the set $\{x_1, \ldots, x_n\}$ is deleted at random, where each point in $\{x_1, \ldots, x_n\}$ has the same probability $1/n$ to be deleted. The deleted point is, say, x_k.

2. A new point is simulated based on the conditional probability density function

$$f(x|x_1, \ldots, x_{k-1}, x_{k+1}, \ldots, x_n) = \frac{f(x_1, \ldots, x_{k-1}, x, x_{k+1}, \ldots, x_n)}{\int_B f(x_1, \ldots, x_{k-1}, y, x_{k+1}, \ldots, x_n)\, dy}. \qquad (5.71)$$

The new configuration $\{x_1, \ldots, x_{k-1}, x, x_{k+1}, \ldots, x_n\}$ is the result after the $(l + 1)^{\text{st}}$ step. As above for the Gibbs hard-core process, the x in substep 2 may be simulated by the rejection method: let M be an upper bound of the nonnormalised conditional density function

$$h(x) = \exp\left(-\sum_{j=1, j \neq k}^{n} \theta(\|x - x_j\|)\right).$$

A random point x is generated uniformly in B along with an independent uniform random number u on $[0, 1]$. The point x is accepted if $h(x) \geq Mu$. Otherwise a new point x is proposed.

The algorithm for the Gibbs hard-core process follows this approach with $h(x) = 1$ if $\|x - x_j\| > D$ for all $j = 1, \ldots, n$, $j \neq k$, and zero otherwise. Illian *et al.* (2008, Section 3.6.2) presented details for the application of the birth-and-death algorithm.

5.5.3 Stationary Gibbs point processes

The methods of Section 5.5.2 cannot be used directly in the case when B is replaced by $\mathbb{R}^d$ and the number of points is infinite. A more sophisticated approach is required, which considers the restrictions of point processes to bounded sets B, and these must be conditioned on the behaviour of the point process outside the bounded region B. Formally one considers probabilities of the form $\pi_B(Y \mid \psi)$. This is the probability that the pattern in B belongs to the configuration set Y, given that outside B the pattern takes on the configuration ψ. So

$$\pi_B(Y \mid \psi) = \mathbf{P}(\Phi \cap B \in Y \mid \Phi \cap B^c = \psi), \qquad (5.72)$$

where B is a bounded Borel set, Y is in $\mathcal{N}$, and ψ belongs to the set $\mathbb{N}_{B^c}$, where

$$\mathbb{N}_{B^c} = \{\varphi \in \mathbb{N} : \varphi(B) = 0\}.$$

These probabilities must satisfy certain measurability and compatibility conditions, which are omitted here. The objective of this section is to show how in many cases the probabilities π_B can be expressed elegantly using a weight process and a pair potential.

The mathematical problems of the above formulation were intensively studied in the 1970s by, for example, Georgii (1976), Preston (1976), Matthes *et al.* (1978, 1979), Nguyen and Zessin (1979b), Glötzl (1980) and Kallenberg (1983a).

Rather than deriving the π_B from the point process Φ, one can attempt to move in the reverse direction. Indeed, this is the more useful procedure for model specification.

Suppose the *local specification* $\{\pi_B\}$ is given for all bounded Borel B (using the terminology of Preston, 1976). A point process Φ with distribution P is said to be a *Gibbs point process with respect to the local specification* $\{\pi_B\}$, if for all bounded Borel B

$$P(Y) = \int \pi_B(Y \mid \varphi \cap B^c) P(d\varphi). \tag{5.73}$$

The equation (5.73) is called the *DLR-equation* after Dobrushin, Lanford and Ruelle. Note that its solution P may not be unique.

The concept of a Gibbs point process is closely connected to the older notion of a *Boltzmann–Gibbs distribution* in statistical physics. The corresponding physical question concerns the existence and uniqueness of equilibrium states for dynamic particle systems given local dynamic specifications analogous to (5.73). See Ruelle (1969) for a discussion of the general physical theory and Georgii (1988) for a general study of uniqueness and phase transition in Gibbs processes.

Formula (5.65) gives an exponential form for the density of P in the bounded case. By analogy the probabilities π_B are considered in exponential form

$$\pi_B(Y \mid \psi) = \frac{1}{Z(B, \psi)} \int 1_Y\big((\varphi \cap B) \cup \psi\big) \exp\big(-\overline{E}(\varphi \cap B, \psi)\big) Q(d\varphi), \tag{5.74}$$

where Q is the distribution of a basic or weight process and $Z(B, \psi)$ is a normalising constant. The function $\overline{E} : \mathbb{N}_{\text{finite}} \times \mathbb{N} \to (-\infty, \infty]$ is the *conditional energy* (here $\mathbb{N}_{\text{finite}} = \{\varphi \in \mathbb{N} : \varphi(\mathbb{R}^d) < \infty\}$). The conditional energy $\overline{E}(\chi, \psi)$ can be interpreted as the energy required to add the point set χ to the configuration ψ.

To complete the analogy with (5.65), Formula (5.74) can be derived as the solution to a maximum entropy problem when the energy is held constant.

Two simplifying conditions are imposed in the subsequent discussion.

First, $\overline{E}$ is assumed to have an additive form expressed in terms of the *local energy* E

$$\overline{E}(\chi, \psi) = E(x_1, \psi) + E(x_2, \psi \cup \{x_1\}) + \cdots + E(x_n, \psi \cup \{x_1, \ldots, x_{n-1}\}), \tag{5.75}$$

where $\chi = \{x_1, \ldots, x_n\}$. Naturally a consistency condition must be imposed on E to ensure that the sum does not depend on the order of the x_i. The term $E(x, \varphi)$ can be interpreted as the energy required to add the point x to the configuration φ and is called *local energy*.

Second, the weight process is supposed to be Poisson with intensity measure Λ_Q. This is the natural generalisation of the 'uniform' weight process used to treat Gibbs processes with nonrandom finite total number of points as in Section 5.5.2.

Clearly it is important to decide when a distribution P leads to a local specification $\{\pi_B\}$ in the form (5.74) with a local energy. Some condition resembling absolute continuity is required. The answer has been given by Georgii (1976), Nguyen and Zessin (1979b) and Glötzl (1980).

Theorem 5.1. *A necessary and sufficient condition for the point process distribution P to have local specification $\{\pi_B\}$ in the form of (5.74) with $\overline{E}$ given by (5.75) is as follows: the reduced Campbell measure $\mathscr{C}^!$ corresponding to P satisfies the absolute continuity condition*

$$\mathscr{C}^! \ll \Lambda_Q \times P, \tag{5.76}$$

where Λ_Q is the intensity measure of the Poisson weight process. Furthermore, if (5.76) holds, then for almost all x and φ (with respect to P)

$$\frac{\mathrm{d}\mathscr{C}^!}{\mathrm{d}(\Lambda_Q \times P)}(x, \varphi) = \exp\bigl(-E(x, \varphi)\bigr). \tag{5.77}$$

Equation (5.77) suggests a plausible argument for the interpretation of $E(x, \varphi)$ as the energy required to add x to φ. For simplicity consider the case when P and Q are both stationary with intensities λ and λ_Q. Then the Radon–Nikodym derivative on the left-hand side of (5.77) can be written informally as

$$\frac{\mathrm{d}\mathscr{C}^!}{\mathrm{d}(\Lambda_Q \times P)}(x, \varphi) = \frac{\mathscr{C}^!(\mathrm{d}x \times \mathrm{d}\varphi)}{\lambda_Q \nu_d(\mathrm{d}x) P(\mathrm{d}\varphi)}.$$

By Formula (4.65), which expresses the reduced Palm measure in terms of the reduced Campbell measure $\mathscr{C}^!$, the infinitesimal quotient can be written as

$$\frac{\mathscr{C}^!(\mathrm{d}x \times \mathrm{d}\varphi)}{\lambda_Q \nu_d(\mathrm{d}x) P(\mathrm{d}\varphi)} = \frac{P_o^!(\mathrm{d}\varphi) \lambda \nu_d(\mathrm{d}x)}{\lambda_Q \nu_d(\mathrm{d}x) P(\mathrm{d}\varphi)},$$

which is proportional to

$$\frac{P_o^!(\mathrm{d}\varphi)}{P(\mathrm{d}\varphi)}.$$

This last quotient is the ratio of the infinitesimal probabilities for Φ under P (in the denominator) and P conditioned to have a point at o (in the numerator). The standard physical interpretation regards the negative logarithm of the quotient as the energy required to make the transition from φ to $\varphi \cup \{o\}$.

The question of uniqueness of the process given by the local specification is of great interest and importance in probability theory, but appears not to be of great relevance in spatial statistics. This is because all practical applications involve limited quantities of data observed through bounded windows. The non-uniqueness phenomenon may manifest itself in increased sensitivity to boundary conditions. However, it seems that non-uniqueness phenomena usually begin to play an important part when the numbers involved approach the thermodynamic ranges of the order of 10^{23} points. On the other hand, when the relatively low numbers of points in statistical problems are being considered, the problem of edge-effects is likely to be of far greater significance (B. D. Ripley, 1984, personal communication).

Together with Formula (4.65), Equation (5.77) implies that

$$\lambda P_o^!(Y) = \lambda_Q \int_Y \exp\big(-E(o, \varphi)\big) P(\mathrm{d}\varphi), \tag{5.78}$$

which in turn implies that the reduced Palm distribution $P_o^!$ is absolutely continuous with respect to P. This enables a quick determination that a point process is *not* a Gibbs process (with respect to a Poisson process, using local energy). For example, the lattice of points of integral coordinates, translated in both coordinates by random shifts uniformly over $(0, 1)$, is a stationary point process which cannot be so expressed. Indeed its reduced Palm probability measure yields simply the lattice with the origin deleted and this is clearly not absolutely continuous with respect to the original law.

In the following it is assumed that $\lambda_Q = 1$, which can be compensated by the freedom to choose the local energy.

An equivalent formulation of Equation (5.78) is

$$\lambda \int g(\varphi) P_o^!(\mathrm{d}\varphi) = \int g(\varphi) \exp\big(-E(o, \varphi)\big) P(\mathrm{d}\varphi) \tag{5.79}$$

for every nonnegative measurable $g : \mathbb{N} \to \mathbb{R}$, which has several applications, for example in the Takacs–Fiksel method.

A more general equation is the *GNZ (Georgii–Nguyen–Zessin) equation*

$$\int \int h(x, \varphi \setminus \{x\}) \varphi(\mathrm{d}x) P(\mathrm{d}\varphi) = \int \int h(x, \varphi) \exp\big(-E(x, \varphi)\big) P(\mathrm{d}\varphi) \mathrm{d}x \tag{5.80}$$

or

$$\mathbf{E}\left(\sum_{x \in \Phi} h(x, \Phi \setminus \{x\})\right) = \int \mathbf{E}\big(h(x, \Phi) \exp(-E(x, \Phi))\big) \mathrm{d}x,$$

where h is a function as in (4.64). The Campbell–Mecke theorem yields (5.79).

An important class of general Gibbs processes corresponding to the class of models specified by the Formulae (5.74) and (5.76) is obtained by requiring that the local energy $E(x, \varphi)$ can be expressed in terms of a *pair potential* θ as in Section 5.5.2:

$$E(x, \varphi) = \alpha + \sum_{y \in \varphi} \theta(\|x - y\|). \tag{5.81}$$

The constant α is called the *chemical activity*. If the pair potential $\theta(r)$ is fixed, then there is the tendency that the Gibbs process intensity λ increases with decreasing α.

Note that the assumption (5.75) implies that the Gibbs process is *hereditary*, that is, for every 'forbidden' point configuration φ and every $x \in \mathbb{R}^d$, the configuration $\varphi \cup \{x\}$ is also forbidden. Dereudre and Lavancier (2009) and Dereudre *et al.* (2012) consider Gibbs processes that are nonhereditary and give a generalised form of Equation (5.80). Such processes appear, for example, in the context of Delaunay and Voronoi tessellations with cells of limited variability (neither too large, too small, nor too flat); see Dereudre and Lavancier (2011).

It is difficult to obtain formulae for the point process characteristics introduced in Chapter 4. However, there are various approximations in the pair potential case. For the intensity λ,

Baddeley and Nair (2012) suggest the following approach: Substituting $g(\varphi) \equiv 1$ in (5.79) gives

$$\lambda = e^{-\alpha} \int \exp\left(-\sum_{y \in \varphi} \theta(y)\right) P(d\varphi).$$

The integral on the right-hand side is a mean value with respect to the Gibbs process. The idea is to approximate it by an analogous mean with respect to the Poisson process of the same intensity. The latter mean is equal to $\exp(-\lambda G)$ with

$$G = \int_{\mathbb{R}^d} \left(1 - \exp(-\theta(x))\right) dx. \tag{5.82}$$

Thus the intensity approximation is the solution of

$$\lambda = e^{-\alpha} \exp(-\lambda G). \tag{5.83}$$

In the particular case of a *Gibbs hard-core process*, which is the Gibbs process with pair potential given by

$$\theta(r) = \begin{cases} \infty, & \text{if } r \leq D, \\ 0, & \text{otherwise,} \end{cases} \tag{5.84}$$

the constant G in (5.82) is given by

$$G = b_d D^d \tag{5.85}$$

and the approximation of λ is then the solution of

$$\lambda = e^{-\alpha} \exp(-\lambda b_d D^d). \tag{5.86}$$

Clearly, for high intensities this approximation may fail; then in the approximation a lattice process may be used instead of a Poisson process.

Mase (1990) suggests an approximation method for calculating means of the form

$$\mathbf{E}\left(\prod_{x \in \Phi} c(x)\right)$$

for nonnegative measurable functions c on $\mathbb{R}^d$. Quantities such as the spherical contact distribution function $H_s(r)$ and the nearest-neighbour distance distribution function $D(r)$ belong to this class. For example, consider

$$c(x) = 1 - \mathbf{1}_B(x)$$

for a Borel set B. Then

$$\mathbf{E}\left(\prod_{x \in \Phi} c(x)\right) = \mathbf{E}\mathbf{1}_{\{\Phi(B)=0\}} = \mathbf{P}\big(\Phi(B) = 0\big).$$

If $\theta(r)$ has a finite *interaction radius* ρ, that is to say, if $\theta(r)$ is zero for $r \geq \rho$, then the following relationship holds:

$$\lambda P_o^!(Y_r) = P(Y_r)e^{-\alpha} \qquad \text{for } r \geq \rho, \tag{5.87}$$

where $Y_r \in \mathcal{N}_{\mathbb{N}_{B(o,r)^c}}$, the trace of $\mathcal{N}$ on the set $\mathbb{N}_{B(o,r)^c}$ of point sequences φ that $\varphi \cap B(o, r) = \emptyset$. In terms of the nearest-neighbour distance distribution function $D(r)$ and the spherical contact distribution function $H_s(r)$, the relationship given in (5.87) becomes

$$\lambda(1 - D(r)) = (1 - H_s(r))e^{-\alpha} \qquad \text{for } r \geq \rho. \tag{5.88}$$

For large windows, when $D(r)$ and $H_s(r)$ can be estimated quite accurately, Formula (5.88) suggests a way to estimate α.

In the case of a Gibbs hard-core process, substituting $r = D$ in Formula (5.88) yields a close relationship between λ, D and α:

$$\lambda = (1 - H_s(D))e^{-\alpha}. \tag{5.89}$$

Formula (6.103) in Section 6.5.3 for the spherical contact distribution function of a random dense hard ball packing can be used to calculate $H_s(r)$ for the Gibbs hard-core process for $d = 3$ and $r > \frac{D}{2}$.

In the physical literature there are ways to derive approximations for the pair correlation function $g(r)$ of stationary Gibbs processes, given λ and $\theta(r)$; see the classical book by Hansen and McDonald (2006). These are acceptable in the case of small intensity λ, as 'low-density' approximations. The simplest approximation is

$$g(r) \simeq \exp(-\theta(r)) \qquad \text{for } r \geq 0, \tag{5.90}$$

(see Hansen and McDonald, 2006, p. 35). Better results are obtained by means of the so-called Percus–Yevick approximation, which yields approximations for $g(r)$ in terms of λ and $\theta(r)$:

$$g(r) \approx \frac{c(r)}{1 - \exp(\theta(r))} \qquad \text{for } r \geq 0, \tag{5.91}$$

(see Diggle *et al.*, 1987, and Hansen and McDonald, 2006, p. 91), where $c(r)$ is a further function called *direct correlation function*, which is defined implicitly by the solution of the so-called Ornstein–Zernicke equation

$$c(r) = h(r) - \lambda(c * h)(r) \qquad \text{for } r \geq 0, \tag{5.92}$$

with $h(r) = g(r) - 1$ and

$$(c * h)(r) = \begin{cases} \displaystyle\int_0^\infty \int_0^{2\pi} p(s)q\left(\sqrt{r^2 + s^2 - 2rs\cos\alpha}\right)s\,d\alpha\,ds, & \text{for } d - 2, \\[3mm] \displaystyle\int_0^\infty \int_0^{2\pi} \int_0^{\pi} p(s)q\left(\sqrt{r^2 + s^2 - 2rs\cos\alpha\sin\beta}\right)s^2\sin\beta\,d\beta\,d\alpha\,ds & \text{for } d = 3, \end{cases} \tag{5.93}$$

following Ornstein and Zernike (1914). Combining (5.91) and (5.92) yields an integral equation for $g(r)$. Torquato (2002, p. 72ff) and Buryachenko (2007, p. 156) discuss its solution by means of Fourier methods. In the case of hard balls for volume fractions up to 0.5 excellent approximations of $g(r)$ are obtained; see for example Hansen and McDonald (2006, p. 94).

The theory of stationary Gibbs processes was extended to the *marked case*; see Coeurjolly *et al.* (2012). This opens the way to important random-set models, for example the quermass models. These form a class of germ–grain models with morphological interaction between the grains, which leads to structures that cannot come from Boolean models.

An approximation for $g(r)$ for three-dimensional Gibbs hard-core processes with low intensity is given in Kanaun and Levin (1994), based on Willis (1978):

$$g(r) = \left(1 - \mathbf{1}_{[0,D)}(r)\right)\left(1 + \left(\frac{2+p}{2(1-p)^2} - 1\right)\cos\left(\frac{2\pi r}{D}\right)\exp\left(4(1 - \frac{r}{D})\right)\right) \qquad \text{for } r \geq 0,$$

(5.94)

where p denotes the volume fraction V_V of the system of hard balls of diameter D centred at the points of the Gibbs hard-core process. Comparison of the graph corresponding to the $g(r)$ in this formula with Figure 4.7 shows that many details of $g(r)$ appearing in the case of high volume fraction p are not reflected by the low-density approximation (5.94).

The planar case is also discussed in Buryachenko (2007).

Methods of statistics for stationary Gibbs processes are described in Illian *et al.* (2008) and Baddeley and Dereudre (2013). A particular method is the Takacs–Fiksel method, which is based on the GNZ equation (5.80); see Coeurjolly *et al.* (2012), where the statistical properties of this method are thoroughly investigated. An alternative may be the approach in Diggle *et al.* (1987), which is based on (5.91) and uses estimates of the pair correlation function. Another popular approach is the 'method' to ignore that the given sample belongs to a stationary process and to apply methods for bounded regions B, where B is identified with the window of observation and it is believed that the estimated pair potential is also valid for the stationary process. This approach opens the way to application of the powerful methods in van Lieshout (2000) and Møller and Waagepetersen (2004).

5.5.4 Spatial birth-and-death processes

A spatial birth-and-death process is a continuous-time Markov process with state space $\mathbb{N}$ the family of point patterns; see Preston (1977), Glötzl (1981), van Lieshout (2000) and Daley and Vere-Jones (2008). These processes provide models of spatio–temporal (or space–time) point processes, which are studied in the books by Daley and Vere-Jones (2008), Illian *et al.* (2008) and Cressie and Wikle (2011). There the term 'birth-and-death' process is sometimes used in a more general sense than in this section, for systems where points randomly appear, exist for a random time and then disappear. The relevance of the birth-and-death processes discussed here lies in the close relationship to Gibbs processes and especially in the way in which they provide a means to simulate Gibbs processes as suggested by Ripley (1977).

A brief description of a spatial birth-and-death process is that it is a point process in space which changes at isolated instants of time either by the *birth* of a new point, which is added to the pattern, or by the *death* of an old point, which is deleted. Moreover, it has a *Markov* property in time: the probability of a change depends only on the current configuration of the process. Births are controlled by a *birth-rate* $\boldsymbol{b}$, a positive measurable function

$$\boldsymbol{b} : \mathbb{R}^d \times \mathbb{N} \to \mathbb{R}$$

with

$$\int_B b(x, \varphi)\, dx < \infty$$

for all bounded Borel sets B and all φ in $\mathbb{N}$. The probability that a birth occurs within the set B in the short time interval $[t, t + s)$, given that the process has configuration φ at time t, is

$$s \int_B b(x, \varphi)\, dx + o(s).$$

The occurrence of a death is controlled by a *death-rate* d, another positive measurable function

$$d : \mathbb{R}^d \times \mathbb{N} \to \mathbb{R}.$$

The probability that the point x is deleted from the pattern $\varphi \cup \{x\}$ in the time interval $[t, t + s)$, given that $\varphi \cup \{x\}$ is the state of the birth-and-death process at time t, is

$$sd(x, \varphi) + o(s).$$

The two functions b and d characterise the birth-and-death process. It is of course necessary to impose conditions on b and d to eliminate various pathological properties. Suitable conditions as well as a formal construction can be found in Preston (1976).

The *equilibrium* or time-stationary distributions of the process are of particular interest. A point process distribution P is an *equilibrium distribution* of the birth-and-death process if when the birth-and-death process is started with distribution P at time 0 its distribution continues as P for all time. The actual configuration may change but the statistics of the configuration have a time-invariant distribution. An equilibrium state is a *time-reversible* state of the birth-and-death process if in addition a realisation of the ensuing evolution cannot be statistically distinguished from its reversal in time. Technically for P to be time-reversible it is required that

$$\int g(\varphi)\Omega h(\varphi) P(d\varphi) = \int h(\varphi)\Omega g(\varphi) P(d\varphi), \tag{5.95}$$

where Ω is the closure of the infinitesimal generator of the birth-and-death process, defined for nonnegative measurable locally bounded g (i.e. there is a bounded Borel set A_g with $g(\varphi) = g(\varphi \cap A_g)$) by

$$\Omega g(\varphi) = \int b(x, \varphi)\big(g(\varphi \cup \{x\}) - g(\varphi)\big)\varphi(dx)$$

$$+ \int d(x, \varphi \setminus \{x\})\big(g(\varphi \setminus \{x\}) - g(\varphi)\big)\varphi(dx).$$

For these definitions to be sensible, it is necessary to assume b and d are well behaved, and in particular that they generate a uniquely defined Markov process.

Glötzl (1981) presents a simple example: Assume $b(x, \varphi) \equiv 1$ and $d \equiv 1$. Then the generated process in any bounded set B is independent of what happens outside. After a random waiting time, which follows an exponential distribution with parameter $v_d(B)$ a new point is born in B at an uniformly distributed location in B, independently of other points. Each point dies after a random lifetime, which is exponentially distributed with parameter 1 and independent of the behaviour of other points. In analogy to queueing theory, this model can be called *spatial M/M/∞ queue*. The shapshots of this process (i.e. the system of all points existing at the same instant), when in equilibrium, are Poisson point processes of intensity 1.

A similar model, but with lifetimes of the points following a general distribution is called *spatial M/G/∞ queue*; see Daley and Vere-Jones (2008). This latter model is not a birth-and-death process in the sense considered here, if the lifetimes are not exponentially distributed.

Under suitable conditions the spatial birth-and-death process will converge to a statistical equilibrium which is the time-reversible distribution P; see Preston (1976, Theorem 7.1). Lotwick and Silverman (1981) prove this by using coupling arguments; see also Møller (1999). If the birth-and-death process has a time-reversible equilibrium distribution P, then, whatever initial distribution is given, it converges to P. This fact forms the basis for a practical simulation procedure of stationary Gibbs processes as described in the next section.

Glötzl (1981) demonstrates the relationship between time-reversible birth-and-death process and Gibbs processes.

Theorem 5.2. *Let the rates b and d satisfy*

$$\frac{b(x, \varphi)b(y, \varphi \cup \{x\})}{d(x, \varphi)d(y, \varphi \cup \{x\})} = \frac{b(y, \varphi)b(x, \varphi \cup \{y\})}{d(y, \varphi)d(x, \varphi \cup \{y\})}$$

for all x and y in $\mathbb{R}^d$ and φ in $\mathbb{N}$.

Then the distribution P is a time-reversible distribution of the spatial birth-and-death process with rates b and d if and only if P is the distribution of a Gibbs process based on a homogeneous Poisson weight process with intensity 1 and local energy

$$E(x, \varphi) = -\ln \left(\frac{b(x, \varphi)}{d(x, \varphi)} \right)$$

for x in $\mathbb{R}^d$ and φ in $\mathbb{N}$.

5.5.5   Simulation of stationary Gibbs processes

Stationary Gibbs processes are also simulated by means of MCMC methods, and there are two widely used methods.

Application of birth-and-death processes

A first method uses spatial birth-and-death processes as the first 'MC' component. The simulation is carried out in a bounded region, usually a rectangle or cuboid W, with periodic boundary conditions. In this way one hopes to obtain point patterns that behave like samples of stationary point processes. The idea is to repeat the point patterns periodically outside of W. Figure 4.9 in Chapter 4 illustrates the periodic continuation for a rectangular window.

The simulation method is as follows:

(1) choose a starting configuration;

(2) choose the rates b and d;

(3) simulate the birth-and-death process with these rates until it is considered sufficiently close to equilibrium.

Then Theorem 5.2 is applied to assert that the final configuration behaves approximately as a sample of the desired Gibbs process.

The starting configuration used in (1) can be of n points generated by the uniform distribution on W, where n is some estimate of the expected mean point number in W, or a lattice satisfying a hard-core condition or the empirical pattern which is under investigation.

The rates chosen in (2) can take many forms. The form which is most easily programmed is to take constant birth rate $b(x, \varphi) = 1$ with new points distributed uniformly over W, and death rate $d(x, \varphi) = \exp(E(x, \varphi))$. However, this leads to slow convergence in many cases, as observed already by Ripley (1977). The reason is that new-born points are frequently removed almost immediately, in particular in the hard-core case. It is usually preferable to take constant death rate $d(x, \varphi) = 1$ and birth rate proportional to the configuration $b(x, \varphi) = \exp(-E(x, \varphi))$. This means new points have to be generated by rejection sampling, as described below, but in practice the convergence to equilibrium is then faster, compensating for the more complicated way of simulating new points.

The simulation step (3) then takes place in continuous time running in variable time-steps from one incident of birth or death to the next, as follows.

Suppose that at time t a birth or death has just happened, and the configuration φ in W consists of n points $x_1, \ldots, x_n$. In the computation of the birth rate $b(x, \varphi) = \exp(-E(x, \varphi))$ the continuation of the point pattern outside of W must be taken into account. It must be determined whether the next incident is actually a birth or a death, when it is to happen, and (if it is a birth) where it is to happen or (if it is a death) which point is to be removed.

First, simulate the time T to the next incident as an exponential random variable with mean

$$\left(n + \int_W \exp(-E(x, \varphi)) \, dx \right)^{-1},$$

evaluating the integral $I(\varphi) = \int_W \exp(-E(x, \varphi)) \, dx$ numerically. Then use simulation to arrange for the next incident to be the death of point x_i, for $i = 1, \ldots,$ or n, each with probability

$$\frac{1}{n + I(\varphi)},$$

and a birth otherwise.

If it is to be a birth, then use rejection sampling to add a new point at x with probability density

$$\frac{\exp(-E(x, \varphi))}{I(\varphi)}.$$

Thus it has been determined that the next birth or death has occurred at time $t + T$, and whether the next incident was a birth or a death, and which point has been affected. Now the whole procedure is repeated with, of course, recalculating $I(\varphi)$ whenever the point pattern φ is altered, and using independent random variables for each simulation step.

Application of the Metropolis–Hastings algorithm

An alternative to the birth-and-death-process simulation is the Metropolis–Hastings algorithm. Usually it is also carried out in a rectangular or cuboidal window W with periodic boundary conditions. Then methods for the simulation of Gibbs processes in bounded windows and random point number are applied. (Use of periodic boundary conditions ensures that the patterns look like realisations of a stationary point process, the random-number models lead to the density fluctuations which appear in stationary processes in bounded windows.)

The general idea of a Metropolis–Hastings simulation algorithm is that in each step a randomly chosen *proposal* for a change of the current configuration is made, and by some *acceptance rule* it is decided to accept the proposal or not. This rule makes that the resulting Markov chain has the target stationary distribution. Whether a proposal is accepted or not depends on chance.

For Gibbs processes it is based on the nonnormalised density

$$f_n(x_1, \ldots, x_n) = \exp\left(-\left(\alpha_f n + \sum_{i=1}^{n-1}\sum_{j=i+1}^{n} \theta(\|x_i - x_j\|)\right)\right) \qquad (5.96)$$

with $x_1, \ldots, x_n \in W$ for $n = 0, 1, \ldots$, where α_f is a parameter which controls the intensity in a way similar to the chemical activity α in (5.81); see p. 198.

The proposals are controlled by the following three simulation parameters:

- $b(\{x_1, \ldots, x_n\})$: the probability that the 'birth' of a new point will be proposed, while $1 - b(\{x_1, \ldots, x_n\})$ is the probability that the 'death' of a point will be proposed;

- $q_{\text{birth}}(\{x_1, \ldots, x_n\}; x)$: the proposal density function for the location of the new point x;

- $q_{\text{death}}(\{x_1, \ldots, x_n\}; x_k)$: the probability for the proposal to delete the point x_k from the set $\{x_1, \ldots, x_n\}$.

Simple concrete choices of the proposal parameters are as follows:

$$b(\{x_1, \ldots, x_n\}) \equiv \frac{1}{2},$$

$$q_{\text{birth}}(\{x_1, \ldots, x_n\}; x) = \frac{1}{v(W)} \qquad \text{for } x \in W,$$

$$q_{\text{death}}(\{x_1, \ldots, x_n\}; x_k) = \frac{1}{n} \qquad \text{for } x_k \in \{x_1, \ldots, x_n\}.$$

The simulation starts with a point pattern $\{x_{01}, \ldots, x_{0m}\}$ for which

$$f_m(\{x_{01}, \ldots, x_{0m}\}) > 0,$$

that is, which does not contradict the basic properties of the process. (For example, it is a hard-core pattern if a hard-core process is to be simulated.)

Suppose that after l iteration steps the configuration is $\{x_1, \ldots, x_n\}$. In the $(l + 1)^{\text{st}}$ step, initially a decision is made as to whether a new point may be added to the pattern (birth) or removed (death), with probability $b(\{x_1, \ldots, x_n\})$ or $1 - b(\{x_1, \ldots, x_n\})$, respectively.

If the $(l+1)^{\text{st}}$ step is a birth, the position of the new point $x \in W$ is proposed from the density function $q_{\text{birth}}(\{x_1, \ldots, x_n\}; x)$ and the proposal is accepted with probability

$$p_{\text{birth}} = \min\{1, \rho_{\text{birth}}\}, \tag{5.97}$$

where

$$\rho_{\text{birth}} = \frac{f_n(\{x_1, \ldots, x_n, x\})\big(1 - b(\{x_1, \ldots, x_n, x\})\big)q_{\text{death}}(\{x_1, \ldots, x_n, x\}; x)}{f_n(\{x_1, \ldots, x_n\})b(\{x_1, \ldots, x_n\})q_{\text{birth}}(\{x_1, \ldots, x_n\}; x)} \tag{5.98}$$

is the so-called Metropolis–Hastings birth ratio. With this ratio and the corresponding death ratio the Markov chain is controlled such that its stationary distribution is the desired point process distribution. If the proposal is accepted, the new point is added to the point configuration. If the proposal is not accepted, the configuration does not change in this step.

If the $(l+1)^{\text{st}}$ step is a death and if $\{x_1, \ldots, x_n\}$ is not empty, a point x_k in $\{x_1, \ldots, x_n\}$ is proposed to be deleted with probability $q_{\text{death}}(\{x_1, \ldots, x_n\}; x_k)$. This deletion is accepted with probability

$$p_{\text{death}} = \min\{1, \rho_{\text{death}}\}, \tag{5.99}$$

where

$$\rho_{\text{death}} = \frac{f_n(\{x_1, \ldots, x_n\} \setminus \{x_k\})\, b(\{x_1, \ldots, x_n\} \setminus \{x_k\})\, q_{\text{birth}}(\{x_1, \ldots, x_n\} \setminus \{x_k\}; x_k)}{f_n(\{x_1, \ldots, x_n\})\big(1 - b(\{x_1, \ldots, x_n\})\big)\, q_{\text{death}}(\{x_1, \ldots, x_n\}; x_k)} \tag{5.100}$$

is now the Metropolis–Hastings death ratio. Here $\{x_1, \ldots, x_n\} \setminus \{x_k\}$ denotes the point pattern $\{x_1, \ldots, x_n\}$ without the point x_k, $k \in \{1, \ldots, n\}$. If the proposal is accepted, the point x_k is removed from the configuration. If the proposal is not accepted, the configuration does not change in this step. (If the configuration was empty, the empty configuration does not change in a death step.)

Note that the Metropolis–Hastings algorithm works also in the nonhereditary case; see Dereudre and Lavancier (2011).

Example 5.11. *Simulation of a planar Strauss process with random number of points (Illian et al., 2008, p. 152)*

The Strauss process with parameters $\alpha_f = -8.0$, $b = -\exp(0.3) = -1.35$ and $\rho = 0.08$ is simulated in W, the unit square. The parameters of the Metropolis–Hastings algorithm are chosen as above.

Suppose the configuration at step l is $\{x_1, \ldots, x_n\}$. At first, a decision has to be made as whether to add or delete a point, either of which happens with probability $\frac{1}{2}$. If a point is to be added to the configuration, the candidate x is chosen uniformly within W. Then the value of the Metropolis–Hastings ratio for a birth is calculated,

$$\rho_{\text{birth}} = \frac{1}{n+1} \exp\left(-\alpha_f + b \sum_{i=1}^{n} \mathbf{1}(0 < \|x - x_i\| \leq \rho)\right).$$

The proposal x is accepted with probability

$$p_{\text{birth}} = \min\{1, \rho_{\text{birth}}\}.$$

This means, a random number u is drawn from $[0, 1]$ and if $u \leq p_{birth}$, the proposal x is accepted and the new configuration is $\{x_1, \ldots, x_n, x\}$. If $u > p_{birth}$, the old configuration $\{x_1, \ldots, x_n\}$ does not change. (Note that a similar mechanism is used above to decide whether the next step is a potential birth or death.)

If a point is to be deleted, a number k is randomly chosen from $\{1, \ldots, n\}$ and x_k is proposed to be removed. The Metropolis–Hastings ratio for a death

$$\rho_{death} = n \exp \left(\alpha_f - b \sum_{i=1, i \neq k}^{n} \mathbf{1}(0 < \|x_k - x_i\| \leq \rho) \right)$$

and $p_{death} = \min\{1, \rho_{death}\}$ are evaluated. The proposal is accepted, that is, x_k is removed, with probability α_{death} as above. If the proposal is accepted, the new configuration is $\{x_1, \ldots, x_{k-1}, x_{k+1}, x_n\}$, otherwise the configuration does not change.

In this simulation example, the initial configuration is a realisation from a binomial process with 120 points. Figure 5.9 on p. 199 shows the simulation results after 100, 5000 and 10 000 iterations, where the numbers of points were by chance 120, 166 and 166, points respectively. (It is only by coincidence that the last two numbers are the same.) Figure 5.10 on p. 200 shows a trace plot of the evolution of the number of points in the 10 000 iteration steps.

In this particular case, the burn-in took around 2000 iterations after which the generated point patterns were considered to be realisations from the Strauss process.

Technical remarks on Gibbs process simulation

(1) In all kinds of simulation, the Gibbs process samples arise as equilibrium limits. In practice it suffices to continue the simulation until the approximation to statistical equilibrium is judged as acceptable. That means, a burn-in period must be accepted, the length of which is not known *a priori*. This implies that the user has to observe the simulation process in trace plots (Møller and Waagepetersen, 2004) and to decide if the algorithm is likely to have converged. In favourable cases the so-called 'perfect simulation' approach can be employed to avoid the burn-in problem: see for example Kendall and Møller (2000) and Huber (2011). Possible indicators are the current point number n or the numbers of births, deaths or moves in fixed numbers of simulation steps. The computing times can be long, in particular for Gibbs processes with repulsion and high point density.

A typical example of such a simulation might lead to the following: after perhaps $10n$ steps the influence of the starting configuration will have largely disappeared, where n is the average number of points in W in the stationary state. Patterns which are separated by $2n$, or perhaps $4n$ steps can then be considered as 'independent'. Practical experience shows that excessively long simulation times do not solve all problems: numerical effects can then make difficulties.

For example, in the case of a hard-core potential and high intensity (very negative α) the burn-in period may be very long and perhaps the packing algorithms as in Section 6.5.3 provide a better approach than MCMC methods.

(2) A difficult problem in the case of stationary Gibbs processes is related to chemical activity α and intensity λ. If the birth-and-death process simulation is used, α can be explicitly built in the algorithm; the intensity may be first unknown and is obtained (approximately) by

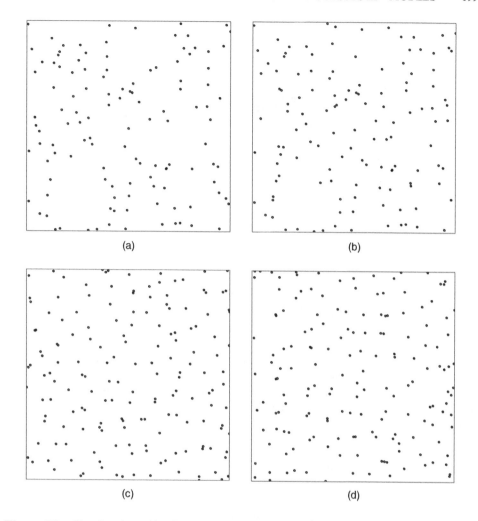

Figure 5.9 Simulated samples from a Strauss process with random number of points generated by the Metropolis–Hastings algorithm on the unit square. The parameters of the process
and the specific implementation of the algorithm are described in Example 5.11. (a) Initial
configuration ($n = 120$), configuration after (b) 100 steps ($n = 120$), (c) 5000 steps ($n = 166$)
and (d) 10 000 steps ($n = 166$). Some tendency towards regularity can be detected but close
pairs are also allowed. Reproduced from Illian *et al.* (2008, Figure 3.21).

statistical analysis of simulated samples. If the intensity is prescribed, α has to be determined
by experimentation.

In the Metropolis–Hastings approach one has to determine that value of α_f which yields
the desired intensity. Samples that behave (approximately) as realisations from a stationary
Gibbs process can be obtained by simulating with fixed point number (defined by the desired
intensity) and periodic boundary conditions but in an extended window W_{ext} containing the
window W in which samples are needed. Then, as it must be, the point number in W shows
random fluctuations, while the number in W_{ext} is fixed.

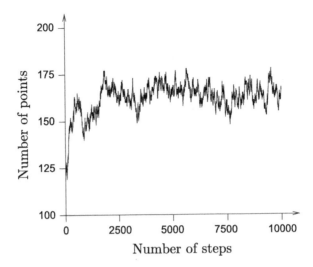

Figure 5.10   A trace plot for the number of points of a run of the Metropolis–Hastings algorithm described in Example 5.11. Reproduced from Illian *et al.* (2008, Figure 3.22(a)).

(3) Some kind of verification of the various approximations is necessary at the end of such simulations. Clearly, one can re-estimate the process parameters and see if the estimated parameters are close to those used in the simulation. Stoyan and Grabarnik (1991) suggested the marking of points of a stationary Gibbs process with 'energy marks': the point x_n has the mark

$$m_n = E(x_n, \Phi \setminus \{x_n\}) = \exp\left(\alpha + \sum_{y \in \Phi \setminus \{x_n\}} \theta(\|x_n - y\|)\right). \tag{5.101}$$

By means of Formula (5.79) it is easy to show that the mean of the marks is $1/\lambda$, where λ is the intensity. Thus a necessary condition for a 'good sample' is that the sum of all energy marks is approximately equal to the window area.

5.6 Shot-noise fields

5.6.1 Definition and examples

A random field $\{Z(x)\}$ is a family of random variables as explained in books on geostatistics. It has values $Z(x)$ in *all* $x \in \mathbb{R}^d$. An important class of random fields is formed by shot-noise fields $\{S_\Psi(x)\}$. Such a random field describes a structure which results from the superposition of (random) *impulses*, also called *responses* or *effects*, which are related to the points of a (marked) point process Ψ. The name origins from the one-dimensional case, where the points are instants of time and the impulses acoustic signals.

The impulses are usually assumed to be homogeneous, that is, they only depend on the mark of the point and the difference $x - x_n$ of the location x of interest and the point x_n. Then

$s(x - x_n, m_n)$ is the contribution of point x_n with mark m_n to $S_\Psi(x)$. Formally, $s(x, m)$ is the contribution of a point at origin o with mark m to the shot-noise field.

In many cases a suitable name for the impulse function $s(x, m)$ is *attenuation function*, as its value decreases with the distance of x from o and really describes an attenuation process. Superposition of the impulses of all points of Ψ yields the *shot-noise field*:

$$S_\Psi(x) = \sum_{[x_n;m_n]\in\Psi} s(x - x_n, m_n) \qquad \text{for } x \in \mathbb{R}^d. \tag{5.102}$$

An important special case is

$$s(x - x_n, m_n) = m_n f(x - x_n), \tag{5.103}$$

where f is a probability density function. In this approach the 'point mass' m_n is continuously distributed around x_n.

However, there are also situations where the superposition follows the max-rule, that is, the value at x is the maximum of the $s(x - x_n, m_n)$,

$$M(x) = \max_{[x_n;m_n]\in\Psi} s(x - x_n, m_n) \qquad \text{for } x \in \mathbb{R}^d; \tag{5.104}$$

multiplicative superposition may also be considered, for example in the context of resource interference in ecology; see Wu *et al.* (1985).

Usually, the main interest is to determine characteristics of the field $\{S_\Psi(x)\}$. Some formulae are given in the next section for the stationary case and independent marks.

Example 5.12. *Competition load (Adler, 1996)*

The points of Ψ represent plants in a (planar) plant community, where each plant competes with its neighbours, and the marks are size parameters such as tree stem diameters. Clearly, the strength of the competition decreases with distance from the reference plant and a given point is influenced by the competition load from all its neighbours. The result is a random competition field $\{S_\Psi(x)\}$, where $S_\Psi(x)$ is the sum of the competition contributions from all plants in the vicinity of location x. In this context, the attenuation function $s(x, m)$ is called 'local competition function'. Here the value of the random field at a deterministic location x is probably less important than at the typical plant.

In Adler (1996) the following attenuation function

$$s(x, m) = m^\alpha \exp\left(-\delta \frac{\|x\|}{m^\beta}\right) \tag{5.105}$$

is considered, where α, β and δ are model parameters.

Example 5.13. *Interference field in wireless communication (Baccelli and Błaszczyszyn, 2009b)*

The points of Ψ are transmitters and the 'impulses' correspond to the signals emitted by antennas. The total power received from all transmitters at a given location x is of interest. The strength of the signals depends on the distances from the antennas and on noise.

The impulse function $s(x, m)$ can have the form

$$s(x, m) = \frac{m}{l(\|x\|)},\tag{5.106}$$

which corresponds to the case of omnidirectional antennas, where l is the so-called *path-loss function*. An example is

$$l(r) = (Ar)^\beta,\tag{5.107}$$

called *simplified path-loss function*, where A is positive and β greater than 2. (In fact, this $l(r)$ is only a mathematical approximation; for small r the values are too large, and the mean of the corresponding random field $\{S_\Psi(x)\}$ is infinite.) An exponential mark distribution is also a popular choice.

Shot-noise fields are also used to construct point process models, the so-called *shot-noise Cox processes*; see Møller and Waagepetersen (2004), Illian *et al.* (2008) and Jalilian *et al.* (2013). In optics and meteorology, as well as in other fields of physics, shot-noise fields have also been successfully used; see Frieden (1983), Rodrigez-Itube *et al.* (1987, 1988) and Cox and Isham (1988).

5.6.2   Moment formulae for stationary shot-noise fields

In the stationary case, mean, variance and variogram (defined in Formula (6.89)) can be derived analytically for the shot-noise random fields, but the relevant equations are simple only for the mean. Despite their complexity the reader may get an impression of the structure of these equations, by considering the volume integrals for the second-order characteristics in this section, taken from Cox and Isham (1980) and Schmidt (1985).

In order to avoid overly complex formulae, assume that the marks are quantitative (real-valued) and independent, that the ground process Φ (Ψ without the marks) is motion-invariant and that the impulse function is rotation-invariant with respect to o, which is used to replace $s(x, m)$ by $s(r, m)$, where r is the distance of x from o.

With these assumptions the field $\{S_\Psi(x)\}$ is stationary and isotropic,

$$S_\Psi(x) = \sum_{[x_n;m_n]\in\Psi} s(x - x_n, m_n) \qquad \text{for } x \in \mathbb{R}^d.\tag{5.108}$$

This means that the value of the random field at location x is a sum of impulses $s(x - x_n, m_n)$ in the points x_n, which depend on their marks m_n.

Mean value:

$$\mathbf{E}\big(S_\Psi(o)\big) = \lambda \int_{\mathbb{R}^d} e(x)\mathrm{d}x\tag{5.109}$$

with

$$e(x) = \int_0^\infty s(x, m)f_{\mathcal{M}}(m)\mathrm{d}m,\tag{5.110}$$

where $f_M(m)$ is the mark probability density function. Due to stationarity, the mean of $S_\Psi(x)$ is the same for all $x \in \mathbb{R}^d$, therefore only the value for $x = o$ has to be considered. Equation (5.109) is a simple consequence of the Campbell theorem for marked point processes.

For the Adler competition field the mean is

$$\mathbf{E}\big(S_\Psi(o)\big) = 2\pi\lambda \int_0^\infty \int_0^\infty m^\alpha r \exp\left(-\delta\frac{r}{m^\beta}\right) f_M(m)\mathrm{d}r\mathrm{d}m = \frac{2\pi\lambda}{\delta^2}\, \mathbf{E}\left(\mathcal{M}^{\alpha+2\beta}\right), \quad (5.111)$$

where $\mathbf{E}\left(\mathcal{M}^{\alpha+2\beta}\right)$ denotes the $(\alpha + 2\beta)^{\text{th}}$ moment of the marks.

Variance:

$$\mathbf{var}\big(S_\Psi(o)\big) = \lambda \int_{\mathbb{R}^d} e^{(2)}(x)\mathrm{d}x + \lambda^2 \int_{\mathbb{R}^d}\int_{\mathbb{R}^d} e(x)e(x+h)\mathrm{d}x\, \mathcal{K}(\mathrm{d}h) - \big(\mathbf{E}(S(o))\big)^2 \quad (5.112)$$

with $e(x)$ defined in (5.110) and

$$e^{(2)}(x) = \int_0^\infty \big(s(x,m)\big)^2 f_M(m)\,\mathrm{d}m,$$

where $\mathcal{K}$ is the reduced second moment measure of the ground process Φ, as introduced in Section 4.5. For a homogeneous Poisson process Formula (5.112) simplifies to

$$\mathbf{var}\big(S_\Psi(o)\big) = \lambda \int_{\mathbb{R}^d} e^{(2)}(x)\mathrm{d}x, \quad (5.113)$$

since in this case $\mathcal{K} = \nu_d$. For the Adler field

$$\mathbf{var}\big(S(o)\big) = \frac{2\pi\lambda}{4\delta^2}\, \mathbf{E}\left(\mathcal{M}^{2(\alpha+\beta)}\right). \quad (5.114)$$

For the interference field similar formulas can be derived; see Baccelli and Błaszczyszyn (2009b). The Laplace transform of $S_\Psi(o)$ is also known in the case of a Poisson process.

Variogram:

$$\gamma(r) = \mathbf{var}\big(S_\Psi(o)\big) + \big(\mathbf{E}(S_\Psi(o))\big)^2 - \lambda \int_{\mathbb{R}^d} e_r(x)\mathrm{d}x - \lambda^2 \int_{\mathbb{R}^d}\int_{\mathbb{R}^d} e(x)e(x+h-\mathbf{r})\mathrm{d}x\mathcal{K}(\mathrm{d}h)$$

$$(5.115)$$

with $e(x)$ defined in (5.110) and

$$e_r(x) = \int_0^\infty s(x,m)s(x-\mathbf{r},m)f_M(m)\mathrm{d}m, \quad (5.116)$$

where $\mathbf{r}$ is any point with distance r from the origin o.

Mean value of the shot-noise field at the typical point

In ecological modelling, the mean competition load at the typical point is of substantial interest, since it reflects the strength of the competition on a typical individual in the community, while

$\mathbf{E}(S_\Psi(o))$ is a spatial mean reflecting the strength of the competition on potential individuals. In the planar case, again based on polar coordinates,

$$\mathbf{E}_o(S_\Psi(o)) = 2\pi\lambda \int_0^\infty rg(r) \int_0^\infty s(r, m) f_\mathcal{M}(m) \, dm dr. \tag{5.117}$$

For the Adler field

$$\mathbf{E}_o(S_\Psi(o)) = 2\pi\lambda \int_0^\infty \int_0^\infty m^\alpha rg(r) \exp\left(-\delta \frac{r}{m^\beta}\right) f_\mathcal{M}(m) dr. \tag{5.118}$$

For a homogeneous Poisson process the pair correlation function $g(r)$ is equal to one and consequently

$$\mathbf{E}(S_\Psi(o)) = \mathbf{E}_o(S_\Psi(o)), \tag{5.119}$$

which is again a special case of the Slivnyak–Mecke theorem.

6

Random closed sets II – The general case

6.1 Basic properties

6.1.1 Introduction

Random closed sets serve as general mathematical models for irregular geometrical patterns, such as porous media. In the planar case they are sometimes called *random area patterns*. Random closed sets play a central rôle in stochastic geometry. Chapter 3 presents an important special case of a random set — the Boolean model. In this chapter a more general theory is considered.

What today are known as random sets occur relatively early in the literature of probability theory. Systematic studies can be found as early as in the decade of the thirties; the famous book by Kolmogorov (1933, p. 41) contains the phrase 'region ... whose shape depends on chance', and Kolmogorov gives there a mean-value formula closely related to Formula (6.23) below. Furthermore, in 1937 he studied a particular case of the Boolean model, in the context of crystallisation; see the KJMA theory in Section 6.6.4 on p. 273. Further early studies of particular random set models can be found in the literature of the 1940s and 1950s; see the references in Section 3.1.2.

Foundations of the modern theory of random closed sets were laid by Choquet (1954) in his work on *capacities*. Matheron (1967, 1975) studied random closed sets in locally compact separable Hausdorff spaces (LCS spaces). A more general theory has been published by Kendall (1974); this derives from work in the 1960s in which Davidson played an important part (see the remarks in Kendall, 1974). Modern references are the books Goutsias *et al.* (1997), Jeulin (1997), Molchanov (1993, 2005), Nguyen (2006) and Schneider and Weil (2008).

Random sets also play a rôle in the theory of control and optimisation. This theory uses so-called multivalued functions, which can be considered to be random sets provided their parameter space is endowed with a certain probability distribution (see Artstein, 1984b; Salinetti

Stochastic Geometry and its Applications, Third Edition.
Sung Nok Chiu, Dietrich Stoyan, Wilfrid S. Kendall and Joseph Mecke.
© 2013 John Wiley & Sons, Ltd. Published 2013 by John Wiley & Sons, Ltd.

and Wets, 1986; Aubin and Frankowska, 1990). Moreover, random set theory provides tools for the statistical inference for the set of parameter values that are observationally equivalent under the given data and maintained assumptions (Beresteanu *et al.*, 2011). The literature also discusses random fuzzy sets (see e.g. Kruse and Meyer, 1987; Bandemer and Näther, 1992; Dozzi *et al.*, 2001; Li *et al.*, 2002; Nguyen, 2005; Krätschmer, 2006; Nguyen and Wu, 2006; and Fu and Zhang, 2008).

In this chapter, the theory of Matheron (1975) will be presented, but only for the particular case in which the random sets lie in $\mathbb{R}^d$. (A reader who is familiar with the theory of locally compact separable Hausdorff spaces will observe that the generalisation is straightforward.) As a matter of convention the random sets considered here are *closed*; that is, the boundaries of the sets are included in the sets themselves.

6.1.2   Random set definition

The hitting σ-algebra $\mathcal{F}$

Random sets can be considered and characterised in many ways, typically by employing some system of deterministic test sets and then analysing the intersections of random sets with these test sets. Thus two choices must be made: first of the system of test sets, and then of the nature of measurement of the intersection. For example, if the random set Ξ might be infinite in extent then the test sets are best chosen to be bounded. On the other hand, as Kendall (1974) points out, some measurements (such as the Lebesgue measure of the intersection $\Xi \cap T$ for test sets T) are not always appropriate. (The measure-related way is briefly discussed in Section 7.3.4.) Rather, to quote Kendall, 'instead of working with some version of the "size" of $\Xi \cap T$ we should merely note whether $\Xi \cap T$ is empty or not'. This means of measurement, the mere observation of whether the intersection is void, is fundamental to the modern random set theories as developed by Matheron and Kendall. The hitting σ-algebra $\mathcal{F}$ is closely related to this means of measurement.

Suppose that $\mathbb{F}$ is the family of all closed subsets of $\mathbb{R}^d$. Then $\mathcal{F}$ denotes the smallest σ-algebra of subsets of $\mathbb{F}$ that contains all the 'hitting sets':

$$\mathbb{F}_K = \{F \in \mathbb{F} : F \cap K \neq \emptyset\} \qquad \text{for } K \in \mathbb{K}.$$

Thus the system of test sets is the system $\mathbb{K}$ of all compact subsets of $\mathbb{R}^d$; and the measurement of the intersection is simply an observation of whether or not the intersection is void. The hitting σ-algebra $\mathcal{F}$, also known as the *Effros σ-algebra* in the context of general topology, is built out of a family of assertions about whether or not the random set hits particular compact subsets.

Important properties of $\mathcal{F}$ can be found in Beer (1993), Molchanov (2005) and Schneider and Weil (2008); it can be shown that $\mathcal{F}$ is the Borel σ-algebra with respect to a suitable topology (the Fell topology) on $\mathbb{F}$.

Random closed sets

A random closed set Ξ is a random variable taking values in $[\mathbb{F}, \mathcal{F}]$. So Ξ is an $(\mathcal{A}, \mathcal{F})$-measurable mapping of a probability space $[\Omega, \mathcal{A}, \mathbf{P}]$ into $[\mathbb{F}, \mathcal{F}]$. This means that probabilities can be assigned to statements about Ξ when the statements are composed out of assertions that Ξ hits compact sets. Thus Ξ generates a distribution P on $[\mathbb{F}, \mathcal{F}]$ called the *distribution* of Ξ, with $P(A) = \mathbf{P}(\Xi \in A)$ for every $A \in \mathcal{F}$.

Simple examples of random sets

The full sophistication of the formal definition of random sets is not needed in order to appreciate many examples, such as:

- random points,
- closed balls of random centre positions and radii,
- closed cubes with random corner positions and edge lengths.

More complicated random sets can be generated by using set-theoretic operations such as $\cup$, $\cap$ and $\oplus$. It can be shown that these operations have suitable measurability properties, so that the resulting structures are indeed still random closed sets. An important example of a random closed set generated by such means is the Boolean model in Chapter 3, the definition of which uses the operations of set union and translation.

6.1.3 Capacity functional and Choquet theorem

Capacity functional

An effective means of characterising the distribution P of a random set Ξ is given by the *capacity functional* T_Ξ. It is defined by

$$T_\Xi(K) = \mathbf{P}(\Xi \cap K \neq \emptyset) = P(\mathbb{F}_K), \tag{6.1}$$

where K is any compact subset of $\mathbb{R}^d$. That means, $T_\Xi(K)$ is the probability that the random set Ξ hits the set K. Considered as a function on $\mathbb{K}$ the capacity functional is actually an *alternating Choquet capacity of infinite order* (Matheron, 1975). That is to say, $T_\Xi(K)$ satisfies

(a) upper semicontinuity: if $K_n \downarrow K$ then $T_\Xi(K_n) \downarrow T_\Xi(K)$,

(b) complete alternation: $S_n(K; K_1, \ldots, K_n) \geq 0,$ for all n,

where

$$S_n(K; K_1, \ldots, K_n) = S_{n-1}(K; K_1, \ldots, K_{n-1}) - S_{n-1}(K \cup K_n; K_1, \ldots, K_{n-1}),$$

and $S_0(K) = 1 - T_\Xi(K)$; here K and $K_1, \ldots, K_n$ are compact subsets of $\mathbb{R}^d$. Note that $S_n(K; K_1, \ldots, K_n)$ is equal to

$$\mathbf{P}(\Xi \cap K = \emptyset, \ \Xi \cap K_1 \neq \emptyset, \ \ldots, \ \Xi \cap K_n \neq \emptyset),$$

and that

(c) $$T_\Xi(\emptyset) = 0,$$

and

(d) $$0 \leq T_\Xi(K) \leq 1 \qquad \text{for } K \in \mathbb{K}.$$

The Choquet theorem

The Choquet theorem on capacities gives a precise characterisation of the distribution of a random closed set: the distribution is completely determined by knowledge of the capacity functional.

Theorem 6.1. *Let T be a functional on* $\mathbb{K}$. *Then there is a (necessarily unique) distribution P on* $\mathcal{F}$ *with*

$$P(\mathbb{F}_K) = T(K)$$

if and only if T is an alternating Choquet capacity of infinite order and such that

$$0 \le T(K) \le 1 \quad \text{for } K \in \mathbb{K},$$

and $T(\emptyset) = 0$, *that is, T satisfies conditions (a)–(d) above.*

The necessity of these conditions is clear from the discussion above. The sufficiency is plausible, since the hitting sets $\mathbb{F}_K$ generate $\mathcal{F}$; a proof can be found in Schneider and Weil (2008, pp. 22–31). It is based on the original proof of Matheron (1975) and uses an idea of Salinetti and Wets (1986). Other proofs are given in Berg *et al.* (1984) and Norberg (1989), who use harmonic analysis on semigroups and lattice theory, respectively; see also Molchanov (2005, pp. 13–18).

If the random closed set satisfies additional properties then it may be possible to determine its distribution using a smaller family of hitting sets; see Cressie and Laslett (1987) and Molchanov (2005, pp. 18–20). For example, the distribution of a *convex* random compact set Ξ is determined by the probabilities

$$\mathbf{P}(\Xi \subset K) \quad \text{for compact convex } K.$$

The Choquet theorem shows that capacity functionals play a describing rôle for random closed sets similar to that distribution functions play for random variables. Several distributional properties of random closed sets can be succinctly expressed in terms of their capacity functionals. For example, convergence in distribution is ensured if and only if the corresponding capacity functionals are convergent; see Norberg (1984), Kallenberg (2002, p. 325) and Molchanov (2005, p. 86).

6.1.4 Distributional properties

Independence

Two random closed sets Ξ_1 and Ξ_2 are said to be *independent* if for arbitrary A_1 and A_2 in $\mathcal{F}$

$$\mathbf{P}(\Xi_1 \in A_1, \ \Xi_2 \in A_2) = \mathbf{P}(\Xi_1 \in A_1)\,\mathbf{P}(\Xi_2 \in A_2). \tag{6.2}$$

By the Choquet theorem this is equivalent to the condition

$$\mathbf{P}(\Xi_1 \cap K_1 \ne \emptyset, \ \Xi_2 \cap K_2 \ne \emptyset) = T_{\Xi_1}(K_1)\,T_{\Xi_2}(K_2) \tag{6.3}$$

for all compact K_1 and K_2. If two random closed sets Ξ_1 and Ξ_2 are independent, then the capacity functional of $\Xi_1 \cup \Xi_2$ is given by

$$T_{\Xi_1 \cup \Xi_2}(K) = T_{\Xi_1}(K) + T_{\Xi_2}(K) - T_{\Xi_1}(K)T_{\Xi_2}(K) \qquad \text{for } K \in \mathbb{K}. \tag{6.4}$$

Stationarity and isotropy

A random closed set Ξ is said to be *stationary* if Ξ and the translated sets $\Xi_x = \Xi + x$ all have the same distribution, whatever x in $\mathbb{R}^d$ may be. Again the Choquet theorem shows that this holds if and only if the capacity functional is translation-invariant, that is,

$$T_\Xi(K) = T_\Xi(K_x) \tag{6.5}$$

for every compact K and every x in $\mathbb{R}^d$. The proof of this uses the equalities

$$\begin{aligned}
T_\Xi(K) &= \mathbf{P}(\Xi \cap K \neq \emptyset) \\
&= \mathbf{P}((\Xi - x) \cap K \neq \emptyset) \\
&= \mathbf{P}(\Xi \cap (K + x) \neq \emptyset) = T_\Xi(K_x).
\end{aligned}$$

A random closed set is said to be *stationary and isotropic* (or *motion-invariant*) if for every rigid motion m the distributions of Ξ and of $m\Xi$ are the same. As before an equivalent condition can be given using capacity functionals:

$$T_\Xi(K) = T_\Xi(mK) \tag{6.6}$$

for all such m and $K \in \mathbb{K}$.

Sometimes stationary random $\mathbb{S}$-sets (random closed sets that are almost surely elements of the extended convex ring $\mathbb{S}$, i.e. almost surely locally polyconvex; see Section 1.8) satisfying a certain integrability condition are called *standard random sets*; see Schneider and Weil (2008, p. 397). For example, Boolean models with convex grains are standard.

Ergodicity

Consider the translated family of sets

$$A_x = \{F_x : F \in A\}$$

for $A \in \mathcal{F}$ and x in $\mathbb{R}^d$. A random closed set Ξ (or its distribution P) is said to be *metrically transitive* (with respect to the family of spatial translations) if the condition on A that

$$P((A \setminus A_x) \cup (A_x \setminus A)) = 0 \qquad \text{for all } x \in \mathbb{R}^d \tag{6.7}$$

always implies that $P(A)$ is zero or one. Condition (6.7) means (if events of probability zero are neglected) that Ξ belongs or fails to belong to A and to A_x simultaneously. Consequently if Ξ is metrically transitive then assertions that are translation-invariant (make no explicit or implicit reference to the location of Ξ) are either almost certainly true or almost certainly false.

A random set Ξ is *ergodic* if

$$\lim_{n \to \infty} \frac{1}{v_d(W_n)} \int_{W_n} P(A_x \cap B) \, dx = P(A)P(B) \tag{6.8}$$

for any two A and B in $\mathcal{F}$ and any convex averaging sequence of windows $\{W_n\}$.

It can be shown (see Daley and Vere-Jones, 2008, Exercise 12.2.3, and Heinrich, 1992b) that a stationary Ξ is ergodic if and only if it is metrically transitive.

A sufficient condition for ergodicity is the *mixing property*:

$$P(A \cap B_x) \to P(A)P(B) \qquad \text{for all } A, B \text{ in } \mathcal{F}, \tag{6.9}$$

or, equivalently, in terms of capacity functional

$$1 - T_\Xi(K \cup K'_x) \to \left(1 - T_\Xi(K)\right)\left(1 - T_\Xi(K')\right) \qquad \text{for all } K, K' \text{ in } \mathbb{K}, \tag{6.10}$$

as $\|x\| \to \infty$. Stationary Boolean models and homogeneous Poisson processes are mixing (see Schneider and Weil, 2008, Section 9.3).

The use of the property of ergodicity may be summarised as follows: statistical averages can be expressed by limits of arithmetic or spatial averages. For example, the classical ergodic theorem states that under metrical transitivity (with respect to iterates of a measure-preserving transformation T) the arithmetic averages

$$\left(f(TX) + \cdots + f(T^n X) \right) \big/ n$$

converge as $n \to \infty$ to $\mathbf{E}\left(f(X) \right)$, when f is a bounded measurable function, T^k the kth iterate of T, and X a random variable. Nguyen (1979) and Nguyen and Zessin (1979a) proved ergodic theorems for 'spatial processes', in which spatial averages replace arithmetic averages.

A simple case of such results concerns spatial processes of the form

$$X_B = h(B \cap \Xi) \qquad \text{for } B \text{ a fixed bounded Borel subset of } \mathbb{R}^d,$$

where h is a suitable function. Such a case arises, for example, in the study of statistics of random sets and in the definition of 'densities' of random sets; see Schneider and Weil (2008, Sections 9.2–9.3) and Section 6.3.6.

In order to discuss this, notation must be introduced to describe the kind of spatial averaging being used. Let C_0 denote the semi-open cube,

$$C_0 = \left\{ (x_1, \ldots, x_d) \in \mathbb{R}^d : -\frac{1}{2} \leq x_i < \frac{1}{2} \right\}.$$

Clearly, the whole space can be divided into translates of C_0.

Theorem 6.2. *(Nguyen and Zessin, 1979a) Suppose that Ξ is a stationary ergodic random closed set and that there is given a measurable mapping $h : \mathcal{B}^d \to \mathbb{R}$ with*

$$h(B + x) = h(B) \qquad \text{for each } B \in \mathcal{B}^d \text{ and all } x \in \mathbb{R}^d \quad \text{(translation-invariance)},$$

and

$$h(B_1 \cup B_2) = h(B_1) + h(B_2) \qquad \text{for all disjoint } B_1, B_2 \quad \text{(additivity)}.$$

Further, let $\{W_n\}$ *be a convex averaging sequence of windows.*

(1) If there exists a positive constant c such that

$$\mathbf{E}\big(|h(K \cap \Xi)|\big) \leq c \qquad \text{for all convex bodies } K \subset C_0,$$

then

$$\lim_{n \to \infty} \mathbf{E}\left(\left|\frac{h(W_n \cap \Xi)}{v_d(W_n)} - \mathbf{E}\big(h(C_0 \cap \Xi)\big)\right|\right) = 0. \tag{6.11}$$

(2) If there exists a nonnegative random variable ξ of finite mean such that

$$|h(K \cap \Xi)| \leq \xi \qquad \textbf{P}\text{-almost surely for all convex bodies } K \subset C_0,$$

then

$$\lim_{n \to \infty} \frac{h(W_n \cap \Xi)}{v_d(W_n)} = \mathbf{E}\big(h(C_0 \cap \Xi)\big) \qquad \textbf{P}\text{-almost surely.} \tag{6.12}$$

Thus the spatial average $h(W_n \cap \Xi)/v_d(W_n)$ converges to the statistical average $\mathbf{E}\big(h(C_0 \cap \Xi)\big)$ under the conditions of stationarity and ergodicity, and the latter average is regarded as the density of h; see Section 6.3.6.

Case (1) of Theorem 6.2 provides a weaker form of convergence than case (2). In later applications, the class $\mathcal{B}^d$ is sometimes replaced by the convex ring $\mathcal{R}$.

6.1.5 Miscellany

Random fields

From any random closed set Ξ a random field $\{Y(x) : x \in \mathbb{R}^d\}$ can be constructed as the indicator function of the set Ξ. That is

$$Y(x) = \mathbf{1}_\Xi(x) \qquad \text{for } x \text{ in } \mathbb{R}^d. \tag{6.13}$$

This correspondence is of use in applying methods of the statistics of random fields to the statistical analysis of random sets. Conversely, from a random field a random closed set can be obtained as an excursion set by truncating the random field at a fixed level; see Section 6.6.3. Buryachenko (2007) tries to circumvent random sets by using indicator functions.

Another constructed random field is

$$V_\Xi(x) = v_d\big(B(x, r) \cap \Xi\big), \tag{6.14}$$

the volume of Ξ in the ball $B(x, r)$ of radius r centred at x; see p. 287.

Random marked sets

A *random marked set* is a random field with a random set as support. That means, it is a pair (Ξ, Z) where Ξ is a random closed set in $\mathbb{R}^d$ and Z a real-valued random mapping defined on

Ξ. Random marked sets were introduced in Ballani *et al.* (2012), where precise definitions are given. Three examples are:

(1) Random field over a level u.

Start with a random field $\{Y(x) : x \in \mathbb{R}^d\}$ which has continuous realisations and a fixed real number u, the level. Take as random set the u-level excursion set $\Xi = \{x \in \mathbb{R}^d : Y(x) \geq u\}$ (discussed in detail in Section 6.6.3) and define Z by

$$Z(x) = Y(x) \qquad \text{for } x \in \Xi.$$

For x lying outside Ξ, $Z(x)$ is not defined. Wrong results will be obtained if the covariance function of $\{Y(x)\}$ is estimated from (Ξ, Z) by the usual geostatistical methods ignoring the fact that outside of Ξ no values of the field can be observed.

(2) Marked point process.

A marked point process $\Psi = \{[x_n; m_n]\}$ can be considered as a random marked set, if $\Xi = \Phi = \{x_n\}$, that is, the ground process is the random supporting set, and $m_n = Z(x_n)$ for $x_n \in \Phi$.

(3) Marked fibre process.

Take a fibre process, that is, a random collection of curves (fibres) in space as considered in Section 8.4, and mark each fibre point x by a real value $Z(x)$. This value may come from an independent random field $\{Y(x)\}$ as in example (1) via $Z(x) = Y(x)$ or may depend on the given fibre, for example as $Z(x) =$ tube thickness in x, if the fibre stands for a tube with variable thickness. See Section 8.6 for further discussion of marked fibre processes and related objects.

The notion of 'stochastic processes with random domains' (Molchanov, 2005, pp. 319–22) is closely related to this concept of random marked sets.

Ballani *et al.* (2012) study second-order characteristics of random marked sets.

Comparison of random closed sets

The random closed set Ξ_2 is *stochastically larger than* Ξ_1, denoted by $\Xi_1 \subseteq_{st} \Xi_2$, if there exist random closed sets N_1 and N_2, distributed respectively as Ξ_1 and Ξ_2, such that $N_1 \subset N_2$ almost surely.

Norberg (1992a,b) showed that $\Xi_1 \subseteq_{st} \Xi_2$ if and only if

$$\mathbf{P}\left(\bigcap_{i=1}^{n}(\Xi_1 \cap K_i \neq \emptyset)\right) \leq \mathbf{P}\left(\bigcap_{i=1}^{n}(\Xi_2 \cap K_i \neq \emptyset)\right) \qquad (6.15)$$

for all n and all compact sets $K_1, K_2, \ldots \in \mathbb{K}$; see also Molchanov (2005, Theorem 4.42).

Formula (6.15) implies

$$T_{\Xi_1}(K) \leq T_{\Xi_2}(K), \qquad (6.16)$$

but the converse is not true; see Norberg (1992b) and Müller and Stoyan (2002, p.247).

Comparisons of Boolean models and of RSA (see Section 6.5.3) and dead leaves models with respect to $\subseteq_{st}$ are briefly discussed in Müller and Stoyan (2002, Section 7.4).

Szekli (1995) considers comparison based on comparison of the empty space distribution function. This is greatly extended in Last and Szekli (2011) to the empty space hazard function, which leads to comparison of clustering degrees for germ–grain models.

Stoyan and Stoyan (1980b) consider some partial orderings, weaker than $\subseteq_{st}$, of random closed sets. Cascos and Molchanov (2003) introduce orderings based on the Aumann expectation; see Formula (6.17) below and Cascos (2010).

Simulation of random sets

Lantuéjoul (2002) presents efficient algorithms for the simulation, including perfect simulation (Kendall and Thönnes, 1999; Møller, 2001), of several random set models described in this book. Simulation of Gaussian random fields, which are needed for generating for example excursion sets or marked random sets, is described in Schlather (2001b) using the R statistical programming language.

6.2 Random compact sets

6.2.1 Definition of means

A random compact set is simply a random closed set that is compact with probability one. Random compact sets are of interest as components of stationary random sets but also as models of particles or small objects of technological, biological or geological origin. Some stochastic models of such sets are considered in Molchanov (1993), Stoyan and Stoyan (1994) and Molchanov and Stoyan (1995). Of particular interest are typical cells of tessellations (see Chapter 9), convex hulls of points (see Reitzner, 2010), and sets resulting from limit procedures.

Beginning with Aumann (1965) several authors have suggested definitions of set-valued means of random compact sets (see Stoyan and Stoyan, 1994, Section 8.3, and Molchanov, 2005, Chapter 2). The typical pattern of such definitions is as follows:

(1) assign to the random compact set Ξ a (family of) random function(s);

(2) use a suitable definition for determining a (family of) mean function(s);

(3) determine a set which corresponds to this (these) mean function(s).

Selection expectation

Aumann's original definition is

$$\mathbf{E}(\Xi) = \{\mathbf{E}(X) : X \text{ is a selection of } \Xi \text{ with } \mathbf{E}(\|X\|) < \infty\}. \qquad (6.17)$$

A *selection* of Ξ is a random point X in $\mathbb{R}^d$ with $\mathbf{P}(X \in \Xi) = 1$. Since $\mathbf{E}(\|X\|)$ is finite for all X, $\mathbf{E}(\Xi)$ itself is compact. Furthermore, $\mathbf{E}(\Xi)$ is convex and coincides with $\mathbf{E}(\operatorname{conv}\Xi)$, with the exception that the probability space is atomic. If Ξ is convex, then the definition can be re-expressed in terms of the support function (see Section 1.6): $\mathbf{E}(\Xi)$ satisfies

$$s(\mathbf{E}(\Xi), u) = \mathbf{E}(s(\Xi, u)) \qquad \text{for } u \in S^{d-1}, \qquad (6.18)$$

which is sufficient to uniquely determine a member in the family of convex compact sets. If Ξ is isotropic, then $\mathbf{E}(\Xi)$ is a ball centred at o. Often $\mathbf{E}(\Xi)$ is called the *Aumann expectation* or *selection expectation* of Ξ.

The average breadth of the Aumann expectation of Ξ coincides with the expected average breadth of Ξ:

$$\mathbf{E}\big(\overline{b}(\Xi)\big) = \overline{b}\big(\mathbf{E}(\Xi)\big), \tag{6.19}$$

and consequently an analogous equality holds for the perimeter in the planar convex case. Thus the Aumann expectation of a random ball centred at o is the ball with mean diameter.

For volume and area such relations do not hold; instead, there is an inequality for area in the planar case:

$$\mathbf{E}\left(\sqrt{A(\Xi)}\right) \le \sqrt{A\big(\mathbf{E}(\Xi)\big)}. \tag{6.20}$$

Many authors have established limit theorems for random compact sets. Artstein and Vitale (1975) prove that for a sequence of i.i.d. random compact sets Ξ, Ξ_1, Ξ_2, ... with $\mathbf{E}(\sup\{\|x\| : x \in \Xi\}) < \infty$,

$$\frac{1}{n}(\Xi_1 \oplus \cdots \oplus \Xi_n) \to \mathbf{E}(\Xi) \qquad \text{almost surely} \tag{6.21}$$

with respect to the Hausdorff metric; see also Artstein (1984a). Weil (1982a) proves a central limit theorem, which corresponds to the strong law of large numbers given in (6.21).

For other limit theorems involving the Minkowski sum, such as the law of iterated logarithm, renewal theorems and ergodic theorems, as well as limit theorems for unions and convex hulls of random compact sets, generalising classical results of extreme value theory; see Molchanov (1993) and Molchanov (2005, Chapters 3–4).

Other expectations

An alternative natural way of defining a mean for a random compact set Ξ is the use of the *coverage function*,

$$p_\Xi(x) = \mathbf{E}\big(\mathbf{1}_\Xi(x)\big) = T_\Xi(\{x\}) \qquad \text{for } x \in \mathbb{R}^d, \tag{6.22}$$

that is, the mean of the indicator function of Ξ. Another form is

$$p_\Xi(x) = \mathbf{P}(x \in \Xi) \qquad \text{for } x \in \mathbb{R}^d. \tag{6.23}$$

In this case the mean is not a set but a function taking values between 0 and 1. Perhaps one may consider $p_\Xi(x)$ as the membership function of a fuzzy set (see e.g. Dubois and Prade, 2000; Nguyen and Wu, 2006). The mean volume of Ξ satisfies

$$\mathbf{E}\big(v_d(\Xi)\big) = \int_{\mathbb{R}^d} p_\Xi(x)\mathrm{d}x. \tag{6.24}$$

If Ξ is isotropic then $p_\Xi(x)$ depends only on $\|x\|$.

A third mean-set definition is the *Vorob'ev mean*. The idea is to take the excursion set of the coverage function $p_\Xi(x)$ such that the volume of the excursion set and the mean volume of Ξ coincide:

$$\mathbf{E}_{\mathrm{V}}(\Xi) = \{x \in \mathbb{R}^d : p_{\Xi}(x) \geq u\}, \tag{6.25}$$

where u is determined by

$$v_d(\{x \in \mathbb{R}^d : p_{\Xi}(x) \geq u\}) = \mathbf{E}(v_d(\Xi)). \tag{6.26}$$

If u is not unique then take the infimum of all such u. Vorob'ev's definition is quite natural: the mean set $\mathbf{E}_{\mathrm{V}}(\Xi)$ satisfies the inequality

$$\mathbf{E}(v_d(\Xi \triangle \mathbf{E}_{\mathrm{V}}(\Xi))) \leq \mathbf{E} v_d(\Xi \triangle B)$$

for all Borel sets B with $v_d(B) = \mathbf{E}(v_d(\Xi))$. Here $\triangle$ denotes the symmetric difference operator:

$$C \triangle D = C \setminus D \cup D \setminus C.$$

In this sense $\mathbf{E}_{\mathrm{V}}(\Xi)$ is that set of area $\mathbf{E}(v_d(\Xi))$ that is best fitted to Ξ. Heinrich *et al.* (2012) proposed a consistent estimator of $\mathbf{E}_{\mathrm{V}}(\Xi)$ from discretised independent realisations of Ξ.

Molchanov (2005, Chapter 3) discusses mean sets defined by various distance functions and, when restricted to star-shaped sets, radius-vector function. Jankowski and Stanberry (2010, 2012) define mean sets and mean boundaries, and construct their confidence regions, using signed distance functions; see Osher and Fedkiw (2003).

There are also attempts to define variances, which are positive numbers characterising variability; see Stoyan and Stoyan (1994, Section 8.3.5).

The application of ideas of mean sets to samples of particles such as sand grains is difficult since for these only shape and size are of interest, while spatial position is disregarded. Stoyan and Molchanov (1997) show how to proceed: the particles are shifted and rotated before determining a mean.

The book Stoyan and Stoyan (1994) studies systematically summary characteristics and, in Section 8.5, models for particles. The papers Stoyan *et al.* (2002) and Ballani and van den Boogaart (2013) present particle models with Gibbs distributions, for pixel sets and polytopes, respectively. Further models are mentioned in Molchanov (2005, pp. 218–223) (Gaussian, p-stable and Minkowski infinitely divisible) and in Jónsdóttir *et al.* (2008). Figure 3.4 can be interpreted as a sample of a random compact set.

6.2.2 Mean-value formulae for convex random sets

In this section the random closed set Ξ is taken to be convex and compact with probability one. Thus the intrinsic volumes $V_k(\Xi)$ can be considered. It is assumed that their first moments are finite:

$$\overline{V}_k = \mathbf{E}(V_k(\Xi)) < \infty \qquad \text{for } k = 0, 1, \ldots, d. \tag{6.27}$$

The *Steiner*, *Crofton*, and other integral-geometric formulae will then be satisfied for Ξ almost surely, and analogous formulae will hold for the mean values. For example, corresponding to the Steiner formula there is

$$\mathbf{E}\left(v_d(\Xi \oplus b(o, r))\right) = \sum_{k=0}^{d} b_{d-k} \overline{V}_k r^{d-k} \qquad \text{for } r \geq 0. \tag{6.28}$$

If Ξ has a distribution invariant with respect to rotations about the origin o, then the important so-called *generalised Steiner formula*

$$\mathbf{E}\big(v_d(\Xi \oplus K)\big) = \frac{1}{b_d} \sum_{k=0}^{d} \frac{b_k b_{d-k}}{\binom{d}{k}} \overline{V}_k V_{d-k}(K) \tag{6.29}$$

holds for every compact convex K. Its proof rests on the Hadwiger characterisation theorem: the functional $h(K)$ appearing in Formula (1.44) is

$$h(K) = \mathbf{E}\big(v_d(\Xi \oplus K)\big),$$

which can be shown to be motion-invariant, C-additive, and monotone in the compact set K. Thus from the Hadwiger theorem $h(K)$ is a linear combination of the $V_k(K)$ and its coefficients can be derived by setting $K = B(o, r)$.

In the important planar and spatial cases Formula (6.29) takes on the forms:

$$\mathbf{E}\big(A(\Xi \oplus K)\big) = \overline{A} + \frac{\overline{L}\, L(K)}{2\pi} + A(K) \qquad \text{if } d = 2 \tag{6.30}$$

and

$$\mathbf{E}\big(V(\Xi \oplus K)\big) = \overline{V} + \frac{\overline{\overline{b}}\, S(K)}{2} + \frac{\overline{S}\, \overline{b}(K)}{2} + V(K) \qquad \text{if } d = 3. \tag{6.31}$$

A similar generalisation can be carried out on Formulae (1.45) for the intrinsic volumes, and also for the generalised intrinsic volumes of sets in the convex ring $\mathcal{R}$.

A useful consequence of Formula (6.31) is that for the volume of the mean *excluded volume* of a random convex compact subset Ξ of $\mathbb{R}^3$:

$$\mathbf{E}\big(V(\Xi \oplus \check{\Xi})\big) = 2\overline{V} + \overline{\overline{b}} \cdot \overline{S}, \tag{6.32}$$

where $\check{\Xi}$ denotes a random closed set of the same distribution as $-\Xi$ but independent of Ξ.

6.3 Characteristics for stationary and isotropic random closed sets

6.3.1 The area or volume fraction

The area or volume fraction of a stationary random set Ξ is the mean volume of Ξ intersected with the unit cube C_0, that is,

$$p = \mathbf{E}\big(v_d(\Xi \cap C_0)\big). \tag{6.33}$$

If Ξ models the pore network of a porous medium, then p is the *porosity*. The volume fraction is the most important random set characteristic and plays a rôle similar to that of the intensity of a point process. Unfortunately, in the literature different symbols are used, for example ϕ or η, and even in this book the notations V_V and A_A are used from time to time as alternatives to p; see Section 10.2.2. The meaning of V_V is 'volume of Ξ per unit volume', and A_A is analogous. These two symbols appear when the dimension of the space is emphasised.

It is the case that

$$p = \mathbf{P}(o \in \Xi) = \mathbf{P}(x \in \Xi) = T_\Xi(\{o\}) \qquad \text{for } x \in \mathbb{R}^d, \tag{6.34}$$

which results from the stationarity of Ξ by the following chain of equalities:

$$\begin{aligned}
\mathbf{E}\big(v_d(\Xi \cap C_0)\big) &= \mathbf{E}\left(\int_{\mathbb{R}^d} \mathbf{1}_{\Xi \cap C_0}(x)\, dx\right) \\
&= \mathbf{E}\left(\int_{C_0} \mathbf{1}_\Xi(x)\, dx\right) \\
&= \int_{C_0} \mathbf{E}\big(\mathbf{1}_\Xi(x)\big)\, dx \\
&= \int_{C_0} \mathbf{P}(x \in \Xi)\, dx = \mathbf{P}(o \in \Xi),
\end{aligned}$$

since by stationarity $\mathbf{P}(x \in \Xi) = \mathbf{P}(o \in \Xi)$ for all x in $\mathbb{R}^d$.

The volume fraction can arise in other ways, which are described as follows.

(a) If the volume (or coverage) measure V_Ξ is defined on $\mathbb{R}^d$ by

$$V_\Xi(B) = v_d(\Xi \cap B) \qquad \text{for all Borel sets } B, \tag{6.35}$$

then V_Ξ is a stationary random measure; see Chapter 7. Its *intensity* is p.

(b) Let v be the translation-invariant measure on $\mathbb{R}^d$ defined by

$$v(B) = \mathbf{E}\big(v_d(\Xi \cap B)\big) \qquad \text{for all Borel sets } B. \tag{6.36}$$

It is finite on bounded Borel sets. Suppose that K is a convex body with inner points. Then the corresponding *density*

$$D_V(\Xi) = \frac{v(K)}{v_d(K)} \tag{6.37}$$

exists and equals p.

(c) Suppose that Ξ is ergodic and K is as in (b). Then the ergodic theorem given in Theorem 6.2 states that the ratio

$$\frac{v_d\big(\Xi \cap (rK)\big)}{v_d(rK)}$$

converges to p almost surely as $r \to \infty$.

6.3.2 The covariance

Just as p describes the first-order structure of Ξ, the *covariance* or *two-point probability function* deals with the second-order structure. (Note that the term is 'covariance' and not 'covariance function'.) As for volume fraction so here different symbols are used, for example $\gamma_2(r)$ (Frisch and Stillinger, 1963) and $S_2(r)$ (Torquato, 2002).

Clearly, for the covariance to be an effective measure of second-order structure the volume fraction p must be positive, since otherwise the covariance will be identically zero. Some random sets, such as point or fibre processes, have $p = 0$ and their second-order structure is described directly by moment measures. (The covariance is essentially a product density corresponding to the random volume measure defined by (6.35).)

The covariance is defined by

$$C(\mathbf{r}) = C_\Xi(\mathbf{r}) = \mathbf{P}(o \in \Xi, \mathbf{r} \in \Xi) = T_\Xi(\{o, \mathbf{r}\}) \qquad \text{for } \mathbf{r} \in \mathbb{R}^d. \tag{6.38}$$

It satisfies the equations

$$C(\mathbf{r}) = \mathbf{P}(o \notin \Xi, \mathbf{r} \notin \Xi) + 2p - 1, \tag{6.39}$$

which is proved as on p. 74, and

$$C(\mathbf{r}) = \mathbf{P}(o \in \Xi \ominus \check{B}), \tag{6.40}$$

where $B = \{o, \mathbf{r}\}$ (a set composed of two points) and $\check{B} = -B$. Thus $C(\mathbf{r})$ is the volume fraction of the set Ξ eroded by the structuring element B. Formula (6.40) is a simple consequence of the definition of the operation of erosion:

$$o \in \Xi \ominus \check{B} \text{ if and only if } o \in \Xi \text{ and } \mathbf{r} \in \Xi.$$

Formula (6.38) is the same as

$$C(\mathbf{r}) = \mathbf{P}\big(o \in \Xi \cap (\Xi - \mathbf{r})\big). \tag{6.41}$$

This provides a basis for statistical estimation of $C(\mathbf{r})$ as a volume fraction.

There is a close relation between $C(\mathbf{r})$ and the *covariance function* of the random field associated with Ξ: the mean of the stationary random field

$$\{\mathbf{1}_\Xi(x) : x \in \mathbb{R}^d\}$$

is given by $\mathbf{E}(\mathbf{1}_\Xi(x)) = p$, and its covariance function $k(\mathbf{r})$ is

$$k(\mathbf{r}) = \mathbf{E}\left((\mathbf{1}_\Xi(o) - p)(\mathbf{1}_\Xi(\mathbf{r}) - p)\right) = C(\mathbf{r}) - p^2, \tag{6.42}$$

as can be seen on expanding the product. Notice that, in contrast to $k(\mathbf{r})$, the covariance $C(\mathbf{r})$ is not centred.

The covariance function is often normalised by dividing by $k(o) = p(1 - p)$, which gives the *correlation function* or *Debye X-ray correlation function*

$$\kappa(\mathbf{r}) = \frac{k(\mathbf{r})}{k(o)} \qquad \text{for } \mathbf{r} \in \mathbb{R}^d, \tag{6.43}$$

with

$$k(o) = p(1 - p),$$

If Ξ is not only stationary but also isotropic then the functions $C(\mathbf{r})$, $k(\mathbf{r})$ and $\kappa(\mathbf{r})$ depend only on the distance $r = \|\mathbf{r}\|$. For simplicity the argument is then written as a scalar r, and the function symbols are retained, for example:

$$C(\mathbf{r}) = C(\|\mathbf{r}\|) = C(r) \qquad \text{for motion-invariant } \Xi.$$

The function

$$g(r) = \frac{C(r)}{p^2}$$

is sometimes called *pair correlation function*. Indeed it is the pair correlation function of the volume measure; see Section 7.3.4. The function

$$\gamma(r) = C(0) - C(r) \qquad \text{for } r \geq 0 \tag{6.44}$$

is also known as the *variogram*.

Note that

$$C(0) = C(o) = p, \tag{6.45}$$

and if Ξ satisfies the mixing property, then

$$C(\infty) = p^2. \tag{6.46}$$

Not every nonnegative function $C(r)$ with $C(0) = p$ and $C(\infty) = p^2$ is the covariance of a random set. A necessary condition is that the corresponding covariance function $k(r)$ is positive definite. Torquato (2006) and Jiao *et al.* (2007, 2010) discussed necessary and sufficient conditions for the reconstructions of random sets from a given covariance. In general, $C(r)$ alone does not determine uniquely the distribution of a random set.

Analysis of the three functions $C(r)$, $k(r)$ and $\kappa(r)$ allows for information to be gained about the structure of Ξ even in the case where no image of Ξ is available as, for example, in the cases of X-ray scattering analysis, or geological exploration by means of boreholes. Figure 6.1 shows possible forms of $C(r)$ by giving three typical images of samples of planar random sets together with corresponding empirical covariances (determined by image analyser). Thus the functions can be used in a descriptive analysis. They also have theoretical uses in determining the accuracy of statistical estimators; see Section 6.4.2. Integrating the correlation function leads to the *integral range a_V* (Matheron, 1971), which is, despite the name, a volume defined by

$$a_V = \int_{\mathbb{R}^d} \kappa(\mathbf{r}) d\mathbf{r}. \tag{6.47}$$

This value gives an approximation of the size of the representative volume element for the determination of volume fraction; see Section 6.4.6.

Finally the three functions can be used directly in estimation of model parameters, for example in the Boolean model as demonstrated in Example 3.3.

Let the random set Ξ be almost surely regular closed (coinciding with the closure of its interior) and suppose that the specific surface area $S_V^{(d)}$ of the boundary $\partial \Xi$ of Ξ is finite (see Section 7.3.4 for a definition of specific surface area). Then the first derivative $C'(0+)$ is related to $S_V^{(d)}$ by

$$S_V^{(d)} = -\frac{d b_d}{b_{d-1}} C'(0+), \tag{6.48}$$

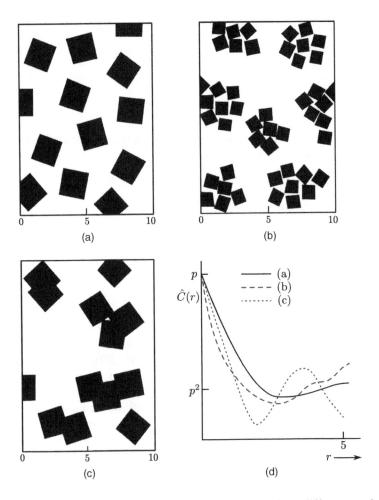

Figure 6.1 Forms of the covariance $C(r)$ for realisations of three different random closed sets sharing the same volume fraction p. The form of $C(r)$ allows conclusions to be drawn concerning the properties of the corresponding random sets, such as the inter-particle distances or the tendency to clustering. See also Figures 6.2 and 6.4.

as discovered by Debye *et al.* (1957) and Guinier and Fournet (1995). In the particular cases of $d = 1, 2$ or 3 the coefficient of $-C'(0+)$ is 2, π or 4, respectively. See Berryman (1987), Kiderlen and Jensen (2003) and Gokhale *et al.* (2005) for the anisotropic case.

Proof of Formula (6.48). The proof is given here for $d = 3$; for $d = 2$ it is analogous. Let $\boldsymbol{g}$ be an oriented line through o. Then the set $\partial\Xi \cap \boldsymbol{g}$ of intersection points of the boundary of Ξ with the line $\boldsymbol{g}$ is a stationary point process on $\boldsymbol{g}$. It has finite intensity P_L, where P_L is connected to $S_V^{(3)} = S_V$ by the stereological formula (see Section 8.5.2)

$$P_L = \frac{S_V}{2}. \tag{6.49}$$

Now suppose that $\{x_n\}$ is the point process of left endpoints of the intervals forming the connected components of $\Xi^{\mathrm{int}} \cap g$ or chords in Ξ (recall that g is oriented, so 'left' has a meaning along g). Attach a mark l_n to each x_n by defining l_n to be the length of the corresponding connected component of $\Xi^{\mathrm{int}} \cap g$ starting in x_n. The marked point process $\Psi = \{[x_n; l_n]\}$ inherits stationarity from Ξ and has intensity $\lambda = S_V/4$, the half of the value given by (6.49).

The random set $\Xi \cap g$ can be viewed as a subset of the real line $\mathbb{R}$. Its covariance $C(r)$ coincides with the covariance of Ξ. Because of (6.40), $C(r)$ is equal to the length fraction of the eroded set

$$(\Xi \cap g) \ominus \{o, r\} = \{x \in \mathbb{R} : \{x, x+r\} \subset \Xi \cap g\},$$

and so it follows that $C(0) - C(r)$ is the length fraction of the set $\{x \in \Xi \cap g : x + r \notin \Xi \cap g\}$. In other words,

$$C(0) - C(r) = \mathbf{E}\big(\nu_1(\{x \in \Xi \cap g : x + r \notin \Xi \cap g\} \cap [0, 1])\big). \tag{6.50}$$

This equation yields upper and lower bounds for $C(0) - C(r)$ for small r ($r < 1$). The upper bound is

$$\mathbf{E}\left(\sum_{n : x_n + l_n \in [0, 1+r]} r \right),$$

and the lower bound is

$$r\mathbf{E}\left(\sum_{n : x_n + l_n \in [r, 1]} \mathbf{1}(l_n > r) \right) - r\mathbf{E}\left(\sum_{n : x_n \in [0, 1]} \mathbf{1}(d_\Psi(x_n) \le r) \right),$$

where $d_\Psi(x_n)$ is the distance from x_n to its nearest neighbour in Ψ.

Applying the Campbell theorem (4.29) for marked point processes to both bounds yields

$$(1 - r)\lambda \mathbf{P}(l > r) - \lambda \mathbf{P}\big(d_\Psi(o) \le r\big) \le \frac{C(0) - C(r)}{r} \le (1 + r)\lambda,$$

where l follows the same distribution as l_n. Letting $r \to 0$ leads to

$$C'(0+) = -\lambda = -\frac{S_V}{4},$$

since $\mathbf{P}\big(d_\Psi(o) \le 0\big) = 0$. ☐

The second derivative $C''(0+)$ has also been studied; see Ciccariello (1995) and Böhm and Schmidt (2003), who showed that it is given by the values at $r = 0$ of the chord length probability density functions (see Section 6.3.4) for Ξ and its complement.

It has proved useful in many applications since Debye et al. (1957) to approximate complicated covariances by a simpler form, the so-called *exponential covariance*

$$C_e(r) = p(1 - p)e^{-\alpha r} + p^2 \qquad \text{for } r \ge 0, \tag{6.51}$$

where α is a positive parameter describing the degree of variability or randomness in the random closed set. The corresponding covariance function and correlation function are given by

$$k_e(r) = p(1 - p)e^{-\alpha r} \tag{6.52}$$

and

$$\kappa_e(r) = e^{-\alpha r} \tag{6.53}$$

both for $r \geq 0$. In the spatial case $(d = 3)$ it can be shown that

$$\alpha = \frac{S_V}{4V_V(1 - V_V)}, \tag{6.54}$$

with

$$V_V = p;$$

see Ohser and Tscherny (1988). This is a useful expression of the theoretical quantity α in terms of quantities that can be easily estimated statistically. Under the assumption of the exponential covariance (6.51), the integral range defined in (6.47) is equal to

$$a_V = \frac{d!\, b_d}{\alpha^d}. \tag{6.55}$$

Sometimes a structure with an exponential covariance is called 'Debye random medium'. A simple example can be constructed as follows. Consider a Poisson line (or plane) tessellation with parameter ϱ as described in Section 9.5. Let each cell of the tessellation be painted independently of the others, painted black with probability p and white with probability $1 - p$. Then the closure of the union of all black cells is a planar (spatial) random closed set with exponential covariance, with parameter $\alpha = \varrho$. Ohanian (1973) gave examples of random sets with exponential covariances but with curved boundaries.

Zachary and Torquato (2011) generalised $C(r)$ by using a function defined as in (6.38), where Ξ is replaced by $\Xi \oplus B(o, \delta)$ or $\Xi \ominus B(o, \delta)$. In the first case also structures with $p = 0$, for example fibre processes, can be analysed.

In some cases it is easy to give the covariance of sets obtained by set-theoretic operations. For the closure of the *complement* $(\Xi^c)^{cl}$ of Ξ it is

$$C_{(\Xi^c)^{cl}}(\mathbf{r}) = C_\Xi(\mathbf{r}) - 2p + 1, \tag{6.56}$$

and for the intersection $\Xi_1 \cap \Xi_2$ of two independent stationary closed random sets Ξ_1 and Ξ_2

$$C_{\Xi_1 \cap \Xi_2}(\mathbf{r}) = C_{\Xi_1}(\mathbf{r}) \cdot C_{\Xi_2}(\mathbf{r}). \tag{6.57}$$

If the two sets Ξ_1 and Ξ_2 are dependent, then their spatial correlation may be described by the *cross-covariance*

$$C_{12}(\mathbf{r}) = C_{\Xi_1, \Xi_2}(\mathbf{r}) = \mathbf{P}(o \in \Xi_1, \mathbf{r} \in \Xi_2). \tag{6.58}$$

An example of a pair of stationary dependent sets is Ξ_1 and Ξ_2, where Ξ_2 is constructed by Ξ_1 and a third set Ξ_3, which are both stationary and independent, by

$$\Xi_2 = (\Xi_3 \setminus \Xi_1)^{\mathrm{cl}}.$$

Here

$$C_{12}(\mathbf{r}) = \mathbf{P}(o \in \Xi_1, \mathbf{r} \in \Xi_2) = \big(p_1 - C_{\Xi_1}(\mathbf{r})\big)p_3,$$

where p_1 and p_3 are the volume fractions of Ξ_1 and Ξ_3, respectively.

Also *n-point probability functions* are of interest; see for example the monograph Torquato (2002). However, the theoretical determination of these functions for random set models is notoriously complicated.

6.3.3   Contact distribution functions

It is frequently the case that random sets must be considered for which identification of single particles or pores is inappropriate. Nevertheless, some kind of measurement of 'size' would be desired. The *contact distribution functions* $H_B(r)$ provide a useful tool for this, as explained in Section 3.1.7. This concept was introduced by Prager (1969), and Delfiner (1972) and Serra (1982, Chapter XIII) discussed its general value. Hug *et al.* (2002a) is an excellent survey on contact distributions.

The following presents a sketch of the theory for the stationary case. A contact distribution function depends on the choice of a *structuring element B*, which is a convex body. If B contains the origin o then $H_B(r)$ is defined by

$$H_B(r) = 1 - \mathbf{P}(o \notin \Xi \oplus r\check{B}|o \notin \Xi), \tag{6.59}$$

or

$$H_B(r) = 1 - \frac{\mathbf{P}(o \notin \Xi \oplus r\check{B})}{1 - p} \qquad \text{for } r \geq 0. \tag{6.60}$$

The numerator of the second term is

$$1 - \text{volume fraction of the set } \Xi \oplus r\check{B}.$$

The function $H_B(r)$ is indeed a distribution function if B is compact and convex with o an inner point. Then even a probability density function $h_B(r)$ exists,

$$h_B(r) = H'_B(r);$$

see Hansen *et al.* (1999). If B is compact and star-shaped (e.g. B is a segment with endpoint at o) then the additional conditions of $p > 0$ and Ξ mixing are necessary; see Heinrich (1993).

Sometimes the *empty space function* $F_B(r)$ is used, defined by

$$F_B(r) = \mathbf{P}(o \in \Xi \oplus r\check{B}) \qquad \text{for } r \geq 0. \tag{6.61}$$

The corresponding distribution has an atom of size p at $r = 0$. It can be interpreted as the distribution function of the random distance $d_B(o, \Xi)$ in the B-metric from a test point (which is represented by o) to Ξ, which is zero if the test point is in Ξ. In some context, the corresponding hazard function $\eta_B(r)$ is also useful:

$$\eta_B(r) = \frac{f_B(r)}{1 - F_B(r)} \qquad \text{for } r > 0, \tag{6.62}$$

where $f_B(r)$ is the derivative (and also the density) of $F_B(r)$ and exists whenever $h_B(r)$ exists.

The empty space function can be extended to a distribution function on the whole real line by

$$F_B(r) = \begin{cases} \mathbf{P}(o \in \Xi \oplus r\check{B}) & \text{for } r \geq 0, \\ \mathbf{P}(o \in \Xi \ominus |r|B) & \text{for } r < 0. \end{cases} \tag{6.63}$$

This is the distribution function of the signed distance in the B-metric from a test point to the boundary of Ξ, which has negative sign if the test point is in Ξ. Note that the negative branch of Formula (6.63) uses $|r|B$, not $|r|\check{B}$ as might be expected; see van Lieshout (1999).

Analogous functions can be defined by replacing $\oplus$ and $\ominus$ by opening $\circ$ and closing $\bullet$; see van Lieshout (1999). The counterpart of the extended $F_B(r)$ is there called 'size distribution function'. Sivakumar and Goutsias (1999) and Goutsias and Batman (2000) discuss a discrete analogue of the corresponding probability density function, called *size density*, defined in Formula (6.143) , which is a valuable tool in practical image analysis; see for example Asano *et al.* (2003), Fletcher and Evans (2005) and Land and Wilkison (2009).

Hug *et al.* (2002a) consider random distances from test sets (replacing the test point o above by a compact set) to random sets. Ohser and Schladitz (2009, Section 5.5.1) consider distances from a second set Ψ, which stands for a third constituent in a sample, to Ξ.

The cases of particular practical importance are those of the two extremes: the *linear contact distribution function $H_l(r)$*, in which B is a segment of unit length (in the case of anisotropy the orientation of this segment is of importance; see Section 6.3.5); and the *spherical contact distribution function $H_s(r)$*, in which B is the unit ball (disc) centred at o in the spatial (planar) case. (Note that the empty space function $F_B(r)$ defined in (6.61) often, if not always, employs a spherical B in practice.) Figure 6.2 shows the empirical $H_s(r)$ for the three patterns shown in Figure 6.1. These curves reveal information on the distributional properties of the size of the particles. Also of practical interest is the case of B a square (cube) or a disc in $\mathbb{R}^3$, leading to $H_q(r)$ and $H_d(r)$; see p. 83.

The quantity $1 - H_s(r)$ is the conditional probability that a test point is the centre of a ball of radius r lying completely outside Ξ, given that the test point in question does not belong to Ξ. Thus the spherical contact distribution function $H_s(r)$ is also referred to as 'the law of first contact'. It can be seen also as the distribution function of the Euclidean distance from o to Ξ,

$$H_s(r) = \mathbf{P}(d(o, \Xi) \leq r \mid o \notin \Xi) \qquad \text{for } r \geq 0. \tag{6.64}$$

In the isotropic case and if Ξ has a sufficiently smooth boundary, the spherical contact distribution function is related to the specific surface area by

$$S_V^{(d)} = (1 - p)H_s'(0+), \tag{6.65}$$

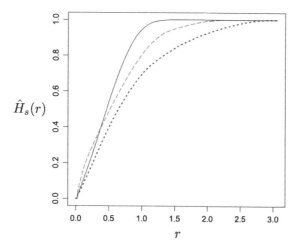

Figure 6.2   Forms of the spherical contact distribution function $H_s(r)$ for the three random set realisations (key: — (a); - - - (b); $\cdots$ (c)) of Figure 6.1. The form of $H_s(r)$ allows conclusions to be drawn concerning the distributional properties of the corresponding random sets. Because of the nearly uniform distribution of the grain positions in (a) the distances between test points and the random set are particularly small, while in (c) large distances are possible. See also Figures 6.1 and 6.4.

see Last and Schassberger (1998) and Hug *et al.* (2002a). Kiderlen and Rataj (2006) showed what happens for general B and stationary Ξ. They used a general theory to explain the Formulae (6.65) and (6.48). Ballani (2006b, 2011) studied two-point spherical contact distribution functions (where the condition '$o \notin \Xi$' is replaced by '$x_1 \notin \Xi, x_2 \notin \Xi$' and two radii r_1 and r_2 appear) and showed that the partial derivatives with respect to r_1 and r_2 are related to the second-order product density of the corresponding surface measure.

The second derivative $H_s''(0+)$ is related to a random geometrical signed measure as shown by Last and Schassberger (2001).

The survey Hug *et al.* (2002a) contains formulae for contact distributions for many random set models, and it considers also the nonstationary case, while Last and Holtmann (1999) give formulae for $H_s(r)$ for various germ–grain models.

The contact distributions introduced above characterise primarily the complement of the random set Ξ of interest: the scaled structuring element rB is placed outside of Ξ. Therefore, the contact distributions could be called 'outer' contact distributions. Of course, analogously, 'inner' contact distributions can be defined by using $(\Xi^c)^{cl}$, the closure of the complement of Ξ.

Rataj (1993) suggests a rather general approach to contact distances and their estimation including edge corrections. There it is shown that distributions of contact distances cannot be derived from the finite-order characteristics of the corresponding random set. If the set is observable only within a compact window, then the observed contact distances are censored; see Rataj (1993) for relevant mathematical models.

Other summarising functional characteristics of random closed sets are the granulometry functions and functions constructed by opening and closing; see Matheron (1975) and Ripley (1988).

6.3.4 Chord length distributions

While Section 1.7.3 considers chords through convex compact sets, this section studies chords through stationary random sets. These are generated by intersecting the random set Ξ of interest by a directed line g. The result is an alternating sequence of random chords inside and outside of Ξ. As for the contact distributions, the chords outside of Ξ are considered in the following. The complementary theory can be simply obtained by interchanging the rôles of Ξ and its complement.

The linear contact distribution function $H_l(r)$ characterises the distribution of chord lengths of the complement of Ξ. In fact, $H_l(r)$ is the distribution function of residual chord length in the direction of a test line segment which begins at a fixed point o, under the condition that o does not belong to Ξ. This is illustrated in Figure 6.3.

This interpretation must be considered in the light of the difference between 'length-biased' and 'unbiased' sampling of chord lengths. For suppose that g is a directed line containing the test segment and consider as on p. 220 the stationary point process $g \cap \partial\Xi$. One may consider $g \cap \partial\Xi$ to be a point process on $\mathbb{R}$ and assume that its intensity is finite. Then from $g \cap \partial\Xi$ one may construct a marked point process $\{[x_n; r_n]\}$. Here $\{x_n\}$ is the sequence of right endpoints of chords in Ξ and r_n is the length of the chord outside of Ξ whose right endpoint is x_n. The corresponding mark distribution function $L(r)$ is the length distribution function of the typical chord of the complement of Ξ, selected in a fashion corresponding to the Palm distribution of the marked point process $\{[x_n; r_n]\}$, which is not length-biased. The distribution functions $L(r)$ and $H_l(r)$ are linked by

$$H_l(r) = \frac{1}{m_L} \int_0^r \left(1 - L(x)\right) dx, \tag{6.66}$$

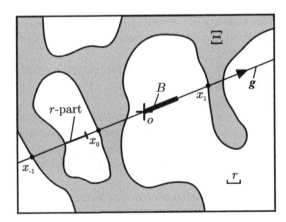

Figure 6.3 Intersection of a closed random set Ξ (shown in grey) and a line g (the direction of the line being defined by a structuring element B that is a segment). The right endpoints of the chords outside of Ξ $\ldots, x_{-1}, x_0, x_1, \ldots$ form a point process. The r-part of a chord is the set of all points of distance more than r from the right endpoint. The symbol $+$ indicates the location of the origin o.

where m_L is the mean typical chord length

$$m_L = \int_0^\infty x \, dL(x) \tag{6.67}$$

with $0 < m_L < \infty$.

Proof of Formula (6.66). First, note that the marked point $[x_n; r_n]$ corresponds to a chord of length r_n with right endpoint x_n. For a chord $[x_n; r_n]$ of length greater than r let the 'r-part' be the subset of points in the chord further than r from the right endpoint. Thus the r-part has length $\max\{0, r_n - r\}$; see Figure 6.3. Then $1 - H_l(r)$ is the conditional probability that o lies in an r-part, given that o does not lie in Ξ. Hence $1 - H_l(r)$ is the quotient of $L_L(r)$ (the length fraction of the union of all r-parts) by L_L. If λ is the intensity of the process $\{x_n\}$ of right endpoints then

$$L_L = \lambda m_L$$

and

$$L_L(r) = \lambda \int_r^\infty (x - r) \, dL(x),$$

so

$$1 - H_l(r) = \frac{L_L(r)}{L_L} = \frac{1}{m_L} \int_r^\infty (x - r) \, dL(x),$$

and Formula (6.66) follows by integration by parts. $\qquad\qquad\qquad\square$

Up to this point attention has focused on chords lying outside of Ξ. As mentioned above, if Ξ has positive volume fraction p, then also chords inside of Ξ make sense. These can be analysed as above simply by exchanging the rôles of Ξ and $(\Xi^c)^{cl}$.

Chords outside of Ξ also make sense if Ξ consists of fragments of curves ($d = 2$) or of surfaces ($d = 3$). Last and Schassberger (1996) express $H_l(r)$ in terms of the Palm distribution of the fibre (or surface) process of the boundary of Ξ.

By means of the stereological Formulae (8.37) and (10.3) it is easy to show that the mean chord length m_L satisfies

$$m_L = \frac{\pi(1 - A_A)}{L_A} \qquad \text{if } d = 2, \tag{6.68}$$

and

$$m_L = \frac{4(1 - V_V)}{S_V} \qquad \text{if } d = 3. \tag{6.69}$$

If Ξ is a Boolean model of convex grains, or a semi-Markovian random set in the sense of Matheron (1975), then the chords of the complement of Ξ have an exponential distribution. In the case of a Boolean model this is a consequence of Formulae (6.66) and (3.58).

6.3.5 Directional analysis of random closed sets

Anisotropy is exhibited by many patterns that may be modelled as random closed sets. Such anisotropy can be quantified by analysis of $C(\mathbf{r})$ and $H_l(r)$, by means of varying the direction of $\mathbf{r}$ and of the linear segment used in the definition of the linear contact distribution. Figure 6.4(a) shows a planar pattern exhibiting clear anisotropy, and Figures 6.4(b) and (c) show $C(\mathbf{r})$ and $H_l(r)$ measured for $(\Xi^c)^{cl}$ in two perpendicular directions. One direction is parallel to the bottom line of Figure 6.4(a), the other perpendicular. The differences resulting from anisotropy are readily observed. Other techniques for studying the anisotropy of random sets are presented in Weil (1988), Stoyan and Beneš (1991) and Molchanov *et al.* (1993). In the second paper the case of germ–grain models is considered.

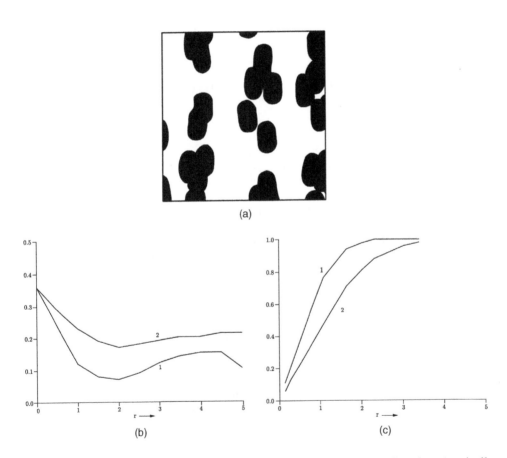

Figure 6.4   (a) An anisotropic area pattern. (b) Estimate of $C(\mathbf{r})$ for two directions (vertically and horizontally). (c) Estimate of $H_l(r)$ for the complement $(\Xi^c)^{cl}$ using the same directions as in (b). Horizontal direction ($\rightarrow$), curves 1; vertical direction $\uparrow$, curves 2.

In the context of surface processes, Section 8.5 provides methods for the statistical analysis of the distribution of surface normals of random closed sets.

6.3.6   Intensities or densities of random closed sets

A given random closed set Ξ is associated with various random measures. One example is the *volume measure* V_Ξ, which is defined by

$$V_\Xi(B) = \nu_d(B \cap \Xi) \qquad \text{for } B \text{ a bounded Borel set.}$$

Other examples include the *curvature measures* discussed in Section 7.3.4.

If such an associated random measure μ_Ξ is stationary then it has an *intensity* $\lambda(\mu_\Xi)$. When $\lambda(\mu_\Xi)$ is finite and positive it is an instructive characteristic of Ξ. The *specific surface* $S_V^{(d)}$ arises in this fashion as the mean surface area of Ξ per unit volume. If $S_V^{(d)}$ is to be well-defined then either assumptions on the smoothness of $\partial\Xi$ must be imposed or the notion of surface area must be modified. Zähle (1982, 1983) discussed suitable modifications of surface area for random closed sets.

Another approach to define densities is via the ergodic theorem. Let h be an additive translation-invariant functional on the Borel sets or on some subsystem of the Borel sets. Suppose that the limit

$$\bar{h}(\Xi) = \lim_{n\to\infty} \frac{\mathbf{E}\big(h(\Xi \cap W_n)\big)}{\nu_d(W_n)} \qquad (6.70)$$

exists (c.f. Theorem 6.2 on p. 210) for all convex averaging sequences $\{W_n\}$. Then $\bar{h}(\Xi)$ is said to be the *h-density* of Ξ; see Schneider and Weil (2008, Section 9.2).

In particular, if $h(\cdot)$ is the kth intrinsic volume $V_k(\cdot)$, its density v_k, defined as

$$v_k = \lim_{n\to\infty} \frac{\mathbf{E}\big(V_k(\Xi \cap W_n)\big)}{\nu_d(W_n)}, \qquad (6.71)$$

is the *specific k^{th} intrinsic volume*, which is the intensity of the corresponding signed curvature measure; see Section 7.3.4.

Last but not least, the *morphological functions* of K. Mecke must be mentioned here; see Mecke (2000) and p. 84 for application in the context of Boolean model. These functions provide a very powerful tool for the statistical description of random sets. (Note that K. Mecke speaks in his papers not about 'morphological functions' but of 'Minkowski functionals'.) The idea is simple and natural, explained here for the three-dimensional case: Consider the dilated random set $\Xi \oplus B(o, r)$ and determine the corresponding densities of intrinsic volumes as functions of the dilation radius r. In the case of volume density or volume fraction, the morphological function is nothing else than the empty space function; in the case of specific surface area it has a close relation to the probability density function of the spherical contact distribution. However, analogous functions related to integral of mean curvature and Euler–Poincaré number yield additional information. The last one, related to the specific connectivity number, is particularly informative. It is also quite useful to show all four functions (of r) in parallel (similarly as in Figure 6.6) and to compare them; see Mecke (2000) and Arns *et al.* (2002).

6.4 Nonparametric statistics for stationary random closed sets

6.4.1 Introduction

This section presents nonparametric methods for the estimation of p, $C(r)$ and $H_B(r)$, and discusses the problem of determining the accuracy of estimation for p. The data are given in an observation window W. Throughout the section the random closed set Ξ under consideration is assumed to be stationary and, in most cases, isotropic. Statistical methods for the particular case of germ–grain models, excursion sets and birth-and-growth models are discussed in Sections 6.5.7, 6.6.3 and 6.6.4, respectively.

Many of the estimators presented are in fact estimators of the capacity functional $T_\Xi(K)$ for special K. If edge-effects are ignored, in the stationary case $T_\Xi(K)$ can be estimated by means of a grid G of n lattice points u by

$$\hat{T}_\Xi(K) = \frac{1}{n} \sum_{u \in G} \mathbf{1}(\Xi \cap K + u \neq \emptyset). \tag{6.72}$$

If the random set Ξ is not stationary, $T_\Xi(K)$ can be estimated from n independent samples $\{\Xi_i\}$ of Ξ:

$$\hat{T}_\Xi(K) = \frac{1}{n} \sum_{i=1}^{n} \mathbf{1}(\Xi_i \cap K \neq \emptyset). \tag{6.73}$$

Little is known of the theoretical properties of the estimators given. If Ξ is ergodic then many of them are consistent for increasing windows. Under certain mixing conditions some of the estimators can be shown to be asymptotically normal.

6.4.2 Estimation of the area or volume fraction p

The methods described here are based on sampling Ξ in the fixed bounded window W of observation by a finite grid of isolated points, a system of parallel lines or planes, or the whole W. The isotropy property is not required for their application.

The point-count method (Thompson, 1930; Glagolev, 1933)

A grid of points $x_1, \ldots, x_n$ in W is used and p is estimated by

$$\hat{p}_p = \frac{1}{n} \left(\sum_{i=1}^{n} \mathbf{1}_\Xi(x_i) \right), \tag{6.74}$$

the fraction of grid points lying in Ξ. This estimator is today frequently applied in image analysis, where the grid points are the centre points of the pixels. (One speaks about 'Gauss digitisation', see e.g. Ohser and Schladitz, 2009.)

The estimator $\hat{p}_p$ is unbiased since

$$\mathbf{E}(\hat{p}_p) = \frac{1}{n} \sum_{i=1}^{n} \mathbf{E}\big(\mathbf{1}_\Xi(x_i)\big) = \frac{1}{n} np = p. \tag{6.75}$$

Its variance σ_p^2 is given by

$$\sigma_p^2 = \frac{1}{n^2}\mathbf{E}\left(np - \sum_{i=1}^{n}\mathbf{1}_\Xi(x_i)\right)^2 = \frac{1}{n^2}\left(np(1-p) + 2\sum_{i>j}k(r_{ij})\right), \tag{6.76}$$

where $r_{ij} = \|x_i - x_j\|$ and $k(r)$ is the covariance function of Ξ. Formula (6.76) follows from expanding the square

$$\left(np - \sum_{i=1}^{n}\mathbf{1}_\Xi(x_i)\right)^2 = \left(\sum_{i=1}^{n}(p - \mathbf{1}_\Xi(x_i))\right)^2.$$

The variance formula (6.76) depends both on the grid geometry and the form of the covariance function $k(r)$. If $k(r)$ decays to zero quickly then approximations are available. A very simple example is

$$\sigma_p^2 \approx \frac{p(1-p)}{n}. \tag{6.77}$$

This would be the exact variance were the indicators $\mathbf{1}_\Xi(x_i)$ independent. Many authors have studied point-count methods; see, for example, Matérn (1986) and Diggle and ter Braak (1982). Also random sampling points are used; if the covariance function $k(r)$ is decreasing then, for fixed n, the estimation variance is greater than for the grid case.

Example 6.1. *Approximation of estimation variance for the point-count method in the planar case*
Suppose the grid is quadratic of mesh width Δ and the covariance function $k(r)$ is of exponential form with $\alpha\Delta \gg 1$. Neglecting edge-effects and only considering up to second-nearest neighbours in the grid lead to

$$\sum_{i \neq j} k(r_{ij}) = 4np(1-p)\exp(-\alpha\Delta)\big(1 + \exp(-(\sqrt{2}-1)\alpha\Delta)\big),$$

and thus

$$\sigma_p^2 \approx \frac{1}{n}p\Big((1-p) + 8(1-p)\exp(-\alpha\Delta)\big(1 + \exp(-(\sqrt{2}-1)\alpha\Delta)\big)\Big). \tag{6.78}$$

The lineal method (Rosiwal, 1898)

An array of N parallel line segments, each of length l, is placed in the window W of observation. The volume fraction p is estimated by

$$\hat{p}_l = \frac{L}{Nl}, \tag{6.79}$$

where L is the (random) total length of line segments intersecting Ξ. As for the point-count method, this estimator is unbiased. Its accuracy has been studied in Stoyan (1979b) and Stoyan and Steyer (1979) for the planar case, with reference also to the interline distance. The particular models used in these studies were either of exponential covariance (6.51) or formed by a Boolean model of spherical grains. In the case of exponential covariance with large α the estimation variance σ_l^2 of $\hat{p}_l$ is approximately

$$\sigma_l^2 \simeq \frac{2p(1-p)\left(1-\dfrac{1}{\alpha l}\right)}{N\alpha l}. \tag{6.80}$$

In the case of a planar sample of a motion-invariant three-dimensional set, σ_l^2 can be expressed in terms of V_V and S_V approximately by

$$\sigma_l^2 \simeq \frac{8V_V^2(1-V_V)^2}{NlS_V}, \tag{6.81}$$

using (6.54). It should be noted that in case of anisotropy the lineal method, based on parallel lines in a fixed orientation, may be affected adversely by the relationship between axes of anisotropy and the common direction of the sampling lines.

The density method

The volume fraction p is estimated in the spirit of volume measure density by the *empirical volume fraction*

$$\hat{p}_v = \frac{v_d(\Xi \cap W)}{v_d(W)}. \tag{6.82}$$

Measurement errors occur only through the bounded nature of the window, if pixelisation errors are ignored. The estimation variance σ_v^2 is given by

$$\sigma_v^2 = \mathbf{E}\big((\hat{p}_v - p)^2\big) = \frac{1}{v_d(W)^2} \int_W \int_W k(\|x - y\|)\, dx\, dy. \tag{6.83}$$

The domain of integration in the variance is W, meaning that the variance depends on the shape and size of the window W of observation. If the window is suitably large (in the sense of extending far in all directions, as for example in the case of a ball or disc of large radius) then an approximate formula is available:

$$\sigma_v^2 \simeq \frac{p(1-p)a_V}{v_d(W)}, \tag{6.84}$$

where a_V is the integral range defined in (6.47). In the particular case of an exponential covariance given in (6.51) with parameter α the variance becomes

$$\sigma_v^2 \simeq \frac{d!\, b_d\, p(1-p)}{\alpha^d\, v_d(W)}, \tag{6.85}$$

see Ohser and Mücklich (2000, Section 5.1) and Kanit *et al.* (2003). Use of Formula (6.54) leads in the three-dimensional case to

$$\sigma_v^2 \simeq \frac{512V_V^4(1-V_V)^4}{S_V^3\, v_d(W)}. \tag{6.86}$$

The formulae given here can also be applied in the context of image analysis if the pixel size is very small in comparison to the details of the random set Ξ investigated.

Heinrich (2005) presents a rigorous study of the limit behaviour of the empirical volume fraction for large windows W.

Example 6.2. *Comparison of estimation variances of the point-count, lineal, and density methods in the planar case*

To illustrate the methods discussed above, consider a random closed set of exponential co-variance with $\alpha = 5\,\text{m}^{-1}$, observed through a square sampling window W of dimension $30 \times 30\,\text{m}^2$. For the point-count method a square grid is used with $31 \times 31 = 961$ points with mesh width $\Delta = 1\,\text{m}$. For the lineal method 31 parallel lines (parallel also to a window side) are employed, each of length $30\,\text{m}$ and with separation distance $1\,\text{m}$.

In the case of area fraction $p = 0.10$ Formulae (6.78), (6.80) and (6.85) yield the following approximate estimation variances, as compared with exact values given in parentheses, obtained by numerical integration:

$$\sigma_p^2 = 0.010 \quad (0.0098),$$

$$\sigma_l^2 = 0.006 \quad (0.0059),$$

$$\sigma_v^2 = 0.005 \quad (0.0046).$$

It is typical that the approximations have small positive biases.

In this case, greater use of information contained in the window is rewarded by lower estimation variance. However, this is not always so; see the interesting discussion in Baddeley and Cruz-Orive (1995).

6.4.3 Estimation of the covariance

Using Formula (6.40) the covariance $C(\mathbf{r})$ can be estimated as the volume fraction of the set $\Xi \ominus \{o, \mathbf{r}\}$. Under the assumption of isotropy the same estimator can be used for $C(r)$ with $\|\mathbf{r}\| = r$. For lattice data this is equivalent to

$$\hat{C}(r) = \frac{\sum_{i=1}^{n-1} \sum_{j=i+1}^{n} \mathbf{1}_\Xi(x_i)\mathbf{1}_\Xi(x_j)\mathbf{1}(\|x_i - x_j\| = r)}{\sum_{i=1}^{n-1} \sum_{j=i+1}^{n} \mathbf{1}_\Xi(x_i)\mathbf{1}_\Xi(x_j)}, \tag{6.87}$$

where x_i and x_j are lattice points of a finite grid. (Of course, the denominator must be positive.) For large samples, which are typical for materials research and physics, Fourier methods are recommended; see Ohser and Schladitz (2009, Chapter 6).

If normalised forms of the covariance are estimated and the sampling window is small, it makes sense to use distance-adapted estimators $\hat{p}(r)$ of p. This was shown in Mattfeldt and Stoyan (2000) for the pair correlation function $g(r) = C(r)/p^2$. In this case it is then natural to use the adapted $\hat{p}(r)$ resulting from the same reduced window from which $C(r)$ is estimated.

6.4.4 Second-order analysis with random fields

The analysis of the spatial variability of random sets should not be limited to analysis of the covariance $C(r)$, which is related to the volume measure; see Section 7.3.4. Arns *et al.* (2005) demonstrate an approach which includes also the other curvature measures $\Phi_{\Xi, k}$ explained in Section 7.3.4.

The idea is to construct for a given random set Ξ random fields $\{Z_k(x)\}$, called the *intrinsic volume fields*, for $k = 0, 1, \ldots, d$ by

$$Z_k(x) = \Phi_{\Xi, k}\big(B(x, R)\big) \qquad \text{for } x \in \mathbb{R}^d, \tag{6.88}$$

that is, the value $Z_k(x)$ is given by the value of the kth curvature measure Φ_k corresponding to Ξ for the ball $B(x, R)$ of radius R centred at x. Here the radius R is a procedure parameter controlling the smoothness of the field. For example, $Z_d(x) = v_d\big(\Xi \cap B(x, R)\big)$ and if $d = 3$ and $k = 2$ then $Z_2(x)$ is the half surface area content of Ξ in $B(x, R)$ (where the surface of the ball $B(x, R)$ is excluded), and its mean is $\frac{1}{2} S_V \cdot \frac{4}{3}\pi R^3$.

These random fields can be constructed numerically and analysed statistically by methods of geostatistics, that is, by statistical methods for random fields. This is explained in Arns *et al.* (2005) for samples of foams and sandstone. Figure 6.5 shows a $(1.4\,\text{mm}^3)$ sample of Fontainebleau sandstone of porosity 13% and the corresponding empirical covariance $C(r)$.

Figure 6.6 shows the empirical normalised variograms of the corresponding random fields. Recall that for a random field $Z_k(x)$ the variogram $\gamma_k(r)$ is given by

$$\gamma_k(r) = \frac{1}{2} \mathbf{E}\left(\big(Z_k(o) - Z_k(\mathbf{r})\big)^2\right). \tag{6.89}$$

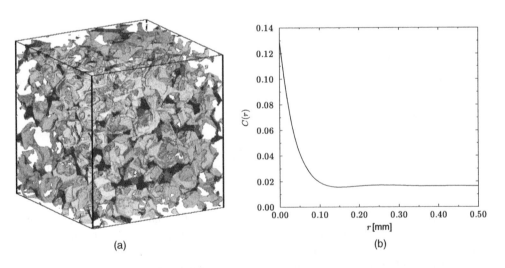

(a)   (b)

Figure 6.5   (a) A $(1.4\,\text{mm}^3)$ sample of Fontainebleau sandstone of porosity 13%, and (b) the corresponding empirical covariance $C(r)$ of the pore space. Reproduced from Arns *et al.* (2005) with permission of Springer.

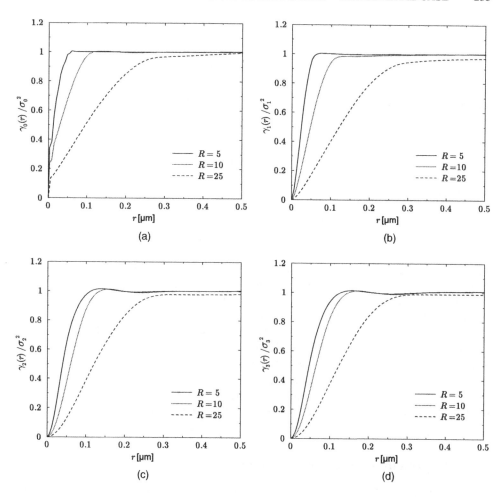

Figure 6.6 Normalised variograms $\gamma_k(r)/\sigma_k^2$ of the intrinsic volume fields for the Fontainebleau sandstone. (a) Euler–Poincaré characteristic, (b) integral of mean curvature, (c) surface area, and (d) volume. Reproduced from Arns *et al.* (2005) with permission of Springer.

(Because of the factor $\frac{1}{2}$, some authors call $\gamma_k(r)$ the *semi-variogram*.) It is natural to normalise it by the variance σ_k^2 of $Z_k(o)$. The (normalised) variograms show that the variability of volume and surface distribution are similar and higher than that of integral of mean curvature and Euler–Poincaré characteristic. The variogram $\gamma_3(r)$, as an alternative to $C(r)$, characterises the variability of volume distribution. The range of correlation for volume is about 0.3 mm.

6.4.5 Estimation of contact distributions

Since $\left(1 - H_B(r)\right)(1 - p)$ is the volume fraction of the stationary random set $\Xi \oplus r \check{B}$, methods in Section 6.4.2 are applicable, where B satisfies the conditions on p. 223. The following considers only the case where area or volume measurement is possible. From the definition of $H_B(r)$ and the minus-sampling estimator (Section 6.5.7) of it is

$$\hat{H}_B(r) = 1 - \frac{1}{1-\hat{p}}\left(1 - \frac{\nu_d\big((W \ominus r\check{B}) \cap (\Xi \oplus r\check{B})\big)}{\nu_d(W \ominus r\check{B})}\right) \qquad \text{for } r \geq 0. \qquad (6.90)$$

For the particular case of the spherical contact distribution $H_s(r)$, the estimator in (6.90) can be written as

$$\hat{H}_s(r) = 1 - \frac{1}{1-\hat{p}}\left(1 - \frac{\nu_d\big((W \ominus B(o,r)) \cap (\Xi \oplus B(o,r))\big)}{\nu_d\big(W \ominus B(o,r)\big)}\right) \qquad \text{for } r \geq 0. \qquad (6.91)$$

An estimator of $H_s(r)$ in the spirit of Hanisch's D-estimator (4.124) is

$$\hat{H}_s(r) = 1 - \frac{1}{1-\hat{p}}\left(1 - \int_W \frac{\mathbf{1}\big(x \in W \ominus B(o,d(x))\big)\mathbf{1}(d(x) \leq r)}{\nu_d\big(W \ominus B(o,d(x))\big)}\,dx\right) \qquad \text{for } r \geq 0. \qquad (6.92)$$

Here $d(x) = d(x, \Xi)$ is the Euclidean distance from x to Ξ; if $x \in \Xi$ then $d(x) = 0$. In practical application the integral becomes a sum, which is based on a grid of test points.

If $\hat{p}$ is unbiased then the above estimators of $H_s(r)$ are ratio-unbiased. Mayer (2004) showed how the estimator (6.90) can be efficiently applied.

Hansen et al. (1996) and Baddeley and Gill (1997) suggested Kaplan–Meier estimators for $H_B(r)(1-p)$ and the corresponding hazard rate. They argued that the latter is a good characteristic for model choice. Chiu and Stoyan (1998) showed that the ideas behind the Hanisch and Kaplan–Meier estimator are similar.

The distance-adapted $\hat{p}(r)$ mentioned in the previous section can of course be employed also here. However, the simulation in Stoyan et al. (2001) suggests that $\hat{p}(r)$ may not be better than the simple estimator $\hat{p}$ for the purposes of estimating $H_B(r)$.

6.4.6   Representative volume elements

A question of great practical importance in the planning of statistical analysis and in calculations of macroscopic properties of structures with random microstructure is the size of the necessary window W of observation or calculation. If this window is too small, edge-effects may play a rôle too great and typical fluctuations of the studied random sets are perhaps overlooked. Using a terminology that first appeared in the context of *homogenisation*, this book speaks of, even in the two-dimensional case, *representative volume elements* (RVEs). The size of the RVE is called *relevant volume* (RV).

In the case of a study of fluid permeability, for example, one takes a sample of the random set representing the porous medium of interest. The flow behaviour in the sample is studied using partial differential equation solvers and the permeability of a homogeneous sample of equal size is determined. If the sample (i.e. observation window W) is small, then edge-effects may play some rôle so that the observed porosity variability cannot be considered as typical. Consequently, the results obtained are not representative. Larger samples deliver estimates which should be (in favourable cases) closer to values that would be obtained for infinitely extended samples, which correspond to the statistically homogeneous case. However, the samples should be kept sufficiently small in order to be considered as volume elements (i.e. differential elements) of the continuum model of the porous medium. The smallest sample that is large enough to guarantee sufficiently small errors is the RVE (Nemat-Nasser and Hori,

1999). That is to say, an RVE and its size RV, which has to be interpreted as a lower bound, depend on a user-prescribed upper bound of statistical error and are only meaningful if this statistical tolerance is given.

The determination of an RVE is, of course, difficult. It depends both on the variability of the random set and on the physical quantity of interest and the summary characteristic being considered (see Freudenthal, 1950, Kanit *et al.*, 2003, Stroeven *et al.*, 2004, and Buryachenko, 2007). If the same summary statistic is considered, random sets with irregularly scattered large components require larger RVEs than random sets with many small and regularly distributed particles. On the other hand, for a fixed microstructure, a larger RVE is required for a precise estimation of the covariance $C(r)$ or spherical contact distribution function $H_s(r)$ than for the estimation of volume fraction p, say.

In classical statistics, sample size calculations require some prior knowledge on the nature of the data that will be analysed, such as their variation. A similar requirement holds in the context of random sets. It is impossible to determine the RVE without any *a priori* knowledge of the distribution of the random set investigated. A straightforward approach for acquiring *a priori* knowledge is a *pilot study* consisting of a preliminary statistical analysis of a small window (or a small number of windows if a series of windows have to be analysed). The expectation is that the pilot study yields a useful yet rough estimate of volume fraction p and some rough information on $C(r)$, perhaps the range of correlation.

If volume fraction p is of interest, the formulae in Section 6.4.2 may be used. In particular Formula (6.84) yields an approximation of RV for a prescribed precision σ_v^2 of the p-estimator. There the integral range plays an important rôle.

For other characteristics either (a) statistical experiments or (b) reconstruction simulations may be used for the determination of RVEs. Method (a) assumes that it is possible to generate a series of nested windows of increasing size. The summary characteristics of interest are estimated from them in order to identify the minimum window size that can be considered representative, such that any further increase of the window size will essentially not change the statistical results further. Method (b) generates random set samples larger than the original pattern by the reconstruction simulation method described in Section 6.7, under the assumption that the pattern given in the single original observation window shows the typical process behaviour. Subsequently method (a) can be applied to the simulated patterns.

Two papers in which RVEs are thoroughly discussed from a statistical standpoint are Kanit *et al.* (2003) and Stroeven *et al.* (2004). The first paper considers as an example two-phase Voronoi tessellations (where the cells are coloured randomly black and white), the second one focuses on granular materials, that is, special germ–grain models.

6.5 Germ–grain models

6.5.1 Basic facts

In a natural generalisation of the Boolean model, (a) the Poisson point process of germs is replaced by a general point process and (b) the grains are permitted to be dependent. This leads to the class of *germ–grain models*. In particular, the generalisation allows for treatment of collections of non-overlapping grains, since independence between germ and grains, as in the Boolean model, may lead to overlapping of grains.

A formal definition starts with a marked point process $\Psi = \{[x_n; \Xi_n]\}$ where the points x_n lie in $\mathbb{R}^d$ and the marks Ξ_n are compact subsets of $\mathbb{R}^d$. A *germ–grain model* Ξ as defined by Hanisch (1981) arises from such a marked point process as the union

$$\Xi = \bigcup_{n=1}^{\infty}(\Xi_n + x_n) = (\Xi_1 + x_1) \cup (\Xi_2 + x_2) \cup \cdots . \tag{6.93}$$

The points x_n are the *germs* and the compact sets Ξ_n are the *grains* of the germ–grain model. Sometimes, Ψ is called a *germ–grain process* (Schneider and Weil, 2008), while Hall (1988) calls $\{\Xi_n + x_n\}$ a *coverage process*.

In the case when Ψ and Ξ are stationary it is useful to introduce the notion of the typical grain Ξ_0, which is a random compact set having the same distribution as the marks of the marked point process Ψ. It is a distribution on the space $\mathbb{K}$ of compact sets endowed with the σ-algebra $\mathcal{F}_{\mathbb{K}}$, the trace of $\mathcal{F}$ on $\mathbb{K}$. There may exist dependences between the grains, as well as between grains and germs. The intensity of the stationary germ process is denoted by λ.

Note that rotating the random set Ξ is not the same as rotating the germ–grain process Ψ, because for the latter, which is a marked point process, the rotation applies only to the germs (points) but not to the grains (marks); see p. 118. Thus, when speaking about rotation and isotropy, one has to be careful.

Every random closed set can be trivially decomposed to become a germ–grain model. Weil and Wieacker (1987) (see also Schneider and Weil, 2008, Section 4.3) provide a meaningful decomposition by showing that for every random closed set Ξ in $\mathbb{R}^d$, there is a point process in $\mathbb{F}$ such that the union of its 'points' (which are closed sets in $\mathbb{R}^d$) is equal to Ξ in such a way that invariance properties under rigid motion (e.g. stationarity and isotropy) of Ξ in $\mathbb{R}^d$ are preserved in the corresponding point process in $\mathbb{F}$. Moreover, if Ξ is a random S-set satisfying a local finiteness condition (see Schneider and Weil, 2008, p. 119), then the corresponding point process takes place in $C(\mathbb{K})$ so that the 'points' are in this case convex bodies. Thus random S-sets are essentially the same as germ–grain models with convex grains.

Random S-sets Ξ have boundaries that are smooth enough for the definition of random curvature measures as in Chapter 7. These are examples of random measures accompanying Ξ, as described in Section 7.3.4. When Ξ is stationary then so are the random curvature measures.

6.5.2 Formulae for p and $C(r)$

Explicit formulae for general germ–grain models are complicated. The capacity functional T_{Ξ} can be expressed in terms of the generating functional G_{Ψ} of the marked point process Ψ,

$$T_{\Xi}(K) = \mathbf{P}(\Xi \cap K \neq \emptyset) = 1 - G_{\Psi}(v_K), \tag{6.94}$$

where K is compact and

$$v_K(x, C) = 1 - \mathbf{1}_{\check{C}\oplus K}(x) \qquad \text{for compact } C \text{ and } x \in \mathbb{R}^d. \tag{6.95}$$

The left-hand side of Formula (6.94) can be expressed in terms of the germ process $\{x_n\}$, denoted by Φ, and the capacity functional T_{Ξ_0} of the typical grain Ξ_0 so that it can be rewritten as

$$T_{\Xi}(K) = 1 - \mathbf{E} \prod_{x \in \Phi} \left(1 - T_{\Xi_0}(K - x)\right). \tag{6.96}$$

When the grains are non-overlapping, which is to say when

$$(\Xi_i + x_i) \cap (\Xi_j + x_j) = \emptyset \qquad \text{for all } i \neq j$$

with probability one, and when, moreover, Ξ is stationary, then the volume fraction is given by

$$p = \lambda \overline{V}, \tag{6.97}$$

where $\overline{V} = \mathbf{E}\left(v_d(\Xi_0)\right)$ is the mean volume of the typical grain Ξ_0. Hall (1988, Section 3.8) gives bounds for p for the case of independent grains and Cox and Neyman–Scott germ processes.

In the case of non-overlapping grains and stationarity the intensities v_k of the intrinsic volume measures satisfy the analogous formula

$$v_k = \lambda \overline{V}_k, \tag{6.98}$$

where $\overline{V}_k = \mathbf{E}\left(V_k(\Xi_0)\right)$ is the mean of the kth intrinsic volume of the typical grain Ξ_0, and the covariance $C(r)$ is the sum of two terms

$$C(r) = C_1(r) + C_2(r), \tag{6.99}$$

where the first term is the probability that the origin o and the second point $\mathbf{r}$, any point of $\mathbb{R}^d$ with $\|\mathbf{r}\| = r$, belong to the same grain, and the second term the probability that these two points belong to different grains. When the non-overlapping grains are independent, isotropic and convex with positive volume, (6.99) can be expressed as

$$C(r) = \lambda \overline{\gamma}_{\Xi_0}(r) + \int_{\mathbb{R}^d} \int_{\mathbb{R}^d} \mathbf{1}_{\Xi_0}(x) p_{\Xi_0}(x - \mathbf{r} + z) \varrho^{(2)}(z) \, dx \, dz, \tag{6.100}$$

where $\overline{\gamma}_{\Xi_0}(r)$ is the isotropised set covariance of the typical grain Ξ_0 and $p_{\Xi_0}(x)$ its coverage function. Furthermore, λ and $\varrho^{(2)}(r)$ are the intensity and second-order product intensity of the (motion-invariant) germ-process; see Hanisch (1984a,b).

Torquato (1991) discusses the form (6.99) of $C(r)$, which is also useful for interpreting empirical covariances. He calls $C_1(r)$ 'two-point cluster function' and $C_2(r)$ 'two-point blocking function'.

6.5.3   Models of mutually non-overlapping balls

Models of random systems of mutually non-overlapping balls are of great practical interest in biology, physics and materials science. The following discusses briefly some models for such structures. The text starts with models of larger density, which are of great practical interest but mathematically difficult to analyse.

The RSA model

RSA is an abbreviation for 'random sequential adsorption' (or 'addition'), a term used in physics and chemistry. The RSA model is also termed SSI model in the statistical literature, an abbreviation of 'simple sequential inhibition'. Since the model is used much more frequently in physics and chemistry than in statistics, the name RSA is used here. Evans (1993) and Talbot *et al.* (2000) are key references.

The RSA model yields a process taking place in a finite region W and is hence a model for a structure composed of finitely many balls. The pattern is constructed by iteratively and randomly placing centres of balls into W with radii following some distribution function. Once a ball is successfully placed, its position is permanently fixed. If a new ball intersects with an already existing ball, the new ball is rejected and another ball with a different centre and perhaps a new radius (or just the same radius) is generated, and so on. There is, furthermore, a model abbreviated CSA, meaning 'cooperative sequential adsorption' (Talbot *et al.*, 2000), where the probability of adsorption of a new ball is proportional to a Boltzmann factor involving interaction between the new ball and the pre-adsorbed balls.

Once it is impossible to place any new ball (then the 'jamming' state is attained) in W the process stops. The pattern formed by the balls is a sample of the random set to be generated.

Stoyan and Schlather (2000) discuss a stationary version of the model. In the case of random radii, the (proposal) distribution used to generate radii of new balls and the (resulting) distribution of the radii of adsorbed balls have to be distinguished, because larger balls are less likely than smaller balls to be adsorbed.

All numerical information for the RSA model (Evans, 1993) has been obtained by simulation, for sufficiently large W such that one could speak about 'stationary case'. The RSA model is simulated along the lines of the model description. For an efficient simulation close to jamming, the search for potential locations for new balls should make use of an efficient search algorithm; see Döge (2001).

The area and volume fraction A_A and V_V for the model with constant radii can be determined by simulation

$$A_A = 0.547 \quad \text{and} \quad V_V = 0.382.$$

Statistical methods for the dynamic RSA model are described in van Lieshout (2006). Provatas *et al.* (2000) consider closely related models where the balls are replaced by fibres.

Packings of hard balls

Random close packings of identical balls were already mentioned in Section 5.4. Following Torquato and Stillinger (2010), instead of 'close' this section uses the term 'jammed'. In jammed packings each particle is in contact with its nearest neighbours, whilst in not-jammed packings 'rattlers' (particles not in contact with the rigid cluster, called the backbone, of particles that are jammed) may exist.

Perhaps it is risky to write in a mathematical book about a structure for which to date there is no mathematical model. The great practical importance of random close packings is the reason for this short section; the reader is also referred to the book by Aste and Weaire (2008) and the survey papers Torquato and Stillinger (2010) and Stachurski (2011), the latter of which discusses applications in the theory of amorphous materials. Comparable structures exist in reality, as real random packings of real hard balls (which approximate ideal hard balls) and have been often experimentally generated since Bernal (1960); see Aste *et al.* (2004, 2005) for more recent reports. Empirical studies by different authors have repeatedly established

(a) volume fraction $V_V = p = 0.64$;

(b) pair correlation function $g(r)$ of the point process of centres as in Figure 4.7.

The following approximation formulae are given for hard ball systems with general volume fraction V_V between 0.5 and 0.7 and radius R. The paper Lochmann *et al.* (2006) shows a sequence of pair correlation functions for different V_V.

Covariance

$$C(r) \approx V_V - \frac{S_V}{4} r + \frac{Z V_V}{4} \left(\frac{r}{2R} \right)^2 + O(r^3) \qquad \text{for } r \geq 0, \tag{6.101}$$

where Z is the mean coordination number, that is, the mean number of contacts of the typical ball with other balls; see Torquato (2002, p. 38), who refers to Frisch and Stillinger (1963). In the case of $V_V = 0.64$, for Z the value 6.05 was found for simulated packings (Bezrukov *et al.*, 2002). For experimental packings smaller values were obtained, for example, between 5 and 5.5 (Delaney *et al.*, 2010). From the standpoint of physicists, the simulated structures are such 'in the limit of zero friction'.

Linear contact distribution function

$$H_l(r) \approx 1 - \exp \left(-\frac{3 V_V}{4(1 - V_V)} \frac{r}{R} \right) \qquad \text{for } r \geq 0, \tag{6.102}$$

see Levitz and Tchoubar (1992), Lu and Torquato (1992) and Stoyan *et al.* (2011).

Density function of spherical contact distribution function

$$h_s(r) \approx \frac{2}{\sqrt{2\pi}\sigma} \exp \left(-\frac{r^2}{2\sigma^2} \right) \qquad \text{for } r \geq 0, \tag{6.103}$$

with

$$\sigma = \frac{(1 - V_V)2R}{3\sqrt{2\pi} V_V}, \tag{6.104}$$

see Stoyan *et al.* (2011).

That these summary characteristics are used by various authors shows that the idea of a motion-invariant hard-ball-packed structure is widely accepted, though properly speaking it can refer only to ensembles made up of infinitely many balls.

Such structures have been simulated by various algorithms (differing only in detail), for example the Lubachevsky–Stillinger algorithm (see Donev *et al.*, 2005) and the force-biased algorithm (see Bezrukov *et al.*, 2002, and Illian *et al.*, 2008, p. 395). Simulation studies have often reproduced the results above. This leads the authors to believe that there does exist an objective mathematical object called 'random jammed packing of hard balls', for which, one day, some genius will propose a mathematical model, which has in some approximation the properties above and will lead to a formula for V_V. Stoyan (1998) *conjectured* that

$$\boxed{V_V = \frac{2}{\pi}}, \tag{6.105}$$

see also Buryachenko (2007, p. 180). Already Scott (1960) gives the value $V_V = 0.6366$, which is very close to $2/\pi$, and it seems to be clear that he, as well as his contemporaries J. D. Bernal and J. L. Finney, believed that the true unknown value is indeed $2/\pi$. (Note that although Song *et al.*, 2008, argue, by interpreting the random jammed packing as the ground state of the ensemble of jammed matter, that the volume fraction V_V cannot exceed 0.634, their value 0.634 is not precise but just an approximate value.) And perhaps one day even a formula for $g(r)$ will be found. Then it may become clear whether at $r = 2R$ there is, in addition to the δ-component (resulting from the direct contacts of balls), also a pole.

This book assumes the existence of the set-theoretic union of the random jammed packing of hard balls as a motion-invariant random closed set, which in this section is denoted by Ξ.

This set Ξ is the base for another model, the *cherry-pit model* $\Xi_{\oplus r}$,

$$\Xi_{\oplus r} = \Xi \oplus B(o, r). \tag{6.106}$$

This means that all balls of Ξ are enlarged in such a way that the radii are increased by the value r, then being $R + r$. The enlargement leads to the possibility of overlapping of the larger balls. (The hard balls are the 'pits' and the enlarged ones the 'cherries'.) Clearly, also $\Xi_{\oplus r}$ is motion-invariant. (Note that in Torquato, 2002, the cherry-pit model is defined with respect to a Gibbs hard-core process.) Clearly, its volume fraction $V_{V,r}$ and specific surface area $S_{V,r}$ satisfy

$$V_{V,r} \to 1 \quad \text{and} \quad S_{V,r} \to 0 \qquad \text{for } r \to \infty.$$

From (6.103) and the fact that the spherical contact distribution function of $\Xi_{\oplus r}$ can be easily expressed by that of Ξ, the following approximations are obtained:

$$V_{V,r} \approx 1 - 2(1 - V_V)\left(1 - \Phi\left(\frac{r}{\sigma}\right)\right), \tag{6.107}$$

and

$$S_{V,r} \approx (1 - V_V)\frac{2}{\sqrt{2\pi}\sigma}\exp\left(-\frac{r^2}{2\sigma^2}\right), \tag{6.108}$$

where V_V denotes the volume fraction of the hard ball packing Ξ, $\Phi(x)$ is the standard Gaussian distribution function (with $\Phi(0) = \frac{1}{2}$), and σ is given by (6.104); see Elsner *et al.* (2009).

The case of balls of random diameters is considered for example in Lochmann *et al.* (2006). Hermann *et al.* (2013) discuss the corresponding cherry-pit model.

Moreover, there are also algorithms to pack ellipsoids, cylinders and polyhedra; see Bezrukov and Stoyan (2006), Bargieł (2008), Torquato and Stillinger (2010) and Jiao and Torquato (2011). Other algorithms try to simulate processes of sedimentation under the influence of gravity, which do not lead to stationary or equilibrium systems; see for example Jodrey and Tory (1979) and the experiment-oriented paper Royall *et al.* (2007).

Figure 6.7 A natural packing of cubical paving-stones. Courtesy of E. Rothe.

Also radial simulations have been used, where one irregular cluster of balls is placed at the origin and is then enlarged by adding successively layers of balls; see p. 165 of SKM95, and Stachurski (2011).

Figure 6.7 shows a natural random packing of cubical paving-stones.

The paper Ballani *et al.* (2006) gives an example for a statistical analysis of a structure which can be modelled by a packing of hard balls, some special kind of concrete. The first step was computerised tomography (CT) and led to pixel data. In the second step, methods of Bayesian image analysis (van Lieshout, 1995, and Winkler, 2003) were applied to segment the grains as non-overlapping ideal balls. For this simulated annealing was used. Having the data of balls, one could then apply methods of point process statistics; see also Ballani (2006a). An alternative approach is described in Thiedmann *et al.* (2012), which is suitable for more noisy data.

The Stienen model

Now two models are considered where some mathematics is possible. Unfortunately, these have densities much smaller than the models discussed above.

For the *Stienen model* (Stoyan, 1990a) the germ process is a homogeneous Poisson process of intensity λ. The grains are balls of random diameters: the diameter d_n of the ball around the germ point x_n is equal to the distance from x_n to its nearest neighbour. Figure 6.8 shows a simulated sample of this model for the planar case. Typically large balls are isolated, and small ones appear sometimes as isolated pairs. The stationary and isotropic marked point process $\{[x_n; d_n]\}$ is an example of a *dependently* marked Poisson process. Clearly, the Stienen model is not a Boolean model. By the way, it can also be interpreted as the system of *in*-balls (centred at the generating points) of the Poisson-Voronoi tessellation; see Section 9.7.

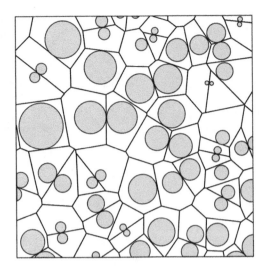

Figure 6.8 A sample of the planar Stienen model. The centres of the discs form a sample of a Poisson process; the disc diameters are equal to the corresponding nearest-neighbour distances. The polygons are the corresponding Dirichlet cells.

The volume fraction p can be calculated by means of Formula (6.97) as

$$p = \lambda b_d \cdot \int_0^\infty r^d \lambda b_d d r^{d-1} \exp(-\lambda b_d r^d) \, dr,$$

which is $1/8$ in the case $d = 3$. Wiencek and Stoyan (1993) give for $d = 3$ the covariance $C(r)$, obtained by simulation. Schlather and Stoyan (1997) express $C(r)$ in integral form for $d \geq 2$. Olsbo (2007) calculates the correlation coefficient between the volume of the typical Poisson-Voronoi cell and the corresponding Stienen ball.

The Stienen model can be generalised by means of a reduction factor α (< 1): the corresponding marked point process is then $\{[x_n; \alpha d_n]\}$.

In the spatial case it is possible to calculate characteristics related to planar sections: the section disc diameter distribution functions, the intensity of the point process of section disc centres and even the pair correlation function of this point process; see Section 10.7.2.

The lilypond model

Häggström and Meester (1996) introduced a model closely related to the Stienen model. Consider a point process of germs in $\mathbb{R}^d$. Centred at these germs, grains start growing radially at the same instant with the same constant speed. A grain stops growing when it touches another grain. This germ–grain model of non-overlapping balls is called the *lilypond model*. A more formal definition (to allow non-Poisson germs) given by Daley and Last (2005) is that the lilypond model is a system of non-overlapping balls such that each ball is in touch with a ball of equal or smaller size. Heveling and Last (2006) showed that there always exists a unique lilypond system of balls for any germ process.

Obviously, the lilypond model contains as a subset the Stienen model of the same set of germs. Consequently, its volume fraction is larger than that of the Stienen model. Daley *et al.*

(1999, 2000) determined the volume fraction of the Poisson lilypond model:

$$
p = \begin{cases} 1 - e^{-1} & \text{for } d = 1, \\ 0.349 & \text{for } d = 2, \\ 0.186 & \text{for } d = 3, \end{cases} \tag{6.109}
$$

the latter two values by simulation. The former paper also gives an exponential upper bound for the tail of the distribution of the size of the typical ball. Subexponential bounds for the tail of the size of the cluster (of touching balls) containing the origin are given in Last and Penrose (2013), where also central limit theorems for the total volume and the number of clusters in expanding windows are proved.

For a homogeneous Poisson germ process, Häggström and Meester (1996) proved that the model does not percolate, that is, there is no infinite cluster of touching balls. Daley and Last (2005) showed that there is also no percolation for a wide range of germ processes, including certain Poisson cluster, Cox and Gibbs processes satisfying some moment conditions. Obviously, for $d > 1$, if the radii of the balls in the Poisson lilypond model are increased by a sufficiently large fixed amount δ, then percolation occurs; the infimum of such δ is called the *critical enhancement* δ_c. Last and Penrose (2013) showed that δ_c is strictly positive, meaning that for sufficiently small $0 < \delta < \delta_c$, the union of balls enlarged by δ still does not percolate. Thus, in a Poisson lilypond, not every frog that is able to jump from one lily pad to another can travel infinitely far without getting into water; if its jump range is too small, it is still unable to do so.

Various

Pelikan *et al.* (1994) study germ–grain models where the grains are balls or cylinders with caps and the positions are given as in the RSA process. Cylindric grains also appear in the filament model in Stoica *et al.* (2010).

Biswal *et al.* (2009) report about a random-field-controlled germ–grain model of RSA type.

Other models come from the field of Gibbs processes; see Baddeley and Møller (1989), Mase (1985), Stoyan (1989), Stoyan and Stoyan (1994, p. 334), van Lieshout (1995), W. S. Kendall *et al.* (1999), Møller and Helisová (2010) and Coeurjolly *et al.* (2012). A simple example is the system of hard balls belonging to the Gibbs hard-core process as in Section 5.5.3.

6.5.4 Shot-noise germ–grain models

An interesting class of germ–grain processes with dependent grains is closely related to shot-noise fields. Its random-set version, the corresponding germ–grain model, is called the *SINR coverage process*, where SINR stands for *signal to interference and noise ratio* (Baccelli and Błaszczyszyn, 2009a, Chapter 7).

The construction starts from a shot-noise field $\{S_\Psi(x)\}$ as in Section 5.6 constructed using a marked point process $\Psi = \{[x_n; m_n]\}$ of intensity λ and a response function $s(x, m)$. The germs of the shot-noise germ–grain process are simply the points x_n, while the grains Ξ_n are constructed as follows:

$$
\Xi_n = \left\{ x \in \mathbb{R}^d : s(x - x_n, m_n) \geq t \cdot \left(w(x) + S_\Psi(x) \right) \right\}. \tag{6.110}
$$

Here t is a nonnegative threshold, which can, in more general models, also depend on n, and $w(x)$ denotes external or thermal noise. Thus Ξ_n is the set where the signal from x_n is t times greater than the relevant noise and interference. The shot-noise germ–grain model Ξ is given by an equation analogous to (6.93).

Baccelli and Błaszczyszyn (2009a, Section 7.4) show that the grains Ξ_n defined in (6.110) are random closed sets, and under boundedness conditions they are also compact. The grains Ξ_n can be empty; they are not necessarily spherical or convex, can overlap and are in general mutually dependent. If $s(x, m)$ is defined by (5.106) with the simplified path-loss function, then each Ξ_n is not empty and contains always x_n, since $s(o, m) = \infty$.

Figure 6.9 shows a simulated shot-noise germ–grain model in a $2 \times 2\,\mathrm{km}^2$ window, with a homogeneous Poisson germ process of intensity $\lambda = 10\,\mathrm{km}^{-2}$, simplified path-loss response

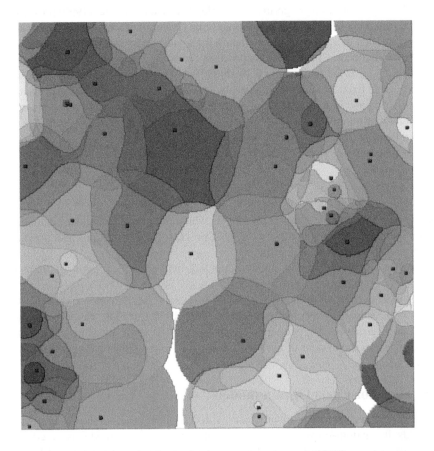

Figure 6.9 A simulated sample of a shot-noise germ–grain model. Different grey-tones mark different grains, the white phase is the complement of the random set Ξ. A grain corresponds to the region where the signal emitted with power of 1 W (Watt) by an antenna located at the corresponding germ can be received with the signal-to-interference-and-noise ratio not smaller than $t = 0.2$; the interference is caused by all other germs emitting also signals of power 1 W each, and modelled by the shot-noise, noise power $w(x) \equiv 10^{-10}$ W. The parameters are given in the text. Courtesy of B. Błaszczyszyn.

function $s(x, m) = L(\|x\|) = L(r)$, where

$$L(r) = (Ar)^{-\beta} \quad \text{(in Watt)} \tag{6.111}$$

with $A = 8000 \, \text{km}^{-1}$ and $\beta = 3$, $w(x) \equiv 10^{-10} \, \text{W}$ (Watt) and $t = 0.2$. These parameters are quite realistic for power attenuation of radio signals.

In dependence on the model characteristics the grains have different shapes. Extremal cases are balls and Voronoi cells.

In the case that Ψ is an independently marked homogeneous Poisson process the distribution of the typical grain, denoted as Ξ_o, can be characterised analytically. The Slivnyak–Mecke formula makes it possible to consider the grain Ξ_o as if based on a germ at the origin o and mark m with the same mark distribution as the m_n and independent of Ψ as

$$\Xi_o = \left\{ x \in \mathbb{R}^d : s(x, m) \geq t \cdot \left(w(x) + S_\Psi(x) \right) \right\}. \tag{6.112}$$

In particular, the mean volume of Ξ_o can be obtained as

$$\overline{V} = \int_{\mathbb{R}^d} p(x) \mathrm{d}x, \tag{6.113}$$

with

$$p(x) = \mathbf{P}(x \in \Xi_o). \tag{6.114}$$

For the planar case of the simplified path-loss function in (6.111) with $A > 0$ and $\beta > 2$, $w(x) \equiv 0$ and exponentially distributed marks, the coverage probability is given by

$$p(x) = \exp(-\lambda \|x\|^2 t^{2/\beta} K), \tag{6.115}$$

where

$$K = \frac{2\pi}{\beta} \Gamma\!\left(\frac{2}{\beta}\right) \Gamma\!\left(1 - \frac{2}{\beta}\right) = \frac{2\pi^2}{\beta \sin(2\pi/\beta)} \tag{6.116}$$

(Baccelli and Błaszczyszyn, 2009a, Example 5.6, and Baccelli and Błaszczyszyn, 2009b, Proposition 16.1). Note that (6.115) and (6.116) do not contain the parameter of the exponential mark distribution and A. The volume fraction p is difficult to obtain, but clearly the qualitative relation

$$p = \lambda \overline{V} - O(\lambda^2) \tag{6.117}$$

holds.

Baccelli and Błaszczyszyn (2009a,b) present many results for this germ–grain process, for example information on the set covariance $\gamma_{\Xi_o}(r)$ of the typical grain Ξ_o and the random number of grains overlapping a given point. Also properties of the random graph with nodes $\{x_n\}$ and edges connecting germs of overlapping grains and percolation are studied; see Sections 3.3.3 and 3.3.4.

6.5.5 Weighted grain distributions

The mark distribution M describes the shape and size of the typical grain of a germ–grain model. It is sometimes useful to weight the grain by a size parameter and to study the weighted mark distribution of the grain. The *volume-weighted mark distribution* $M_{(v)}$ is an important special case. It is defined by

$$\int f(K)M_{(v)}(\mathrm{d}K) = \int f(K)v_d(K)M(\mathrm{d}K)\Big/\overline{V} \qquad (6.118)$$

for nonnegative measurable functions $f(K)$ on $\mathbb{K}$, where $\overline{V}$ is the mean volume of the typical (unweighted) grain. A small number of large grains contribute to this distribution as much as a large number of small grains.

Such volume-weighted mark distributions can be of more interest than the unmodified or *number-weighted* mark distribution M. In the case of balls the term 'sieving distribution' is often used, since a series of sieving operations using sieves of different mesh sizes will produce an approximation to a volume-weighted distribution of balls.

Weighting can be carried out by using other functionals such as surface area rather than volume.

A series of moment formulae are implied by (6.118):

$$\int \big(v_d(K)\big)^n M_{(v)}(\mathrm{d}K) = \int \big(v_d(K)\big)^{n+1} M(\mathrm{d}K)\Big/\overline{V} \qquad \text{for } n \geq 1. \qquad (6.119)$$

6.5.6   Intersection formulae

Many measurements tend to be of the form $h(\Xi \cap W)$ for a given test set or observation window W and suitable functionals $h(\cdot)$. Consequently formulae for the expectations of such quantities are important in statistical applications. Schneider and Weil (2008, Section 9.4) give formulae for such means in the case of stationary standard random sets (see p. 209), which can be decomposed into germ–grain models with convex grains (see p. 238).

There are two important special cases:

$$\mathbf{E}\big(v_d(\Xi \cap W)\big) = p v_d(W) \qquad (6.120)$$

for the volume and

$$\mathbf{E}\big(S(\Xi \cap W)\big) = p S(W) + S_V^{(d)} v_d(W) \qquad (6.121)$$

for the surface area. Formula (6.120) is clearly true for general stationary random closed sets under the sole assumption that $v_d(W)$ is finite. In Formula (6.121) S denotes the boundary length measure (surface area measure in the spatial case), while $S_V^{(d)}$ denotes the intensity of the fibre process (surface process) formed by the boundary (surface) $\partial \Xi$ of Ξ and is equal to the mean boundary length per unit area (mean surface area per unit volume in the spatial case). The first term on the right-hand side of Formula (6.121) results from the intersection of Ξ with the boundary of W.

If the standard random set Ξ is motion-invariant, and if W is convex and compact, then the following general formula holds for all intrinsic volumes of $\Xi \cap W$ (see e.g. Schneider and Weil, 2008, p. 416):

$$\mathbf{E}\big(V_k(\Xi \cap W)\big) = \frac{1}{k!d!b_k b_d} \sum_{i=k}^{d} i!(d-i+k)!b_i b_{d-i+k}\, v_i\, V_{d-i+k}(W) \quad \text{for } k = 0, 1, \ldots, d,$$

$$(6.122)$$

where v_i is the specific i^{th} intrinsic volume (see p. 229). In stereological notation, as in Section 10.2,

$$
\begin{array}{llll}
v_0 = N_A, & v_1 = \dfrac{L_A}{2}, & v_2 = A_A & \text{if } d = 2, \\[2mm]
v_0 = N_V, & v_1 = \dfrac{M_V}{\pi}, & v_2 = \dfrac{S_V}{2}, \quad v_3 = V_V & \text{if } d = 3.
\end{array}
$$

Formula (6.122) follows from the Hadwiger characterisation theorem (1.44) since the functionals $h_k(\cdot) = \mathbf{E}\big(V_k(\Xi \cap \cdot)\big)$ are finite, monotone, C-additive, and motion-invariant.

In the planar case of $d = 2$ a particularly important version of (6.122) is that of $k = 0$. In that case

$$\mathbf{E}\big(\chi(\Xi \cap W)\big) = \mathbf{E}\big(V_0(\Xi \cap W)\big)$$

$$= \frac{2\pi v_0 V_2(W) + 4v_1 V_1(W) + 2\pi v_2 V_0(W)}{2\pi}$$

$$= A_A + \frac{L_A L(W)}{2\pi} + N_A A(W), \qquad (6.123)$$

where $\chi(\cdot)$ denotes the connectivity number or Euler–Poincaré characteristic, $A_A = p$, and L_A, N_A and $A(W)$ are respectively the mean boundary length per unit area, the specific connectivity number and the area of W.

Proof of Formula (6.123). Consider the case of a germ–grain model with convex disjoint grains with rotation-invariant distribution. Since the grains are disjoint, the number of grains to hit W is given by $\chi(\Xi \cap W)$. Then the Campbell theorem (4.29) yields for the corresponding marked point process

$$\mathbf{E}\big(\chi(\Xi \cap W)\big) = \lambda \mathbf{E}\big(v_d(\check{\Xi}_0 \oplus W)\big), \qquad (6.124)$$

where Ξ_0 denotes the typical grain and λ is the intensity of the germ process (so $\lambda = N_A$). Formula (6.124) is true because of

$$\mathbf{E}\big(\chi(\Xi \cap W)\big) = \lambda \int_{C(\mathbb{K})} \int_{\mathbb{R}^d} \mathbf{1}_{H(W)}(x, K)\, dx\, M(dK)$$

$$= \lambda \int_{C(\mathbb{K})} \int_{W \oplus \check{K}} dx\, M(dK)$$

$$= \lambda \int_{C(\mathbb{K})} v_d(\check{K} \oplus W) M(dK)$$

$$= \lambda \mathbf{E}\big(v_d(\check{\Xi}_0 \oplus W)\big),$$

where M is the distribution of Ξ_0, $H(W) = \{(x, K) \in \mathbb{R}^d \times C(\mathbb{K}) : (x + K) \cap W \neq \emptyset\}$ and K is a dummy variable of integration ranging over the system of convex bodies $C(\mathbb{K})$.   □

The mean volume $\mathbf{E}(\nu_d(\check{\Xi}_0 \oplus W))$ can be calculated by methods of integral geometry. If Ξ_0 is rotation-invariant then the generalised Steiner formula (6.29) can be used. In the case $d = 2$ Formula (6.30) yields the result.

6.5.7 Statistics for motion-invariant germ–grain models

Practical measurement procedures typically involve a sample of a germ–grain model Ξ via a convex compact sampling window W.

The fundamental statistical problems are estimation of the intensity λ of the germ process and some description of the grains. These problems are nontrivial because of edge-effects; some grains will only be partially observed as they do not lie completely in the window W. Measurements on these grains will be difficult or even impossible; the possibility of the presence of such grains introduces potential bias in counting procedures.

Determination of λ

(1) Counting rules
Baddeley and Jensen (2005, Section 3.3.5 and Chapter 10) describe various counting rules for the unbiased estimation of the germ density λ. The most natural procedure assigns to each grain a unique 'associated' point, such as the extremal point in a fixed direction or the centre of gravity. An unbiased estimator of λ is given by the number of these points in W divided by $\nu_d(W)$.

Another popular method uses the *Gundersen counting frame* or *Gundersen's tiling rule*, illustrated by Figure 6.10 for the planar case. Here all grains are counted which hit the rectangular window W but not the area on the left-hand side of the 'forbidden line' including the window's left vertical and lower horizontal sides; it must be possible to decide which grain fragments in W belong to the same grain and also to decide for each grain of interest whether or not it hits the 'forbidden line'. The number n_W of grains hitting W but not the 'forbidden lines' leads to the estimator

$$\hat{\lambda} = \frac{n_W}{A(W)}. \tag{6.125}$$

It is unbiased even if the germ–grain model is only stationary.

Proof of the unbiasedness of $\hat{\lambda}$. Consider

$$n_W = \sum_{[x_n; \Xi_n] \in \Psi} \mathbf{1}_{\mathcal{T}(\Xi_n)}(x_n),$$

where $\mathcal{T}(K) = \{x \in \mathbb{R}^2 : x + K \text{ is counted by the frame}\}$. Geometrical arguments show that for every K the set $\mathcal{T}(K)$ is a rectangle, which is congruent to W. Thus the Campbell theorem (4.29) for marked point processes gives

$$\mathbf{E}(n_W) = \lambda \int \int \mathbf{1}_{\mathcal{T}(K)}(x)\mathrm{d}x \, M(\mathrm{d}K) = \lambda A(W);$$

see Schwandtke *et al.* (1988).   □

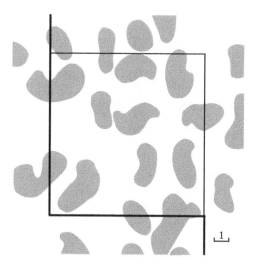

Figure 6.10   A sample of a germ–grain model in a window, to be considered together with Example 6.3 on p. 253, in which the length scale shown is used. The window W is the inner square; the additional space shown is needed for getting full information on the grains hitting the Gundersen counting frame, which is given for the inner square; the three bold lines are its 'forbidden lines'.

Note that the Gundersen frame uses more information than given in W: it is possible that grains intersect both W and the frame outside of W; see Figure 6.10.

In the three-dimensional case the Gundersen frame is constructed analogously; see Baddeley and Jensen (2005, Chapter 11).

(2) Disector

A simple three-dimensional method of measurement for the estimation of λ is the *disector*, which was invented by Sterio (1984); see also Baddeley and Jensen (2005, Section 3.3.5 and Chapter 10). It uses two parallel section planes of known distance t. The geometrical setup is shown in Figure 6.11. The lower plane is called the reference plane and the upper the look-up plane. The window of observation in the reference plane is W.

Let n_W be the number of grains which intersect the reference plane in W *but not* the look-up plane (which is deemed to be infinite in extent). Then λ is estimated by

$$\hat{\lambda} = \frac{n_W}{v_2(W)t}. \tag{6.126}$$

This is clearly an unbiased estimator of λ: assign to each particle (as associated point) its highest tangent point. These points form a stationary point process of intensity λ, and the disector finds the points which are between the two planes.

For purposes of estimation the phrase 'intersect the window' has to be made exact. An unbiased counting rule has to be used, such as described above; it can be implemented, for example, by the Gundersen frame.

Clearly, the distance t should be reasonably small, in order to minimise the possibility of there being grains which lie completely between the two planes without hitting either one. If

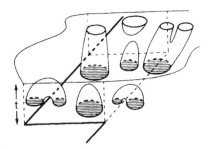

Figure 6.11 A schematic illustration of the disector. The rectangle below is the window in the reference plane, shown together with the Gundersen counting frame (the bold lines). The grains having contact with the window but not hitting the area on the left-hand side of the forbidden lines of the frame or the (upper) look-up plane are counted. In the illustration there are three such grains.

topologically complicated grains are considered, it is sometimes necessary to look below the reference plane in order to check connectivity properties of grain fragments.

Ohser and Mücklich (2000, Section 3.2.5) explained how the specific Euler–Poincaré characteristic N_V can be estimated by serial sections.

Simultaneous estimation of all intrinsic volume densities

In the planar case let $L(W)$ and $A(W)$ be the boundary length and area of W, and assume the grains are convex. The following quantities are measured:

$A(\Xi \cap W)$ the area of Ξ in W,

$L(\Xi \cap W)$ the boundary length of Ξ inside the interior of W plus the length of $\partial W \cap \Xi$,

$\chi(\Xi \cap W)$ the connectivity number of $\Xi \cap W$.

They satisfy the following system of equations:

$$\mathbf{E}\big(A(\Xi \cap W)\big) = A_A A(W), \tag{6.127}$$

$$\mathbf{E}\big(L(\Xi \cap W)\big) = A_A L(W) + L_A A(W), \tag{6.128}$$

$$\mathbf{E}\big(\chi(\Xi \cap W)\big) = A_A + \frac{L_A L(W)}{2\pi} + N_A A(W), \tag{6.129}$$

which is the particular case of (6.122) for $d = 2$ and $k = 0, 1, 2$. If the expectations on the left-hand side are replaced by the corresponding random quantities $A(\Xi \cap W)$, $L(\Xi \cap W)$ and $\chi(\Xi \cap W)$ then solving for A_A, L_A and N_A produces unbiased estimators for these fractions. The same approach is possible in the spatial case.

However, the above presentation is somewhat naïve. The areas, lengths, and volumes do not appear from nowhere but must be somehow determined, a procedure which is not at all trivial if data are given as pixels. This book does not consider these problems; the methods needed in this case are described in Ohser and Schladitz (2009). The spatial case uses ideas of stereology, which are based on the Crofton formula.

A similar approach appears in Schmidt and Spodarev (2005), which is based on counting a topological index function of Ξ within the window W. Its algorithmic realisation is described in Klenk *et al.* (2006) and Guderlei *et al.* (2007). An alternative approach is presented in Mrkvička and Rataj (2008), where a local version of the Steiner formula is applied for various dilation radii r_i and linear regression then leads to estimates of the intrinsic volume densities.

Example 6.3. *Determination of* λ, N_A, A_A *and* L_A *for the pattern of Figure 6.10*
Figure 6.10 shows a sample of a planar motion-invariant germ–grain model Ξ observed through the window W, which is the *inner* square of side-length 10.5. The measurements made only through the window can be compared with others obtained by observing all of the grains that happen to intersect the window.

Image analyser measurements through the window give

$$A(\Xi \cap W) = 41.2,$$

$$L(\Xi \cap W) = 97.3,$$

$$\chi(\Xi \cap W) = 14,$$

using the unit of length of Figure 6.10. Thus the equation system (6.127) to (6.129) yields, assuming convexity (which is not given, but affects only N_A), the estimates

$$\hat{A}_A = 0.37,$$
$$\hat{L}_A = 0.74,$$
$$\hat{N}_A = 0.08.$$

All individual grains have connectivity number one. So the value $\hat{N}_A$ can be interpreted as estimating the mean number λ of grains per unit area. This number can also be estimated by counting associated points, for example the lower left boundary points of grains falling in W, which comes to 10, and so leads to an estimate of N_A as 0.09. The Gundersen frame method applied to the inner square W yields the estimate $\hat{N}_A = \hat{\lambda} = 9/A(W) = 0.08$. Note that the upper left particle must not be counted though it hits the window because it hits also the 'forbidden line'. However, both these measurement procedures utilise information exterior to the measurement window W.

The specific convexity number N_A^+ can be estimated as 0.11, by counting the lower tangent points with respect to the bottom line of Figure 6.10.

Schwandtke *et al.* (1987, 1988) study the estimation variances of these and other estimators of N_A and show that the estimator of N_A which results from the equation system above is more accurate than those arising from methods using associated points.

Tests of goodness-of-fit are carried out following the pattern of such tests in Sections 3.4.2 and 4.7.10, using summary characteristics and deviation tests.

Estimation of mean values for convex grains

Let $M(\cdot)$ be the distribution of the typical grain: so $M(\cdot)$ is a probability measure on the space $C(\mathbb{K})$ of convex bodies. Given a measurable nonnegative functional $f(K)$ the problem is to estimate the quantity

$$M_f = \int_{C(\mathbb{K})} f(K)M(dK), \qquad (6.130)$$

where as before K is a dummy variable of integration running through the space of convex bodies. Particular cases are

$$f(K) = v_d(K)$$

and

$$f(K) = \mathbf{1}(\mathrm{diam}(K) \le r).$$

In the first case M_f is the mean grain volume and in the second for solid spherical grains the value of the distribution function of diameter at r.

There are two possible methods for the estimation of M_f.

(1) *Plus-sampling* (Miles, 1974b; Weil, 1982b). The values of $f(\Xi_i)$ are determined for all grains Ξ_i hitting the sampling window W. This may mean that information from outside of the window W must be gathered: hence the phrase 'plus-sampling'. An unbiased estimator for λM_f is $\hat{M}_f$:

$$\hat{M}_f = \sum_{[x_n;\Xi_n]\in\Psi} f(\Xi_n)\frac{\mathbf{1}_{W\oplus\check{\Xi}_n}(x_n)}{v_d(W\oplus\check{\Xi}_n)}. \qquad (6.131)$$

The sum is taken only over those grains $[x_n;\Xi_n]$ that actually hit W as described above, since otherwise $\mathbf{1}_{W\oplus\check{\Xi}_n}(x_n) = 0$. The unbiased nature of M_f follows by an application of the Campbell theorem (4.29) for marked point processes to the germ–grain process $\Psi = \{[x_n;\Xi_n]\}$.

Formulae in integral geometry help simplify the evaluation of $v_d(W\oplus\check{\Xi}_n)$.

It is clear that in many cases plus-sampling will be unrealistic, since only the interior of the window W will be available for observation and so insufficient information for the determination of all $f(K)$-values is given.

(2) *Minus-sampling or Miles–Lantuéjoul sampling* (Miles, 1974b; Lantuéjoul, 1978a,b). Only the grains lying completely within W are considered. Analogous to $\hat{M}_f$ in (6.131), the quantity

$$\overline{M}_f = \sum_{[x_n;\Xi_n]\in\Psi} f(\Xi_n)\frac{\mathbf{1}_{W\ominus\check{\Xi}_n}(x_n)}{v_d(W\ominus\check{\Xi}_n)} \qquad (6.132)$$

is an unbiased estimator for λM_f. Again the sum is restricted to particular grains, since $\mathbf{1}_{W\ominus\check{\Xi}_n}(x_n) = 0$ unless $x_n + \Xi_n \subset W$: only the grains fully contained in W are counted.

Matheron (1978) and Weil (1982b) discuss means of determination of $v_d(W\ominus\check{\Xi}_n)$.

Baddeley and Jensen (2005, p. 272) consider the case with associated points.

The simulation of germ–grain models follows the pattern of simulation of point processes and Boolean models as presented in Chapters 3, 4 and 5.

6.6 Other random closed set models

6.6.1 Gibbs discrete random sets

Usually in this book random sets are defined in the spirit of Euclidean geometry, as subsets of $\mathbb{R}^d$. However, random set data often appear in the form of discrete sets, as pixel structures. Therefore, it makes sense to build also models in this geometry. The following aims to give the reader some idea of such sets, following the paper Sivakumar and Goutsias (1997b). The exposition is restricted to the planar case, but generalisation to d dimensions is possible.

The space in which the sets are existing is the lattice W,

$$W = \{(m, n) : 1 \leq m \leq M, 1 \leq n \leq N\},$$

where M and N are natural numbers. The points $w = (m, n)$ are called *sites*. Subsets of W are denoted by X, $\mathcal{P}(W)$ is the power set of W, that is, the collection of all X, and $|X|$ is the cardinality of X, that is, number of sites of X. Assume that to each site $w \in W$ a corresponding random variable $x(w)$, called its *state*, is assigned, which takes the value 0 or 1. Then the set

$$\Xi = \{w \in W : x(w) = 1\}$$

is a discrete random set, whose distribution is given by the probabilities

$$\mathbf{P}(\Xi = X) \qquad \text{for } X \subset W,$$

saying that the random set Ξ is equal to the set X.

The particular case of discrete random sets considered in this section are *Gibbs discrete random sets*, the distribution of which is given by the elegant formula (6.133). These sets can be simulated in a well-established way and have found many successful applications. Goutsias and Sivakumar (1998) summarise advances of models based on Gibbs distributions. These are often natural choices, leading to a wide spectrum of distributions and making it possible to incorporate constraints and to model local interactions in a simple way.

A Gibbs distribution is given by

$$\mathbf{P}(\Xi = X) = \frac{1}{Z} \exp\left(-\beta U(X)\right) \qquad \text{for } X \subset W, \tag{6.133}$$

where Z, called *partition function*, is a normalising constant equal to

$$Z = \sum_{X \subset W} \exp\left(-\beta U(X)\right). \tag{6.134}$$

Furthermore, $U(X)$ is the *energy* of X and β a positive model parameter, called *inverse temperature*. Though it is notoriously difficult to calculate Z and numerical characteristics of the distribution such as the mean number $\mathbf{E}(|\Xi|)$ of sites in Ξ or the probability $\mathbf{P}(w \in \Xi)$ for a given site w, this model is very popular.

The inverse temperature β somehow controls the variability of a Gibbs discrete random set Ξ. If $\beta \to 0$ then Ξ becomes uniform, that is,

$$\lim_{\beta \to 0} \mathbf{P}(\Xi = X) = p \qquad \text{for } X \subset W, \tag{6.135}$$

where

$$p = 2^{-MN} \tag{6.136}$$

since 2^{MN} is the total number of all subsets X of W, that is, the number of elements of the power set $\mathcal{P}(W)$. If $\beta \to \infty$ then

$$\lim_{\beta \to \infty} \mathbf{P}(\Xi = X) = \begin{cases} q & \text{for } X \in \mathcal{U}, \\ 0 & \text{otherwise,} \end{cases} \tag{6.137}$$

with

$$q = |\mathcal{U}|^{-1},$$

in which $\mathcal{U}$ is the set of all $X \subset W$ with energy $U(X) = u$, where u is the global minimum of $U(X)$. The members in $\mathcal{U}$ are often called *ground states*.

Gibbs discrete random sets are usually simulated by the Markov chain Monte Carlo technique: A homogeneous ergodic Markov chain $\{X_n\}$ is constructed, the states of which are subsets of W, whose equilibrium distribution is just the Gibbs distribution (6.133). After sufficiently many jumps of the Markov chain any of its states can be approximately considered as a sample drawn from the Gibbs distribution.

Example 6.4. *The Ising model*
The energy is defined as

$$U(X) = \sum_{\substack{(v,w) \subset W \\ |v-w|=1}} \mathbf{1}\big(x(v) \neq x(w)\big), \tag{6.138}$$

where $|v - w| = |k - m| + |l - n|$ for $v = (k, l)$ and $w = (m, n)$. The energy $U(X)$ is high if X consists of many small components.

In the form given here the model is called the 'ferromagnetic' Ising model. Also the 'anti-ferromagnetic' case with $\beta < 0$ is of interest; see for example Georgii *et al.* (2001). Sivakumar and Goutsias (1997a,b, 1999) consider a more general energy function. Georgii (2000) and Georgii *et al.* (2001) show how to define Ising models on infinite lattices.

For the simulation of the Ising model here a quite simple form is described, the *Metropolis algorithm with single-site updating*. It works as follows.

Let X_n be the state of the Markov chain in the nth iteration. The following state X_{n+1} is obtained in two steps:

(i) In W a site w is uniformly chosen. For X_{n+1} a proposal Y is made by changing the state of site w: either w is added to X_n (if not in X_n) or w is deleted from X_n (if in X_n); all other sites remain unchanged.

(ii) The proposal Y is accepted (i.e. $X_{n+1} = Y$) with probability

$$\min\{1, \exp(-\beta\Delta U)\},$$

where

$$\Delta U = U(Y) - U(X_n).$$

If Y is not accepted, then $X_{n+1} = X_n$.

It is clear that ΔU is completely given by the elements in $(B + w) \cap W$, where B is the *rhombus*

$$B = \{(0, -1), (0, 0), (0, 1), (-1, 0), (1, 0)\}.$$

Statistics for the Ising model is considered in Sherman (2011, Section 4.2), where also an example discussing cancer rates in the eastern USA is presented. In statistical applications, the Ising model often arises as a prior in Bayesian statistics.

Example 6.5. *Morphologically constrained discrete random sets*
These Gibbs discrete random sets, which were first used by Chen and Kelly (1992), have energies that lead to samples which tend to have prescribed morphological properties. A simple example is

$$U(X) = |X \setminus X \circ B| \qquad \text{for } X \subset W, \tag{6.139}$$

where $\circ$ denotes the opening operation on the lattice W and B is some structuring element. (Sivakumar and Goutsias, 1997c, present a theory of discretised morphological operators; see also Soille, 2003, Chapter 4.) The underlying idea is to have ground states that are B-open, in order to favour realisations X that are B-open: X and $X \circ B$ will differ only for a small number of sites, that is, X is 'smooth' in a sense determined by B. This class of Gibbs discrete random sets is thoroughly studied in a series of papers including Sivakumar and Goutsias (1999). Instead of defining the energy with opening, it can be defined also with respect to closing, as

$$U(X) = |X \bullet B \setminus X| \qquad \text{for } X \subset W, \tag{6.140}$$

or even

$$U(X) = |X \setminus X \circ B| + |X \bullet B \setminus X| \qquad \text{for } X \subset W. \tag{6.141}$$

The ground states are such sets that fulfill some morphological conditions. In the case of (6.139) this means that there are no components of a 'size' smaller than the 'size' of the structuring element B. Sivakumar and Goutsias (1999) present many interesting examples and extend the theory to the case of grey-tone-value fields. For simulation they use some multi-state updating Metropolis method. Finally, they show how to fit Gibbs discrete random sets to image data and use a maximum likelihood technique where the size density (6.143) plays an important rôle.

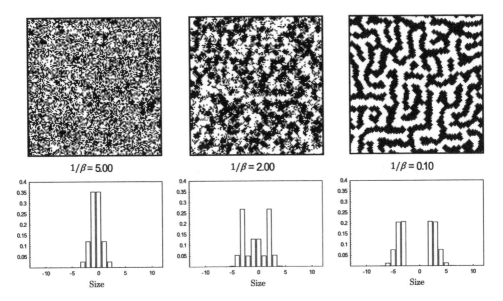

Figure 6.12 Realisations of the morphologically constrained Gibbs discrete random set as in Example 6.5 at three inverse temperatures β. The format of each figure is 128×128 pixels. The histograms show the corresponding size densities (6.143). Courtesy of K. Sivakumar.

Figure 6.12 shows three simulated morphologically constrained Gibbs discrete random sets as in Sivakumar and Goutsias (1999, Example 3.1). The energy function is

$$U(X) = \sum_{i=0}^{1} \left\| \gamma_i(X) \setminus \gamma_{i+1}(X) \right\|_W + \sum_{j=1}^{2} \left\| \phi_j(X) \setminus \phi_{j-1}(X) \right\|_W, \qquad (6.142)$$

where

$$\gamma_k(X) = X \circ kB,$$

$$\phi_k(X) = X \bullet kB,$$

and

$$\|Y\|_W = \sum_{w \in W} Y(w)$$

for

$$Y(w) = \begin{cases} 0 & \text{if white,} \\ 1 & \text{if black,} \end{cases}$$

with B being the rhombus, so that $2B = B \oplus B$. The size density $s(k)$ of X is

$$
s(k) = \begin{cases} \dfrac{1}{NM}\mathbf{E}\left(\left\|\gamma_k(X) - \gamma_{k+1}(X)\right\|_W\right) & \text{for } k = 0, 1, 2, \ldots, \\[3mm] \dfrac{1}{NM}\mathbf{E}\left(\left\|\phi_{|k|}(X) - \phi_{|k+1|}(X)\right\|_W\right) & \text{for } k = -1, -2, \ldots. \end{cases}
\tag{6.143}
$$

6.6.2 Dilated fibre and surface processes

Interesting random sets are obtained when the fibres (surfaces) of a fibre (surface) process are enlarged by set-theoretic dilation. An example is the *Boolean cylinder model*, which is obtained when the lines of a Poisson line process are dilated by the ball $B(o, r)$. Its basic characteristics are given in Ohser and Schladitz (2009, pp. 251–2) and Spiess and Spodarev (2011); see also Heinrich and Spiess (2009). A similar model is obtained by dilating the planes of a Poisson plane process. Redenbach (2011) studies the system of dilated Poisson-Voronoi tessellation faces.

6.6.3 Excursion sets

Random fields and u-level excursion sets

Excursion sets of random fields form another important class of random set models. The most well-studied case is that of excursion sets of stationary Gaussian and Gaussian-related random fields. A thorough mathematical treatment of the theory of random fields and excursion sets can be found in Adler (1981, 2000), Piterbarg (1996), Adler *et al.* (2010), Adler and Taylor (2007, 2011) and Vanmarcke (2010). This section presents a short introduction.

Excursion sets are a 'continuous alternative' to germ–grain models: the reader should compare Figures 3.2 and 6.13 to see the difference. The boundaries of excursion sets, typically, are 'smoother' than those of germ–grain models, but, by suitable choice of the covariance function of the underlying random field, can also be fractals; see the discussion of the smoothness parameter in Formulae (6.150) and (6.151) on p. 261. Random fractals will not be discussed here and the interested reader is referred to Stoyan and Stoyan (1994) and Mörters (2010).

A real-valued spatial process $\{Y(x) : x \in \mathbb{R}^d\}$, or $\{Y(x)\}$ for short, is called a *Gaussian (random) field* if the finite-dimensional distributions

$$
\mathbf{P}(Y(x_1) \in B_1, \ldots, Y(x_n) \in B_n) \qquad \text{for Borel sets } B_1, \ldots, B_n \qquad \text{for } n \geq 1 \tag{6.144}
$$

are multivariate normal. Hence a Gaussian field is completely characterised by its *mean function*

$$
m(x) = \mathbf{E}(Y(x)) \qquad \text{for } x \in \mathbb{R}^d, \tag{6.145}
$$

and *covariance function*

$$
k(x_1, x_2) = \mathbf{E}\big((Y(x_1) - m(x_1))(Y(x_2) - m(x_2))\big) \qquad \text{for } x_1, x_2 \in \mathbb{R}^d. \tag{6.146}
$$

A Gaussian field is stationary and isotropic, which is assumed throughout this section, if its $m(x)$ is a constant μ and $k(x_1, x_2)$ is a function only of the distance $r = \|x_1 - x_2\|$ of the points x_1 and x_2. In this case, with some abuse of notation, write

$$k(x_1, x_2) = k(\|x_1 - x_2\|) = k(r), \qquad r \geq 0.$$

The value $\sigma^2 = k(0)$ is the *variance* of the random field $\{Y(x)\}$. Any Gaussian field can be normalised to have zero mean and unit variance. Hence, without loss of generality, hereinafter only normalised Gaussian fields, denoted by $\{Z(x)\}$, are considered.

If $k(r)$ is continuous, then the Gaussian field is *mean square continuous* in the sense that

$$\mathbf{E}\big(Z(x) - Z(x + \mathbf{r})\big)^2 = 2\big(k(0) - k(r)\big) \to 0 \qquad \text{for } r = \|\mathbf{r}\| \to 0, \qquad (6.147)$$

but the realisations, known as *sample functions* or *sample paths*, are not necessarily continuous. A sufficient condition on $k(r)$ for almost surely continuous sample functions over $I \subset \mathbb{R}^d$ is that for some positive constants c and ε,

$$k(0) - k(\|x\|) \leq \frac{c}{|\log \|x\||^{1+\varepsilon}} \qquad \text{for all } x \in I \qquad (6.148)$$

(Adler, 1981, p. 62).

Not any function can be used as a covariance function, because for any finite set of locations $\{x_1, \ldots, x_n\}$, the covariance matrix of the finite dimensional distribution (6.144) has to be nonnegative definite; functions that fulfil this condition are called *positive definite*; see Adler and Taylor (2007, Theorem 5.7.2) and Gneiting and Guttorp (2010, pp. 20–3) for necessary and sufficient conditions for positive definiteness.

Perhaps the most important class of covariance functions is the *Matérn* class (Matérn, 1986, p. 18), given by

$$k(r) = \frac{2^{1-\nu}}{\Gamma(\nu)} \left(\frac{r}{\theta}\right)^\nu K_\nu \left(\frac{r}{\theta}\right) \qquad \text{for } \nu, \theta > 0, \qquad (6.149)$$

where $K_\nu(\cdot)$ is the modified Bessel function of the second kind, θ a scale parameter and ν a smoothness parameter such that the sample functions are m times differentiable if and only if $m < \nu$. Members of the *Matérn* class include

$$\nu = \tfrac{1}{2}, \quad k(r) = \exp\left(-\frac{r}{\theta}\right),$$

$$\nu = 1, \quad k(r) = \frac{r}{\theta} K_1 \left(\frac{r}{\theta}\right),$$

$$\nu = \tfrac{3}{2}, \quad k(r) = \left(1 + \left(\frac{r}{\theta}\right)\right) \exp\left(-\frac{r}{\theta}\right),$$

$$\nu = \tfrac{5}{2}, \quad k(r) = \left(1 + \frac{r}{\theta} + \frac{1}{3}\left(\frac{r}{\theta}\right)^2\right) \exp\left(-\frac{r}{\theta}\right).$$

Other important classes of covariance functions are the *powered exponential family*

$$k(r) = \exp\left(-\left(\frac{r}{\theta}\right)^\alpha\right) \qquad \text{for } 0 < \alpha \leq 2, \qquad (6.150)$$

the *Cauchy family*

$$k(r) = \left(1 + \left(\frac{r}{\theta}\right)^{-\alpha}\right)^{-\beta/\alpha} \qquad \text{for } 0 < \alpha \le 2 \text{ and } \beta > 0, \qquad (6.151)$$

and the *spherical covariance function*

$$k(r) = \begin{cases} 1 + \dfrac{3}{2}\dfrac{r}{\theta} + \dfrac{1}{2}\left(\dfrac{r}{\theta}\right)^3 & \text{if } r \le \theta, \\ 0 & \text{otherwise,} \end{cases} \qquad (6.152)$$

where $\theta > 0$ is a scale parameter. The spherical covariance function is closely related to the set covariance of a three-dimensional ball with diameter θ. By the way, covariance functions related to balls of other dimensions are also used.

The long-memory parameter β in the Cauchy family (6.151) controls the decay rate of $k(r)$ such that the smaller the value of β, the stronger the long-range dependence.

Fractal structures are possible, and they could result from choosing small values for α in (6.150) and (6.151). The smoothness parameter α governs the behaviour of $k(r)$ at the origin, and the fractal dimension of the sample functions is equal to $d + 1 - \frac{\alpha}{2}$. Therefore, the larger the value of α, the smoother the sample functions. However, note that even the sample functions can be made smoother by increasing α, they are still not differentiable if $\alpha < 2$. Nevertheless, when $\alpha = 2$, the sample functions become infinitely differentiable.

Note that whilst the Matérn, the powered exponential and the Cauchy family are valid covariance functions for any $d \ge 1$, the spherical function is valid only for $d \le 3$. For more possible forms; see Schlather (1999) and Gneiting and Guttorp (2010, pp. 24–6).

Gaussian fields may serve as building blocks for other random fields. An important example is the χ^2 *field with parameter n*:

$$\chi_n^2(x) = \sum_{i=1}^n Z_i(x)^2 \qquad \text{for } x \in \mathbb{R}^d, \qquad (6.153)$$

where the $\{Z_i(x)\}$ are independent normalised Gaussian fields with the common covariance function $k(r)$. The mean and the covariance function of $\{\chi_n^2(x)\}$, denoted by $\mu^{*(n)}$ and $k^{*(n)}(r)$ respectively, are given by

$$\mu^{*(n)} = n, \qquad (6.154)$$

$$k^{*(n)}(r) = 2nk(r)^2 \qquad \text{for } r \ge 0. \qquad (6.155)$$

The *u-level excursion set* of a random field $\{Y(x)\}$, denoted by $\Xi_u(Y)$, results from truncating $\{Y(x)\}$ at level u, that is,

$$\Xi_u(Y) = \{x \in \mathbb{R}^d : Y(x) \ge u\}. \qquad (6.156)$$

The closure of $\Xi_u(Y)$ is a random closed set, and if the sample functions are continuous, then $\Xi_u(Y)$ itself is a random closed set. Figure 6.13 on next page shows realisations of such random sets. These have boundaries 'smoother' than Boolean models, compare with Figure 3.2 on p. 66.

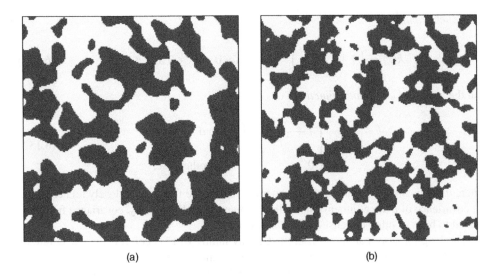

(a)   (b)

Figure 6.13   Simulated 0-level excursion sets in $[0, 20]^2$ of normalised Gaussian fields with covariance functions (a) $k(r) = e^{-r^2}$ and (b) $k(r) = (1 + r^2)^{-2}$, $r \geq 0$. Note that for these models at the 0 level, because of symmetry, volume fraction is 0.5 and the number of holes and the number of connected components are, on average, the same and hence the specific connectivity number is zero. The structural difference results from different covariance functions: for case (a) the covariance function decreases more slowly for small r than for (b).

For a normalised Gaussian field $\{Z(x)\}$ the volume fraction p of $\Xi_u(Z)$ is clearly given by

$$p = \mathbf{P}\big(Z(0) \leq u\big) = 1 - \Phi(u), \qquad (6.157)$$

where $\Phi(u)$ denotes the standard normal distribution function.

Also, for the planar case $d = 2$, the (noncentred) covariance $C_u(r)$ of $\Xi_u(Z)$ can be calculated for a normalised Gaussian field $\{Z(x)\}$ with covariance function $k(r)$:

$$C_u(r) = p^2 + \frac{1}{2\pi} \int_0^{k(r)} \frac{e^{-\frac{u^2}{1+z}}}{\sqrt{1 - z^2}}\, dz \qquad \text{for } r \geq 0. \qquad (6.158)$$

In particular, when $u = 0$,

$$C_0(r) = \frac{1}{4} + \frac{1}{2\pi} \sin^{-1} k(r) \qquad \text{for } r \geq 0. \qquad (6.159)$$

For a χ^2 field, the planar u-level excursion set $\Xi_u(\chi_n^2)$ has the following covariance:

$$C_u^*(r) = p^2 + \frac{1}{\pi} \int_0^{|k(r)|} \frac{e^{-\frac{u^2}{1+z}} - e^{-\frac{u^2}{1-z}}}{\sqrt{1 - z^2}}\, dz \qquad \text{for } r \geq 0, \qquad (6.160)$$

where $k(r)$ is the common covariance function of the underlying normalised Gaussian fields and p the volume fraction of $\Xi_u(\chi_n^2)$.

The linear contact distribution function $H_l(r)$ of $\Xi_u(Y)$ can sometimes be computed explicitly. It is related to the first passage time distribution of a certain associated process. For a normalised Gaussian field $\{Z(x)\}$ with twice differentiable covariance function $k(r)$, Nott and Wilson (1997) show that

$$H_l(r) = \mathbf{P}(T_u < r), \tag{6.161}$$

in which

$$T_u = \inf\{t > 0 : \xi_u(t) = u\} \tag{6.162}$$

is the first passage time of the process $\{\xi_u(t)\}$, defined as

$$\xi_u(t) = uk(t) - \eta\frac{k'(t)}{\lambda_2} + \zeta(t) \qquad \text{for } t > 0,$$

where λ_2 is the so-called *second spectral moment*, that is,

$$\lambda_2 = -k''(0), \tag{6.163}$$

which is assumed to be finite and positive throughout this section, η is a Rayleigh random variable with probability density function

$$f_\eta(t) = \lambda_2^{-1} t e^{-t^2/(2\lambda_2)},$$

and $\{\zeta(t)\}$ is a zero mean nonstationary Gaussian process, independent of η, with covariance function

$$k_\zeta(s, t) = k(t - s) - k(t)k(s) - \frac{k'(t)k'(s)}{\lambda_2}.$$

There is no closed form expression for the distribution of T_u, and Lindgren and Rychlik (1991) suggest a regression approximation for this distribution.

For $S_V^{(d)}$, the specific surface area, Formula (6.48) gives

$$S_V^{(d)} = -\frac{db_d}{b_{d-1}} C_u'(0+). \tag{6.164}$$

Ballani *et al.* (2012) show that whenever $k(r)$ is twice differentiable, then

$$C_u'(0+) - -\frac{\sqrt{\lambda_2}}{2\pi} e^{-u^2/2}, \tag{6.165}$$

where λ_2 is the second spectral moment defined in (6.163) above.

One-dimensional random sets are important when one considers linear sections, and such random sets are always unions of disjoint closed intervals. For a u-level excursion set in $\mathbb{R}$, the number of disjoint closed intervals will be the number of *upcrossings*, where an upcrossing happens whenever the sample function (which has been assumed to be continuous) crosses the level u from below.

Denote by N_u the number of upcrossings of the level u on $[0, 1]$ by a normalised one-dimensional Gaussian process with covariance function $k(r)$. Rice (1944, 1945) shows that

$$E(N_u) = \frac{\sqrt{\lambda_2}}{2\pi} e^{-u^2/2}, \tag{6.166}$$

which agrees with (6.164) and (6.165). Formula (6.166) immediately suggests unbiased estimators for λ_2, which is the only parameter for the limiting distribution of the maximum of $\{Z(x)\}$; see Adler (1981, pp. 68–9).

The number of upcrossings N_u in $[0, 1]$ is in fact the one-dimensional case of the specific connectivity number $v_0(u)$, which is also known as the *mean differential topology characteristic* per unit volume in the context of excursion sets. For a normalised Gaussian field $\{Z(x)\}$ under some smoothness and nondegeneracy regularity conditions,

$$v_0(u) = \frac{e^{-u^2/2}\sqrt{\lambda_2^d}}{(2\pi)^{(d+1)/2}} \sum_{j=0}^{d-1} \frac{(d-1)!(-1)^j u^{d-2j-1}}{j!(d-2j-1)!2^j} \tag{6.167}$$

(Adler, 1981, Theorem 5.3.1, and Adler, 2000, Theorem 3.2.2). For $d = 1$, Formula (6.167) is the same as the Rice formula (6.166), whilst for the planar and spatial case, it leads to

$$N_A = \frac{ue^{-u^2/2}\lambda_2}{(2\pi)^{3/2}}, \tag{6.168}$$

$$N_V = \frac{(u^2-1)e^{-u^2/2}\sqrt{\lambda_2^3}}{(2\pi)^2}, \tag{6.169}$$

the specific connectivity number. Note that it is not the same as the mean Euler–Poincaré characteristic of the intersection of the excursion set and a unit cube, because of the complication at the intersection between the boundary and the excursion set. The interested reader is referred to Adler and Taylor (2007, pp. 140–1 and 289–98).

The chord lengths of sections of both Ξ_u and its complement are studied in Estrade *et al.* (2012). In general subsequent chords have dependent lengths, different to the case of a Boolean model. For the length distribution functions, bounds in terms of the random field characteristics can be given. The paper by Demichel *et al.* (2011) discusses the tails of chord length distributions and shows that the decay is always faster than for any negative power function.

Formula (6.169) shows that N_V as a function of the level u has zeros at $u = -1$ and $u = +1$. The presence of zeros can be explained intuitively by the relationship that the Euler–Poincaré characteristic in three dimensions is

$$\#\text{components} - \#\text{tunnels} + \#\text{holes};$$

see Section 1.8. Tunnels and holes, if any, are very rare for high u (positive N_V), tunnels dominate for u around 0 (negative N_V), and for deeply negative u there is one component and many holes (positive N_V). Thus there are zeros of N_V, and it is a matter of mathematical beauty that these are just at -1 and $+1$.

Worsley (1997) and Adler and Taylor (2011, Chapter 5) describe the applications of the Euler–Poincaré characteristic of the excursion set in astrophysics and medical imaging. Astrophysicists were able to construct a map of (standardised) galaxy density. At each level u, the empirical Euler–Poincaré characteristic of the u-level excursion set was plotted to obtain a function of u. Such a plot on one hand suggested a good agreement between the empirical Euler–Poincaré characteristics and the theoretical ones from the excursion sets of a Gaussian field, and on the other hand revealed the structure of the universe: at high levels of u, the topology of the universe is like a meatball, at medium levels a sponge and at low levels a bubble. Another example came from the cosmic microwave background radiation data across the full sky. These data are directional and hence can be modelled by a random field on the sphere. Again, the empirical Euler–Poincaré characteristics as a function of the level u were compared with the theoretical ones from the fitted Gaussian model. Some evidence of non-Gaussianity was found. This approach is quite common in comparing competing theories in cosmology.

For medical imaging, paired data on brain activities, taken by positron emission tomography or functional magnetic resonance imaging, were collected when the participants were performing a task and were at rest. Under the random noise hypothesis, the sample mean of the pointwise differences at each voxel inside the brain follows a normal distribution. Hence, these sample means, after normalisation, can be modelled by a Gaussian random field on a compact set, representing the brain, in $\mathbb{R}^3$. Consider its u-level excursion set. When u is sufficient high, only the global maximum will survive and the Euler–Poincaré characteristics is 1. Thus, for sufficient high levels, the mean Euler–Poincaré characteristics can approximate the tail distribution of supremum of the Gaussian field (Adler and Taylor, 2007, Formula (14.0.2)). Consequently a critical level u_0 can be obtained, exceeding which will lead to rejection of the random noise hypothesis; if the hypothesis is rejected then it is possible to identify the brain region associated with the task performed.

Arns *et al.* (2002) considered Mecke's morphological functions for excursion sets. Vogel (2002) discussed applications in soil research, in which excursion sets were used to model porous media. For three-dimensional reconstructions of porous networks using correlated Gaussian fields (and other characteristics of porous materials); see Schüth *et al.* (2002).

For a χ_n^2 field in which the component normalised Gaussian fields satisfy some smoothness and nondegeneracy regularity conditions, the specific connectivity number for a general d is also explicitly known (Worsley, 1994). The formulae for the planar and spatial cases are

$$N_A = \frac{u^{(n-2)/2}e^{-u/2}\lambda_2}{2^{n/2}\pi\Gamma(n/2)}\left(u - (n-1)\right), \tag{6.170}$$

$$N_V = \frac{u^{(n-3)/2}e^{-u/2}\sqrt{\lambda_2^3}}{(2\pi)^{3/2}2^{(n-2)/2}\Gamma(n/2)}\left(u^2 - (2n-1)u + (n-1)(n-2)\right). \tag{6.171}$$

Statistics

It is easy to see that in random-set statistics it suffices to use normalised Gaussian fields. If so, only the level parameter u and the covariance function $k(r)$ have to be estimated. The level u can be easily estimated from the relationship given in (6.157), where for p the observed area or volume fraction is plugged in. The covariance function can then be estimated nonparametrically, with the help of Formula (6.158), from an estimate of the random set covariance $C(r)$, obtained from the binary image using the method described in Section 6.4.3. However, this way is difficult since inverting the integral in Formula (6.158) does not necessarily yield the positive definite function property required for a valid covariance function. Thus, again the two-step method is recommended, where in the first step a rough estimate of $k(r)$ is determined and in the second step parametric statistical methods are applied.

For the second step, parametric analysis, various approaches exist.

(1) *Method of moments*

If $k(r)$ contains one or two parameters, then plugging in the empirical specific surface area and specific connectivity number into Formulae (6.164), (6.165), and (6.168) or (6.169), gives estimates for the parameters.

(2) *Minimum contrast*

Another possibility is to minimise the distance between the theoretical form of some characteristic function and its empirical values. For example, the theoretical random set covariance $C_{\hat{u}}(r)$ of the excursion set at the estimated level $\hat{u}$ can be compared with the empirical covariance. However, there is a potential structural defect in this approach. Whilst $C_u(r)$ in Formula (6.158) is at least p^2, the empirical covariance can be smaller, due to for example holes around connected components.

A better candidate may be the linear contact distribution function $H_l(r)$, which can be easily estimated using the methods described in Section 6.4.5. Their theoretical counterparts given in Formulae (6.161) and (6.162) can be approximated efficiently (Rychlik and Lindgren, 1993). Thus, minimum contrast estimators for the parameters can be computed; see Nott and Wilson (1996).

(3) *Pairwise likelihood*

Denote by $\{z(x)\}$ a given binary image. Instead of full likelihood, Nott and Rydén (1999) considered pairwise likelihood

$$\prod_{i<j} p\big(\mathbf{1}(z(x_i) > u), \mathbf{1}(z(x_j) > u); \Theta\big)$$

where $p(\cdot, \cdot; \Theta)$ is the joint probability density function of the two binary random variables $\big(\mathbf{1}(Z(x_i) > u), \mathbf{1}(Z(x_j) > u)\big)$ and Θ the vector of parameters.

In practical implementation, the pairwise likelihood takes the product over distinct locations

$$L\big(\Theta; \Xi_u(z)\big) = \sum_{x\in\mathcal{L}} \prod_{x'\in\mathcal{L}'} p\big(\mathbf{1}(z(x) > u), \mathbf{1}(z(x + x') > u); \Theta\big), \qquad (6.172)$$

where $\mathcal{L}$ and $\mathcal{L}'$ are two finite sets of sites in the sampling window W such that $\mathcal{L} \subset \{x : x + \mathcal{L}' \in W\}$. The maximiser $\hat{\Theta}$ of $L\big(\Theta; \Xi_u(z)\big)$ is the maximum pairwise likelihood estimator for Θ. Nott and Rydén (1999) showed that $\hat{\Theta}$ is consistent and asymptotically follows the normal distribution.

Example 6.6. *Modelling heather pattern by excursion set*

Figure 6.14(a) shows a data set which has been already often analysed by random-set methods. The dark areas in the figure show areas covered by heather over a $10 \times 20\,\mathrm{m}^2$ rectangular

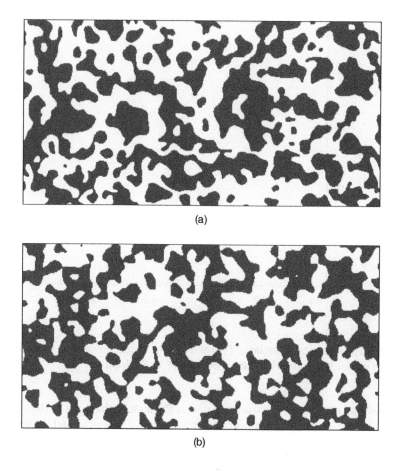

(a)

(b)

Figure 6.14   (a) The heather data in a $10 \times 20 \, \text{m}^2$ rectangle (available from the `spatstat` package in R); (b) a simulated realisation of a 0-level excursion set of a normalised Gaussian field in 10×20 rectangle, with covariance function $k(r) = e^{-(r/0.4)^2}$.

window at Jädraås, Sweden. Since heather appears in bushes, it is natural to try a statistical fit with a germ–grain model, where the 'grains' are perhaps the heather bushes. In this spirit Diggle (1981) and Møller and Helisová (2010) try to fit a Boolean model with random discs or more general germ–grain models to the data. However, their results do not confirm the germ–grain model assumption. In contrast, visual inspection already raises doubt whether a model of this type is appropriate. Comparison between Figures 6.13 and 6.14(a) suggests that an excursion set, perhaps of a Gaussian field, may be a good model. The questions why the heather pattern sample looks like an excursion set and how the excursion set hypothesis can be tested against the germ–grain model hypothesis will not be discussed.

The estimates of the fundamental random-set characteristics are

$$\hat{A}_A = 0.50, \quad \hat{L}_A = 1.77 \, \text{m}^{-1} \quad \text{and} \quad \hat{N}_A = 0.185 \, \text{m}^{-2}.$$

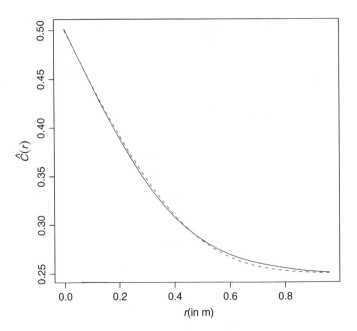

Figure 6.15   The empirical covariance (—) and the theoretical covariance (- - -) of the fitted excursion set model for the heather data. Courtesy of K. S. Helisová.

Figure 6.15 shows the empirical covariance $\hat{C}(r)$. (The authors thank K. S. Helisová for the numerical values of $\hat{A}_A$ and $\hat{L}_A$, estimated by the method of Mrkvička and Rataj, 2008, and for the values of $\hat{C}(r)$.)

These values can be used to estimate Boolean model parameters, by means of Formulae (3.50) to (3.52). Under the assumption that the grains are discs, for the mean radius the value 0.41 m is obtained and for the second moment of radius 0.162 m². This yields for the radius a negative variance, which seems to show that a Boolean model with discoidal grains is not suitable.

As an alternative, the u-level excursion set of a normalised Gaussian field is fitted to the heather data. For the covariance function $k(r)$ of the random field the simplest *ad hoc* choice,

$$k(r) = \exp\left(-\left(\frac{r}{\theta}\right)^2\right) \qquad \text{for } r \geq 0, \tag{6.173}$$

is made, in order to get smooth boundaries as observed.

Since (by good luck) $\hat{A}_A = 0.5$, u is estimated as $\hat{u} = 0$. Formulae (6.164) and (6.165) yield

$$L_A = -\pi C'(0+) = \frac{1}{\sqrt{2}\theta}.$$

This leads to the estimate

$$\hat{\theta} = 0.40\,\text{m},$$

which is very close to the estimate 0.38 obtained by the pairwise likelihood approach (Nott and Rydén, 1999) and the estimate 0.39 obtained by minimising the L^2-norm of theoretical and empirical linear contact distribution of the right half of the data (Nott and Wilson, 1997).

Figure 6.15 shows, in addition to the empirical $\hat{C}(r)$, also the theoretical covariance of the excursion set model with covariance function given in (6.173) and $\theta = 0.4$, and a simulated realisation of the fitted model is given in Figure 6.14(b).

To check whether the model offers a good fit of the data, a parametric bootstrap test (see Sections 3.4.2 and 4.7.10) can be used. A summary characteristic such as the linear contact distribution function $H_l(r)$ or the covariance is chosen. Measured in, for example the L^2- or L^∞-norm, the deviations from the theoretical values of the empirical values in the given data and in 999 independent simulated realisations of the fitted models are then calculated. If the deviation calculated from the given data is not among the top 5% of these 1000 deviations, then the model is not rejected. Alternatively, simulated confidence envelopes can also be employed for visual inspection of the fit. Nott and Rydén (1999) has, using the covariance, adopted the latter approach, and found that such an excursion set model for the heather data was not rejected.

6.6.4 Birth-and-growth processes

Fundamentals

A birth-and-growth process $\{\Xi_t\}$ is a dynamic germ–grain model used for modelling situations in which germs (nuclei) are born at random instants at random spatial locations. Each germ x_i is the origin of a grain $\Xi_{t,i}$, which evolves in time according to a given growth law (Villa and Rios, 2010, and Aletti *et al.*, 2011). At time t the random closed set Ξ_t can be written as

$$\Xi_t = \bigcup_i (\Xi_{t,i} + x_i),$$

where the union is taken over all germs existing at time t. Symbolically, the evolution is written as

$$\Xi_t = \left(\Xi_0 \oplus \int_0^t G_\tau d\tau\right) \cup \bigcup_{s \in [0,t]} \left(dB_s \oplus \int_s^t G_\tau d\tau\right), \tag{6.174}$$

where $\{G_\tau\}$ is an increasing compact convex set-valued process representing growth, and $\{B_s\}$ is an increasing closed set-valued process representing nucleation. The infinitesimal dB_s can be understood as $\lim_{\delta s \to 0} (B_{s+\delta s} - B_s^{int})$. Here growth is nonlocal since the same Minkowski-addend $G_t dt$ is added to every $x \in \Xi_t$ to represent the growth between $(t, t + dt]$.

In some applications the growth depends on the 'ages' of the grains, so that at time t the model is given by

$$\Xi_t = \left(\Xi_0 \oplus \int_0^t G_\tau d\tau\right) \cup \bigcup_{s \in [0,t]} \left(dB_s \oplus \int_0^{t-s} G_\tau d\tau\right). \tag{6.175}$$

An analogous model with discrete time is considered in Aletti *et al.* (2009), where also statistical problems are discussed.

This book considers only spatially stationary (or spatially statistically homogeneous) structures, where Ξ_t forms a stationary random closed set at each time t. Particular cases where the germs are located at single planes or lines (in $\mathbb{R}^3$) are considered in Cahn (1956) and Villa and Rios (2010). Bounded growing sets are considered in Cressie and Hulting (1992), Deijfen (2003) and Jónsdóttir et al. (2008); see also the references to random compact set models on p. 215.

The following construction is closely related to the Boolean model. The germ-birth rate is denoted by $\lambda(t)$. Births can happen anywhere in $\mathbb{R}^d$ and all germs born in the time interval $[0, t]$ form a homogeneous Poisson process in $\mathbb{R}^d$ of intensity λ_t,

$$\lambda_t = \int_0^t \lambda(u)\mathrm{d}u. \tag{6.176}$$

An important special case is

$$\lambda(t) = ct^{m-1} \qquad \text{for } t \geq 0, \tag{6.177}$$

where c and m are positive constants. (This intensity leads to the so-called Weibull theory; see below.) Important is also the so-called site-saturated case:

$$\lambda(t) = \lambda\delta(t) \qquad \text{for } t \geq 0, \tag{6.178}$$

in which $\delta(t)$ is the Dirac delta function and λ a positive constant, meaning that all germs, forming a homogeneous Poisson process of intensity λ, start to grow at time $t = 0$.

The grains grow according to some growth rate or velocity $v(t)$. If a grain with germ x grows spherically and born at time s, then the grain is at time t the ball $B(x, r_t)$ with radius

$$r_t = \int_s^t v(u)\mathrm{d}u. \tag{6.179}$$

In this case, if in addition $v(t)$ decays exponentially and $m = -d$ in (6.177), the resultant structure of such a birth-and-growth process is a fractal (Chiu, 1995c).

The definitions here have been given in terms of 'time t'. However, in some (mechanical) applications the rôle of time is played by some load σ. The germs then appear with increasing load and may lead to fractures.

In this book interactions of growing grains are excluded. In reality, for example, it may happen that a growing segment grain stops growing once it contacts another grain.

The main aim of this section is to calculate the volume fraction $V_V(t)$ of Ξ_t at time t. It is given by

$$V_V(t) = \mathbf{P}(o \in \Xi_t), \tag{6.180}$$

for which formulae will be given.

Birth-and-growth processes have found applications in the contexts of polymerisation (Capasso, 2003) and of nucleation of materials (Cahn, 1956).

Evolution processes for random closed sets are also modelled by stochastic differential equations; see Lorenz (2010, Section 3.7), Kloeden and Lorenz (2011) and Aletti et al. (2011). Lorenz and Kloeden speak about 'stochastic morphological evolution equations' and use ideas from set-valued analysis as in Aubin (1999). By means of an extended form of differential equations beyond the body of vector spaces they could develop a general theory without the convexity assumption as, for example, in Malinowski and Michta (2010). It is to be expected that this approach will permit the pure geometrical modelling to be enriched with physical ideas.

The Weibull model

In this example, which follows Jeulin (1994), instead of time t some mechanical load σ is considered, which is increasing like time. According to a birth rate $\lambda(\sigma)$ germs are activated in the whole $\mathbb{R}^d$. The birth rate has the form

$$\lambda(\sigma) = c\sigma^{m-1} \qquad \text{for } \sigma \geq 0, \tag{6.181}$$

with $m \geq 1$; c is some positive constant. (The corresponding inhomogeneous Poisson process is sometimes called Weibull-Poisson process.)

A Borel set B of volume $v_d(B)$ is considered. It is said that B fails under load when the first germ is born in B. Let S be the random load under which B fails. Its distribution function $F(\sigma)$ is given by

$$F(\sigma) = \mathbf{P}\big(\Phi_B([0, t]) > 0\big),$$

where Φ_B is the one-dimensional point process of germ birth instants for B. This is a Poisson process of intensity $\lambda(\sigma)v_d(B)$. Consequently

$$\mathbf{P}\big(\Phi_B([0, t]) > 0\big) = 1 - \mathbf{P}\big(\Phi_B([0, t]) = 0\big) = 1 - \exp\left(-v_d(B) \int_0^t \lambda(u)du\right),$$

and for the birth rate given in (6.181),

$$F(\sigma) = 1 - \exp\left(-v_d(B) \frac{c\sigma^m}{m}\right),$$

or

$$F(\sigma) = 1 - \exp\left(-\left(\frac{\sigma}{\sigma_0}\right)^m\right) \qquad \text{for } \sigma \geq 0. \tag{6.182}$$

This is the well-known Weibull distribution. The parameter m is called *Weibull modulus*.

Length distribution of growing segments

In this example, t and 'time', instead of σ and 'load', are used, though the latter could also be applied. The planar case is considered. Each grain is a segment with one endpoint at its germ and the other endpoint moves with constant velocity v and without any interaction with other grains.

The birth rate is

$$\lambda(t) = ct^{m-1} \qquad \text{for } t \geq 0, \tag{6.183}$$

which is the same as (6.181) with $\sigma = t$.

The problem is the determination of the length probability density function $f(l)$ of the segments existing at time t_0. Only segments initiated before t_0 count and clearly older segments are longer than younger ones.

The probability density function $g(t)$ of the instants of initiation is

$$g(t) = \frac{\lambda(t)}{\displaystyle\int_0^{t_0} \lambda(u)du} \qquad \text{for } 0 \leq t \leq t_0.$$

The length of a segment initiated at time t is $l = v(t_0 - t)$ and thus the probability density function of segment length is

$$f(l) = \frac{g\left(t_0 - \dfrac{l}{v}\right)}{v}.$$

With (6.183) this yields

$$f(l) = \frac{m\left(t_0 - \dfrac{l}{v}\right)^{m-1}}{vt_0^m} \qquad \text{for } 0 \leq l \leq vt_0. \tag{6.184}$$

This gives $f(0) = m/(vt_0)$ and $f(vt_0) = 0$ and between 0 and vt_0 the length density $f(l)$ is decreasing.

Such a distribution is often observed in statistical analyses of the lengths of geological faults; see Stoyan and Gloaguen (2011). Figure 6.16 shows the empirical and the fitted density $f(l)$ of the fault lengths in Figure 8.1 on p. 298.

The causal cone

The causal cone $\mathcal{C}(t, x)$ helps calculate the volume fractions of birth-and-growth processes. It is the space–time region in which at least one birth event has to take place in order to cover the point x by the time t; see Villa and Rios (2010). In the case of spherical growth the causal cone is

$$\mathcal{C}(t, x) = \left\{ (u, y) : 0 \leq u \leq t, y \in B\left(x, \int_u^t v(z)dz\right) \right\} \qquad \text{for } t \geq 0 \text{ and } x \in \mathbb{R}^d. \tag{6.185}$$

If (spatially) stationary processes are studied (as here) it suffices to consider $x = o$, because of (6.180).

Let furthermore Φ_{st} (st = spatial–temporal) be the point process of germs in $\mathbb{R}^d \times \mathbb{R}_+$, where each germ is characterised by both location x and instant of initiation t. Under the

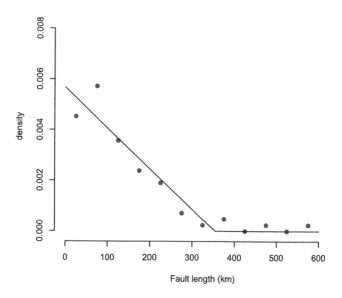

Figure 6.16 Empirical and fitted probability density function of the lengths of the geological faults of the NNE–SSW system in Figure 8.1. See Stoyan and Gloaguen (2011) for details and the estimation method. It is assumed that for very short faults there is some censoring.

assumptions above Φ_{st} is a Poisson process with intensity function $\lambda(t, x) = \lambda(t)$. The probability $\mathbf{P}(x \in \Xi_t)$ satisfies

$$\mathbf{P}(x \in \Xi_t) = \mathbf{P}\big(\Phi_{st}(\mathcal{C}(t, x)) > 0\big),$$

and so putting $x = o$, one obtains

$$V_V(t) = 1 - \exp\left(-\int_0^t v_d\big(\mathcal{C}(t, o)\big)\lambda(u)\mathrm{d}u\right). \tag{6.186}$$

The KJMA theory

Kolmogorov (1937), Johnson and Mehl (1939) and Avrami (1939) developed a theory for the volume fractions $V_V(t)$ of birth-and growth processes as above, where the germs form a Poisson process in the whole $\mathbb{R}^d$. In the site-saturated case (6.178), Formula (3.45) for the volume fraction of a Boolean model can be used to obtain

$$V_V(t) = 1 - \exp\left(-b_d\lambda\left(\int_0^t v(u)\mathrm{d}u\right)^d\right) \qquad \text{for } t > 0, \tag{6.187}$$

since Ξ_t is a Boolean model with intensity λ and solid spherical grains of the same radius

$$r_t = \int_0^t v(u)\mathrm{d}u.$$

In the case of time-dependent $\lambda(t)$ again Ξ_t is a Boolean model, now with intensity $\lambda_t = \int_0^t \lambda(u)\mathrm{d}u$ and solid spherical grains with variable radii. Using the causal cone and (6.186) yields

$$V_V(t) = 1 - \exp\left(-b_d \int_0^t \lambda(s) \left(\int_s^t v(u)du\right)^d ds\right) \qquad \text{for } t \geq 0. \qquad (6.188)$$

Balls on lines and planes

Cahn (1956) and Villa and Rios (2010) studied birth-and-growth processes where the germs are not scattered in the whole space but only on some planes or lines. In the spirit of these papers here two cases are discussed where the germs lie on the lines of a Poisson line process and on the planes of a Poisson plane process, both in $\mathbb{R}^3$. (These processes are explained in Section 8.2.2 and 8.2.3. Stationarity is always assumed.) For simplicity the case of constant growth velocity v and site-saturation (6.178) is considered.

The point process of all germs is the union of independent homogeneous one-dimensional (two-dimensional, respectively) Poisson processes of intensity ϱ on the lines (planes) of the line (plane) process. The result is a stationary Cox process in $\mathbb{R}^3$ of intensity ϱL_V (ϱS_V), where L_V (S_V) is the length (area) density of the line (plane) process.

According to the spirit of the causal cone, $V_V(t)$ is nothing else than the complement of the void-probability of the Cox process of germs:

$$V_V(t) = 1 - v_{B(o,vt)}, \qquad (6.189)$$

as discussed in Section 5.2. Of course, the origin o is not covered by one of the balls $B(x, vt)$ centred at the germs if and only if the Cox process has no point within $B(o, vt)$.

The following gives the result for the line process, for the case known in the materials scientific context as 'grain edge nucleation', see Cahn (1956). The case of a plane process goes similarly.

The random number of lines of the line process hitting $B(o, vt)$ has a Poisson distribution of mean $\mu = \pi v^2 t^2 L_V$, which follows from (8.14). The lengths of the chords generated by intersection of lines with the ball $B(o, vt)$ are independent and have the probability density function

$$f(l) = \frac{l}{2(vt)^2} \qquad \text{for } 0 \leq l \leq 2vt;$$

see Formula (10.61). Since the germs on each line form a homogeneous Poisson process of intensity ϱ, the probability that on an intersecting line there is no point within $B(o, vt)$ is

$$p_0 = \int_0^{2vt} \exp(-\varrho\lambda) f(l)dl = \frac{1 - (2vt\varrho + 1)\exp(-2vt\varrho)}{2(vt\varrho)^2}.$$

Then the void-probability is

$$v_{B(o,vt)} = \sum_{i=0}^{\infty} \frac{\mu^i}{i!} e^{-\mu} p_0^i,$$

and

$$V_V(t) = 1 - \exp(-\mu(1 - p_0)). \qquad (6.190)$$

For the plane process μ is $2rS_V$ and the radius distribution of the section discs of planes with the ball $B(o, vt)$ is the distribution given by Formula (10.42).

The case where the germs are scattered on the faces of tessellations is considered in Saxl et al. (2003).

Process duration

Suppose the birth-and-growth process takes place in a finite region, then the time until, if possible, the volume fraction becomes 1 (and hence the process stops) is of practical interest. Consider the model with birth rate given by (6.177) and spherical growth at a positive constant rate v, operating in a cube of side length L. The duration T_L of the process until the volume fraction in the cube reaches 1 can be approximated, when L is sufficiently large, by

$$T_L \simeq A_d^{-1} k c^{-1} b_d^{-1} v^{-d} \left(\frac{d+m}{m} \right)^{-1/(d+m)}, \qquad (6.191)$$

where

$$A_d = \sum_{i=0}^{d} \binom{d}{i} \frac{(-1)^i}{i+m},$$

$$k = \log \left(\frac{1}{cb_d v^d} \left(\frac{cL^d}{m} \right)^{(d+m)/m} \right);$$

see Chiu (1995c), in which the limiting distributions of T_L for various forms of the birth rate are given.

Estimation of the birth rate and the growth speed

If the births of germs are observable only when they are born in places not occupied by any grains, then the actual birth rate $\lambda(t)$ will be different from the rate $\eta(t)$ of observable births.

At each time s, where $s \leq t$, the process gives a (spatially) stationary Boolean model. Hence, for a model with spherical growth at a constant speed v, the two rates are related by

$$\eta(t) = \lambda(t) \exp \left(-b_d v^d \int_0^t (t-s)^d \lambda(s) ds \right), \qquad (6.192)$$

in which the term after $\lambda(t)$ on the right-hand side is the probability that a germ born is observable. Molchanov and Chiu (2000) suggest that by considering $\lambda(t)$ in (6.192) to be a constant λ_i in $[t_i, t_{i+1})$, where $t_i = i\delta$ for small $\delta > 0$, $\lambda(t)$ can be approximated by the solutions of

$$\lambda_i = \eta(t) \exp \left(\frac{b_d v^d \delta^{d+1}}{d+1} \sum_{k=0}^{i-1} \lambda_k \left((i-k)^{d+1} - (i-k-1)^{d+1} \right) \right) \qquad \text{for } i \geq 1, \quad (6.193)$$

with the initial value $\lambda_0 = \eta(\delta)$. A stepwise constant estimate of $\lambda(t)$ can then be obtained by replacing $\eta(t)$ in (6.193) by a kernel smoothed estimate of $\eta(t)$, based on the observed birth instants. If $\lambda(t)$ comes from some parametric family, minimising the distance between the corresponding parametric form of $\eta(t)$ and a smoothed estimate of it yields the minimum contrast estimates of the parameters.

If v is unknown, the maximum likelihood approach suggested by Chiu *et al.* (2003) can be used. The likelihood $L(\lambda, v)$ of observing births at $\{(t_1, x_1), \ldots, (t_n, x_n) \in [0, \infty) \times \mathbb{R}^d\}$ is proportional to the product of the likelihood that there are births at t_i and the probability that, apart from those observed, there is no other birth in the causal cones $C(t_i, x_i)$, that is,

$$L(\lambda, v) \propto \left(\prod_{i=1}^{n} \lambda(t_i) \right) \exp \left(- \int_{\bigcup_{i=1}^{n} C(t_i, x_i)} \lambda(t) \mathrm{d}(t, x) \right). \tag{6.194}$$

Note that the larger the speed v, the bigger the causal cones. Thus, no matter what $\lambda(t)$ is, the likelihood in (6.194) increases with v and so the maximum likelihood estimator $\hat{v}$ of the speed v is the largest possible v such that none of the points (t_i, x_i) is contained in the interior of a causal cone.

When a parametric form of $\lambda(t)$ is given, maximising $L(\lambda, \hat{v})$ leads to the maximum likelihood estimators of the parameters of $\lambda(t)$.

Since the birth locations are stationary, even if only birth times but not birth locations can be observed, one still can express the likelihood in an integral form and consequently maximising it, often by numerical methods, to estimate the growth speed and the parameters of the birth rate; see Chiu *et al.* (2003).

6.7   Stochastic reconstruction of random sets

Often one is confronted with the problem of providing a quantitative description of the pore-space geometry and topology of a random-set-type structure. The starting point is data from a (small) sample, which may be only lower-dimensional, resulting from a planar (or even linear) section. The aim is to find a way to construct many samples of arbitrary size of the corresponding random set (typically one constituent of a random structure, e.g. the pore space of a porous medium) in full dimensionality. If a fully specified mathematical model is explicitly given, one can generate by simulation as many samples as one likes. Even if no model is specified, given such samples in a computer, as a pixel structure, one still can carry out calculations for the determination of characteristics of connectivity, flow and transport properties, etc. Of course, in the construction information from the original data is used and therefore one speaks about **re**construction.

The following sketches an approach for the case of a two- or three-dimensional stationary and isotropic random set Ξ, which is represented as a pixel structure, such that 0-pixels and 1-pixels, respectively, stand for pore and solid. The structure of Ξ is characterised by its volume fraction p and some summary characteristics $F_1(r), \ldots, F_m(r)$, for example $F_1(r) = C(r)$ (co-variance) and $F_2(r) = L(r)$ (chord length distribution function). The aim is to obtain a pixel structure with periodic boundary conditions (to approximate stationarity) with summary characteristics $\hat{F}_1(r), \ldots, \hat{F}_m(r)$ which are as close as possible to the prescribed $F_1(r), \ldots, F_m(r)$. Thus in least-squares thinking the following minimisation problem has to be solved:

$$E = \alpha_1 \int \left(F_1(r) - \hat{F}_1(r) \right)^2 \mathrm{d}r + \ldots + \alpha_m \int \left(F_m(r) - \hat{F}_m(r) \right)^2 \mathrm{d}r \to \min! \tag{6.195}$$

The minimisation is by choosing an optimal pixel structure, where only such structures having a fraction p of 1-pixels are permitted. The coefficients α_i are weights, which should be chosen in such a way that the summands in (6.195) are of similar size order. Each integral is in practice often replaced by a summation over r going over suitable inter-pixel distances, which may by different for the different $F_i(r)$.

The term 'reconstruction' is appropriate, since the starting point is fragmentary information on Ξ and the above procedure is applied to build samples of Ξ that are in good agreement with these fragments. In the case of simulations from a parametric stochastic model one would not speak about reconstruction, even when the parameters come from a sample of Ξ.

The summary characteristics should be chosen in such a way that they describe possibly different aspects of the pixel distribution. If, for example, $F_1(r) = C(r)$ for the solid phase, then $F_2(r) = L(r) =$ chord length distribution function for the pore phase is a good choice, if $m = 2$ is wanted. In addition to the characteristics of 'one-dimensional nature' such as $C(r)$, $L(r)$ and $H_l(r)$, it may be valuable to use also 'higher-dimensional' ones, such as $H_d(r)$ and $H_s(r)$ or n-point probability functions, if available. However, it may happen that some of the summary characteristics prescribed are contradicting — there is perhaps no random set that possesses all of them as its summary characteristics.

The usual minimisation method is *simulated annealing* (Hazlett, 1997; Yeong and Torquato, 1998a,b; Hilfer, 2000; Torquato, 2002, 2010). This works as follows (Talukdar et al., 2002,a,b): In the beginning the pixels are randomly marked as 0 (void) and 1 (solid), such that volume fraction is p. The start configuration then changes stepwise, at the kth iteration step a void and a solid pixel are randomly chosen and their marks are interchanged. The change slightly modifies the functions $\hat{F}_i(r)$, which describe the actual pixel structure. Therefore, also the 'energy' E defined by (6.195) changes from E_k to E_{k+1}, while volume fraction p is preserved. A pixel change is accepted with probability p_k given by the Metropolis rule

$$p_k = \min\left\{1, \exp\left(-\frac{E_{k+1} - E_k}{T_k}\right)\right\}, \tag{6.196}$$

where T_k is a control parameter, called 'temperature'. It decreases as k increases with the goal that a global minimum is achieved as quickly as possible. Talukdar et al. (2002) sketches the choice of the temperature schedule. Experience shows that the simpler form

$$p_k = \begin{cases} 1 & \text{if } E_k > E_{k+1}, \\ 0 & \text{otherwise,} \end{cases} \tag{6.197}$$

meaning that one accepts 'improvements only', is also sufficient; see Tscheschel and Stoyan (2006).

For an efficient use of the reconstruction method it is important to calculate the functions $\hat{F}_i(r)$ in a way that uses the fact that the pixel interchanges are only local and thus one can reuse the functions $\hat{F}_i(r)$ measured in the kth step, instead of recomputing them fully, at the $(k+1)^{st}$ step.

Examples for the application of the reconstruction method can be found in Manwart et al. (2000), Torquato (2002) and Talukdar et al. (2002,a,b), and in p. 418 in the context of stereology, when for a three-dimensional set only one- or two-dimensional information is given. For a realisation of Ξ observed in W, reconstruction of samples in $W \oplus B(o, r) \setminus W$ for

$r \geq 0$ enables one to apply the plus-sampling (Section 6.5.7) for the estimation of summary characteristics of Ξ. Tscheschel and Chiu (2008) apply such an approach, called *quasi-plus-sampling*, to point patterns and show empirically that it can yield more accurate estimates than many other edge-correction methods (see Section 4.7.2).

A reconstruction method in the spirit of germ–grain models is described in Singh *et al.* (2006). There the grains are empirically given, and the approach is to model the germs first and then place the grains.

As a by-product, the reconstruction method may also help characterise the information content of summary characteristics: those which enable a precise reconstruction can be considered as highly informative; see the discussion in Jiao *et al.* (2009).

7

Random measures

7.1 Fundamentals

7.1.1 Introduction

This chapter presents a brief discussion of random measures on $\mathbb{R}^d$. Random measures are an elegant tool of stochastic geometry often leading to a deeper understanding. In particular curvature measures and marked measures are very useful.

Random measures are here considered as special random set functions taking finite values on bounded sets. Attention is confined mainly to the *nonnegative* case; however, there is a general theory for signed random measures, which closely follows the theory for the non-negative case. Signed random measures are of importance for the study of curvatures (see Section 7.3) as well as for providing a unifying theoretical framework.

This chapter only provides an introduction to the theory of random measures, being a short sketch of the theory in so far as it is relevant to stochastic geometry. A more detailed account can be found in Mecke (1972), Kallenberg (1976b, 1986, 2002) and Schneider and Weil (2008).

Random measures are mostly auxiliary entities in stochastic geometry, giving alternative, often elegant descriptions of random structures. Random closed sets, fibre processes, and surface processes all have various random measures associated with them. This allows the use of notation and methods from measure and integration theory, for example Palm distributions or moment measures.

The following example illustrates the concept of a random measure. Consider a random collection Ξ of fibres in space. For a Borel set B let $\Phi(B)$ be the total length of the fibre pieces within B. For each realisation of Ξ the function Φ has the properties of a measure: value zero for $B = \emptyset$ and σ-additivity. So Φ is genuinely a *random* measure.

Further (partly more abstract, but important) examples of random measures are given in Section 7.3.

Stochastic Geometry and its Applications, Third Edition.
Sung Nok Chiu, Dietrich Stoyan, Wilfrid S. Kendall and Joseph Mecke.
© 2013 John Wiley & Sons, Ltd. Published 2013 by John Wiley & Sons, Ltd.

7.1.2 Definitions and facts

The random measures considered here are all measures on Euclidean spaces. It is straightforward to extend the discussion to more general spaces.

Let $\mathbb{M}$ be the set of all locally finite measures φ on $[\mathbb{R}^d, \mathcal{B}^d]$. Let $\mathcal{M}$ denote the smallest σ-algebra of subsets of $\mathbb{M}$ making all mappings $\varphi \mapsto \varphi(B)$ measurable, where B runs through the bounded Borel sets.

A *random measure* Φ on $\mathbb{R}^d$ is a random element of $[\mathbb{M}, \mathcal{M}]$. So Φ is a measurable mapping of a probability space $[\Omega, \mathcal{A}, \mathbf{P}]$ into the measurable space $[\mathbb{M}, \mathcal{M}]$. For each given fixed B, $\Phi(B)$ is a real-valued random variable. The *distribution* of Φ is the probability measure P given by

$$P(Y) = \mathbf{P}(\Phi \in Y) \qquad \text{for } Y \in \mathcal{M}. \tag{7.1}$$

Here Y is a configuration set, for example $Y = \{\varphi : \varphi\big(B(o, r)\big) = 0\}$, the set of all measures giving the ball $B(o, r)$ the measure zero. For a measure $\varphi \in \mathbb{M}$, let $\varphi_x = T_{-x}\varphi$ be the *translated measure* defined by

$$\varphi_x(B) = (T_{-x}\varphi)(B) = \varphi(B - x) \qquad \text{for all Borel sets } B.$$

This notion makes it possible to define stationarity: a random measure Φ and its distribution P are said to be *stationary* if Φ_x has the same distribution as Φ for all $x \in \mathbb{R}^d$. This can be rewritten as

$$P(Y_x) = P(Y) \qquad \text{for all } x \in \mathbb{R}^d \text{ and all } Y \in \mathcal{M}, \tag{7.2}$$

with the notation

$$Y_x = \{\varphi \in \mathbb{M} : \varphi_{-x} \in Y\}. \tag{7.3}$$

The definition of *isotropy* is analogous to that of stationarity, instead of translation- now rotation-invariance is considered.

The standard first-order characteristic of a random measure is the (deterministic) measure Λ on $[\mathbb{R}^d, \mathcal{B}^d]$, which is defined by

$$\Lambda(B) = \mathbf{E}\big(\Phi(B)\big) \qquad \text{for all Borel sets } B. \tag{7.4}$$

It is called the *intensity measure* of Φ. This measure is not necessarily locally finite, but in the following this will be assumed. In addition, the null measure will be excluded in the discussion so that $\Lambda(B) \neq 0$ for some B. It can be shown that in the stationary case Λ is a multiple of Lebesgue measure ν_d and there is some λ in $(0, \infty)$ such that

$$\Lambda(B) = \lambda\nu_d(B) \qquad \text{for all Borel sets } B. \tag{7.5}$$

The constant λ is called the *intensity* of Φ. Its interpretation is similar to that in the case of point processes: λ is the mean measure per volume unit or the mean measure of a set of volume one.

The notion of ergodicity is analogous to that in the case of point processes in Section 4.1.6, and there is an ergodic theorem analogous to (4.14); see Kallenberg (2002, Corollary 10.19).

There are important *operations* that transform given random measures into others, similar to the operations on point processes given in Section 5.1. By way of example, the operations of superposition and compounding are described here.

Suppose that Φ_1 and Φ_2 are two random measures defined on the same probability space. Then Φ, given by

$$\Phi(B) = \Phi_1(B) + \Phi_2(B) \qquad \text{for all Borel sets } B, \tag{7.6}$$

is again a random measure, the *superposition* of Φ_1 and Φ_2.

If Φ is a random measure and $\{Z(x)\}$ a nonnegative random field independent of Ψ, then the integral

$$Z \circ \Phi(B) = \int_B Z(x)\Phi(\mathrm{d}x) \qquad \text{for all Borel sets } B \tag{7.7}$$

allows construction of a further random measure, the so-called *Z-compound* of Φ (or the measure *derived* from Φ by Z). Some properties of such derived measures are studied by Karr (1978).

The *support* of a locally finite measure φ is the set

$$\mathrm{supp}(\varphi) = \{x \in \mathbb{R}^d : x \in G \text{ implies } \varphi(G) > 0 \text{ for an open } G \subset \mathbb{R}^d\}. \tag{7.8}$$

The set $\mathrm{supp}(\varphi)$ is closed. In the case of a random measure Φ the support $\mathrm{supp}(\Phi)$ is a random closed set, whose distribution is uniquely determined by the distribution of Φ (Molchanov, 2005, p. 116).

7.1.3 Palm distributions

Just as in the case of point processes, for random measures it is possible to define Palm distributions. These are still conditional distributions, which can be interpreted analogously to Palm distributions of point processes. Their definition is based on Campbell measures, as in the point process case.

Consider a random measure Φ with distribution P. Its *Campbell measure* $\mathscr{C}$ is given by

$$\mathscr{C}(B \times Y) = \mathbf{E}\big(\Phi(B)\mathbf{1}_Y(\Phi)\big) = \int_{\mathbb{M}} \varphi(B)\mathbf{1}_Y(\varphi)P(\mathrm{d}\varphi) \tag{7.9}$$

for Borel B and for $Y \in \mathcal{M}$.

In analogy with the case of point processes, because the Campbell measure $\mathscr{C}$ is absolutely continuous with respect to the intensity measure Λ, and the latter is a locally finite measure, there is a family $\{P_x\}$ of distributions on $(\mathbb{M}, \mathcal{M})$ such that

$$\mathscr{C}(B \times Y) = \int_B P_x(Y)\Lambda(\mathrm{d}x) \qquad \text{for all Borel } B \text{ and } Y \in \mathcal{M}. \tag{7.10}$$

The P_x are called *Palm distributions*. If Ψ is stationary then

$$P_x(Y_x) = P_o(Y) \qquad \text{for } Y \in \mathcal{M} \text{ and for } Y_x \text{ defined as in (7.3)}.$$

So in the stationary case it is sufficient to consider P_o, the Palm distribution with respect to the origin. It satisfies

$$\lambda P_o(Y) = \int_M \int_{\mathbb{R}^d} g(x) \mathbf{1}_Y(\varphi_{-x}) \varphi(\mathrm{d}x) P(\mathrm{d}\varphi) = \mathbf{E} \left(\int_{\mathbb{R}^d} g(x) \mathbf{1}_Y(\Phi_{-x}) \Phi(\mathrm{d}x) \right) \quad (7.11)$$

for every $g : \mathbb{R}^d \to [0, \infty)$ with $\int_{\mathbb{R}^d} g(x)\,\mathrm{d}x = 1$.

Mecke (1967) proved the following properties of the Palm distribution P_o of a stationary random measure Φ.

Theorem 7.1. *(a) For all nonnegative measurable functions f on $\mathbb{R}^d \times M$,*

$$\lambda \int_M \int_{\mathbb{R}^d} f(x, \varphi)\,\mathrm{d}x\, P_o(\mathrm{d}\varphi) = \int_M \int_{\mathbb{R}^d} f(x, \varphi_{-x}) \varphi(\mathrm{d}x) P(\mathrm{d}\varphi). \quad (7.12)$$

(b) For all nonnegative measurable functions u on M such that u is zero if evaluated at the null measure,

$$\int_M u(\varphi) P(\mathrm{d}\varphi) = \lambda \int_M \int_{\mathbb{R}^d} h(x, \varphi_x) u(\varphi_x)\,\mathrm{d}x\, P_o(\mathrm{d}\varphi), \quad (7.13)$$

where h is an arbitrary nonnegative measurable function on $\mathbb{R}^d \times M$ with

$$\int_{\mathbb{R}^d} h(x, \varphi) \varphi(\mathrm{d}x) = 1 \qquad \text{for all } \varphi \text{ except the null measure.}$$

(c) For all f as in (a)

$$\int_M \int_{\mathbb{R}^d} f(-x, \varphi_{-x}) \varphi(\mathrm{d}x) P_o(\mathrm{d}\varphi) = \int_M \int_{\mathbb{R}^d} f(x, \varphi) \varphi(\mathrm{d}x) P_o(\mathrm{d}\varphi). \quad (7.14)$$

Remarks (1) Formula (7.12) corresponds to Formula (4.64). There is only one distribution P_o satisfying (7.12) for any given stationary P.

(2) Formula (7.13) allows the determination of P for given P_o, so it is the basis for the so-called *inversion formulae.*

(3) U. Zähle (1984d) showed that, just as in the point process case (see pp. 48 and 130), a local interpretation of the Palm distributions for random measures corresponding to stationary random surface and fibre processes is possible.

This book makes much use of the term 'typical', for example 'typical fibre point', 'typical cell' or 'typical line'. Underlying this language there is always a Palm distribution, which, however, is often not explicitly defined. The idea is explained in detail in Chapter 4 in the point process context. Below, in Section 7.3.4, the Palm distribution of the volume measure is explained, where the typical point is, so to say, 'any' point in the random set considered. In the case of fibre and surface processes, the Palm distribution is nowhere explicitly defined in the book. Nevertheless, the authors hope that the reader can follow: heuristically, the typical fibre (surface) point could be thought of as the result of random uniform sampling on the fibres (surfaces) of the fibre (surface) process.

The case of 'typical cell' or 'typical line' is however different. The underlying Palm distribution is that of a point process in some representation space, and 'typical cell' or 'typical line' then corresponds to the typical point of this point process. Note that it is not customary to speak about 'typical fibre' (with some effort this also could be defined — using some representation space), but only about 'typical fibre point', in the sense of random uniform point sampling on fibres.

7.1.4 Marked random measures

Marked (or weighted) random measures are to ordinary random measures as marked point processes are to ordinary point processes. Marks give additional information about the random structure investigated. A simple example is a fibre process, where at each fibre point an additional value is given. This value may characterise the direction of the fibre in that point, or, when the fibres help to model a system of tubes, the tube diameter at that point.

Let $\mathbb{W}$ be the space of possible *marks*, with corresponding σ-algebra $\mathcal{W}$. (In the example with the fibres, $\mathbb{W}$ is simply $\mathbb{R}$. Also note that there is a change of notation here; $(\mathbb{M}, \mathcal{M})$ in earlier chapters denotes the measurable space of marks but in this chapter denotes that of locally finite measures, and hence $(\mathbb{W}, \mathcal{W})$, originated from the word 'weight', is used for the marks here.) Let $\mathbb{M}_\mathbb{W}$ be the set of all measures ψ on $[\mathbb{R}^d \times \mathbb{W}, \mathcal{B}^d \otimes \mathcal{W}]$ with $\psi(B \times \mathbb{W}) < \infty$ for all bounded Borel sets B. Let $\mathcal{M}_\mathbb{W}$ be the smallest σ-algebra on $\mathbb{M}_\mathbb{W}$ making all mappings $\psi \mapsto \psi(B \times U)$ measurable for B a bounded Borel set and U in $\mathcal{W}$. A marked random measure on $\mathbb{R}^d \times \mathbb{W}$ is then a random variable Ψ on $[\mathbb{M}_\mathbb{W}, \mathcal{M}_\mathbb{W}]$.

Examples of marked random measures occur in Section 8.6, in the context of fibre and surface processes. The random measures that arise are the length or surface measures; and the marks are the directions of tangents or normals, curvatures or other quantities. Random marked sets (see p. 211) are closely related to marked random measures.

Translations of marked random measures are defined to be shifts which leave the marks unchanged:

$$T_x A = A - x = \{(a, b) : (a - x, b) \in A\} \qquad \text{for } A \in \mathcal{B}^d \otimes \mathcal{W}.$$

If $\psi \in \mathbb{M}_\mathbb{W}$ then $\psi_x = T_{-x}\psi$ is the translated marked measure:

$$\psi_x(A) = T_{-x}\psi(A) = \psi(A - x).$$

Using this notion of translation, stationary marked random measures can be defined.

It is possible to define for marked random measures an intensity λ, which is simply the intensity of the corresponding unmarked random measure Φ where $\Phi(B) = \Psi(B \times \mathbb{W})$. Also moment measures (see Stoyan and Ohser, 1984) and Palm distributions can be defined for marked random measures. As with the mark distribution for a marked point process, so one can define a mark distribution M for a marked random measure Ψ by

$$M(U) = \frac{1}{\lambda}\mathbf{E}\big(\Psi([0, 1]^d \times U)\big) \qquad \text{for } U \in \mathcal{W}, \tag{7.15}$$

and there is a Campbell theorem which is analogous to Formula (4.29):

$$\mathbf{E}\left(\int_{\mathbb{R}^d \times \mathbb{W}} f(x, w)\Psi\big(\mathrm{d}(x, w)\big)\right) = \lambda \int_{\mathbb{R}^d} \int_{\mathbb{W}} f(x, w)M(\mathrm{d}w)\,\mathrm{d}x. \tag{7.16}$$

7.2 Moment measures and related characteristics

7.2.1 The Laplace functional

In the theory of random measures the Laplace functional plays a rôle similar to that of the Laplace transform in the theory of random variables. Each random measure Φ on $\mathbb{R}^d$ is connected with a functional L_Φ defined on the set $\mathbf{U}$ of all bounded nonnegative measurable functions u of bounded support. The Laplace functional is given by

$$L_\Phi(u) = \mathbf{E}\left(\exp\left(-\int_{\mathbb{R}^d} u(x)\Phi(\mathrm{d}x)\right)\right) \qquad \text{for } u \in \mathbf{U}. \tag{7.17}$$

Note that $0 < L_\Phi(u) \le 1$. The distribution of a random measure is uniquely determined by its Laplace functional; see Mecke (1972). If Φ is a random measure and $B_1, \ldots, B_n$ are bounded Borel sets, then the Laplace transform of the random vector $(\Phi(B_1), \ldots, \Phi(B_n))$ can be determined from knowledge of L_Φ:

$$\mathbf{E}\big(\exp(-s_1\Phi(B_1) - \cdots - s_n\Phi(B_n))\big) = L_\Phi(s_1\mathbf{1}_{B_1} + \cdots + s_n\mathbf{1}_{B_n}) \tag{7.18}$$

for $s_1, \ldots, s_n \ge 0$. If Φ_1 and Φ_2 are independent then the Laplace functional L_Φ of their superposition $\Phi = \Phi_1 + \Phi_2$ is given by

$$L_\Phi = L_{\Phi_1} \cdot L_{\Phi_2}. \tag{7.19}$$

7.2.2 Moment measures

Consider as before a random measure Φ on $\mathbb{R}^d$ with distribution P. The nth *moment measure* $\mu^{(n)}$ of Φ is a measure on $\mathcal{B}^{nd}$ given by

$$\int_{\mathbb{R}^{nd}} f(x_1, \ldots, x_n)\mu^{(n)}\big(\mathrm{d}(x_1, \ldots, x_n)\big)$$

$$= \mathbf{E}\left(\int_{\mathbb{R}^d} \cdots \int_{\mathbb{R}^d} f(x_1, \ldots, x_n)\Phi(\mathrm{d}x_1) \cdots \Phi(\mathrm{d}x_n)\right)$$

$$= \int_{\mathbb{M}}\left(\int_{\mathbb{R}^d} \cdots \int_{\mathbb{R}^d} f(x_1, \ldots, x_n)\varphi(\mathrm{d}x_1) \cdots \varphi(\mathrm{d}x_n)\right) P(\mathrm{d}\varphi) \tag{7.20}$$

for all nonnegative measurable functions $f(x_1, \ldots, x_n)$ on $\mathbb{R}^{nd}$. In particular

$$\mu^{(n)}(B_1 \times \cdots \times B_n) = \mathbf{E}\big(\Phi(B_1) \cdot \ldots \cdot \Phi(B_n)\big)$$

$$= \int_{\mathbb{M}} \big(\varphi(B_1) \cdot \ldots \cdot \varphi(B_n)\big) P(\mathrm{d}\varphi)$$

and $\mu^{(n)}(B^n) = \mathbf{E}\big(\Phi(B)^n\big)$. Thus the first moment measure coincides with the intensity measure Λ introduced by (7.4).

As in the case of point processes, densities with respect to the Lebesgue measure may exist. They are also called *product densities*. In the motion-invariant case the second-order product density is denoted by $\varrho^{(2)}(r)$. Normalisation yields the pair correlation function $g(r)$,

$$g(r) = \frac{\varrho^{(2)}(r)}{\lambda^2}. \tag{7.21}$$

The rôle of the moment measures of random measures is similar to that of moments of random variables. They can be obtained from the Laplace functional by differentiation:

$$\mu^{(n)}(B_1 \times \cdots \times B_n) = (-1)^n \lim_{s_1,\dots,s_n \downarrow 0} \frac{\partial}{\partial s_1} \cdots \frac{\partial}{\partial s_n} L_\Phi(s_1 \mathbf{1}_{B_1} + \cdots + s_n \mathbf{1}_{B_n}) \tag{7.22}$$

for bounded Borel sets $B_1, \dots, B_n$.

Conversely, under certain conditions the Laplace functional can be determined by the moment measures. Suppose all moment measures $\mu^{(n)}$ corresponding to a distribution P are locally finite, and the right-hand side of (7.23) converges. Then for u in $\mathbf{U}$ with $0 \le u(x) \le 1$ for all x in $\mathbb{R}^d$,

$$L_\Phi(u) = 1 + \sum_{n=1}^{\infty} \frac{(-1)^n}{n!} \int_{\mathbb{R}^{nd}} u(x_1) \cdot \ldots \cdot u(x_n) \mu^{(n)}\big(\mathrm{d}(x_1 \ldots x_n)\big). \tag{7.23}$$

Zessin (1983) gave conditions to ensure uniqueness of a random measure when the moment measures are given. All this is in close analogy to the *moment problem* for random variables.

Again as in the case of point processes (Section 4.5), it is possible to *reduce* the moment measures. In the stationary and isotropic case, for a diffuse measure a K-function can be used which is defined by

$$\lambda K(r) = \int \varphi\big(B(o, r)\big) P_o(\mathrm{d}\varphi) \qquad \text{for } r \ge 0. \tag{7.24}$$

Note that for point processes, which are atomic, in an analogous equation the reduced Palm distribution $P_o^!$ appears.

The variance of $\Phi(B)$ is given in the stationary and isotropic case by the following formula (compare with Formula (4.106))

$$\mathbf{var}\big(\Phi(B)\big) = \lambda^2 \int \overline{\gamma}_B(r) \, \mathrm{d}K(r) - \big(\lambda v_d(B)\big)^2, \tag{7.25}$$

where $\overline{\gamma}_B(r)$ is the isotropised set covariance of B.

Cross-correlation measures and reduced cross-correlation measures can be defined for any two random measures that are based on the same probability space. They can be used to study correlations between two different random structures such as for example a point process and a fibre process. Stoyan and Ohser (1982, 1984) studied examples of this, for instance when the point process describes tree locations and the fibre process the courses of small brooks in a forest. The case of correlation between a random set and a point process is considered in Foxall and Baddeley (2002).

7.3 Examples of random measures

7.3.1 Random measures constructed from point processes

Each point process on $\mathbb{R}^d$ can be interpreted as a random integer-valued measure, since Φ, given by

$$\Phi(B) = \text{number of points in } B, \quad \text{for all Borel sets } B, \tag{7.26}$$

is a random measure. It is clearly nonnegative and σ-additive. For such a random measure, with probability one the realisations are so-called *counting measures*: integer-valued locally finite measures on $[\mathbb{R}^d, \mathcal{B}^d]$. Consequently each point process corresponds to a random counting measure. If the notion of a point process is generalised to allow for multiple points then the correspondence is one-to-one.

The Laplace functional of the random measure Φ is connected with the generating functional G of the point process Φ as follows:

$$L_\Phi(u) = G(e^{-u}) \quad \text{for } u \in \mathbf{U}. \tag{7.27}$$

For a marked point process with nonnegative marks, the mark sum measure S_m as introduced in Formula (4.33) is a further random measure, associated to the point process.

7.3.2 Random measures constructed from random fields

Let $\{Z(x)\}$ be a nonnegative random field on $\mathbb{R}^d$. Then

$$\Phi_Z(B) = \int_B Z(x)\,\mathrm{d}x \quad \text{for all Borel sets } B \tag{7.28}$$

defines a set function, with the tacit assumption that $Z(x)$ is integrable over all Borel sets B. In fact Φ_Z is a random measure on $[\mathbb{R}^d, \mathcal{B}^d]$, which is almost surely diffuse. Under continuity assumptions, each realisation of Φ_Z delivers almost surely the corresponding realisation of the random field $\{Z(x)\}$ as the density of Φ_Z with respect to Lebesgue measure ν_d.

7.3.3 Completely random measures

Let Φ be the random counting measure corresponding to a Poisson process on $\mathbb{R}^d$. Then (by definition) for every family of pairwise disjoint bounded Borel sets, $B_1, \ldots, B_n$, the random variables $\Phi(B_1), \ldots, \Phi(B_n)$ are independent; see Section 2.3.1. It is natural to ask: which other random measures have also this independence property. Such random measures are called *completely random*.

It is clear that the mark sum measure of an independently marked Poisson process is completely random, because sums of independent random variables are involved. More generally, suppose that H is a Borel measure on $\mathbb{R}^d \times (0, \infty)$, with

$$\int_{B \times (0,\infty)} (1 - e^{-a}) H\big(\mathrm{d}(x, a)\big) < \infty$$

for all bounded Borel sets B. Let Φ_H be the random counting measure corresponding to a Poisson process on $\mathbb{R}^d \times (0, \infty)$ with intensity measure H. (If H has an atomic part then Φ_H

can have multiple points.) The measure Φ_H given by

$$\Phi_H(B) = \int_{B \times (0,\infty)} a\,\Phi_H(d(x,a)) \qquad \text{for all Borel sets } B \qquad (7.29)$$

is also completely random. The general structure of completely random measures is described in Kingman (1993).

The Laplace functional of Φ_H is

$$L_{\Phi_H}(u) = \exp\left(-\int_{\mathbb{R}^d \times (0,\infty)} \left(1 - \exp(-au(x))\right) H(d(x,a))\right) \qquad \text{for } u \in \mathbf{U}. \quad (7.30)$$

Its intensity measure Λ_{Φ_H} is given by

$$\Lambda_{\Phi_H}(B) = \int_{B \times (0,\infty)} aH(d(x,a)) \qquad \text{for all bounded Borel sets } B. \qquad (7.31)$$

The higher moment measures are given by

$$\mu^{(2)}(B_1 \times B_2) = m_1(B_1) \cdot m_1(B_2) + m_2(B_1 \cap B_2)$$

$$\begin{aligned}
\mu^{(3)}(B_1 \times B_2 \times B_3) = {}& m_1(B_1) \cdot m_1(B_2) \cdot m_1(B_3) + m_1(B_1) \cdot m_2(B_2 \cap B_3) \\
& + m_1(B_2) \cdot m_2(B_3 \cap B_1) + m_1(B_3) \cdot m_2(B_1 \cap B_2) \\
& + m_3(B_1 \cap B_2 \cap B_3) \qquad \text{for } B_1, B_2, B_3 \text{ Borel sets,}
\end{aligned}$$

etc. Here

$$m_n(B) = \int_{B \times (0,\infty)} a^n H(d(x,a)) \qquad \text{for Borel } B.$$

The nth moment measure is locally finite if and only if the measure m_n is locally finite.

7.3.4 Random measures generated by random closed sets: Curvature measures

In geometrical studies an important rôle is played by measures, in particular curvature measures. These measures generalise and replace the functionals and characteristics introduced in Chapter 1. A simple example is the following.

Volume measure V_Ξ

Given a random closed set Ξ, its volume measure or coverage measure is defined by

$$V_\Xi(B) = \nu_d(B \cap \Xi) \qquad \text{for all Borel sets } B; \qquad (7.32)$$

it is always well-defined.

By definition V_Ξ ignores lower-dimensional parts of Ξ, for example fibres or surface pieces if in $\mathbb{R}^3$. If Ξ is regular closed (the closure of an open set) then $\mathrm{supp}(V_\Xi) = \Xi$. Therefore, V_Ξ is well-fitted to situations where Ξ is regular closed, as it is the case for usual models of porous media. Ayala *et al.* (1991) discussed the problem of recovering the distribution of Ξ from that of the measure V_Ξ. In the following it is assumed that Ξ is almost surely regular closed.

If Ξ is stationary and of volume fraction p, then V_Ξ is likewise stationary and its intensity is p.

It is mathematically interesting to consider the Palm distribution of V_Ξ. This may help to understand the idea of Palm distributions for random measures and sheds light to the nature of contact distribution functions.

It holds that

$$P_o(Y) = \mathbf{P}(V_\Xi \in Y | o \in \Xi) \qquad \text{for any configuration set } Y \text{ in } \mathcal{M}. \tag{7.33}$$

Proof of (7.33) (Ballani, 2011). Note that $\lambda = p = \mathbf{P}(o \in \Xi)$. By definition (7.11) it holds for any nonnegative $g(x)$ with $\int_{\mathbb{R}^d} g(x)\mathrm{d}x = 1$,

$$\lambda P_o(Y) = \mathbf{E}\left(\int_{\mathbb{R}^d} g(x)\mathbf{1}_Y(\Phi_{-x})\Phi(\mathrm{d}x)\right)$$

$$= \mathbf{E}\left(\int_{\mathbb{R}^d} g(x)\mathbf{1}_Y(V_{\Xi-x})\mathbf{1}_\Xi(x)\mathrm{d}x\right)$$

$$= \mathbf{E}\left(\int_{\mathbb{R}^d} g(x)\mathbf{1}_Y(V_{\Xi-x})\mathbf{1}_{\Xi-x}(o)\mathrm{d}x\right)$$

$$= \int_{\mathbb{R}^d} \mathbf{E}\left(\mathbf{1}_Y(V_{\Xi-x})\mathbf{1}_{\Xi-x}(o)\right)g(x)\mathrm{d}x$$

$$= \int_{\mathbb{R}^d} \mathbf{E}\left(\mathbf{1}_Y(V_\Xi)\mathbf{1}_\Xi(o)\right)g(x)\mathrm{d}x$$

$$= \mathbf{E}\left(\mathbf{1}_Y(V_\Xi)\mathbf{1}_\Xi(o)\right) = \mathbf{P}(V_\Xi \in Y, o \in \Xi),$$

and (7.33) follows. $\square$

The statement given in (7.33) can be re-expressed in terms of configuration sets in the trace $\mathcal{F}_{\mathrm{rc}}$ of the hitting σ-algebra on the set of regular closed sets. Let Y_{set} be the counterpart of $Y \in \mathcal{M}$ in $\mathcal{F}_{\mathrm{rc}}$. Then

$$P_o(Y) = \mathbf{P}(\Xi \in Y_{\mathrm{set}} | o \in \Xi). \tag{7.34}$$

In order to discuss the spherical contact distribution, the case $Y = \{\varphi : \varphi(B(o, r)) = b_d r^d\}$ is considered, to which $Y_{\mathrm{set}} = \{X : B(o, r) \subset X\}$ belongs, where φ denotes a measure and X a regular closed set. That is, Y_{set} is the set of all regular closed sets that contain the ball $B(o, r)$. Equation (7.34) yields

$$P_o(Y) = \mathbf{P}(B(o, r) \subset \Xi | o \in \Xi). \tag{7.35}$$

The term on the right-hand side can be considered as a function of r, and

$$F(r) = 1 - \mathbf{P}(B(o, r) \subset \Xi | o \in \Xi)$$

is a distribution function. Comparison with (6.59) shows that $F(r)$ is the 'inner' spherical contact distribution function of Ξ. When the Ξ considered here is the closure of the complement of a given random closed set, then $F(r)$ is the spherical contact distribution function of that set.

This shows that contact distributions of random sets with positive p have the nature of Palm distributions, as also do the nearest-neighbour distance distribution functions $D(r)$ for point processes, as explained in Section 4.4.4.

The second-order characteristics of the volume measure V_Ξ in the case $p > 0$ are closely related to the set-theoretic second-order characteristics of Ξ. In particular, $C(r)$ is the second-order product density of V_Ξ and $C(r)/p^2$ its pair correlation function.

Measure-theoretic ideas play some rôle in statistics of porous media. A powerful method to characterise statistically the variability of random sets is the study of *local porosity*; see Hilfer (1991, 2000). The idea is simple and described as follows.

Take a cube C_L of side-length L and consider the 'volume fraction' of Ξ in C_L, that is, the random variable $p_L = V_\Xi(C_L)/v_d(C_L)$. Here Ξ is assumed to be motion-invariant; otherwise the location and orientation of C_L also plays a rôle. If Ξ serves as a model for the empty part of a porous medium, then p_L is just the local porosity of Ξ in C_L. The random variable is rather difficult to analyse theoretically, even for Boolean models, but it can be easily measured and analysed statistically. Clearly, it is

$$\mathbf{E}(p_L) = p, \tag{7.36}$$

and the variance can be calculated if the covariance of Ξ is known, as σ_v^2 in (6.83). For large L, the local volume fraction p_L is approximately constant, but for smaller L its distribution is rather complicated. For example, $\mathbf{P}(p_L = 0) > 0$ is possible. Hilfer (1991, 2000) showed how the distribution of p_L can be used in the characterisation of porous media.

Hilfer (2000) also demonstrated the use of so-called *local percolation probabilities*. He considered the probability that in the cube C_L there is a path lying in Ξ and connecting a given side of the cube with its opposite side.

Surface area measure S_Ξ

Given a random closed set Ξ in $\mathbb{R}^d$, its surface area measure is defined by

$$S_\Xi(B) = h_{d-1}(B \cap \partial\Xi) \qquad \text{for all Borel sets } B, \tag{7.37}$$

where h_{d-1} is the $(d-1)$-dimensional Hausdorff measure and $h_{d-1}(B \cap \partial\Xi)$ (not $h_{d-1}(B \cap \Xi)$!) is the surface area of all pieces of the boundary $\partial\Xi$ in B. It is well-defined if $\partial\Xi$ has certain smoothness properties, as in the case that Ξ is a regular closed random $\mathbb{S}$-set.

If Ξ is stationary then so is S_Ξ. Likewise S_Ξ inherits isotropy from Ξ. The intensity of S_Ξ, denoted by $S_V^{(d)}$, is called the *specific surface area* of Ξ. In the linear, planar and spatial cases special symbols adopted from stereology are used:

$$S_V = S_V^{(3)},$$
$$L_A = S_V^{(2)},$$
$$P_L = S_V^{(1)}.$$

In the linear or one-dimensional case Ξ is a union of non-overlapping closed intervals, and P_L is the intensity of the point process of all interval endpoints.

Pair correlation functions and reduced second moment functions of S_Ξ for some models are considered in Sections 8.3.1 and 8.5.2.

Curvature measures

To each random $\mathbb{S}$-set Ξ in $\mathbb{R}^d$ there correspond related random measures $\Phi_{\Xi,k}$, where k runs through $0, 1, \ldots, d$. Two of them have already been described:

$$V_\Xi = \Phi_{\Xi,d},$$

$$S_\Xi = 2\Phi_{\Xi,d-1}.$$

For $k = 0, 1, \ldots, d - 1$ the measures $\Phi_{\Xi,k}$ are concentrated on $\partial\Xi$. (Note that the connection with S_Ξ only holds if Ξ is a regular closed set. The comment on (7.41) below indicates what will happen if Ξ possesses lower-dimensional parts.)

The following exposition follows Schneider and Weil (1992); see also Schneider (1993) (but note that there are notational differences between these books and the present text). First curvature measures for deterministic convex bodies are explained, then there is an account of their extension to deterministic polyconvex sets, and finally there is a description of random curvature measures corresponding to random $\mathbb{S}$-sets.

Convex sets

Let K be a convex body of $\mathbb{R}^d$ and suppose that $x \in \mathbb{R}^d$. Then $p(x, K)$ denotes the (uniquely determined) nearest point to x in K. For any $B \subset \mathbb{R}^d$ the *local parallel set* $M_r(K, B)$ of K at distance r is defined by

$$M_r(K, B) = \{x \in \mathbb{R}^d : \|x - p(x, K)\| \le r, p(x, K) \in B\} \qquad \text{for } r \ge 0.$$

If B is a Borel set, then so is $M_r(K, B)$. Finite Borel measures are defined for each positive r by

$$\mu_r(K, B) = v_d\big(M_r(K, B)\big) \qquad \text{for all Borel sets } B. \tag{7.38}$$

These measures $\mu_r(K, \cdot)$ satisfy the *local Steiner formula*

$$\mu_r(K, B) = \sum_{k=0}^{d} b_{d-k}\Phi_k(K, B)r^{d-k} \qquad \text{for } r \ge 0 \text{ and } B \text{ a Borel set;} \tag{7.39}$$

compare this with Formula (1.39). The measures $\Phi_k(K, \cdot)$ appearing here are called *curvature measures* of K. For $k = d$ and $k = d - 1$,

$$\Phi_d(K, B) = V_K(B) = v_d(B \cap K) \tag{7.40}$$

and

$$2\Phi_{d-1}(K, B) = S_K(B) = h_{d-1}(B \cap \partial K) \tag{7.41}$$

for all Borel sets B. The formula for the surface measure is true if K is d-dimensional. If K is $(d - 1)$-dimensional, that is, a subset of a $(d - 1)$-dimensional flat, then

$$\Phi_{d-1}(K, B) = h_{d-1}(\partial K \cap B),$$

as with $S(K)$ on p. 12.

Under smoothness conditions the values $\Phi_k(K, B)$ for $k < d$ can be obtained as integrals with respect to curvatures over $B \cap \partial K$. This explains the name 'curvature measure'; see the explanation of Φ_0 for a planar polyconvex set below.

For $B = \mathbb{R}^d$ the measures yield the intrinsic volumes,

$$\Phi_k(K, \mathbb{R}^d) = V_k(K) \qquad \text{for } k = 0, 1, \ldots, d.$$

Polyconvex sets

As the intrinsic volumes and Minkowski functionals, the curvature measures can be extended to sets in the convex ring $\mathcal{R}$, with the additivity property

$$\Phi_k(A_1 \cup A_2, B) + \Phi_k(A_1 \cap A_2, B) = \Phi_k(A_1, B) + \Phi_k(A_2, B)$$

for A_1 and A_2 in $\mathcal{R}$. This *additive extension* can be achieved by means of the theory of extensions of valuations; see Schneider (1980), Schneider and Weil (1992) and Schneider (1993).

These theories justify the assertion that the extended Minkowski and curvature measures satisfy, analogously to Formula (1.78), the relation

$$\Phi_k(A, B) = \sum_i \Phi_k(K_i, B)$$

$$- \sum_{i_1 < i_2} \sum \Phi_k(K_{i_1} \cap K_{i_2}, B) + \cdots + (-1)^{n-1} \Phi_k(K_1 \cap \cdots \cap K_n, B) \quad (7.42)$$

for any Borel set B, if A is given in the form $A = \bigcup_{i=1}^n K_i$ where the K_i are convex bodies.

Note that the measures $\Phi_k(A, \cdot)$ are finite *signed* measures for $k = 0, 1, \ldots, d - 2$; they are not necessarily positive.

As in the convex case, for any Borel set B,

$$\Phi_d(A, B) = V_A(B), \quad (7.43)$$

and for any regular closed A,

$$2\Phi_{d-1}(A, B) = S_A(B). \quad (7.44)$$

Consider now the measure Φ_0. For convex K it is

$$\Phi_0(K, B) = \frac{1}{db_d} h_{d-1}\big(\sigma(K, B)\big), \quad (7.45)$$

where $\sigma(K, B)$ is the 'spherical image' of K, that is, the set of all outward-pointing unit normal vectors of $B \cap \partial K$ in the sphere S^{d-1}; see Schneider (1993).

In the planar case the value of $\Phi_0(A, B)$ for a sufficiently regular polyconvex set A (e.g. having a boundary of a finite number of vertices and being smooth between the vertices) is obtained as follows for a regular set B (e.g. a rectangle). As with all curvature measures Φ_k with $k < d$, it is concentrated on the boundary ∂A of A. Because of the assumptions on A, the curvature $\kappa(x)$ is finite at almost all points x, and at almost all points of ∂A the interior of

A is on one side of ∂A. The inner angle at a vertex x is denoted by $\omega(x)$. Then it holds

$$2\pi\Phi_0(A, B) = \int_{B\cap\partial A} \kappa(x)\,l(dx) + \sum_{\text{vertices } x \text{ of } \partial A \text{ in } B} \left(\pi - \omega(x)\right), \qquad (7.46)$$

where the integral is taken over all curved segments of ∂A in B, with $l(\cdot)$ being the corresponding length measure. The orientation of the curves is taken so that the interior of A is on the left-hand side of an observer moving in an anticlockwise direction along the curves of ∂A. Thus for a circle belonging to ∂A and lying completely in B, the integration yields the value 2π. If ∂A has a vertex at the point x, then x gives a separate contribution to the integral. It is determined by the inner angle $\omega(x)$ of A at x, which will lie between 0 and 2π. If $0 < \omega(x) < \pi$ then at x there is an outward vertex; if $\pi < \omega(x) < 2\pi$ then there is an inward vertex. The contribution is then $\pi - \omega(x)$, and is positive if the vertex is outward.

The measures Φ_k are homogeneous of degree k, that is,

$$\Phi_k(\alpha A, \alpha B) = \alpha^k \Phi_k(A, B) \qquad \text{for } \alpha > 0,$$

and they are *motion-covariant*, that is,

$$\Phi_k(mA, mB) = \Phi_k(A, B) \qquad \text{for all rigid motions } m.$$

Random S-sets
The local character of the curvature measures enables to extend their definition to sets in the extended convex ring $\mathbb{S}$, that is, to sets A with the property that $A \cap K \in \mathcal{R}$ for every convex body K. For such a set A and any bounded Borel set B define

$$\Phi_{A, k}(B) = \Phi_k\big(A \cap B(o, R), B\big), \qquad (7.47)$$

where R is large enough that B is contained in the open ball $B^{\text{int}}(o, R)$, and by definition, $A \cap B(o, R)$ belongs to the convex ring $\mathcal{R}$.

In the same way the random curvature measures $\Phi_{\Xi, k}$ are defined for random S-sets Ξ:

$$\Phi_{\Xi, k}(B) = \Phi_k\big(\Xi \cap B(o, R), B\big) \qquad \text{for all Borel sets } B. \qquad (7.48)$$

Federer (1959) defined curvature measures for sets of positive reach. Zähle (1984a,b, 1986, 1987a) and Weiss and Zähle (1988) extended this to unions and to locally finite unions of such sets.

Positive or absolute curvature measures

There is still another extension of the curvature measures to the polyconvex sets and S-sets, which leads to proper (positive) measures. However, this other extension is not additive. These absolute curvature measures were introduced by Matheron (1975, p. 119) and, in a refined form, by Schneider (1979a,b, 1980).

In the following the construction of Matheron is sketched for any $A \in \mathcal{R}$.

Let $\Pi_A(x)$ be the set of all projections of x in $\mathbb{R}^d$ onto A, that is, the set of all points $x' \in A$ for which there exists an open set O with $x' \in O$ and

$$\|x - y\| > \|x - x'\|$$

for all $y \in O \cap A$ with $y \neq x'$. This set is finite. (If A is convex then Π_A is the singleton set $\{p(x, A)\}$, with $p(x, A)$ defined as on p. 290.)

Then

$$n(A, B, r; x) = \sum_{z \in \Pi_A(x) \cap B^{\text{int}}(x,r)} \mathbf{1}_B(z) \qquad \text{for } r \geq 0 \text{ and } B \text{ a Borel set}$$

is the number of distinct points in $\Pi_A(x) \cap B^{\text{int}}(x, r) \cap B$, and satisfies

$$\int_{\mathbb{R}^d} n(A, B, r; x) \, dx < \infty \qquad \text{for any bounded Borel set } B. \tag{7.49}$$

Matheron (1975) showed that there exist measures $W_k^+(A, \cdot)$ called *positive Minkowski measures* yielding the expansion

$$\int_{\mathbb{R}^d} n(A, B, r; x) \, dx = \sum_{k=0}^{d} \binom{d}{k} W_k^+(A, B) r^k \tag{7.50}$$

for all Borel sets B and $k = 0, 1, \ldots, d$. These can be transformed in *positive curvature measures* $\Phi_k^+(A, \cdot)$ by

$$b_{d-k} \Phi_k^+(A, \cdot) = \binom{d}{k} W_{d-k}^+(A, \cdot). \tag{7.51}$$

The positive curvature measure Φ_0^+ is related to the so-called convexity number; see Matheron (1975, p. 122) and Schneider (1993, p. 225), who defined positive curvature measures directly.

In the three-dimensional case the measure Φ_1^+ has a natural interpretation in terms of curvature. Let ∂A^+ be the subset of the boundary ∂A in which all curvatures are nonnegative. Since A is a polyconvex set, $\partial A \setminus \partial A^+$ consists of curves generated by the intersections of the surfaces of the constituent convex bodies; see Figure 7.1. Then, if $m(x)$ is the mean curvature at x, it can be shown that

$$\pi \Phi_1^+(A, B) = \int_{B \cap \partial A^+} m(x) \, dS \qquad \text{for all Borel sets } B, \tag{7.52}$$

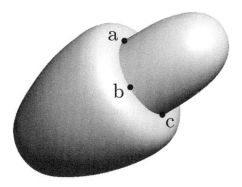

Figure 7.1   Intersection curve ($-$ a $-$ b $-$ c $-$) of two convex bodies.

where dS corresponds to the surface area. Hence, $\Phi_1^+(A, \cdot)$ is called the *mean positive curvature measure*.

A further (positive) curvature measure is related to the *absolute curvature* described in Baddeley and Averback (1983). The absolute curvature $m^a(x)$ of a smooth surface at point x is given by

$$m^a(x) = \begin{cases} |m(x)|, & \text{if } \kappa_1(x) \text{ and } \kappa_2(x) \\ & \text{have the same sign,} \\ \left(\dfrac{4}{\pi}\theta_0(x) - 1\right) m(x) + \dfrac{1}{\pi}(\kappa_1(x) - \kappa_2(x)) \sin 2\theta_0(x), & \text{otherwise,} \end{cases}$$

where $\kappa_1(x)$ and $\kappa_2(x)$ are the principal curvatures and

$$\theta_0(x) = \tan^{-1}\sqrt{\dfrac{-\kappa_1(x)}{\kappa_2(x)}},$$

choosing the branch of arctangent so that $0 \le \theta_0(x) \le \pi/2$; see Baddeley (1980). The definition of the absolute curvature measure is entirely analogous to that of the mean positive curvature measure of (7.52), but the integration is performed over $B \cap \partial A$ rather than $B \cap \partial A^+$ and $m(x)$ is replaced by $m^a(x)$.

Absolute curvature measures can also be defined for $\mathbb{S}$-sets. Such measures for unions of sets of positive reach are studied in Zähle (1989) and Rother and Zähle (1992).

The random positive curvature measures belonging to a random $\mathbb{S}$-set Ξ are denoted by $\Phi_{\Xi,k}^+$. For stationary (motion-invariant) Ξ the random measures are also stationary (motion-invariant).

Intensities of stationary curvature measures

If the underlying random closed set Ξ is stationary, then so are the corresponding curvature measures. Thus it makes sense to introduce the corresponding intensities v_k and v_k^+, $k = 0$, $1, \ldots, d$. These intensities play an important rôle in practical applications as first-order characteristics of Ξ, in particular in the planar and spatial case. Therefore a special notation is reserved for them: in $\mathbb{R}^3$ v_3 is denoted as V_V and $2v_2$ as S_V. On p. 249 there is a complete list of the symbols for the v_k. For the null-dimensional intrinsic volumes these are

$$N_A = v_0 \quad \text{and} \quad N_A^+ = v_0^+ \quad \text{in the planar case} \tag{7.53}$$

and

$$N_V = v_0 \quad \text{and} \quad N_V^+ = v_0^+ \quad \text{in the spatial case.} \tag{7.54}$$

It is natural to call N_A the *specific planar connectivity number* and N_A^+ the *specific planar convexity number*. Furthermore, it is convenient to use N_A^- to denote the intensity of Φ_0^+ of the closure of the complement of Ξ. It is

$$N_A = N_A^+ - N_A^-. \tag{7.55}$$

Formula (7.46) suggests the equation

$$C_A = 2\pi N_A \tag{7.56}$$

for the *integral of mean curvature per area unit*. The intensities of the curvature measures for $d = 3$ are M_V for $\Phi_{\Xi, 2}$ and M_V^+ for $\Phi_{\Xi, 2}^+$.

Note that the intensities discussed here coincide with the densities introduced in Formulae (3.37) and (6.71). Thus, for example, S_V can also be understood as

$$S_V = \lim_{n \to \infty} \frac{\mathbf{E}\big(S(\partial \Xi \cap W_n)\big)}{v_3(W_n)}$$

where $\{W_n\}$ is a convex averaging sequence of windows.

To aid understanding, some formulae are given for germ–grain models of non-intersecting grains. Suppose the germ point process has intensity λ. If $\overline{\chi}$ is the mean connectivity number of the typical grain, then

$$N_A = \lambda \overline{\chi}, \quad \text{and likewise} \quad N_V = \lambda \overline{\chi}.$$

If $\overline{M}$ is the mean of the integral of mean curvature of the typical grain, then

$$M_V = \lambda \overline{M}.$$

Finally, C_A satisfies

$$C_A = 2\pi \lambda \overline{\chi}.$$

In the planar case, N_A can be determined by tangent count. Adler and Taylor (2007, Theorem 6.1.2) gave the theoretical foundation. This can be used statistically in a method suggested already in 1967 by Cahn, DeHoff and Haas *et al.*, which is described as follows.

In the stationary and isotropic case, N_A^+ is the intensity of the point process of lower convex tangent points; see Figure 7.2. The value N_A^+ can be estimated by counting all those lower tangent points that fall in a given window W of observation. This count $n^+(W)$ yields

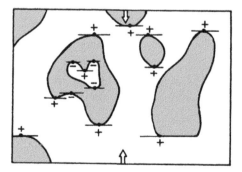

Figure 7.2 Tangent count with respect to the bottom line. There are six lower convex tangent points corresponding to tangents parallel to the bottom line and three lower concave tangent points; there are four upper convex tangent points and one upper concave tangent point. Thus, assuming that the set shown belongs to a planar stationary and isotropic random closed set and that the window area $v_2(W) = 1$, one obtains $\hat{N}_A^+ = 5$, $\hat{N}_A^- = 2$ and $\hat{N}_A = 3$.

the estimator

$$\hat{N}_A^+ = \frac{n^+(W)}{v_2(W)}.$$ (7.57)

It is useful to determine $\hat{N}_A^+$ for various orientations (instead of using only the upward direction) and then to average. Baddeley (1984) considered the accuracy of the tangent count method, which is of course sensitive to image noise.

Analogously, estimates of N_A^- are obtained by counting the lower concave tangents (or upper convex tangents of the complement); see also Figure 7.2. Then finally N_A is estimated via

$$\hat{N}_A = \hat{N}_A^+ - \hat{N}_A^-.$$ (7.58)

The estimation of the other intensities v_k is described in Section 6.5.7.

8

Line, fibre and surface processes

8.1 Introduction

This chapter is concerned with the study of systems of fibres and of systems of surfaces or fragments of surfaces, distributed at random on the plane or in space. Interest is focused on general results that hold in the cases of stationarity or motion-invariance and on simple particular models. In this introductory section some illustrations and examples are discussed in order to motivate basic definitions.

Stochastic models of these systems are called *fibre processes* and *surface processes*; as in the term 'point process' no dependence on time is implied here. The theory of fibre and surface processes can be considered variously as belonging either to the theory of random closed sets or to the theory of random measures. As elsewhere in this book, in the case of stationarity one basic characteristic is the *intensity*, the mean fibre length per unit area (volume) for a planar (spatial) fibre process or the mean surface area per unit volume for a spatial surface process. Moreover, the constituent points of fibres and surfaces have associated directions (of tangents or normals) and these directions lead to a further important characteristic: the *rose of directions*.

Figures 8.1 and 8.2 display two realisations of irregular systems of lines in the plane, to be considered as samples of planar fibre processes. Figure 8.3 displays a spatial system of vessels which might be modelled by means of a spatial fibre process. The intensity here is the mean total fibre length per unit area or volume. The rose of directions summarises the spatial distribution of the collection of tangents to the fibres.

Planar fibre systems can also arise as planar sections of spatial surface systems such as the boundary of a three-dimensional Boolean model (see Chapter 3). There are stereological methods which enable estimation of characteristics of the original three-dimensional structure. The most important of these are the intensity (the mean surface area per unit volume) and the rose of directions.

Statistical investigation begins with estimation of intensity and rose of directions. A further stage considers the suitability of various fibre or surface process models for the sample. On

Stochastic Geometry and its Applications, Third Edition.
Sung Nok Chiu, Dietrich Stoyan, Wilfrid S. Kendall and Joseph Mecke.
© 2013 John Wiley & Sons, Ltd. Published 2013 by John Wiley & Sons, Ltd.

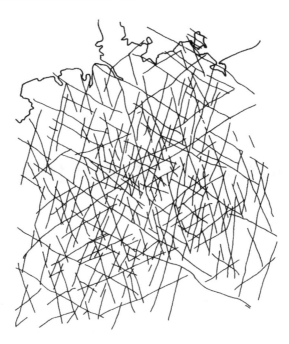

Figure 8.1   Distribution of linear fault zones in Central Europe (from Weber, 1977). The system of lines can be interpreted as a sample of a planar fibre process. A line process could also serve at least as a rough approximation; the pattern can be also interpreted as a sample of a geometrical network. Note that some of the fault zones end at other fault zones – so clearly there is interaction between them. The structure is clearly not isotropic. There are four dominating directions, ENE–WSW, NNE–SSW, NW–SE, ESE–WNW. The apparent lower intensity of fault zones in the northern region of the figure is due to incomplete information about that region. See the statistical analysis of fault lengths in Section 6.6.4, Figure 6.16.

occasion such a description can be suggestive of the manner in which the structures originated. Some examples of models are given below.

Figure 8.7 on p. 309 results from the simulation of a Poisson line process. This is an example of the class of *line processes* discussed in Section 8.2, which also can be considered as fibre processes. There is a considerable theory concerning such processes, which uses ideas of the theory of point processes. Higher-dimensional analogues are *plane, hyperplane* and general *flat processes* in $\mathbb{R}^3$ and $\mathbb{R}^d$. Figure 8.4 shows a sample of a plane process in $\mathbb{R}^3$. Also these structures are well understood as point processes in suitable representation spaces.

Random tessellations of space, as described in Chapter 9, yield fibre and surface processes, formed by their edge and facet systems.

Another particular class of fibre and surface processes comes from Boolean models.

Example 8.1. *Boolean fibre and surface process*
(a) *Planar Boolean fibre process*. Boolean models are characterised by the intensity of the underlying Poisson process and the distribution of the typical grain Ξ_0. In this example Ξ_0 is a random segment defined by a random point ξ on the unit circle and a positive random

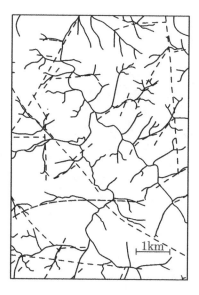

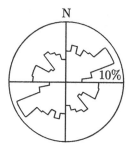

Figure 8.2 A part of the drainage network in the Eastern Erzgebirge (Saxony) and the corresponding rose of directions (derived from a larger region). The main flow direction in that region is SE–NW but the rose has its maximum at SW–NE. This is because of the large number of tributary rivers. The broken lines are fault lines.

variable l, representing orientation and length, respectively. The random set Ξ_0 is the line segment between o and $l\xi$. The corresponding Boolean model forms a fibre process in the plane. A simulated realisation of such a Boolean segment process with anisotropy is illustrated in Figure 8.5(a). To produce fibres from line segments, Kärkkäinen *et al.* (2012) deformed the line segments by dividing each segment into many short pieces and then 'shaking' them so that the angles between adjacent pieces follow a multivariate von Mises distribution (circular normal). Figure 8.5(b) shows a system of *von Mises fibres*, resulting from deforming the segments in Figure 8.5(a).

(b) *Boolean disc process.* If the typical grain Ξ_0 is a random isotropic two-dimensional disc in $\mathbb{R}^3$ of random radius R, then the corresponding Boolean model can be interpreted as a surface process in $\mathbb{R}^3$. This model and its generalisations are used in various papers in the context of discontinuities in engineering geology, such as structures arising from joints in

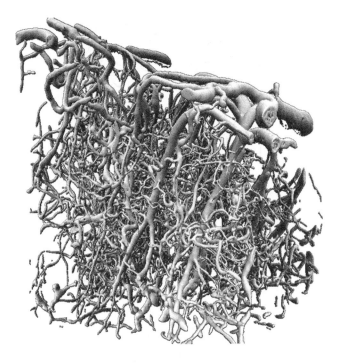

Figure 8.3 Scanning electron microscopic image of vasculature at the cortical surface. This may be regarded as a sample of a spatial fibre process with thick fibres. See Heinzer *et al.* (2006). Courtesy of R. Müller.

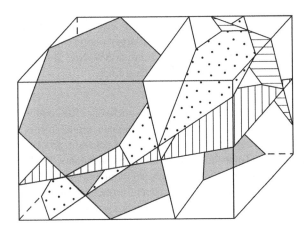

Figure 8.4 A sample of a plane process. This figure is also an example of a parallelepiped intersected by random planes. The intersection figures can be seen to have three, four, five, or six angles.

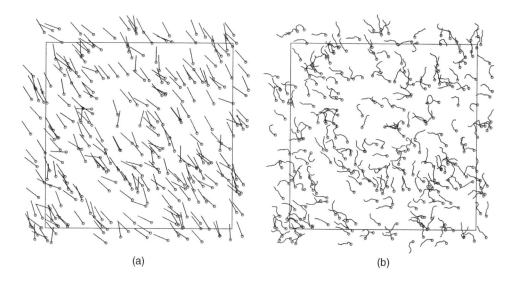

(a)   (b)

Figure 8.5 (a) A simulated realisation of an anisotropic Boolean segment process. All segments are of equal length. (b) A simulated system of von Mises fibres. The fibres result from the segments in part (a) by dividing the segments in 17 equal-length pieces the directions of which follow a multivariate von Mises distribution. See Kärkkäinen *et al.* (2012). Courtesy of S. Kärkkäinen.

rocks or from cracks in soil masses. Planar sections are visible on slopes or on outcrops and so *stereological* methods can be used to determine statistical characteristics of the underlying process; see Koschitzki (1980), Molek *et al.* (1981), Baecher (1983), Adler and Thovert (1999) and Thovert and Adler (2004), who speak about *fracture networks*. Berkowitz and Adler (1998) also study the secondary segment process consisting of the segments resulting from intersection of discs, which they call 'needles'.

This book confines itself to processes of *smooth* fibres and surfaces. Of course, there are more general theories. For example, surfaces are considered which are the boundaries of locally polyconvex sets, as in Weil (1997). Zähle (1982, 1983) and since then many others considered fibre and surface processes (and higher-dimensional generalisations) under much more relaxed regularity assumptions; Zähle (1982) is today a classic reference. The relatively stringent assumptions of Sections 8.3–8.5 are replaced in Zähle's work by the requirement that the sets involved are *Hausdorff-rectifiable* closed sets. This means that (together with a technical regularity condition) for such a set X the Hausdorff k-dimensional measure $h_k(X \cap K)$ is finite for any bounded subset K of $\mathbb{R}^d$; $k = 1$ for fibres, $k = d - 1$ for (hyper) surfaces. This more general theory is developed to give stereological formulae generalising (8.38) and (8.83) and to define direction distributions analogous to the roses of directions mentioned above and defined in Sections 8.3–8.5. Moreover, the formulae for intersections of fibre and surface processes can be generalised.

Zähle (1982, 1983) discussed a further avenue of generalisation: the case of fibre and surface processes on homogeneous Riemannian spaces. The most immediately applicable case of this is of fibre processes on the sphere.

Just as line and plane processes in $\mathbb{R}^2$ and $\mathbb{R}^3$ can be generalised to plane and flat processes in $\mathbb{R}^d$, so fibre and surface processes can be generalised to the *manifold processes* of Mecke (1981b). These are random systems of fragments of *manifolds* of fixed dimension embedded in some higher-dimensional Euclidean space $\mathbb{R}^d$. Here a manifold is a generalisation of a fibre or of a surface: it can be thought of as the solution set of some system of equations $f_1(x) = 0, \ldots, f_r(x) = 0$. Mecke studies the intersections of manifold processes with other manifold processes, and also intersections with fixed manifolds. Related second moments are considered in Jensen *et al.* (1990a,b) and Zähle (1990).

8.2   Flat processes

8.2.1   Introduction

This section considers flat processes as introduced in Section 4.8, that is, random systems of k-dimensional planes in $\mathbb{R}^d$. The particular case of $k = 1$ and $d = 2$, of lines in the plane, is considered in detail, as an example of a point process on a special representation space, which is here a cylinder in $\mathbb{R}^3$. Thus the theory of planar line processes becomes a special chapter in the general theory of point processes. The corresponding geometric structure leads to remarkable constraints on the variety of possible regular planar line processes. In the following this geometric structure is explored, and at the end of the section higher dimensional cases are also considered.

Point processes in other parameter spaces are considered in Small (1996), D. G. Kendall *et al.* (1999) and Kendall and Le (2010), in the context of statistics of shape; see also Chapter 8 of SKM95.

Practical applications tend to concern the special case of *Poisson line processes*. These generate tessellations of the plane and thus random polygons (as described in Section 9.5), and also provide useful models for random line probe sampling in stereology.

Before the formal introduction of the theory, the following remark on the construction of Poisson line processes in the plane may be enlightening.

One might seek to construct a Poisson line process as a particular 'Boolean model' with lines as grains. However, this does not lead to a reasonable structure. From the formal standpoint, a Boolean model has compact grains, but lines are unbounded sets and therefore not compact. The condition (3.2) is of course not satisfied for all K. A more constructive explanation of the problem is as follows. Assume that a Boolean model with lines as grains is given, where the probability for a line to be parallel to x_1-axis is zero. Let the intensity of the germ process be λ. Then almost all lines intersect the x_1-axis. For each integer n, the lines with germs in the strip $\{(x_1, x_2) : n \le x_2 < n + 1\}$ produce a point process Φ_n of intensity λ on the x_1-axis. Consequently, the ensemble of all lines of the Boolean model generates a point process $\Phi = \bigcup_n \Phi_n$ on the x_1-axis of intersection points of *infinite* intensity.

8.2.2   Planar line processes

A representation space for lines in the plane

A *directed line* is a line together with a preferred direction along the line. The family of all undirected lines in the plane is denoted by $A(2, 1)$, and the family of all directed lines

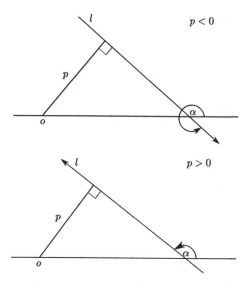

Figure 8.6   Construction of the representation space for directed lines in the plane. The two diagrams illustrate the cases of negative and positive signed perpendicular distance p.

by $A^*(2, 1)$. There is an obvious $2 : 1$ correspondence between elements of $A^*(2, 1)$ and of $A(2, 1)$, obtained by ignoring direction.

Directed lines in the plane can be put into $1 : 1$ correspondence with the set of points on the surface of a cylinder in $\mathbb{R}^3$. To see this, note that a convenient set of coordinates for a directed planar line ℓ is based on its perpendicular distance from the origin o and the angle that it makes with the x_1-axis.

In this approach, the first coordinate p is the *signed* perpendicular distance of ℓ from o; the sign is positive if o lies to the left of ℓ when looking in the direction of ℓ. The second coordinate α is the angle between ℓ and the x_1-axis, measured in an anticlockwise direction. Thus $p \in \mathbb{R}$ and $\alpha \in (0, 2\pi]$. The coordinatisation is illustrated by Figure 8.6. This supplies a $1 : 1$ correspondence relating $A^*(2, 1)$ to the cylinder

$$C^* = \{(\cos \alpha, \sin \alpha, p) \in \mathbb{R}^3 : p \in \mathbb{R}, \alpha \in (0, 2\pi]\}.$$

To each *undirected line* in $A(2, 1)$ there corresponds a pair of directed lines, and thus a pair of points in C^*. The two points of every pair are reflections of each other in the origin. As representative of the undirected line the point is selected which is in the half-cylinder lying to one side of a fixed plane including the cylinder axis:

$$C = \{(\cos \alpha, \sin \alpha, p) : p \in \mathbb{R}, \alpha \in (0, \pi]\}.$$

Note that as the angle α passes through 0 so the representative point jumps from one edge of the half-cylinder to the other, while the value of the p-coordinate is multiplied by -1.

Symmetries of the representation space

The representations described above depend on the particular choice of origin o and x_1-axis for the plane. The stochastic geometry of line processes must take account of such arbitrary choices, since they correspond to important symmetries in the representations.

The *motion-symmetry group* on C^* (or C) is generated by those transformations of $A^*(2, 1)$ (or of $A(2, 1)$) that are induced by plane translations and rotations. The *translation-symmetry group* is composed of those transformations that are induced by plane translations alone. Point processes whose properties are invariant and measures in C^* (or C) which are invariant under these groups are described as being respectively *motion-symmetric* or *translation-symmetric*. Up to a constant multiple there is only one measure on the representation cylinder C^* which is invariant under the motion-symmetry group, and measures which are invariant under the translation-symmetry group must have a special form. Line processes arising in stochastic geometry typically have properties which are at least invariant under the translation-symmetry group.

Discussion of the symmetries entails study of the transformations induced on the representation cylinder C^* by translations and rotations of the plane. From the construction of C^* it follows that a rotation through the angle β about the origin of the plane induces a rotation through the same angle of C^* around the cylinder axis, $(p, \alpha) \mapsto (p, \alpha + \beta)$. The effect of translations is seen by considering a translation through distance s parallel to the x_1-axis. Directions, and thus α, are unchanged but the perpendicular distance p of a line ℓ from the origin is changed to $p + s \sin \alpha$. Thus a translation induces a *shear* on the cylinder C^*. More generally, consider the transformation $T_{(s,\gamma)}$ which corresponds to a translation through distance s in a direction that makes an angle γ with the x_1-axis. This will act on C^* by

$$T_{(s,\gamma)}(p, \alpha) = \left(p + s \sin(\alpha - \gamma), \alpha\right). \tag{8.1}$$

To see this, note that when r is a rotation about the origin through γ, then $rT_{(s,\gamma)}r^{-1}$ is a translation parallel to the x_1-axis.

So the symmetries of the rigid motions correspond to *rotations* of the cylinder and *vertical shears* along the cylinder's axis. Note that the rigid translational symmetries of C^* do not arise from the motion-symmetries of the plane underlying $A^*(2, 1)$.

Measures on the representation space

Line processes give rise to, and are controlled by, measures on the representation space C^*, and analogously on C. For example, Poisson directed-line processes can be defined as Poisson processes on C^*; so their distributions are defined by their intensity measures on C^*. Invariance properties holding for a line process lead to invariance properties for the corresponding measures, so the study of invariant measures is a necessary preliminary to the study of motion-invariant and stationary line processes.

Measures on C^* invariant under the translation-symmetry group induced by planar translations must be of a specific form. If invariant under the motion-symmetry group, they are unique up to a multiplicative factor.

Theorem 8.1. *(a) Suppose that μ is a locally finite measure on C^* which is invariant under the translation-symmetry group (plane-translation-invariant). Then μ is of the form*

$$\mu(dp \, d\alpha) = \nu_1(dp) \cdot \kappa(d\alpha) = dp \cdot \kappa(d\alpha) \tag{8.2}$$

for some finite measure κ on $(0, 2\pi]$, where ν_1 denotes as usual the one-dimensional Lebesgue measure.

(b) Suppose that μ is a locally finite measure on $\boldsymbol{C}^*$ which is invariant under the motion-symmetry group. Then μ is a constant multiple of Lebesgue measure on $\boldsymbol{C}^*$

$$\mu(\mathrm{d}p\,\mathrm{d}\alpha) = m\,\mathrm{d}p\,\mathrm{d}\alpha \tag{8.3}$$

for some constant m with $0 \le m < \infty$. If $m = 1$ then μ is the surface measure on $\boldsymbol{C}^*$ (regarding $\boldsymbol{C}^*$ as a cylinder in $\mathbb{R}^3$).

The measure with $m = 1$ is called *standard invariant measure.*

Theorem 8.1 can be proved by using differential forms (see Santaló, 1976). The famous encyclopaedia article by Crofton (1885) gives a heuristic argument which can be made precise as follows.

Proof. Let $T_{(s,\gamma)}$ be the translation through a distance s in a direction γ. For given α and β in $(0, 2\pi]$ consider the measure defined on intervals of $\mathbb{R}$ by

$$[a, b] \mapsto \mu([a, b] \times [\alpha, \beta]).$$

By translation-invariance of μ

$$\mu([a, b] \times [\alpha, \beta]) = \mu\big(T_{(s,\gamma)}([a, b] \times [\alpha, \beta])\big)$$

and by an approximation argument, it follows that

$$\mu([a, b] \times [\alpha, \beta]) = \int_\alpha^\beta \mu\big(T_{(s,\gamma)}([a, b] \times \mathrm{d}\gamma)\big)$$

$$= \int_\alpha^\beta \mu([s+a, s+b] \times \mathrm{d}\gamma)$$

$$= \mu([s+a, s+b] \times [\alpha, \beta]).$$

So μ is invariant under vertical shifts of the cylinder $\boldsymbol{C}^*$.

Formula (8.2) follows from general results on the uniqueness of translation-invariant measures on $\mathbb{R}$, while (8.3) follows from the implications of rotational invariance on the circle which is the base of $\boldsymbol{C}^*$. $\qquad\square$

The invariant measures for $A(2, 1)$ can be derived (via the $2 : 1$ correspondence) from those for $A^*(2, 1)$. Formulae (8.2) and (8.3) remain valid except that κ is now a measure on $(0, \pi]$.

Geometric arguments can also be used to derive formulae for the invariant measure of hitting sets. Define the *hitting set* of a compact subset K by

$$\mathcal{L}_K^* = \{\ell \in A^*(2, 1) : \ell \cap K \text{ is not empty}\}.$$

Theorem 8.2. *If μ is a motion-symmetric measure on C^*, then*

$$\mu(\mathcal{L}_K^*) = 2mL(K) \qquad \text{for planar compact convex sets } K, \tag{8.4}$$

where $L(K)$ is the perimeter of K and m is the constant factor in (8.3).

Sketch of proof. If S is a line segment, then invariance considerations show that $\mu(\mathcal{L}_S^*)$ is proportional to the length of S, and does not depend on the orientation nor on the location of S. If K is a convex polygon of sides $S_1, \ldots, S_n$, then any ℓ in $\mathcal{L}_K^*$ intersects exactly 2 of the S_i, unless ℓ intersects the boundary of K at a vertex. Thus, using indicator functions,

$$2\mathbf{1}_{\mathcal{L}_K^*} \equiv \sum_i \mathbf{1}_{\mathcal{L}_{S_i}^*},$$

where $\equiv$ denotes equality of the indicator functions except on a set that is the finite union of hitting sets of points. Points p are segments of zero length and so $\mu(\mathcal{L}_p^*) = 0$. Thus

$$2\mu(\mathcal{L}_K^*) = \sum_i \mu(\mathcal{L}_{S_i}^*) = \sum_i s_i \mu_1,$$

where s_i is the length of S_i and $\mu_1 = \mu(\mathcal{L}_{S_0}^*)$, in which S_0 is any segment of unit length. Integration shows that $\mu_1 = 2m$. A limiting argument yields Equation (8.4) for all convex compact K.

If K is the unit disc $B(o, 1)$, then

$$\mu(\mathcal{L}_{B(o,1)}^*) = 4\pi m.$$

An analogue of Theorem 8.2 holds for undirected lines, but $\mu(\mathcal{L}_K^*)$ has to be replaced by $\mu(\mathcal{L}_K)$, the set of all undirected lines hitting K, and the factor m by another constant, representing the ratio of the motion-symmetric μ to the Lebesgue measure on C. The theorem for the undirected case can be generalised to higher-dimensional cases, where the lines are replaced by flats to obtain 'flat processes', see Schneider and Weil (2008, around Equation (4.27)).

One can also consider the evaluation of invariant μ on the hitting set of a finite union of line segments (the *Buffon–Sylvester problem*) and on general hitting sets of finite unions of convex bodies. Ambartzumian (1982) (note also there the appendix by A. Baddeley) described the full solution of this problem and generalisations.

Line processes as point processes on the representation space

A *line process* is a random collection of lines in the plane which is *locally finite*; only finitely many lines hit each compact planar set. It is convenient to consider directed line processes: results transfer easily to the undirected case. Formally, a directed line process is defined as a random collection of points in the representation space C^*. The line process is locally finite exactly when the representing random collection is locally finite, hence a point process, on C^*. Such point processes are particular cases of point processes on $\mathbb{R}^2$ because, as suggested by the parametrisation (p, α) of C^*, the cylinder can be cut and embedded as the subset $\mathbb{R} \times (0, 2\pi]$ of $\mathbb{R}^2$. So the theory of line processes can be regarded as a special case of the theory of planar point processes as described in Chapters 2, 4 and 5.

The definitions of *stationary* and *motion-invariant* line processes are entirely analogous to those of stationary and motion-invariant point processes. The difference lies in the systematic replacement of the usual translation and motion groups in $\mathbb{R}^2$ by the translation-symmetry or motion-symmetry groups of C^* (or C). Thus a line process $\Phi = \{\ell_1, \ell_2, \ldots\}$ is *stationary* if $\Phi_T = \{T\ell_1, T\ell_2, \ldots\}$ has the same distribution (considered as a line process) for every translation T of the plane, and this is to say that in the C^* representation the point process

$$\left\{ \left(p(\ell_1) + s \cdot \sin(\alpha(\ell_1) + \gamma), \alpha(\ell_1) \right), \left(p(\ell_2) + s \cdot \sin(\alpha(\ell_2) + \gamma), \alpha(\ell_2) \right), \ldots \right\}$$

has the same point process distribution as

$$\left\{ \left(p(\ell_1), \alpha(\ell_1) \right), \left(p(\ell_2), \alpha(\ell_2) \right), \ldots \right\} \tag{8.5}$$

for each s in $\mathbb{R}$ and γ in $(0, 2\pi]$.

Motion-invariant line processes have in addition the property that

$$\left\{ \left(p(\ell_1), \alpha(\ell_1) + \gamma \right), \left(p(\ell_2), \alpha(\ell_2) + \gamma \right), \ldots \right\}$$

also has the same point process distribution as (8.5) for each γ in $(0, 2\pi]$, where the additions of angles $\alpha(\ell_k) + \gamma$ are interpreted modulo 2π.

The intensity measure and line density

A directed line process Φ, when regarded as a point process Φ_{C^*} on C^*, yields an *intensity measure* Λ_{C^*} on C^*:

$$\Lambda_{C^*}(A) = \mathbf{E}\left(\#\{(p, \alpha) \in \Phi_{C^*} \cap A\} \right) \qquad \text{for all Borel sets } A \text{ of } C^*. \tag{8.6}$$

If Φ is stationary then Λ_{C^*} is translation-symmetric, and if Λ_{C^*} is locally finite then Formula (8.2) can be applied to obtain

$$\Lambda_{C^*}\left(d(p, \alpha) \right) = \lambda\, dp \cdot \mathcal{R}(d\alpha), \tag{8.7}$$

where λ is a constant, the interpretation of which will become clear below by Formula (8.11), and $\mathcal{R}$ is a probability measure on $(0, 2\pi]$, called the *rose of directions* of Φ. The rose of directions can be interpreted as the distribution of direction of the typical (or, so to speak, a randomly selected) line of Φ.

In the case of motion-invariance, Formula (8.3) can be applied to obtain

$$\Lambda_{C^*}\left(d(p, \alpha) \right) = \lambda\, dp \cdot \frac{d\alpha}{2\pi}. \tag{8.8}$$

The constant $\lambda/(2\pi)$ is then the intensity of the representing point process Φ_{C^*} with respect to the standard invariant measure $dp \cdot d\alpha$.

Now the meaning of the λ in (8.7) and (8.8) is explained as follows. If the line process Φ is stationary, then invariance arguments can be applied to the line length measure

$$\Lambda(B) = \mathbf{E}\left(\sum_{\ell \in \Phi} v_{1,\ell}(\ell \cap B) \right) \qquad \text{for all planar Borel sets } B, \tag{8.9}$$

where $v_{1,\ell}$ is the Lebesgue measure on the line ℓ. That is to say, $\Lambda(B)$ is the mean total length of all line pieces of Φ in B. Because Λ is a translation-invariant measure on $\mathbb{R}^2$, there is a constant L_A such that

$$\Lambda(B) = L_A \, \nu_2(B) \qquad \text{for all planar Borel sets } B. \tag{8.10}$$

Thus, L_A is the mean line length per unit area.

The relationship of L_A to λ can be determined by inserting in Formula (8.10) a special set B, namely, $B = B(o, 1)$, which yields, using the Campbell theorem,

$$\pi L_A = \Lambda\big(B(o, 1)\big) = \mathbf{E}\left(\sum_{\ell \in \Phi} \nu_{1,\ell}\big(\ell \cap B(o, 1)\big)\right) = \mathbf{E}\left(\sum_{(p,\alpha) \in \Phi_{C^*} : |p| \leq 1} 2\sqrt{1 - p^2}\right)$$

$$= 2 \int_{(0,2\pi]} \int_{[-1,1]} \sqrt{1 - p^2} \, \Lambda_{C^*}\big(\mathrm{d}(p, \alpha)\big) = 2\lambda \int_{-1}^{1} \sqrt{1 - p^2} \, \mathrm{d}p = \lambda\pi,$$

since the length of the chord in $B(o, 1)$ of a line of distance p from o is $2\sqrt{1 - p^2}$. Thus

$$\lambda = L_A \tag{8.11}$$

is obtained.

Undirected line processes satisfy similar formulae, but subject to replacement of $\mathrm{d}p \cdot \mathrm{d}\alpha / (2\pi)$ in Formula (8.8) by $\mathrm{d}p \cdot \mathrm{d}\alpha / \pi$, and conversion of the rose of directions $\mathcal{R}$ into a distribution on $(0, \pi]$. Also here Equation (8.11) holds.

Since line processes can be regarded as point processes on the representation space, the definition of second and higher *moment measures* is straightforward. These concepts, though playing an important rôle in the general theory of line processes (see e.g. Ambartzumian, 1990, pp. 256–61, and Daley and Vere-Jones, 2008, pp. 477–82), will not be discussed further here.

Poisson line processes

A Poisson line process is the line process produced by a Poisson process on C^*. Consequently it is characterised completely by its intensity measure Λ_{C^*}. Thus stationary Poisson line processes are characterised by the line density L_A and the rose of directions $\mathcal{R}$. Moreover, stationary Poisson line processes are motion-invariant precisely when the rose of directions is the uniform probability measure on $(0, 2\pi]$.

Poisson line processes provide examples of stationary line processes which are free of parallel lines, since a Poisson point process on C^* with intensity $L_A \, \mathrm{d}p \cdot \mathcal{R}(\mathrm{d}\alpha)$ can have no more than one point per vertical line when the rose of directions $\mathcal{R}$ is diffuse. All this remains true for the case of undirected lines except that $\mathcal{R}$ is then a distribution on $(0, \pi]$.

Example 8.2. *The motion-invariant Poisson line process*
Let Φ be a motion-invariant directed Poisson line process of line density L_A. Then the number of lines of Φ hitting a convex body K is of Poisson distribution of mean

$$\mu_K = \frac{L_A}{\pi} \cdot L(K), \tag{8.12}$$

where $L(K)$ is the perimeter of K.

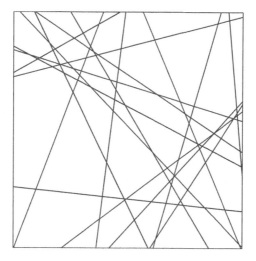

Figure 8.7   A simulated realisation of a Poisson line process of intensity $L_A = 15$ in $[0, 1]^2$.

The Poisson distribution is established by noting that the number of lines hitting K is precisely the number of representing points lying in a certain subset of C^*. The value for the mean is derived by applying Theorem 8.2 to the intensity measure of Φ_{C^*}, which is $\Lambda_{C^*}(\mathrm{d}(p, \alpha)) = L_A \, \mathrm{d}p \cdot \mathrm{d}\alpha/(2\pi)$.

For undirected lines Formula (8.12) is also true.

The *simulation* of a motion-invariant undirected Poisson process of line density L_A usually consists in determining random lines that hit some convex window W. The number of lines is obtained by sampling a Poisson variable with mean $\mu_W = L_A \cdot L(W)/\pi$ and the lines then are obtained as points uniformly distributed in the counterpart of W in C^*. One can also take a disc $B(o, r)$ containing W, simulate random lines hitting $B(o, r)$ and reject those lines that do not hit W.

A simulated realisation of a motion-invariant Poisson line process is shown in Figure 8.7.

Cox line processes

Cox line processes are line processes produced by Cox processes on C^*. They are characterised by their driving random measures Λ_{Cox} as in Section 5.2. Conditional on the realisation Λ of the random measure Λ_{Cox}, the point process Φ_{C^*} in C^* is Poisson of intensity measure Λ. If the driving random measure is translation-symmetric, then so is the Cox line process. Thus a complete characterisation of stationary Cox line processes depends on the analysis of translation-symmetric random measures on C^*.

It is a remarkable fact that if a translation-symmetric random measure Λ_{Cox} on C^* has locally finite second moment measure, and gives zero measure to the set of lines parallel to the cylinder axis, then it must factorise into the form

$$\Lambda_{\mathrm{Cox}}(\mathrm{d}(p, \alpha)) = \mathrm{d}p \cdot \Lambda^{\Theta}(\mathrm{d}\alpha),$$

where Λ^Θ is a random measure on $(0, 2\pi]$. This was observed by Davidson (1974a) for translation-symmetric random measures with second moment density. The density condition was lifted by Krickeberg (1972, 1974) using disintegration of the second moment measure.

This factorisation implies that (under suitable regularity conditions) stationary Cox line processes must be mixtures of stationary Poisson line processes; in effect the line density L_A and the rose of directions $\mathscr{R}$ are randomised. A motion-invariant Cox line process must satisfy the further condition that the random measure Λ^Θ on $(0, 2\pi]$ is invariant under translation modulo 2π.

The central position occupied by mixed Poisson line processes is also indicated by a class of results concerning intersections of the lines of a Cox line process. These intersections form a point process that is stationary if the line process is stationary. The intensity of the intersection process (suitably normalised) is maximised by the isotropic mixed Poisson line processes; see Davidson (1974b).

Davidson's conjecture

The problem posed by Davidson (1974a) was whether or not there existed motion-invariant line processes of locally finite second moment measure which were neither Cox nor contained parallel lines. (Note that a motion-invariant line process has either no or infinitely many pairs of parallel lines or antiparallel lines; see Davidson, 1974c.) Davidson conjectured that there were no such line processes. Indeed, the geometrical properties of C^* place strong restrictions on the characteristics of stationary line processes. Under suitable regularity conditions these restrictions imply the Cox property. However, Kallenberg (1977) showed that in general there *do* exist counterexamples to Davidson's conjecture.

Example 8.3. *A stationary line process with no parallel lines which is not Cox*
The description of this line process is best given using an alternative coordinatisation to the cylindrical parametrisation. Each line ℓ in the plane is parametrised by a pair (a, b) where $a(\ell) = \cot \alpha(\ell)$ (using $\alpha(\ell)$ as in the cylindrical representation), and $b(\ell)$ is the signed distance between the origin and the intercept of ℓ on the x_1-axis. So the representation space is a plane. This coordinatisation breaks down if ℓ is parallel to the x_1-axis. However, in this example that eventuality is of probability zero.

Translations $T_{(s,\gamma)}$ of $\mathbb{R}^2$ correspond to linear shears of the representation plane:

$$T_{(s,\gamma)}(a, b) = \big(a, \ b - s(a \sin \gamma - \cos \gamma)\big)$$

and so translation-symmetries will send a linear lattice on the representation plane into another linear lattice.

Thus the corresponding class of patterns of lines (patterns that correspond under the above planar parametrisation to lattices of points) will be left invariant under the group of translations. Kallenberg constructed a probability measure on the space of lattices which is invariant under the action of the group of linear shears and hence is translation-symmetric. This example suffers the imperfection that a lattice on (a, b)-space corresponds to a line pattern that is not locally finite (there are infinitely many lines of arbitrarily small angle γ hitting each compact set). However, this imperfection is removed by a thinning depending only on the angular parameter, for example by deleting all points of the lattice lying outside a region in (a, b)-space bounded by two lines of the form $a = $ constant. The resulting process has a finite second moment measure.

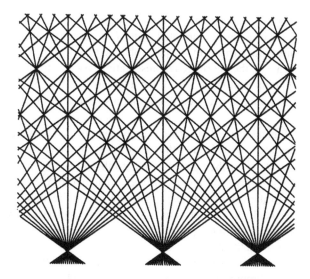

Figure 8.8 A simulated realisation of a Kallenberg lattice process. This is clearly not a Poisson line process: many intersections involve more than two lines and the pattern is very regular.

A simulation of the resulting lattice process is displayed in Figure 8.8. It is clearly not a Cox process; the lattice structure means that there is positive probability of observing more than two lines meeting at some intersections. This failure to have the *bundle-free* property follows immediately from the observation that lines meeting in a common point in the (a, b)-representation correspond exactly to collinear representing points. One might hope that if a stationary line process is both free of parallel lines and bundle-free, then it must also be Cox. Kallenberg (personal communication) has pointed out that this cannot be so; if his example is modified by independent parallel translations of the lines over random distances which are uniform on $[0, \varepsilon]$, then no bundles can occur. The modification cannot be Cox for small $\varepsilon > 0$, since this would not be compatible with the influence of the underlying lattice structure on the point representation.

The Kallenberg lattice-type line process is completely defined by a fundamental cell of the lattice in the (a, b)-space representation; Sukiasian (1987) took parabolic inversion of the lattice to construct further examples. Mecke (1979) constructed the process using only five real-valued random variables.

8.2.3 Spatial line and plane processes

The theory presented in the preceding section can be generalised. Of most practical interest are the cases of planes and lines in $\mathbb{R}^3$.

Parametric representations in $\mathbb{R}^3$

Planes in $\mathbb{R}^3$: If the planes are oriented, then they may be represented by points $(x, \boldsymbol{u})$ lying in a generalised cylinder $\mathbb{R} \times S^2$. Consider an oriented plane P. Let $\boldsymbol{u} \in S^2$ be the unit vector normal to P defined by its orientation. Then the point on P closest to the origin is given by $x\boldsymbol{u}$ for some scalar x in $\mathbb{R}$. So the oriented plane P is parametrised by $(x, \boldsymbol{u}) \in \mathbb{R} \times S^2$. Unoriented planes are parametrised by a space obtained by identifying points $(x, \boldsymbol{u})$ and $(-x, -\boldsymbol{u})$. Away from $x = 0$ the unoriented representation space is given by $(0, \infty) \times S^2$, but there are topological complications at $x = 0$. An alternative is $\mathbb{R} \times S^2_+$, where S^2_+ consists of the points of S^2 with nonnegative x_3-coordinate. See also Baddeley and Jensen (2005, p. 92).

Lines in $\mathbb{R}^3$: For topological reasons, this case is more complicated than the cases considered until now. If the lines are oriented, then they may be represented by points $\big(x, (\boldsymbol{u}, \boldsymbol{v})\big) \in [0, \infty) \times T(S^2)$, where $T(S^2)$ is the tangent bundle to the sphere S^2, namely the collection of pairs $(\boldsymbol{u}, \boldsymbol{v})$, in which $\boldsymbol{u} \in S^2$ and $\boldsymbol{v}$ is a unit vector tangent to S^2 at $\boldsymbol{u}$. The point on the line which is closest to the origin is $(x, \boldsymbol{u})$, while the line's direction is $\boldsymbol{v}$. However, this representation behaves badly at $x = 0$, where $T(S^2)$ has to be replaced by S^2. If the lines are unoriented, then again pairs of points $\big(x, (\boldsymbol{u}, \boldsymbol{v})\big)$ and $\big(x, (\boldsymbol{u}, -\boldsymbol{v})\big)$ have to be identified. See also Baddeley and Jensen (2005, p. 95).

Again there is an alternative parametrisation, sometimes called the phase representation. Fix a reference plane, and represent each line by $\big((x, y), \boldsymbol{u}\big)$, where (x, y) is the intersection point of the line on the reference plane and $\boldsymbol{u}$ is the (three-dimensional) direction of the line. As before, this representation breaks down for lines parallel to the reference plane.

Decomposition of invariant measures

In each case of the above representations, expressions may be obtained for the general form of measures which are invariant under translations or under rigid motions of $\mathbb{R}^3$, analogous to Theorem 8.1; see for example Santaló (1976, II.12.7, Note 5) or Ambartzumian (1990).

Poisson line and plane processes

Poisson line processes and *Poisson plane processes* in space can now be defined simply as Poisson point processes on the corresponding representation spaces, with distributions determined by their intensity measures. As in the case of line processes on the plane, stationary and motion-invariant processes in space can be defined by requiring the point processes on the representation spaces to be translation-symmetric and motion-symmetric with respect to the corresponding induced symmetry groups.

In the stationary case the intensity parameters of main interest are the *area* or *surface density* S_V and the *line density* L_V, representing the mean total area of all plane pieces intersecting a unit cube and the mean total length of all line pieces intersecting a unit cube, respectively. As in the case of lines in $\mathbb{R}^2$ the intensity of the point process in the representative space coincides with S_V and L_V, respectively.

In the Poisson process case, the random number of planes and lines hitting a compact set K has a Poisson distribution. In the motion-invariant case and for convex K the corresponding mean μ_K satisfies

$$\mu_K = S_V \, \overline{b}(K) \qquad \text{for plane process,} \tag{8.13}$$

and

$$\mu_K = \frac{L_V}{4} S(K) \qquad \text{for line process,} \tag{8.14}$$

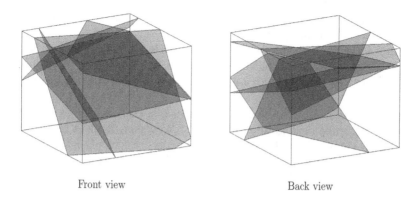

Front view   Back view

Figure 8.9 A simulated piece of a realisation of a Poisson plane process with five planes hitting a parallelepiped.

where $\bar{b}(K)$ and $S(K)$ are the average breadth and the surface area, respectively, of K. Note that there are no mutual intersections of the lines of a Poisson line process in $\mathbb{R}^3$. Figure 8.9 shows a sample of a Poisson plane process.

A hyperplane process in $\mathbb{R}^d$ generates *secondary networks* or *intersection processes* by intersections. The order-k secondary network is the process of flats each of which is the intersection of k hyperplanes from the original process. Thus under suitable regularity conditions (for example in the Poisson case) the order-d network is a point process in $\mathbb{R}^d$, while the order-1 network is the hyperplane process itself. If the hyperplane process is stationary, then so is the order-k network. Similar constructions can be given for a process of flats. The full theory for such networks is presented in Schneider and Weil (2008, Section 4.4).

Generalising work by Miles (1971a, 1974a) and Matheron (1975), Schneider and Weil (2008, Section 4.4) presented a general theory of flat processes without explicitly using representation spaces, which are, however, needed for example for simulations; see Lantuéjoul (2002).

Extending Davidson (1974b) and his treatment of the problem of maximising the intersection rate for a planar line process, Janson and Kallenberg (1981) and Mecke (1983, 1984a) used Fourier methods to produce similar results for the secondary networks of Poisson hyperplane and flat processes. Thomas (1984) applied results from convex geometry concerning the Steiner compact (see Section 8.3.1) and a generalisation of the isoperimetric inequality. This work is continued in Mecke (1986, 1988a,b, 1991) and Mecke and Thomas (1986).

Ambartzumian (1982) generalised his solution of the Buffon–Sylvester problem to the case of higher dimensions. Kallenberg (1976a, 1980, 1981) discussed generalisations to higher dimensions suggested by the Davidson problem. Kallenberg has also applied his work to the study of systems of non-interacting particles.

8.2.4 Applications of line and plane processes

Poisson and Cox line processes can be used as simple mathematical models of random systems of very long fibres of weak curvature (for example, in paper or woven materials; see Lücke and Tittel, 1993) or of trajectories of moving particles. Such models are not intended as a correct description of the global behaviour of the real structures, but can provide a useful local description. Formulae for the point process of self-intersection points and for the induced polygons (see Section 9.5) are of interest. Kallenberg (1983b) discusses the approximation of segment processes by Cox line processes.

In another application, line processes serve as models for traffic networks; see Baccelli and Zuyev (1997), Aldous and Kendall (2008), Voss *et al.* (2009, 2010, 2011) and Kendall (2011). Fairclough and Davies (1990) use line processes in simulating the structure of fibre membranes.

Line and plane processes can be used as construction elements just as point processes are used for constructing random set models. Thus one obtains Boolean cylinder and plate models: the lines and planes are dilated to obtain cylinders or thick plates; see Section 6.6.2.

Finally, the planar Poisson line process and the Poisson plane process play an important rôle in the construction of random tessellations; see Chapter 9.

8.3 Planar fibre processes

8.3.1 Fundamentals

As mentioned above, planar fibre processes model random collections of curves in the plane $\mathbb{R}^2$. Figures 8.1 and 8.2 give examples of systems of curves or fibres that could be interpreted as samples of fibre processes. The discussion below is based on the papers of Mecke and Nagel (1980) and Mecke and Stoyan (1980a). Earlier references on fibre processes are Ambartzumian (1973, 1977), Ambartzumian and Ohanian (1975) and Cowan (1979). In these papers the fibres are usually segments and the processes are usually motion-invariant. The following treatment is more general, but theories of yet greater generality are possible.

Fibres and fibre systems

A fibre is a sufficiently smooth simple curve of finite length in the plane. More precisely, a fibre γ is a subset of $\mathbb{R}^2$ which is the image of a curve $\gamma(t) = \big(\gamma_1(t), \gamma_2(t)\big)$ such that

(i) $\gamma : [0, 1] \to \mathbb{R}^2$ is once continuously differentiable,

(ii) $|\gamma'(t)|^2 = |\gamma_1'(t)|^2 + |\gamma_2'(t)|^2 > 0$ for all t, and

(iii) the mapping γ is one-to-one, so that a fibre does not intersect itself.

Here is a deliberate ambiguity; γ also stands for the measure

$$\gamma(B) = h_1(\gamma \cap B) = \int_0^1 \mathbf{1}_B\big(\gamma(t)\big)\sqrt{\gamma_1'(t)^2 + \gamma_2'(t)^2}\, dt \qquad \text{for planar Borel sets } B,$$

in which the Hausdorff measure h_1 is used, so that $h_1(\gamma \cap B)$ is simply the length of the fibre γ in the set B. Note that the *measure* γ depends only on the curve $\gamma([0, 1])$ and not on the precise representation map γ.

The definition of a fibre system is a straightforward extension of the definition of a fibre. A *fibre system* ϕ is a closed subset of $\mathbb{R}^2$ which can be represented as a union of countably many fibres $\gamma^{(i)}$, with the property that any compact set is intersected by only a finite number of the fibres, and such that distinct fibres have either nothing or only end-points in common:

$$\gamma^{(i)}\big((0, 1)\big) \cap \gamma^{(j)}\big((0, 1)\big) = \emptyset \qquad \text{whenever } i \neq j.$$

The length measure corresponding to the fibre system ϕ is then defined in terms of the measures $\gamma^{(i)}$ by

$$\phi(B) = \sum_{\gamma^{(i)} \in \phi} \gamma^{(i)}(B) \qquad \text{for Borel sets } B.$$

Despite the condition on end-points, ϕ actually may consist of intersecting curves, since fibres can end at intersection points. All the definition requires is that such curves can be dissected into a locally finite collection of fibres as above; of course, such a representation of ϕ as a union of fibres is not unique. The condition prevents fibre systems from having locally dense accumulations of self-intersection points (because of the condition of local finiteness). The local finiteness and smoothness conditions ensure that the measure ϕ is locally finite; it depends only on the closed set that is the union of all the fibres.

The family of all planar fibre systems is denoted by $\mathbb{D}$ and is endowed with a σ-algebra $\mathscr{D}$ generated by sets of the form

$$\{\phi \in \mathbb{D} : \phi(B) < x\}$$

for planar Borel sets B and positive numbers x. It can be shown that $\mathscr{D}$ is the same as the trace $\mathcal{F}_{\mathbb{D}}$ of $\mathcal{F}$ on $\mathbb{D}$, where $\mathcal{F}$ is Matheron's hitting σ-algebra for the family of all closed subsets of $\mathbb{R}^2$ as in Section 6.1.2.

Fibre processes

A (planar) *fibre process* Φ is a random variable taking values in $[\mathbb{D}, \mathscr{D}]$, that is to say, a measurable mapping from an underlying probability space $[\Omega, \mathcal{A}, \mathbf{P}]$ to $[\mathbb{D}, \mathscr{D}]$. The same symbol Φ is also used to denote the corresponding *random length measure* as well as the union set of all fibres. The theory of fibre processes is thus a special part of the theory of random measures, as well of the theory of random sets.

The *distribution* of the fibre process is the measure P generated on $[\mathbb{D}, \mathscr{D}]$ by Φ.

Stationarity and isotropy

The fibre process Φ is said to be *stationary* if it has the same distribution as the translated fibre process Φ_x for all x in $\mathbb{R}^2$. Thus

$$P(Y) = P(Y_x) \qquad \text{for all } Y \in \mathscr{D} \text{ and all } x \in \mathbb{R}^2,$$

where, as in other chapters, $Y_x = \{\phi \in \mathbb{D} : \phi_{-x} \in Y\}$. Likewise, it is *isotropic* if the distribution remains invariant under rotations about the origin. (These definitions are special cases of the general definitions for random measures or random sets.)

Intensity and the rose of directions

The *intensity measure* Λ of a fibre process is given by

$$\Lambda(B) = \mathbf{E}\big(\Phi(B)\big) = \mathbf{E}\left(\sum_{\gamma \in \Phi} h_1(\gamma \cap B)\right) \qquad \text{for Borel sets } B \text{ of } \mathbb{R}^2, \qquad (8.15)$$

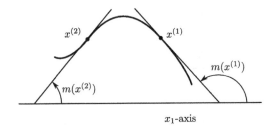

x_1-axis

Figure 8.10   Definition of the tangent direction $m(x)$.

which is the mean length of all fibre pieces in the set B. If Φ is stationary, then invariance considerations show that there exists a constant L_A ($0 \leq L_A \leq \infty$), the *intensity* or *line density* of Φ, such that

$$\Lambda(B) = L_A \cdot \nu_2(B) \qquad \text{for Borel sets } B \text{ of } \mathbb{R}^2, \tag{8.16}$$

where ν_2 is the planar Lebesgue measure. The notation L_A stands for 'mean line length per unit area'. In fact L_A is precisely the intensity of Φ regarded as a stationary random measure. In the following L_A is assumed to be positive and finite.

The *rose of directions* arises from consideration of fibre tangents. Let $m(x)$ denote the direction of the fibre tangent at x, presuming that precisely one fibre of Φ passes through x. Then $m(x)$ is some number between 0 and π; see Figure 8.10. In the case when more than one fibre passes through x, or an end-point occurs at x, let $m(x)$ equal π; such points form a subset of h_1-measure zero by the conditions on fibre systems. Denote by $\mathcal{B}^2$ and Π the families of Borel sets of $\mathbb{R}^d$ and $(0, \pi]$, respectively, then a marked (or weighted) random measure Ψ can be defined on $\mathcal{B}^2 \times \Pi$ by

$$\Psi(B \times L) = \int_B \mathbf{1}_L \big(m(x)\big)\Phi(\mathrm{d}x) \qquad \text{for } B \in \mathcal{B}^2 \text{ and } L \in \Pi. \tag{8.17}$$

Thus $\Psi(B \times L)$ is the total length of all fibre pieces of Φ in B that have tangent directions lying in L.

If Λ_Ψ is the intensity measure of Ψ, so that

$$\Lambda_\Psi(B \times L) = \mathbf{E}\big(\Psi(B \times L)\big), \tag{8.18}$$

and if Φ is stationary, then invariance considerations dictate that Λ_Ψ can be decomposed as

$$\Lambda_\Psi(B \times L) = L_A \cdot \nu_2(B)\mathcal{R}(L), \tag{8.19}$$

where $\mathcal{R}$ is a distribution on Π, called the *rose of directions* of the fibre process Φ. Reference to Section 7.1.4 shows that this is a mark distribution. In analogy to similar constructions in the theory of point processes, $\mathcal{R}$ can be interpreted as the distribution of the direction of the tangent to a fibre at the typical fibre point. (The term 'typical fibre point' is related to the Palm distribution of the random measure associated with the fibre process; see Chapter 7.) Figure 8.2 includes an example of an empirical rose of directions.

This analogy carries further: a *Campbell theorem* also holds for stationary fibre processes, which is stated as follows.

Theorem 8.3. *Suppose Φ is a stationary fibre process with distribution P, intensity L_A and rose of directions $\mathscr{R}$. For any measurable $f : \mathbb{R}^2 \times (0, \pi] \to [0, \infty)$,*

$$\mathbf{E}\left(\int f\big(x, m(x)\big)\Phi(\mathrm{d}x)\right) = \int_{\mathbb{D}}\int_{\mathbb{R}^2} f\big(x, m(x)\big)\phi(\mathrm{d}x)P(\mathrm{d}\phi)$$

$$= L_A \int_{\mathbb{R}^2}\int_{(0,\pi]} f(x, \alpha)\mathscr{R}(\mathrm{d}\alpha)\,\mathrm{d}x. \qquad (8.20)$$

By definition neither L_A nor $\mathscr{R}$ changes if the underlying fibre process is translated. Furthermore L_A does not vary if the fibre process is rotated; but, of course, $\mathscr{R}$ will then change. If r is an anticlockwise rotation about o through an angle θ, then the rotated fibre process has rose of directions $r(\mathscr{R})$ where

$$r(\mathscr{R})(L) = \mathscr{R}\left(r^{-1}(L)\right) \qquad \text{for all Borel sets } L \text{ of } (0, \pi].$$

The inverse rotation r^{-1} is a clockwise rotation through θ.

Finite measures on $(0, \pi]$ can be represented in terms of their *Steiner compacts S*, first used for this purpose by Matheron (1975). (Schneider and Weil, 2008, use the term 'associated zonoid'.) These are centrally symmetric compact convex sets. For a stationary fibre process Φ the modified support function $s_m(S, \beta)$ of the Steiner compact S is given by

$$s_m(S, \beta) = \frac{1}{2}L_A \int_{(0,\pi]} |\sin(\alpha - \beta)|\mathscr{R}(\mathrm{d}\alpha) \qquad \text{for } 0 < \beta \leq \pi. \qquad (8.21)$$

This concept aids the presentation of the important Formula (8.42). The notion of the modified support function for a symmetric convex set is described in Section 1.6. The rose of directions of the boundary ∂S of S is equal to the rose of directions $\mathscr{R}$ of Φ if rotated by $\pi/2$.

If the rose of directions $\mathscr{R}$ is the uniform distribution on $(0, \pi]$, then the resulting Steiner compact S is the ball

$$S = B\left(o, \frac{L_A}{\pi}\right). \qquad (8.22)$$

By construction, the Steiner compact is homogeneous of degree -1 with respect to dilations of the process, and thus its radius, if S is a disc, has dimension length^{-1}.

The Steiner compact is an auxiliary convex body used in the study of intersections of fibre processes with lines; see Section 8.3.2. It is also a valuable tool in statistics. Two further relationships with stationary fibre processes are:

(1) if S is the Steiner compact of Φ, then $L_A = L(S)/2$, where $L(S)$ denotes the perimeter of S;

(2) if Φ_1 and Φ_2 are independent stationary fibre processes with Steiner compacts S_1 and S_2 respectively, then the Steiner compact of the union $\Phi_1 \cup \Phi_2$ is $S = S_1 \oplus S_2$.

Instead of tangent directions, fibre normals also can be considered. Sometimes it makes sense to consider the normals as directed, for example when the fibres are boundary lines of random sets and outer (outward-pointing) normals can be defined. Then one speaks about *oriented direction distributions*; see Rataj (1996).

The second moment measure and the reduced second moment measure

Moment measures for fibre processes can be defined as for general random measures. In particular the *second moment measure* $\mu^{(2)}$ of Φ is given by

$$\mu^{(2)}(B_1 \times B_2) = \mathbf{E}\big(\Phi(B_1)\Phi(B_2)\big) \qquad \text{for } B_1, B_2 \in \mathcal{B}^2. \tag{8.23}$$

In the stationary case this measure can be decomposed just as for random measures in Section 7.2.2. Given $\mu^{(2)}$ for a stationary fibre process a measure $\mathcal{K}$ can be defined on the plane by

$$\mu^{(2)}(B_1 \times B_2) = L_A^2 \int \int \mathbf{1}_{B_1}(x)\mathbf{1}_{B_2}(x+h)\,\mathrm{d}x\,\mathcal{K}(\mathrm{d}h) \qquad \text{for } B_1, B_2 \in \mathcal{B}^2. \tag{8.24}$$

This $\mathcal{K}$ is the *reduced second moment measure* of Φ. Its interpretation follows the point process case: if x is the typical fibre point of Φ, then $L_A \mathcal{K}(B)$ is the mean total length of all fibre pieces of Φ in B_x. This is further discussed in Stoyan (1981), Schwandtke (1988), Jensen *et al.* (1990a,b) and Zähle (1990) for the isotropic case (and also for more general, not necessarily fibrous structures) and in Beneš (1994) for the anisotropic case.

In the case of stationarity and isotropy it is sufficient to consider the *reduced second moment function* $K(r)$ given by

$$K(r) = \mathcal{K}\big(B(o, r)\big) \qquad \text{for } r \geq 0.$$

If $K(r)$ is differentiable then, just as in the point process case, one can consider the *pair correlation function* $g(r)$ given by

$$g(r) = \frac{1}{2\pi r}\frac{\mathrm{d}K(r)}{\mathrm{d}r} \qquad \text{for } r \geq 0. \tag{8.25}$$

Example 8.4. *Basic characteristics of two simple planar fibre processes*
(a) Let Φ be a motion-invariant Poisson line process of intensity L_A. Its reduced second moment function $K(r)$ is given by

$$L_A K(r) = 2r + L_A \pi r^2 \qquad \text{for } r \geq 0. \tag{8.26}$$

The pair correlation function is therefore

$$g(r) = 1 + \frac{1}{L_A \pi r} \qquad \text{for } r \geq 0. \tag{8.27}$$

To make a heuristic argument for Formula (8.26), note that $L_A K(r)$ has two components: one is the contribution from the line containing the typical point and the other is formed from the remainder of the process. Clearly the first component is $2r$. By the Slivnyak–Mecke theorem the remainder of the process has a distribution equal to that of the original process and thus it yields the contribution $L_A v_2\big(B(o, r)\big)$.

(b) Let Φ be a Boolean segment process as in Example 8.1 on p. 298. If the length of the typical segment has distribution function $L(x)$ with mean $\bar{l}$, then the intensity of Φ is $L_A = \lambda \bar{l}$, where λ is the intensity of the germ process. The reduced second moment function $K(r)$ is

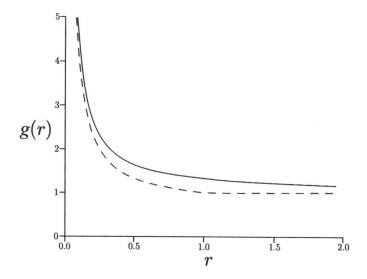

Figure 8.11 Pair correlation function $g(r)$ of a Poisson line process (—) with unit intensity and of a Boolean segment process (- - -) with segments of constant unit length and unit intensity. Note that the curve for the segment process is horizontal for $r \geq 1$.

given by

$$L_A K(r) = L_A \pi r^2 + \frac{1}{l}\left(\int_0^r x^2 \, dL(x) + \int_r^\infty (2xr - r^2) \, dL(x) \right) \qquad \text{for } r \geq 0, \quad (8.28)$$

as in Stoyan (1983). Again the Slivnyak–Mecke theorem explains this; this time the second term arises from the segment on which the typical point lies. The distribution of the length of the typical segment is length-biased and the typical point lies uniformly at random on this segment.

Figure 8.11 shows the pair correlation functions for the two fibre process models in Example 8.4. As r tends to infinity, they tend to 1, but for $r \downarrow 0$ they diverge to infinity. This divergence is typical for fibre processes. It is explained by noting that even in very small discs about the typical fibre point there must still be pieces of at least one fibre. Thus dK/dr does not become small for small r.

The corresponding quantity for point processes uses the second-order *factorial* moment measure, deleting the typical point itself. A similar approach for fibre processes using, for example, $K_0(r) = K(r) - 2r/L_A$ does not in general dispose of the infinity at zero. Note also that even in the point process case there are examples with pair correlation function having pole at zero; see the discussion in Stoyan (1994).

Weiss *et al.* (2010) gave formulae for the pair correlation and K-functions of the systems of edges of some motion-invariant tessellations, the so-called STIT tessellations; see p. 354. They showed that these second-order characteristics are identical to these of a closely related Boolean segment process.

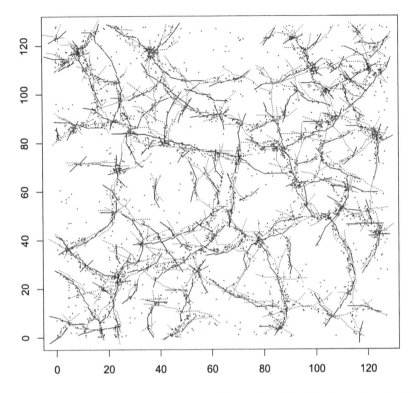

Figure 8.12   Three (—, — and ---), superposed, samples of the Candy model, simulated by a Metropolis–Hastings algorithm with three different sets of model parameters. The aim is to describe filaments of galaxies, which are shown as points, and the unit of measurement is h^{-1} Mpc. The three sets of parameters give almost equally good results. See Stoica *et al.* (2005). Courtesy of R. Stoica.

Candy model

The paper of van Lieshout and Stoica (2003) studies the properties of the so-called *Candy model*. (Its name comes from dubbing a segment as 'candy'.) This is a (planar) finite fibre process with a Gibbs distribution, where the energy function discourages free and singly connected segments as well as sharp crossings and disagreements in orientation of close segments. This model has its origin in statistics for networks of rivers and roads. Figure 8.12 shows simulated samples in an astronomical study.

Still more complicated Gibbs fibre processes have been studied in the context of polymers, based on theories of Flory (e.g. Flory, 1969) and de Gennes (e.g. de Gennes, 1979); see den Hollander (2009) and Caravenna *et al.* (2011). Hill *et al.* (2012) used rather degenerate fibre processes as part of an MCMC algorithm to reconstruct curvilinear structures from point patterns.

Undirected line processes regarded as fibre processes

In Example 8.4 above a Poisson line process is considered as a particular kind of fibre process. This is correct since every line can be considered as a union of line segments, which can be considered as fibres. The union of all lines can also be considered as a random set, and finally a random measure can be associated with the lines, namely, the line length measure as on p. 307 and the line density L_A considered on p. 308 is the line density in the fibre process interpretation.

If the direction $\alpha(l)$ of each line l in a line process Φ is taken into account, then a marked random measure can be introduced. The direction of an undirected line lies in $(0, \pi]$, so one can define

$$\Psi(B \times L) = \sum_{\{l \in \Phi : \alpha(l) \in L\}} h_1(B \cap l) \qquad \text{for } B \in \mathcal{B}^2 \text{ and } L \in \Pi, \tag{8.29}$$

where $\mathcal{B}^2$ and Π are the families of Borel sets of $\mathbb{R}^2$ and $(0, \pi]$, respectively. In the stationary case the intensity measure Λ of Ψ factorises as

$$\Lambda(B \times L) = L_A \cdot v_2(B)\mathcal{R}(L) \qquad \text{for } B \in \mathcal{B}^2 \text{ and } L \in \Pi, \tag{8.30}$$

where $\mathcal{R}$ is a probability measure on $(0, \pi]$, the rose of directions. This rose has a double interpretation: as the rose of directions in the fibre-process sense and as the distribution of the direction of the typical line l of Φ.

Contact distribution functions

Up to this point the exposition has concentrated on random-measure characteristics of fibre processes. However, random-set characteristics are also of interest. Of course, area fraction p and covariance $C(r)$ are simply equal to zero. In contrast, the contact distributions make sense and give distributional information that L_A and the rose of directions cannot give.

In the case of a motion-invariant Poisson line process with line density L_A, the contact distribution function $H_B(r)$ with respect to a convex compact structuring element B is

$$H_B(r) = 1 - \exp\left(-\frac{L_A}{\pi} L(B) r\right) \qquad \text{for } r \geq 0. \tag{8.31}$$

In particular, the spherical contact distribution function is

$$H_s(r) = 1 - \exp(-2L_A r) \qquad \text{for } r \geq 0. \tag{8.32}$$

For a Boolean segment process as in Example 8.4 above it is

$$H_s(r) = 1 - \exp\left(-\lambda r(2\bar{l} + \pi r)\right) \qquad \text{for } r \geq 0. \tag{8.33}$$

These formulae simply follow from the fact that the random number of lines or segments hitting $B(o, r)$ has a Poisson distribution, and $H_s(r)$ is just the complement of probability that no hit happens.

Last and Schassberger (2000) studied the Palm inversion formula of the fibre (surface) process Φ of the edges (cell boundaries) of a general stationary tessellation in $\mathbb{R}^2$ (in $\mathbb{R}^d$), associated with a random vector field that is smooth in $\mathbb{R}^d \setminus \Phi$ but discontinuous on Φ. A formula for certain generalised contact distribution functions of Φ was obtained.

8.3.2   Intersections of fibre processes

Intersections with lines

Important methods of measurement for fibre processes use enumeration of intersections with test systems of lines, circles, or other curves. Therefore it is helpful to derive formulae for the random point patterns which arise from such intersections and, more generally, from the intersections of fibre processes with other fibre processes.

Let Φ be a stationary fibre process with distribution P, intensity L_A, and rose of directions $\mathcal{R}$ satisfying $\mathcal{R}(\{\pi\}) < 1$. Consider the intersection of Φ with a fixed line e. To be definite e is taken to be the x_1-axis.

The process $\Phi \cap e$ is a point process on e and is marked (weighted) by the angles of intersection. This marked point process $\Psi = \{[y_n; m(y_n)]\}$ was first considered by Davidson (1974b), and later discussed in Ambartzumian (1974a, 1977, 1982) and Mecke and Stoyan (1980a). The rose $\mathcal{R}$ can be recovered from the distribution of Ψ, as can L_A if $\mathcal{R}(\{\pi\}) = 0$. As will be shown below, if the fibre process Φ is not just stationary but also isotropic, and if no fibres contain linear segments, then the reduced second moment function $K(r)$ can be determined from knowledge of the distribution of Ψ.

Let P_L be the intensity of Ψ and H its mark distribution on $(0, \pi]$. The mark distribution of Ψ is the distribution of the intersection angle at the typical point of the intersection point process $\Phi \cap e$.

The following formula connects the characteristics of Φ and Ψ:

$$P_L \int_{\mathbb{R}} \int_{(0,\pi]} h(z, \alpha) H(d\alpha) \, dz = L_A \int_{\mathbb{R}} \int_{(0,\pi]} h(z, \alpha) \sin \alpha \, \mathcal{R}(d\alpha) \, dz, \qquad (8.34)$$

where h is any nonnegative and measurable function on $\mathbb{R} \times (0, \pi]$. Formula (8.34) can be proved as follows.

Proof. Let ϕ be any fibre system, and let w be any nonnegative measurable function on $\mathbb{R}^2 \times (0, \pi]$. Furthermore, let $m(x)$ be the direction of the fibre tangent of Φ at x. Simple geometry shows that

$$\int_{\mathbb{R}^2} w(x, m(x)) \sin m(x) \phi(dx) = \int_{\mathbb{R}} \sum_{x_1 : (x_1, x_2) \in \phi} w((x_1, x_2), m(x_1, x_2)) \, dx_2. \qquad (8.35)$$

If g is a nonnegative measurable function with $\int_{\mathbb{R}} g(z) \, dz = 1$, then the Campbell theorem (8.20) can be employed on Φ for

$$f(x, \alpha) = h(x_1, \alpha) \, g(x_2) \sin \alpha,$$

leading to

$$\mathbf{E}\left(\int_{\mathbb{R}^2} h(x_1, m(x)) g(x_2) \sin m(x) \Phi(dx) \right) = L_A \int_{\mathbb{R}^2} \int_{(0,\pi]} h(x_1, \alpha) g(x_2) \sin \alpha \, \mathcal{R}(d\alpha) \, dx.$$

The right-hand side equals

$$L_A \int_{\mathbb{R}} \int_{(0,\pi]} h(z, \alpha) \sin \alpha \, \mathcal{R}(d\alpha) \, dz,$$

and by Formula (8.35) for $w\big(x_1, x_2, m(x)\big) = h\big(x_1, m(x)\big)g(x_2)$ and stationarity, the left-hand side becomes

$$\mathbf{E}\left(\int_{\mathbb{R}} \sum_{x_1:(x_1,\,x_2)\in\Phi} h\big(x_1, m(x)\big)\,g(x_2)\,\mathrm{d}x_2\right) = \int_{\mathbb{D}}\int_{\mathbb{R}} \sum_{x_1:(x_1,\,x_2)\in\phi} h\big(x_1, m(x)\big)\,g(x_2)\,\mathrm{d}x_2\,P(\mathrm{d}\phi)$$

$$= \int_{\mathbb{R}}\int_{\mathbb{D}} \sum_{x_1:(x_1,\,x_2)\in\phi-(0,\,x_2)} h\big(x_1, m(x_1,0)\big)\,P(\mathrm{d}\phi)\,g(x_2)\,\mathrm{d}x_2$$

$$= \int_{\mathbb{D}} \sum_{x_1:(x_1,\,0)\in\phi} h\big(x_1, m(x_1,0)\big)\,P(\mathrm{d}\phi)$$

$$= \mathbf{E}\left(\sum_{[y;\,m(y)]\in\Psi} h\big(y, m(y)\big)\right).$$

The proof is completed by applying the Campbell theorem (4.20) for the marked point process Ψ to this last formula.   □

From (8.34) one can derive some important formulae, relating characteristics of Φ to those of Ψ. For example, the intensity of Ψ is related to the intensity and rose of directions of Φ by

$$P_L = L_A \int_{(0,\pi]} \sin\alpha\,\mathscr{R}(\mathrm{d}\alpha), \tag{8.36}$$

as can be seen by putting $h(z, \alpha) = \mathbf{1}_{[0,1]}(z)$.

If $h(z, \alpha) = \mathbf{1}_{[0,1]}(z)\mathbf{1}_{(0,\beta]}(\alpha)$ for some β in $(0, \pi]$, then a more refined relation is obtained that uses also the mark distribution H

$$P_L H\big((0, \beta]\big) = L_A \int_{(0,\beta]} \sin\alpha\,\mathscr{R}(\mathrm{d}\alpha) \qquad \text{for } 0 < \beta \le \pi. \tag{8.37}$$

It follows that the orientation distribution function $F_H(\beta)$ for H is given by

$$F_H(\beta) = H\big((0, \beta]\big) = \frac{\int_{(0,\beta]} \sin\alpha\,\mathscr{R}(\mathrm{d}\alpha)}{\int_{(0,\pi]} \sin\alpha\,\mathscr{R}(\mathrm{d}\alpha)} \qquad \text{for } 0 < \beta \le \pi.$$

When $\mathscr{R}$ is the uniform distribution on $(0, \pi]$, as in the case of isotropy, then

$$P_L = \frac{2}{\pi}L_A \tag{8.38}$$

and

$$F_H(\beta) = \frac{1}{2}(1 - \cos\beta) \qquad \text{for } 0 < \beta \le \pi, \tag{8.39}$$

with the corresponding density function

$$f_H(\beta) = \frac{1}{2} \sin \beta \qquad \text{for } 0 < \beta \leq \pi.$$

This density also arises in the case of intersection with curves; see Morton (1966).

Let $\mathcal{R}(\{\pi\}) = 0$ so that a typical point on a fibre has fibre tangent almost surely not parallel to the sampling line e. Then the characteristics of Φ can be expressed in terms of those of Ψ:

$$\mathcal{R}\left((0, \beta]\right) = \frac{P_L}{L_A} \int_{(0,\beta]} (\sin \alpha)^{-1} H(d\alpha) \qquad (8.40)$$

and

$$L_A = P_L \int_{(0,\pi]} (\sin \alpha)^{-1} H(d\alpha). \qquad (8.41)$$

A further formula for $\mathcal{R}$ involves the so-called *rose of intersections* $P_L(\beta)$, which is the intensity of the point process of intersection points of Φ with a sampling line at angle β to e. Up to a factor this is the same as the modified support function of the Steiner compact S related to $L_A \mathcal{R}$, as described above in Section 8.3.1.

Consider the point process of intersections of Φ with a straight line forming the angle β with the reference axis. The fibre system rotated by r has rose of directions $r(\mathcal{R})$ as explained in Section 8.3.1. Thus if r rotates through an angle $-\beta$ sending the intersection line to e, then Formula (8.36) can be applied to the rotated fibre process. This shows that the point process of intersection points has intensity

$$P_L(\beta) = L_A \int_{(0,\pi]} |\sin(\alpha - \beta)| \, \mathcal{R}(d\alpha) = L_A \int_0^\pi |\sin(\alpha - \beta)| \, f_{\mathcal{R}}(d\alpha). \qquad (8.42)$$

This formula was first given by Hilliard (1962). Here one speaks about the *sine transform* (of the density function $f_{\mathcal{R}}(\alpha)$ corresponding to the rose of directions).

Formulae (8.21) and (8.42) together yield

$$P_L(\beta) = 2s_m(S, \beta) \qquad \text{for } 0 < \beta \leq \pi. \qquad (8.43)$$

Hence there is a direct connection between the rose of intersections and the modified support function $s_m(S, \beta)$ of the Steiner compact S of Φ. If $\mathcal{R}$ has a continuous density function $f_{\mathcal{R}}(\beta)$, then Formula (8.42) can be differentiated to show

$$\frac{d^2 P_L(\beta)}{d\beta^2} + P_L(\beta) = 2L_A f_{\mathcal{R}}(\beta) \qquad \text{for } 0 < \beta \leq \pi. \qquad (8.44)$$

The case where a density does not exist is considered in Berg (1969) and Mecke (1981a). In principle the formula allows for estimation of $f_{\mathcal{R}}(\beta)$ if the rose of intersections $P_L(\beta)$ is calculated empirically. In practice one must deal with the serious problems of numerical differentiation of $P_L(\beta)$.

A simple example of an anisotropic fibre process is the 'pressed' fibre process discussed in Stoyan *et al.* (1980). This is produced from an isotropic fibre process by the 'pressing' mapping

$$p_c : (x_1, x_2) \rightarrow (x_1, c \cdot x_2) \qquad \text{for some } c \text{ with } 0 < c < 1.$$

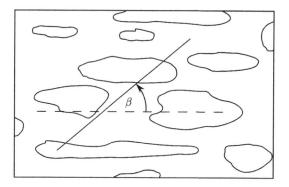

Figure 8.13   A sample of a pressed fibre process together with a 40° measurement line.

Characteristics of the pressed fibre process can be expressed in terms of the original isotropic fibre process. For example, its intensity satisfies in good accuracy and independently of c

$$L_A \approx \frac{\pi}{2} P_L(\beta) \qquad \text{for } \beta = 40°, \tag{8.45}$$

see Figure 8.13. This result is obtained by numerical calculation of the elliptic integral which appears in the formula for $P_L(\beta)$ for the pressed fibre process.

For a motion-invariant fibre process Φ, the reduced second moment function $K(r)$ can be determined in terms of the marked intersection point process Ψ together with certain additional marks. If $[y; m(y)]$ belongs to Ψ, then the additional mark on y required is

$$s_r(y) = L(y) + 2\pi \sum_{\substack{(x_1, 0) \in \Phi, \\ y \leq x_1 \leq y+r}} \frac{x_1 - y}{\sin \alpha(x_1, 0)},$$

where $\alpha(x_1, 0)$ is the intersection angle between the fibre at $(x_1, 0)$ and the x_1-axis (so $0 < \alpha(x_1, 0) \leq \pi$), and $L(y)$ is the total length of all straight line pieces of fibres lying on lines radial to $(y, 0)$ within the ball $B((y, 0), r)$.

Note that if Φ has no straight line pieces of fibre, then $s_r(y)$ is determined entirely by the information contained within the marked point process Ψ. However, the additional information represented by the lengths $L(y)$ is required if there are straight pieces of fibre in Φ.

Consider the new marked point process $\{[y; m(y), s_r(y)]\}$ and let $\mathscr{S}$ be the joint mark distribution of the mark pair $(m(y), s_r(y))$. Then it holds that

$$L_A K(r) = \frac{2}{\pi} \int_{[0,\infty)} \int_{(0,\pi]} \frac{s}{\sin \alpha} \mathscr{S}(\mathrm{d}(\alpha, s)) \qquad \text{for } r \geq 0, \tag{8.46}$$

as shown by Stoyan (1981). Ambartzumian (1981) studied the particular case of segment processes and gave formulae for the product density, which agrees with the pair correlation function up to a factor L_A^2. Formula (8.46) is refined and generalised in Schwandtke (1988). The general theory behind these results is presented in Jensen *et al.* (1990a,b) and Zähle (1990), where also higher-dimensional analogues are considered.

The properties of intersection of fibres processes having segments, triangles, and circles as fibres are studied in Ohanian (1978) and Sukiasian (1978, 1980).

Intersections with fibre systems

Suppose that Φ is a stationary fibre process of intensity L_A and rose of directions $\mathscr{R}$, and that ψ is a nonrandom fibre system with finite total length l. Mecke (1981a) and Ohser (1981) studied the intersection process $\Phi \cap \psi$. This intersection is a point process with only a finite number of points because ψ has finite total length. The fibre system ψ controls the basic characteristics of $\Phi \cap \psi$ by means of two functional parameters:

(a) the angle distribution η_ψ, a probability measure on $(0, \pi]$ given by

$$\eta_\psi(L) = \frac{h_1(\{x \in \psi : m_\psi(x) \in A\})}{l} \qquad \text{for Borel sets } L \text{ of } (0, \pi], \qquad (8.47)$$

where $m_\psi(x)$ is the ψ-fibre tangent angle at x;

(b) the total length $\varrho_\psi(\beta)$ (counted by multiplicity) of the projection of ψ in the direction β, given by

$$\varrho_\psi(\beta) = \int_{l_\beta^\perp} \#\{\psi \cap (l_\beta - y)\} \, dy \qquad \text{for } 0 < \beta \leq \pi,$$

where l_β is a line of direction β and $l_\beta^\perp$ is a line perpendicular to l_β.

Here are two examples of these characteristics for particular ψ:

(1) if ψ is a circle of radius R, then η_ψ is uniform on $(0, \pi]$ and $\varrho_\psi(\beta) = 2l/\pi$ for all $\beta \in (0, \pi]$ (note that $l = 2\pi R$);

(2) if ψ is the union of two segments of equal lengths $l/2$ that make angles α and $\alpha + \pi/2$ with the x_1-axis respectively, then η_ψ is the obvious two-point distribution on $\{\alpha, \alpha + \pi/2\}$ and

$$\varrho_\psi(\beta) = \frac{l}{2} (|\sin(\beta - \alpha)| + |\cos(\beta - \alpha)|) \qquad \text{for } 0 < \beta \leq \pi.$$

The mean number of points $\overline{N}_\psi$ of $\Phi \cap \psi$ satisfies

$$\overline{N}_\psi = L_A l \int_{(0,\pi]} \int_{(0,\pi]} |\sin(\alpha - \beta)| \, \mathscr{R}(d\alpha) \eta_\psi(d\beta). \qquad (8.48)$$

If ψ is a circle of radius R, then (independently of $\mathscr{R}$)

$$\overline{N}_\psi = 4RL_A. \qquad (8.49)$$

If $\overline{N}_\psi(L)$ denotes the mean number of points in $\Phi \cap \psi$ for which the Φ-fibre direction is in the Borel set L of $(0, \pi]$, then

$$H_\psi(L) = \frac{\overline{N}_\psi(L)}{\overline{N}_\psi} \qquad \text{for all } L \in \Pi$$

defines a distribution on $(0, \pi]$ which can be interpreted as the angle distribution for the intersections. It satisfies

$$\overline{N}_\psi H_\psi(L) = L_A \int_L \varrho_\psi(\alpha) \mathcal{R}(\mathrm{d}\alpha), \tag{8.50}$$

and if, in particular, ψ is a circle, then

$$H_\psi = \mathcal{R}. \tag{8.51}$$

Intersection of two fibre processes

Let Φ_1 and Φ_2 be two independent stationary fibre processes of characteristics $L_{A,i}$ and $\mathcal{R}_i$ for $i = 1$ and 2. The intersection $\Phi_1 \cap \Phi_2$ is a stationary point process, the intensity of which is denoted by λ. Then

$$\lambda = L_{A,1} L_{A,2} \int_{(0,\pi]} \int_{(0,\pi]} |\sin(\alpha_1 - \alpha_2)| \mathcal{R}_1(\mathrm{d}\alpha_1) \mathcal{R}_2(\mathrm{d}\alpha_2). \tag{8.52}$$

If either of the roses $\mathcal{R}_2$ or $\mathcal{R}_2$ is actually uniform on $(0, \pi]$ (and this will occur for example if either of the Φ_i is isotropic), then Formula (8.52) simplifies to

$$\lambda = \frac{2}{\pi} L_{A,1} L_{A,2}. \tag{8.53}$$

A formula for the reduced second moment measure of $\Phi_1 \cap \Phi_2$ is given in Hanisch (1985).

8.3.3 Basic statistical methods for planar fibre processes

General

In general, estimation of basic characteristics of fibre processes follows the same path as the case of point processes. However, there is an important difference in the process of model fitting as there is no fibre process model holding a central position analogous to that of the Poisson point process.

There seem to be only a few references on statistics for particular fibre process *models*. Fellous *et al.* (1978) and Rasson and Hermans (1988) wrote on statistics for line processes. Laslett (1982a,b), Gill (1994), van de Laan (1995), Wijers (1995) and van Zwet (2004) studied the estimation of segment length distributions for segment processes.

In this section estimation procedures are discussed for the basic characteristics L_A, $\mathcal{R}$ and K. In every case it is assumed that the data arise as a sample of the fibre process Φ observed in a compact window of observation W. It can be useful to approximate the observed image by systems of segments; see Scheidegger (1979) and Stoyan and Stoyan (1983). In an analysis without the aid of an image analyser this can be helpful, as it simplifies length and angle measurements. More generally it may serve as a useful smoothing procedure when the fibres are irregular.

Estimation of L_A

The most natural unbiased estimator of L_A is

$$\hat{L}_A = \frac{\Phi(W)}{\nu_2(W)}, \tag{8.54}$$

where $\Phi(W)$ is the total length of all fibre pieces in W, which in many situations can be measured by an image analyser. Otherwise, intersections with sampling lines or circles can be used and then Formulae (8.38) and (8.49) be applied. If Φ is motion-invariant, an unbiased estimator is given by

$$\hat{L}_A^{\text{segment}} = \frac{\pi \cdot \#\{T \cap \Phi\}}{2h_1(T)}, \tag{8.55}$$

where T is a test system of segments (such as an array of parallel segments) in W of total length $h_1(T)$ and $\#\{A\}$ denotes the number of points in A. In the general stationary case a test system C composed of n circles of fixed radius R can be used. Formula (8.49) ensures that an unbiased estimator for L_A is

$$\hat{L}_A^{\text{circle}} = \frac{\#\{C \cap \Phi\}}{4nR}. \tag{8.56}$$

The application of these estimators is demonstrated in Example 8.6 on p. 341. Rancoita *et al.* (2011) describe a way practically realise this method. Beneš *et al.* (1994) and Beneš and Rataj (2004, pp. 61–92) propose further estimators of L_A in the anisotropic case and give their estimation variances and other properties.

Lauschmann and Mrkvička (2009) study two regression-based methods of individual perimeter measurement for systems of closed planar curves. Ambrosio *et al.* (2009) show that the intensity can be approximated by a multiple of the volume fraction of the dilated set $\Phi_{\oplus r}$ for small r.

Estimation of the rose of directions $\mathcal{R}$

There are three main approaches for estimating $\mathcal{R}$ or the corresponding density function $f_{\mathcal{R}}(\alpha)$, which can be systematised according to the data used. The reader should note that estimation of $\mathcal{R}$ based on sections and stereological ideas (as in the first method below) is an ill-posed problem of a difficulty similar to that of Wicksell's corpuscle problem (described in Section 10.4.2). As there, a *two-step method* is recommended: determine first by nonparametric methods a reasonable type of distribution and then apply parametric methods to data that can be measured in high precision. Starting from a formula for $f_{\mathcal{R}}(\beta)$ by means of (8.42), a formula for $P_L(\beta)$ is derived and then the parameters of the latter are estimated.

Rose of intersection

Smoothing methods are used in Marriott (1971) (cardioidal approximation of either $P_L(\beta)$ or $\mathcal{R}$), Serra (1975) (elliptic approximation of $P_L(\beta)$) and Louis *et al.* (2011) (using mollifiers to invert the cosine transform, which is equivalent to the sine transform in (8.42)).

Hilliard (1962) and Kanatani (1984) used Fourier methods. They compared Fourier coefficients of the two roses $P_L(\beta)$ and $\mathcal{R}$. Beneš and Gokhale (2000) showed that the obtained estimators do not have good statistical properties; see also the cautionary remark given in Hilliard and Lawson (2003, footnote on p. 143).

Methods based on the Steiner compact S are applied in Rataj and Saxl (1989) for planar fibre processes (see also Beneš and Rataj, 2004, pp. 143–52). The first step is the construction of a polygonal estimator $\hat{S}$ of S starting from values of the rose of intersections $P_L(\beta)$ and using Formula (8.43). Then the rose of directions of the boundary $\partial\hat{S}$ of $\hat{S}$, which can be easily obtained because of the polygonal shape, is an estimator of $\mathcal{R}$. The graphical construction of $\hat{S}$ can be replaced by an analytical one; see Beneš and Gokhale (2000).

In particular, for the Poisson line process, Kiderlen (2001) proposed a maximum likelihood estimator, whose optimisation procedure can be implemented by the EM algorithm, whilst Prokešová (2003) suggested a Bayesian approach, where MCMC can be used to approximate the posterior distribution. Both of these methods were shown to yield consistent estimators of $\mathcal{R}$, even when applied to more general stationary fibre processes.

Angle-marked section process

The intersection $\Phi \cap T$ is analysed for T being a test system of circles of radius R lying in the observation window W. If $m(x)$ is the direction of the fibre tangent at x, then

$$\hat{\mathcal{R}}\big((\alpha_1, \alpha_2]\big) = \frac{1}{\#(\Phi \cap T)} \sum_{x \in \Phi \cap T} \mathbf{1}_{(\alpha_1, \alpha_2]}\big(m(x)\big) \tag{8.57}$$

is a ratio-unbiased estimator of $\mathcal{R}\big((\alpha_1, \alpha_2]\big)$; see Ohser (1981). This is because the angles at the intersection points form a sample from a population of angles of distribution $\mathcal{R}$.

Grey-tone images

When only a degraded grey-tone image of the fibre process is available, the rose of intersections $P_L(\beta)$ cannot be estimated directly. Kärkkäinen *et al.* (2001) suggest approximating its values through evaluating the intersection intensity for a given test line by grey-tone values along the line using a scaled variogram. They assume a parametric model for the rose of directions. Altendorf and Jeulin (2009) and Wirjadi *et al.* (2009) completely avoid the use of test lines.

In parametric statistics, the von Mises distribution for the planar case and the von Mises–Fisher distribution for the spatial case (see Fisher *et al.*, 1987) may be the first choice. For other parametric families, see Mardia and Jupp (2000), Jammalamadaka and SenGupta (2001), León *et al.* (2006) and Oualkacha and Rivest (2009).

Estimation of the reduced second moment function $K(r)$ and the pair correlation function

The estimation of $K(r)$ is a rather complicated matter. Various methods will be sketched. In the following Φ is always assumed to be motion-invariant; for the anisotropic case see Beneš (1994) and Weiss and Nagel (1994).

A first approach uses intersections with test systems. Let T be a system of parallel lines in a distance larger than the expected range of correlation, that is, larger than the value r_0 that it is expected that $g(r) = 1$ for all $r \geq r_0$. A simple approximation, which can give acceptable results for values of r not too small, can be obtained by assuming that the fibre-line intersection angles are independent. Then $g(r)$ and the pair correlation function of the one-dimensional

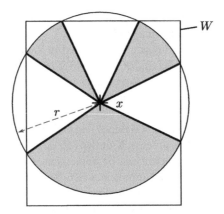

Figure 8.14   From the centre x, draw lines to the intersections between the circle $\partial B(x, r)$ and W. The set $S(x, r, W)$ is the union of the (shaded) sectors corresponding to arcs lying in W.

point process of intersection points are nearly equal, and so estimators of the latter lead to estimators of the former.

A more sophisticated method also starts with the points of $\Phi \cap T$. Formula (8.90) below provides the unbiased estimator $\hat{l}(r)$ of $L_A K(r)$

$$\hat{l}(r) = \frac{1}{\#\{\Phi \cap T\}} \sum_{x \in \Phi \cap T} k(x, r) \Phi\big(S(x, r, W)\big), \tag{8.58}$$

where $S(x, r, W)$ is sketched in Figure 8.14 and $k(x, r) = 2\pi/\alpha_{x, r}$, in which $\alpha_{x, r}$ is the sum of all angles of the arcs in $\partial B(x, r) \cap W$. This estimator is based on a 'weighting' of the fibre points x by the total length $S_r(x)$ of all fibre pieces within the disc $B(x, r)$. Here fibre length measurement is necessary.

The perhaps most elegant approach is the Cox-process approach, probably first used by Hanisch (1985). For a simple example of this idea, consider first a line process. On each line, independently of the others, generate a Poisson point process with some intensity λ. The union of all these points is a sample of a Cox process with driving measure $\lambda\Phi$. By Formula (5.33) the K-function of the Cox point process coincides with the K-function of the line process, and so estimation of one yields the other. The same is true for the pair correlation functions.

The case of a process of general fibres can be treated analogously. This method can be easily realised with the help of an image analyser. If the fibre pixels are black and the rest white, then sample randomly from the black pixels and estimate the K-function or the pair correlation function for the point pattern resulting from the sample.

8.4 Spatial fibre processes

The definitions of spatial line and fibre processes follow the pattern of their planar counterparts, with occasional extra complexities; see Mecke and Nagel (1980) and Nagel (1983). This section is restricted to some brief remarks on stationary fibre processes.

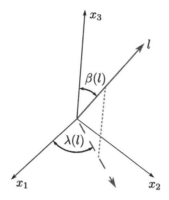

Figure 8.15   A diagram to illustrate the system of spherical polar coordinates.

As with planar fibre processes, so in this case the *intensity* L_V of a stationary spatial fibre process Φ is the mean fibre length per unit volume, and

$$L_V \nu_3(B) = \mathbf{E}\big(\Phi(B)\big) \qquad \text{for all Borel sets } B \text{ of } \mathbb{R}^3. \tag{8.59}$$

The *rose of directions* $\mathscr{R}$ is now defined as a distribution on the space $\mathbb{L}$ of all straight lines through the origin; $\mathbb{L}$ is actually the projective plane $\mathbb{RP}^2$. The corresponding Borel σ-algebra $\mathcal{L}$ is the system of all spatial Borel sets that are unions of lines through o. The quantity $\mathscr{R}(A)$ is defined by

$$L_V \nu_3(B)\mathscr{R}(A) = \mathbf{E}\left(\int_B \mathbf{1}_A\big(l(x)\big)\Phi(\mathrm{d}x) \right) \qquad \text{for } A \in \mathcal{L}, \tag{8.60}$$

where $l(x)$ is the fibre tangent line through x.

If Φ is isotropic, then $\mathscr{R}$ is the rotation-invariant, or uniform, probability distribution U_1 on $\mathbb{L}$. To describe this distribution more explicitly, introduce a coordinate system on $\mathbb{L}$. A line l in $\mathbb{L}$ makes an angle $\beta(l)$ with the x_3-axis, the *azimuthal angle*; and the projection of $l \cap \{x_3 > 0\}$ onto the plane $x_3 = 0$ makes an angle $\lambda(l)$ with the x_1-axis, the *geographical longitude*, see Figure 8.15. These angles lie in the regions

$$0 \le \beta(l) \le \frac{\pi}{2} \qquad \text{and} \qquad -\pi \le \lambda(l) \le \pi,$$

although if $\beta(l) = \pi/2$, then $\big(\beta(l), \lambda(l)\big)$ and $\big(\beta(l), \pi - \lambda(l)\big)$ refer to the same line. (This kind of duplication is inescapable, when a projective plane is coordinatised.) Under this coordinate system the measure U_1 can be expressed as

$$U_1\big(\mathrm{d}(\beta, \lambda)\big) = \frac{\sin \beta \, \mathrm{d}\beta \, \mathrm{d}\lambda}{2\pi}. \tag{8.61}$$

The intersection of a spatial fibre process with a plane e (consider for simplicity this plane e to be the plane $x_3 = 0$) generates a point process on e. A point x of this point process can be marked with the fibre tangent line $l(x)$ through x in $\mathbb{L}$. Thus the intersection generates a stationary marked point process

$$\Psi = \{[x; l(x)]\}.$$

If P_A is the intensity of Ψ and Θ its mark distribution, then by analogy with Formula (8.34), these can be connected to the characteristics of the fibre process using a formula of Mecke and Nagel (1980)

$$P_A \int_{\mathbb{R}} \int_{\mathbb{R}} \int_{\mathbb{L}} h(x_1, x_2, l)\, \Theta(dl)\, dx_1 dx_2$$

$$= L_V \int_{\mathbb{R}} \int_{\mathbb{R}} \int_{\mathbb{L}} h(x_1, x_2, l) \cos \beta(l)\, \mathcal{R}(dl)\, dx_1 dx_2, \qquad (8.62)$$

where h is any nonnegative measurable function on $\mathbb{R}^2 \times \mathbb{L}$. If $\mathcal{R}$ is the uniform distribution U_1 on $\mathbb{L}$, then

$$\int \cos \beta(l) \mathcal{R}(dl) = \frac{1}{2}$$

and so

$$L_V = 2P_A. \qquad (8.63)$$

In this uniform case the mark distribution Θ is given by

$$\Theta(dl) = 2 \cos \beta(l)\, \mathcal{R}(dl) = \sin 2\beta\, \frac{d\beta\, d\lambda}{2\pi} \qquad \text{for } 0 \le \beta < \frac{\pi}{2}. \qquad (8.64)$$

As in the planar case a *rose of intersections* $P_A(l)$ can be defined: $P_A(l)$ is the intensity of the point process $\Phi \cap e_l$, where e_l is the plane perpendicular to l. Mecke and Nagel (1980) give an integral formula connecting $\mathcal{R}$ and $P_A(l)$, involving Legendre functions; see also Hilliard and Lawson (2003, pp. 157–70).

Further results on spatial fibre processes in the context of stereology can be found in Sections 10.2.2 and 10.7, and in Nagel (1983) and Zähle (1984c). Also there it is useful to apply the Steiner compact. Kiderlen and Pfrang (2005) give an example where the rose of directions and the Steiner compact are estimated for a sample of carbon fibres in pyrolytic carbon matrix. The consistency of the estimator used is shown in Kiderlen (2001) and more generally in Gardner *et al.* (2006).

A particular case is that of line processes, in particular Poisson line processes. The corresponding contact distributions are given by

$$H_B(r) = 1 - \exp\left(-\frac{L_V S(B)}{4} r^2\right) \qquad \text{for } r \ge 0 \qquad (8.65)$$

for convex compact structuring element B containing the origin, and

$$H_s(r) = 1 - \exp(-\pi L_V r^2) \qquad \text{for } r \ge 0. \qquad (8.66)$$

Bosan *et al.* (2003) illustrate the statistical use of the spherical contact distribution function $H_s(r)$ in the context of capillaries.

Boolean models with segments as grains are studied in many papers, in particular in the context of the material 'paper'. For applications of fibre processes to the manufacture of paper, nonwoven textiles and glass mats; see Dodson and Sampson (1999) and Sampson (2001). Provatas *et al.* (2000) leave the field of Boolean models and consider other fibre deposition models that lead to clustering of fibres.

Laslett *et al.* (1982) use a spatial segment process to model the occurrence of fission particle tracks within apatite crystals. The objective is to measure the length distribution of the tracks, and in particular to gain information about whether the distribution is unimodal or not. Since most particle tracks are revealed by etching treatment only when exposed by a sheared plane of the crystal, this is a complicated stereological problem. Numerical problems of the kind referred to in the section on planar fibre processes prevail, and it is found to be more effective to sample only the infrequent complete tracks that are revealed because of an intersection with another exposed track. The book Galbraith (2005) presents the corresponding statistical theory. These tracks provide important information about geological history.

Hanisch *et al.* (1985) use spatial fibre processes as models for systems of dislocation lines and show the value of the pair correlation function $g(r)$ for the description of such systems; see also Section 10.7.3, Stoyan and Stoyan (1986) and El-Azab *et al.* (2007). These models are Boolean models, modified Poisson line processes, and planar and spatial lattices of lines.

Fibre process models are used in the context of statistics of roots of plants. Examples include Grabarnik *et al.* (1998) and Fleischer *et al.* (2006) for the cases of maize and forest trees. Thread-like objects, so-called soil hyphae, are studied in Schack-Kirchner *et al.* (2000).

Analysis of vascular systems via fibre process is carried out in Heinzer *et al.* (2006); see also Capasso and Micheletti (2006) and Capasso *et al.* (2008).

To conclude this section, some remarks on practical *fibre length measurement* are given. There are two basic methods, which are described as follows.

(1) Use of pixel data, that is, of binary images of three-dimensional objects. Klette and Rosenfeld (2004, pp. 409–13) is a good source; see also Barber *et al.* (2003).

(2) Use of a randomly rotated and shifted (IUR, standing for 'isotropic, uniformly random') periodic grid of surfaces. The total fibre length L is estimated by

$$\hat{L} = \frac{2N}{S_V},\tag{8.67}$$

where N is the number of intersections of fibres with surfaces and S_V the surface area of the grid per volume unit; see Baddeley and Jensen (2005). Janáček and Kubínová (2010) discuss the variance of this estimator. Čebašek *et al.* (2010) present an interesting application in microvascular research; the grid of surfaces is realised by a stack of optical sections acquired by a confocal microscope.

8.5 Surface processes

8.5.1 Plane processes

Plane processes are used as stochastic models for systems of planes in space, and can be regarded as special cases of spatial surface processes. Just as Section 8.3 used line processes to introduce fibre processes, so here the theory of plane processes is briefly reviewed as a prelude to the general theory of surface processes. As with line processes, so also plane processes can be defined as point processes in a suitable representation space; see p. 311. A Poisson plane process corresponds then to a Poisson point process in the representation space.

As in Section 8.3 the theory of random closed sets can be used to provide a theoretical framework and to introduce the notions of *stationary* and *motion-invariant* plane processes. These notions, and measure theory, tie up in a satisfactory manner with the representation space approach.

A plane process $\Phi = \{e_1, e_2, \ldots\}$ can also be described as a random measure

$$\Phi(B) = \sum_{e \in \Phi} h_2(e \cap B) \qquad \text{for all Borel sets } B \text{ of } \mathbb{R}^3, \tag{8.68}$$

where h_2 is the two-dimensional Hausdorff measure. If Φ is stationary, then one can define the *intensity* S_V of Φ to be the mean plane area per unit volume. Moreover, denote by $\mathbb{L}$ as in Section 8.4 the set of lines in $\mathbb{R}^3$ passing through the origin, a marked random measure Ψ can be defined on $\mathbb{R}^3 \times \mathbb{L}$ by

$$\Psi(B \times A) = \sum_{\{e \in \Phi : e^\perp \in A\}} v_{2,e}(e \cap B) \qquad \text{for all Borel sets } B \text{ of } \mathbb{R}^3 \text{ and } A \text{ of } \mathbb{L}, \tag{8.69}$$

where the sum is over e in Φ with normal $e^\perp$ lying in A and $v_{2,e}$ is the Lebesgue measure on the plane e. If Φ is stationary, then invariance arguments provide a factorisation of the intensity measure Λ of Ψ

$$\Lambda(B \times A) = S_V \cdot v_3(B) \mathscr{R}(A), \tag{8.70}$$

where $\mathscr{R}$ is a distribution on $[\mathbb{L}, \mathcal{L}]$, the *rose of (normal) directions* of Φ. If Φ is isotropic, then $\mathscr{R}$ is the uniform distribution U_1 on $[\mathbb{L}, \mathcal{L}]$. The rose of normal directions $\mathscr{R}$ can be interpreted as the distribution of the normal line to the typical plane in Φ.

If Φ is isotropic, then the mean number $\bar{n}_K$ of planes of Φ intersecting a compact convex set K is proportional to the average breadth $\bar{b}(K)$ of K:

$$\bar{n}_K = S_V \bar{b}(K). \tag{8.71}$$

As mentioned on p. 312, in the case of a Poisson plane process the number of planes intersecting K has a Poisson distribution. The contact distribution functions are

$$H_B(r) = 1 - \exp(-S_V \bar{b}(B)r) \qquad \text{for } r \geq 0 \tag{8.72}$$

and

$$H_s(r) = 1 - \exp(-2 S_V r) \qquad \text{for } r \geq 0. \tag{8.73}$$

As in the case of planar fibres, it is useful to consider the *Steiner compact* (or *associated zonoid*) S corresponding to the measure $S_V \mathscr{R}$ on $\mathbb{L}$. The Steiner compact is the symmetric compact convex set with modified support function $s_m(S, \cdot)$ given by

$$s_m(S, l) = \frac{1}{2} S_V \int_L |\langle l, l' \rangle| \mathscr{R}(\mathrm{d}l') \qquad \text{for } l \in \mathbb{L}, \tag{8.74}$$

where $\langle l, l' \rangle = \cos \alpha$, for α the angle between the lines l and l'.

If Φ is isotropic, then its Steiner compact S is the ball

$$S = B\left(o, \frac{S_V}{4}\right),$$ (8.75)

whose radius, as explained after (8.22), has dimension length^{-1}.

As in the case of planar fibre process the Steiner compact is a useful concept in the study of intersections of Φ with planes and lines. Consider for example the intersection of Φ with a fixed line l in $\mathbb{L}$. Let $P_L(l)$ be the intensity of the point process $\Phi \cap l$, also called the *rose of intersections* of Φ if considered as a function of l. Then

$$P_L(l) = 2s_m(S, l) \qquad \text{for } l \in \mathbb{L},$$ (8.76)

where S is the Steiner compact of Φ.

If e is a fixed plane, then $\Phi \cap e$ is a line process with Steiner compact S_e, which is given by

$$S_e = p_e^{\perp} S,$$ (8.77)

where $p_e^{\perp}$ is the orthogonal projection onto e.

In the case of isotropy Formulae (8.76) and (8.77) give

$$P_L = \frac{1}{2} S_V,$$ (8.78)

where P_L is the intensity of $\Phi \cap l$, and

$$L_A = \frac{\pi}{4} S_V,$$ (8.79)

where L_A is the intensity of $\Phi \cap e$. Note that the left-hand sides do not depend on the particular choices of l and e.

8.5.2 General surface processes

Basic characteristics

The definition of the class of *surface processes* (or *random surfaces*) follows closely the definition of fibre processes; see Pohlmann *et al.* (1981). A *surface system* is a set ϕ contained in $\mathbb{R}^3$ which can be represented as the boundary (that is, the surface) of an $\mathbb{S}$-set A, that is, a set $A \subset \mathbb{R}^3$ in the extended convex ring. Since A is locally a finite union of convex bodies, its surface ϕ has a well-defined area measure using h_2, the two-dimensional Hausdorff measure. For a surface-like set A, which is of zero volume, $h_2(A)$ is the surface area and not the double surface area (contrast the way in which the conventions of Section 1.6 apply to spatial convex sets lying on planes). Thus the following defines, with a slight abuse of notation, for the surface system ϕ a locally finite measure, also denoted by ϕ, in $\mathbb{R}^3$:

$$\phi(B) = h_2(\phi \cap B) \qquad \text{for all Borel sets } B \text{ of } \mathbb{R}^3.$$

Let $\mathbb{Z}$ be the class of all surface systems, endowed with the σ-algebra $\mathcal{Z}$ generated by

$$\{\phi \in \mathbb{Z} : \phi(B) < x\} \qquad \text{for positive numbers } x \text{ and spatial Borel sets } B.$$

A *surface process* Φ is a random variable defined on a probability space $[\Omega, \mathcal{A}, \mathbf{P}]$ and taking values in $[\mathbb{Z}, \mathcal{Z}]$. The probability measure generated by Φ on $[\mathbb{Z}, \mathcal{Z}]$ is called the *distribution* of Φ. The surface process Φ corresponds to a random measure, the *random surface measure*; see Section 7.3.4.

Stationarity and isotropy are defined as for random closed sets or for random measures.

Stationary surface processes have two basic characteristics: a scalar, the *intensity* S_V, and a measure, the *rose of normal directions* $\mathcal{R}$. These are defined in a manner entirely analogous to the corresponding quantities for fibre processes, generalising the definitions for plane processes given in Section 8.5.1. In particular $\mathcal{R}$ gives the distribution of the normal to the typical surface point. These characteristics satisfy a *Campbell theorem* (which can be used to recover their definitions)

$$\mathbf{E}\left(\int_{\mathbb{R}^3} f\big(x, l(x)\big)\Phi(\mathrm{d}x)\right) = S_V \int_{\mathbb{R}^3} \int_{\mathbb{L}} f(x, l)\mathcal{R}(\mathrm{d}l)\,\mathrm{d}x \qquad (8.80)$$

for every nonnegative measurable function f on $\mathbb{R}^3 \times \mathbb{L}$, where $l(x)$ is the surface normal at x.

If Φ is isotropic, then $\mathcal{R}$ is the uniform distribution U_1 on $\mathbb{L}$. The converse is, of course, not true. Consider, for example, a randomised lattice of non-intersecting balls and take the union of all spheres as a surface process. Then the rose of normal directions is the uniform distribution, but the surface process is anisotropic if the lattice is anisotropic. As in Sections 8.3.1 and 8.5.1, the rose of normal directions $\mathcal{R}$ can be characterised by the associated Steiner compact.

It makes sense to refine the approach by using directions of the normals, to use inward-pointing and outwarding-pointing normals. In this context one speaks about the *oriented mean normal measure*; see Kiderlen (2008) and Schneider and Weil (2008, p. 147).

Intersections by lines and planes

As in the case of fibre processes, it is of interest to consider intersections of surface processes with lines and planes. Intersection with a line generates a point process on the line; intersection with a plane produces a fibre process on the plane. (It suffices to consider plane intersections since linear intersections can be regarded as produced by first a planar intersection, and then a linear intersection of the ensuing planar fibre process. Linear intersections of planar fibre processes are dealt with in Section 8.3.2.)

Planar and linear sections of surface processes are also considered in Section 10.2, which is devoted to stereology.

Suppose that Φ is a stationary surface process and $\Phi \cap e$ is the planar fibre process resulting from the intersection of Φ with the plane e, where e is the plane $x_3 = 0$. Furthermore, let Θ be the rose of tangent directions for $\Phi \cap e$. Then Θ is connected with the rose $\mathcal{R}$ of normal directions of Φ by

$$L_A \int_{\mathbb{R}} \int_{\mathbb{R}} \int_{(0,\pi/2]} h(x_1, x_2, \lambda)\Theta(\mathrm{d}\lambda)\,\mathrm{d}x_1\mathrm{d}x_2$$

$$= S_V \int_{\mathbb{R}} \int_{\mathbb{R}} \int_{\mathbb{L}} h\big(x_1, x_2, \lambda(l)\big)\sin\beta(l)\,\mathcal{R}(\mathrm{d}l)\,\mathrm{d}x_1\mathrm{d}x_2 \qquad (8.81)$$

for every nonnegative measurable function h defined on $\mathbb{R}^2 \times (0, \pi]$, where $\beta(l)$ is the azimuthal angle of l in $\mathbb{L}$, and $\lambda(l)$ is the angle of a line in e orthogonal to the projection of l onto e, and so $-\pi < \lambda(l) \leq \pi$.

If $h(x_1, x_2, \lambda) = \mathbf{1}_{[0,1]}(x_1)\mathbf{1}_{[0,1]}(x_2)$, then Formula (8.81) gives

$$L_A = S_V \int \sin \beta(l) \, \mathscr{R}(dl). \tag{8.82}$$

If $\mathscr{R}$ is the uniform distribution (for example, if Φ is isotropic), then by means of Formula (8.61)

$$L_A = S_V \cdot \frac{1}{2\pi} \int_{-\pi}^{\pi} \int_0^{\pi/2} \sin^2 \beta \, d\beta \, d\lambda = \frac{\pi}{4} S_V. \tag{8.83}$$

For the case of intersection with a line the following formula holds:

$$P_L = \frac{1}{2} S_V, \tag{8.84}$$

as in the case of a plane process. This follows from (8.83) and (8.38).

Formulae (8.76) and (8.77) for Steiner compacts remain true for general surface processes.

Intersections with fibre processes and surface processes

Hilliard (1962) discussed intersections of fibre and surface processes in $\mathbb{R}^3$ (see also Hilliard and Lawson, 2003, pp. 170–1). In particular he derived a formula for the mean number P_V of intersections per unit volume of a surface process of intensity S_V with an independent fibre process of intensity L_V, in the case of motion-invariance. The formula is

$$P_V = \frac{1}{2} L_V S_V. \tag{8.85}$$

Denote by Φ_i, $i = 1, 2$ and 3, three independent stationary and isotropic surface processes with intensities $S_{V,i}$. The intersection $\Phi_1 \cap \Phi_2$ gives a spatial fibre process with intensity

$$L_V = \frac{\pi}{4} S_{V,1} S_{V,2} \tag{8.86}$$

(Miles, 1972a; see also Hilliard and Lawson, 2003, p. 181). Combining (8.85) and (8.86) yields the intensity of the spatial point process resulted from the intersection $\Phi_1 \cap \Phi_2 \cap \Phi_3$ of three independent motion-invariant surface processes:

$$P_V = \frac{\pi}{8} S_{V,1} S_{V,2} S_{V,3}. \tag{8.87}$$

General intersection formulae are given in Wieacker (1989, Section 4) and, for the non stationary Poisson case, in Hoffmann (2007), where the Steiner compact plays an important rôle.

Second moment measure

The definition of the *second moment measure* of a surface process follows that of a fibre process. If the surface process is stationary, then its second-order behaviour is summarised in a reduced second moment measure, and in the isotropic case a reduced second moment function $K(r)$ suffices. The interpretation of $S_V K(r)$ is that it is the mean area of all surface pieces within a ball of radius r centred at the typical surface point.

Example 8.5. *Basic characterisations of two simple surface process models*
(a) Let Φ be an isotropic Poisson plane process with intensity S_V. By isotropy the rose of normal directions is the uniform distribution on $\mathbb{L}$. As in Example 8.4 it can be shown that the reduced second moment function $K(r)$ satisfies

$$S_V K(r) = \pi r^2 + \frac{4}{3}\pi r^3 S_V \qquad \text{for } r \geq 0.$$

(b) Let Φ be a three-dimensional Boolean disc model as in Example 8.1, with constant disc radii R. Then

$$S_V = \lambda \pi R^2,$$

where λ is the intensity of the germ process of disc centres. The reduced second moment function $K(r)$ has a more complicated form:

$$S_V K(r) = \frac{4}{3}\pi r^3 S_V + F(r) \qquad \text{for } r \geq 0,$$

where

$$F(r) = \begin{cases} \pi R^2, & \text{if } r > 2R, \\ \dfrac{2}{R^2} \displaystyle\int_0^R tf(t)\,dt, & \text{if } r \leq 2R, \end{cases}$$

and $f(t) = v_2(B(o, r) \cap B(t, R))$ for $\|t\| = t$, that is, $f(t)$ is the intersection area of two discs of radii r (which comes from the ball $B(o, r)$) and R and centre distance t.

The second term comes from the disc holding the typical point while the first arises from contribution by other discs; this is reminiscent of Example 8.4.

Rough random surfaces are studied in some papers. An example is Stroeven (2000), where the surface is obtained by a planar section through a random system of balls, which generates dome-like caps and indentations of spherical particles; see Figure 8.16.

A very important tool in the context of modelling (liquid) surfaces is the *Surface Evolver*, which was developed by A. K. Brakke. It can be used to generate physically founded surfaces by energy minimisation. Information about this program can be obtained from the internet. A paper in which it is used for generating random foams is Kraynik *et al.* (2003).

The two basic methods for fibre length measurement discussed on p. 333 are also applicable for the *surface area measurement* as follows.

(1) Use of voxel data, that is, of binary images of three-dimensional objects. To estimate the surface area of such objects one determines the occurrences of certain configu-rations of black and white voxels in $2 \times 2 \times 2$ cubes. To each configuration type a

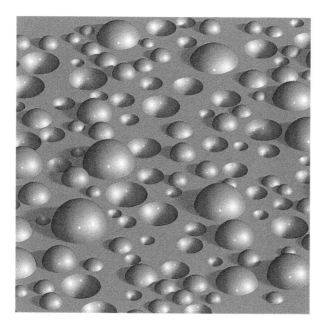

Figure 8.16 Simulated surface composed of inter-connected portions of a dividing planar section and indenting and protruding dome-like caps; see Stroeven (2000). Courtesy of M. Stroeven.

weight is assigned and the total surface area is obtained by summation of weights. See Ohser and Schladitz (2009) and Ziegel and Kiderlen (2010a,b).

(2) Use of a randomly rotated and shifted (IUR) periodic grid of lines. The total surface area S is estimated by

$$\hat{S} = \frac{2N}{L_V},$$ (8.88)

where N is the number of intersections of the surface system with lines and L_V is the line length of the grid per volume unit; see Baddeley and Jensen (2005). Janáček and Kubínová (2010) discuss the variance of this estimator.

8.6 Marked fibre and surface processes

Just as for point processes, it is useful for fibre and surface processes to attach *marks (weights)* to the fibre and surface points constituting the process. The mark might be a real number (for example, the curvature of the fibre at the point, or its thickness if 'thick fibres' are being studied), or it might be more general; at any rate it gives further information about the structure being studied. The space of possible marks will be denoted here by $\mathbb{W}$, and it must be endowed with a σ-algebra $\mathscr{W}$.

For the case where the fibres stand for microvessels, a stereological method for the determination of fibre thickness distribution is described in Krasnoperov and Gerasimov (2003).

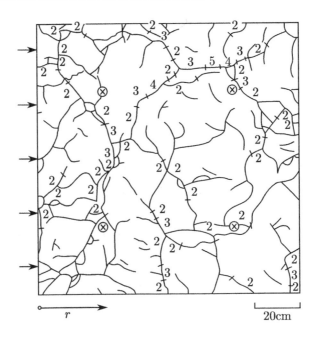

Figure 8.17 Crack pattern in a clay soil, taken from a horizontal section at depth of 30 cm (Doležal, 1982). The numbers indicate the widths of the cracks in the section plane; unlabelled cracks all have width 1 mm.

Figure 8.17 gives an example of a marked fibre process. The fibre system is the system of cracks in a clay soil on plane section. The fibres are marked by the width ('thickness') of the cracks. Note that this is indeed an intersection of a marked surface process with a plane!

As in aspects of point process theory, it can be useful to define marks for an unmarked fibre or surface process so that the marks describe distributional aspects of the process; see Stoyan (1984d). For example, the reduced second moment function $K(r)$ of a planar fibre process can be interpreted as follows: $L_A K(r)$ is the mean of a mark distribution when the mark at the typical point x is the total length $S_r(x)$ of all fibre pieces within the disc $B(x, r)$.

Mecke and Stoyan (1980a), Pohlmann *et al.* (1981) and Stoyan (1984d) gave important formulae for marked fibre and surface processes. In this section, a particular case is given a brief consideration: the case of stationary planar marked fibre processes. (Note in passing that stationarity and isotropy are defined just as in the case of marked random measures.)

The basic characteristics of a stationary planar marked fibre process are its *intensity L_A* (simply the intensity of the fibre process when the marks are disregarded) and the *joint distribution $\mathscr{J}$ of tangent direction and mark*. This distribution $\mathscr{J}$ can be interpreted as the joint distribution of tangent direction and mark for the typical fibre point. For L in the Borel σ-algebra Π of $(0, \pi]$ and C in $\mathscr{W}$,

$$L_A \cdot \mathscr{J}(L \times C) = \text{the mean length per unit area of all fibre fragments for which}$$
$$\text{the constituent points have tangent direction in } L \text{ and mark}$$
$$\text{in } C$$

(its strict definition follows closely the definition of the rose of directions, which itself can be thought of as a mark distribution). Of course,

$$\mathcal{J}(L \times \mathbb{W}) = \mathcal{R}(L) \qquad \text{for all Borel sets } L \text{ of } (0, \pi].$$

The *unconditional mark distribution* M is given by

$$M(C) = \mathcal{J}\big((0, \pi] \times C\big) \qquad \text{for all } C \text{ in } \mathcal{W}.$$

If the marked fibre process is isotropic, then $\mathcal{R}$ is uniform and $\mathcal{J}$ factorises as

$$\mathcal{J}(L \times C) = U_1(L)\, M(C) \tag{8.89}$$

Also a Campbell theorem holds, analogous to Theorem 8.3.

Intersection of a stationary marked fibre process with the x_1-axis yields a marked point process Ψ. The intersection points are marked by pairs of tangent directions and marks. Let P_L be the intensity of Ψ and $\mathcal{J}$ the joint distribution as above. Mecke and Stoyan (1980a) show that for any nonnegative measurable function h on $\mathbb{R} \times (0, \pi] \times \mathbb{W}$, it holds that

$$P_L \int_{\mathbb{R}} \int_{(0,\pi] \times \mathbb{W}} h(x, \alpha, w) \mathcal{J}\big(\mathrm{d}(\alpha, w)\big)\, \mathrm{d}x$$
$$= L_A \int_{\mathbb{R}} \int_{(0,\pi] \times \mathbb{W}} h(x, \alpha, w) \sin \alpha\, \mathcal{J}\big(\mathrm{d}(\alpha, w)\big)\, \mathrm{d}x. \tag{8.90}$$

In the case of isotropy

$$\mathcal{J}\big(\mathrm{d}(\alpha, w)\big) = \frac{1}{2} \sin \alpha\, \mathrm{d}\alpha\, M(\mathrm{d}w). \tag{8.91}$$

Thus in the isotropic case the mark distribution $\mathcal{J}$ also factorises. One factor is the mark distribution M and the other the angle distribution given by Formula (8.39).

Example 8.6. *Cracks in clay soil (Doležal, 1982; Sandau, 1993; Sandau and Vogel, 1993)*
Figure 8.17 shows a pattern of cracks in soil in planar section. Doležal (1982) uses stereological methods to draw inferences about the spatial crack structure. Here the planar fibre system is considered as a marked fibre process, with marks being the crack widths as shown in Figure 8.17.

Length measurements yield an estimate for L_A of $\hat{L}_A = 0.134\,\mathrm{cm}^{-1}$, using Formula (8.54). Line and circle section methods were also used to estimate L_A. The arrows in

Table 8.1 Empirical (mark) distributions for the crack-width.

Width (mm)	Exhaustive measurement	Line section method	Circle section method
1	0.786	0.783	0.770
2	0.163	0.174	0.203
3	0.039	0.043	0.027
4	0.009	0	0
5	0.004	0	0

Figure 8.17 define five line segments parallel to the bottom line of the window, and if these are used as test set and if isotropy is assumed, then Formula (8.55) yields $\hat{L}_A^{\text{segment}} = 0.120\,\text{cm}^{-1}$. The circle section method used four circles of radius $r = 30\,\text{cm}$ and centres at the points marked in Figure 8.17 by $\otimes$. Formula (8.56) yields $\hat{L}_A^{\text{circle}} = 0.154\,\text{cm}^{-1}$. This example may be instructive in illustrating actual errors of measurement.

Using both line and circle methods, marks were measured for the intersection points. The corresponding empirical distributions are tabulated in Table 8.1, together with the distributions of marks obtained by exhaustive length measurement.

This approach is greatly refined in Sandau (1993), who considers the three-dimensional problem and studies a stationary marked surface process. By means of vertical sections Sandau (1993) and Sandau and Vogel (1993) determine the joint distribution of spatial direction and width of the cracks.

9

Random tessellations, geometrical networks and graphs

9.1 Introduction and definitions

A *tessellation* or *mosaic* is a division of the plane into polygons, or of space into polyhedra. Figures 9.1, 9.4 and 9.17 show patterns that can be interpreted as the result of *tessellating* a plane. Such geometrical patterns occur in many natural situations. Examples include crystalline structures, crack patterns, and foam structures. Various models are described in Section 9.2.

This chapter mainly considers the case of random tessellations composed of *convex* polygons or polyhedra, although some of the results given are also true in the nonconvex case; see the references on p. 370. Precise mathematical definitions are given here for the case of planar tessellations. The generalisation to the spatial case and the *d*-dimensional case is straightforward.

There is a vast mathematical literature on random tessellations. There are specialised books such as Møller (1994), Okabe *et al.* (2000) and van de Weygaert *et al.* (2014), while other books contain chapters on tessellations, for example Matheron (1975), Schneider and Weil (2008) and Ohser and Schladitz (2009).

The system of edges of a tessellation is an example of a network. Geometrical properties of such and other *random geometrical networks* are often studied without reference to tessellations. For random geometrical networks the edge lengths and positions of vertices play a rôle, whilst for *random graphs* only the topological properties such as connectivity are of interest. There is a rapidly growing literature on random graphs, which will be reviewed briefly in Section 9.12.

Stochastic Geometry and its Applications, Third Edition.
Sung Nok Chiu, Dietrich Stoyan, Wilfrid S. Kendall and Joseph Mecke.
© 2013 John Wiley & Sons, Ltd. Published 2013 by John Wiley & Sons, Ltd.

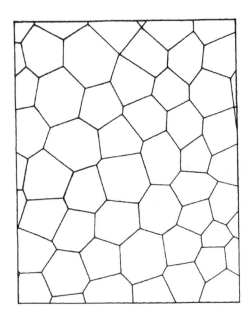

Figure 9.1   Plan view of basaltic columnar jointing at Burg Stolpen (Saxony) as measured by Professor R. A. Koch and his students (see Koch *et al.*, 1983). This pattern can be interpreted as a part of a realisation of a stationary isotropic tessellation. It is not a Dirichlet tessellation; see Stoyan and Stoyan (1980a).

Planar random tessellations

Let $\mathbb{C}$ be the set of all planar compact convex sets C with polygonal boundaries. A subset of polygons $\theta \subset \mathbb{C}$ is said to be a *tessellation* if the interiors of the constituent polygons are pairwise disjoint, if their union fills the plane, and if the family is locally finite. These three conditions can be expressed formally by:

(a) $C_1^{\text{int}} \cap C_2^{\text{int}}$ is empty if C_1 and $C_2 \in \theta$ and $C_1 \neq C_2$,

(b) $\bigcup_{C \in \theta} C = \mathbb{R}^2$,

(c) if B is a bounded planar set then the set $\{C \in \theta : C \cap B \neq \emptyset\}$ is finite.

The polygons $C \in \theta$ are the *cells* of the tessellation θ. The terminology suggested in Weiss and Cowan (2011) is adopted here. *Vertices* and *edges* are the primitive elements of the planar graph of a tessellation (also called, respectively, 0- and 1-*faces* of the tessellation), whilst *corners* and *sides* are 0-faces and 1-faces of the constituent polygons; see Figure 9.11 on p. 361. Because of convexity, the interior angle at a corner of a polygon is strictly less than π. If a vertex appears at an inner point of a polygonal side instead of at a corner, then it is called a π-*vertex*. An edge has a vertex at each of its two endpoints but no vertex in its interior; a side is the union of edges which it contains, and its endpoints are not π-vertices but polygonal corners. Tessellations without π-vertices are called *regular* or *face-to-face*.

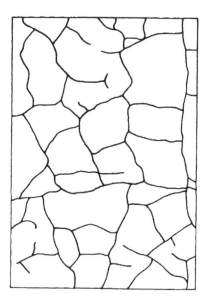

Figure 9.2   Mud-cracks in a ditch. The pattern shows some anisotropy caused by the geometry of the ditch. It can be interpreted as an intermediate stage in the formation of a tessellation with nonconvex cells. As given, the pattern forms a geometrical network.

Denote by E_θ the set-theoretic union of all edges of θ. Clearly, this *edge system E_θ* is a fibre system, more precisely a segment system, and thus a random closed set. Evidently, θ can be recovered from E_θ. Thus ideas of the theories of random closed sets and of fibre processes can be applied to characterise tessellations.

Let $\mathbb{T}$ denote the class of all tessellations and let $\mathscr{T}$ denote the σ-algebra on $\mathbb{T}$ generated by sets of the form

$$\{\theta \in \mathbb{T} : E_\theta \cap K \neq \emptyset\},$$

where K runs through all compact subsets of $\mathbb{R}^2$. If tessellations are identified with their edge systems then $\mathscr{T}$ is the trace of the hitting σ-algebra $\mathcal{F}$ on $\mathbb{T}$.

A *planar random tessellation* (or simply *tessellation* for brevity) is a random variable Θ with values in $[\mathbb{T}, \mathscr{T}]$. Its *distribution* is the induced probability measure on $[\mathbb{T}, \mathscr{T}]$. The definitions of stationarity and isotropy for random tessellation follow the usual form: A tessellation Θ and its distribution P are said to be *stationary* if for all x in $\mathbb{R}^2$ the translated tessellation

$$\Theta + x = \{C + x : C \in \Theta\} = \{C : C - x \in \Theta\}$$

has the same distribution as the original tessellation Θ, or

$$P(Y_x) = P(Y), \qquad \text{for all } x \in \mathbb{R}^2 \text{ and } Y \in \mathscr{T},$$

where $Y_x = \{\theta \in \mathbb{T} : \theta_{-x} \in Y\}$ and Y belongs to $\mathscr{T}$. If there is invariance with respect to rotations about the origin then Θ and P are called *isotropic*; isotropic and stationary tessellations are also called *motion-invariant*.

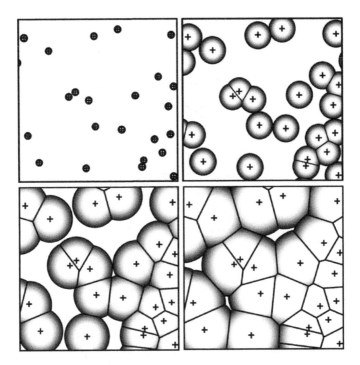

Figure 9.3   Four steps of the growth process which leads to a Dirichlet tessellation. Courtesy of L. Muche.

As noted above, the definition in the three-dimensional and d-dimensional cases is directly analogous; see Schneider and Weil (2008, Chapter 10).

Random tessellations of other spaces, for example the sphere, are also of interest. Miles (1971b) and Arbeiter and Zähle (1994) study such tessellations and give mean-value formulae as well as formulae for intersections with subspheres. Isokawa (2000) investigates tessellations in three-dimensional hyperbolic space.

Another avenue of generalisation is the study of random cell complexes (Zähle, 1987b, 1988; Leistritz and Zähle, 1992). An example is a system of groups of cells randomly scattered in the plane without overlappings and with gaps between them (as in Figure 9.3).

9.2   Mathematical models for random tessellations

In this section a representative selection of models is discussed and a partial summary of the literature is given. The purpose is to acquaint the reader with various types and methods of construction.

Line and plane tessellations

A nondegenerate line process Ψ in $\mathbb{R}^2$ generates a tessellation Θ, whose edge system E_Θ is simply the union of all lines of Ψ. Figure 8.7 on p. 309, showing part of a line process, can therefore also be interpreted as part of a *line tessellation*. In an analogous fashion hyperplane processes provide *hyperplane tessellations* of $\mathbb{R}^d$.

The distributional properties of Θ depend on those of Ψ. In particular, if Ψ is stationary then so is Θ. In general little is known about general line tessellations, with the exception of some simple assertions derived from geometrical considerations, such as that in the planar case the mean number of sides of a randomly chosen polygon is 4 if no line intersections involve more than two lines.

Much work has been done for the special case of *Poisson* line and plane tessellations built on Poisson line and plane processes. Some important formulae and results are presented Section 9.5 below.

Voronoi tessellation and related models

Voronoi tessellations

Dirichlet (1850) and Voronoi (1908) consider regular tessellations of planes and higher-dimensional spaces, motivated by problems in number theory. The application of Dirichlet and Voronoi tessellations to irregular and random point patterns appear to have arisen independently in applications in meteorology (Thiessen and Alter, 1911), metallurgy and crystallography (Kolmogorov, 1937; Johnson and Mehl, 1939) and ecology (Pielou, 1977; Matérn, 1986); see Okabe *et al.* (2000) for a historical sketch of the development of ideas.

Let $\varphi = \{x_1, x_2, \ldots\}$ be a locally finite system of points in $\mathbb{R}^d$. Each location in $\mathbb{R}^d$ is associated to its *nearest point(s)* belonging to φ. The *neighbourhood* or *Voronoi cell* $C(x_i, \varphi)$ of a point x_i, called the *nucleus* or *generator*, of φ is defined by

$$C(x_i, \varphi) = \{y \in \mathbb{R}^d : \|y - x_i\| \le \|y - x_j\| \text{ for all } j \ne i\}. \tag{9.1}$$

The $C(x_i, \varphi)$ are all convex polygons (with the x_i as inner points) but it is possible for some polygons to be unbounded. If this possible deficiency does not arise then the $T(y)$ constitute a tessellation of $\mathbb{R}^d$, the *Voronoi tessellation* $\mathscr{V}(\varphi)$ relative to (or generated by) φ. Some authors refer to the $d = 2$ case as the *Dirichlet* or *Thiessen* tessellation; here the first of these terms is used. When φ is a lattice (as it was in Dirichlet's and Voronoi's cases), physicists and metallurgists call the cells Wigner–Seitz cells or zones.

The Voronoi tessellation can be also interpreted as a result of a growth process. The points x_i of φ play the rôle of nuclei, in which growth begins at the same instant. The speed of growth is uniform in all directions, so that the nuclei grow into discs (or balls in the spatial case) while empty space is available. When two discs come in contact, growth stops at the contact point and a facet begins to develop, while in other places growth continues. Eventually, the whole space is divided into cells. Figure 9.3 shows some steps of the growth process in the planar case. Muche (1993) and Schulz *et al.* (1993) study characteristics of the 'incomplete' tessellation obtained after time t.

This simple growth model makes the Voronoi tessellation attractive as a model for microstructures which result from crystallisation. Surprisingly, it seems to be also a good model for the universe; see Zaninetti (2006), van de Weygaert (2007) and van de Weygaert *et al.* (2014). The idea is that the 'empty space' grows while the galaxies tend to lie on the cell facets, edges and vertices. The so-called Abell clusters of galaxies are believed to be positioned in vertices. This greatly refines the Neyman–Scott model mentioned in Section 5.3.

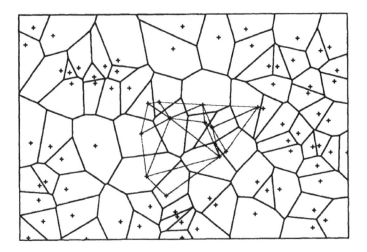

Figure 9.4 A typical realisation of the Dirichlet tessellation $\mathscr{V}$ relative to a sample of a homogeneous Poisson point process and some triangles of the corresponding Delaunay tessellation.

The regular square tessellation is a particular example of a Voronoi tessellation. It arises from a regular square grid of nuclei. This tessellation is degenerate; if by contrast a set of nuclei is in general quadratic position, meaning that no three nuclei lie on the same line and no four nuclei on the same circle, then this set must give rise to a tessellation with vertices at each of which only three cells meet. In the d-dimensional case 'general quadratic position' means that almost surely never $k + 1$ points lie in a $(k - 1)$-dimensional affine subspace of $\mathbb{R}^d$ for $k = 2, \ldots, d$ and never $d + 2$ points lie on the surface of a sphere. Then at each vertex $d + 1$ Voronoi cells meet.

If the nuclei form a stationary point process Φ of finite positive intensity ϱ then almost surely all the $C(x_i, \Phi)$ are bounded so $\mathscr{V}(\Phi)$ is indeed a random tessellation. If Φ is stationary then $\mathscr{V}(\Phi)$ inherits this property.

Known formulae for $\mathscr{V}(\Phi)$ are almost entirely confined to the case when Φ is a homogeneous Poisson process, and are presented in Section 9.7. Figure 9.4 shows a realisation of $\mathscr{V}(\Phi)$ relative to a sample of a homogeneous Poisson process Φ in the planar case. Furthermore, Figure 9.5 shows a simulated cell of a spatial Voronoi tessellation.

Voronoi tessellations relative to non-Poisson point processes are studied by simulation in Hermann *et al.* (1989), Lorz (1990), Kohutek and Saxl (1993), Lorz and Hahn (1993), van de Weygaert (1994), Saxl and Poniżil (2001, 2002) and Dereudre and Lavancier (2011). Heinrich (1998) considers contact and chord length distribution for the case of Poisson cluster and Gibbs processes. See Okabe *et al.* (2000, Section 5.12) for a survey.

The Voronoi tessellation has been often used as a model for natural phenomena in a variety of fields. Here are some now classical papers:

- *agriculture and forestry*: Klier (1969), Fischer and Miles (1973), Gavrikov *et al.* (1993), Kessler and Werner (2003);
- *astrophysics*: Kiang (1966), Icke and van de Weygaert (1987), van de Weygaert and Icke (1989);

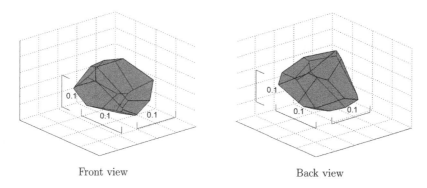

Front view   Back view

Figure 9.5   A cell of a spatial Poisson-Voronoi tessellation of intensity 1000 (so that the mean volume of the typical cell is 0.001). It has 15 facets, 39 edges and 26 vertices.

- *cell biology*: Lewis (1946), Honda (1983);

- *communication theory*: Shannon (1949);

- *crystal growth and aggregates*: Kolmogorov (1937), Meijering (1953), Gilbert (1962);

- *geography*: Mardia *et al.* (1977), Boots (1987);

- *geology*: Stiny (1929), Smalley (1966), Crain (1976), Gray *et al.* (1976), Stoyan and Stoyan (1980a);

- *metallography*: Boots (1982, 1984), Hermann (1991);

- *physics*: Rahman (1966), Ogawa and Tanemura (1974), Finney (1979), Boots (1982);

- *protein structure*: Richards (1974), Finney (1975), Poupon (2004);

- *zoology and ecology*: Hamilton (1971), Hasegawa and Tanemura (1976, 1980), Hasegawa *et al.* (1981).

See also the detailed description of applications of the Voronoi tessellation in the encyclopedic work of Okabe *et al.* (2000).

In addition to physical objects, the Voronoi tessellation can also model abstract concepts in the so-called conceptual spaces (Gärdenfors, 2000; Douven *et al.*, 2013) of cognitive science.

Voronoi tessellations are also used in data analysis of geometrical structures. Finney (1979) uses the term 'polyhedral statistics'. Sibson (1980, 1981) describes the use of Voronoi tessellations relative to given point patterns as a basis for 'natural neighbour interpolation': interpolating a smooth function for data located at irregularly distributed points. Thiessen and Alter had this in mind when they originally suggested this tessellation. A modern method in this spirit is discussed in Bernardeau and van de Weygaert (1996); see also Schaap and van de Weygaert (2000). Their 'Delaunay tessellation field estimator' yields estimates of random field values and point process intensities. To estimate a (unknown) bounded Borel set A in the unit cube $(0, 1)^d$ in $\mathbb{R}^d$, Khmaladze and Toronjadze (2001) and Penrose (2007) show that the union of Voronoi cells

$$A_n = \bigcup_{x_i \in A \cap \Phi_n} C(x_i, \Phi_n) \cap (0, 1)^d, \tag{9.2}$$

where Φ_n is a set of n i.i.d. random test points in the unit cube, is a consistent estimator of A in the sense that

$$\nu_d(A \triangle A_n) \to 0 \qquad \text{almost surely, as } n \to \infty. \tag{9.3}$$

The moments of this symmetric difference are studied in Heveling and Reitzner (2009) and Reitzner *et al.* (2012). For the problem of estimating, in the presence of noise, the region where the intensity of a bounded planar point process is nonzero, such as the detection of a minefield from an aerial image containing mine locations corrupted by clutter, Allard and Fraley (1997) propose a nonparametric maximum likelihood estimator in the form of a union of Voronoi polygons. In planar point pattern analysis, Voronoi tessellation is also used to estimate contour lines of the density of points (Picard and Bar-Hen, 2000), and to construct uniformity measures (Ong *et al.*, 2012) and test statistics for complete spatial randomness (Chiu, 2003).

Delaunay tessellations

If $\mathcal{V}(\Phi)$ has the property that almost surely each vertex is touched by exactly three cells (in the planar case) or by exactly four cells (in the spatial case) then a further tessellation can be constructed, the *Delaunay tessellation* $\mathcal{D}(\Phi)$. In the planar case it is constructed out of the triangles formed by points of whose cells share the same vertex. Section 9.7.4 contains some formulae for $\mathcal{D}(\Phi)$ in the Poisson process case. Early applications of the Delaunay network (the system of edges of the Delaunay tessellation) are given in Priolo *et al.* (1992) and Kumar and Singh (1995) in the context of conductance in disordered matter.

The construction of Voronoi and Delaunay tessellations, which is basic for simulations of random tessellations, is discussed in Preparata and Shamos (1985), Guibas and Stolfi (1988) Aurenhammer (1991), Fortune (1992), van de Weygaert (1994), Boissonnat and Yvinec (1998), Okabe *et al.* (2000), Hjelle and Dæhlen (2006) and de Berg *et al.* (2008). Efficient algorithms are implemented in, for example, the R packages such as `deldir` and `tripack` and the MATLAB toolbox MPT.

Generalised/higher-order Voronoi tessellations

The Voronoi tessellation has been generalised in many ways; see Okabe *et al.* (2000, Chapter 3) and Gavrilova (2008).

Miles (1970) defines the *generalised Voronoi tessellation*, also called n^{th}-*order Voronoi tessellation*, $\mathcal{V}_n$, which uses for cells the sets of positions sharing the same n nearest neighbours of the generating point pattern. Of course, $\mathcal{V}_1 = \mathcal{V}$, giving the original Voronoi tessellation. Details can be found in Miles (1970, 1972b) and Miles and Maillardet (1982). The latter paper gives pictures of $\mathcal{V}_n$ for $n = 4, 16, 64$ and 256 relative to a homogeneous Poisson process. See also Edelsbrunner (1987, Sections 13.3–13.5) and Okabe *et al.* (2000, Section 3.2) for elaborated surveys.

Non-Euclidean norm Voronoi tessellations

Ohser and Lorz (1994) show that interesting tessellations are obtained if the Euclidean norm is replaced by other norms for which the unit ball is not a ball in the usual sense but, for example, a square or cube. Scheike (1994) considers an elliptic distance to study tessellations resulting from elliptic, instead of spherical, growth.

Weighted Voronoi tessellations

Tessellations of particular interest are obtained if the Euclidean distance in (9.1) is replaced by other, weighted, distance definitions, where the weights may represent some measures of actual physical sizes of the nuclei. The following gives briefly two examples; the geometrical ideas behind are discussed in Okabe *et al.* (2000, Section 3.5.3).

The definitions are based on marked point processes. In the given context, the marks $w_i \in \mathbb{R}$ are called weights or powers, and instead of the set φ of unmarked nuclei, the set $\psi = \{[x_i; w_i]\}$ is considered.

(a) *Laguerre tessellations*   The first definition of the distance between the marked point $[x_i; w_i]$ and a point y of $\mathbb{R}^d$ is as follows:

$$\text{dist}_L([x_i; w_i], y) = \|x_i - y\|^2 - w_i. \tag{9.4}$$

The corresponding cells are then defined as

$$C([x_i; w_i], \psi) = \{y \in \mathbb{R}^d : \text{dist}_L([x_i; w_i], y) \leq \text{dist}_L([x_j; w_j], y) \text{ for all } j \neq i\}. \tag{9.5}$$

The resultant tessellation is called *Laguerre tessellation*, also known as *radical* or *additively weighted power Voronoi tessellation*, or similarly. If the powers w_i are identical then the Laguerre tessellation coincides with the Voronoi tessellation.

The cells of a Laguerre tessellation are convex. It can happen that a nucleus produces an empty cell (therefore, it must be said more precisely that the tessellation consists of the non-empty cells only) or that a nucleus is not an element of its cell. There is a tendency for points x_i with large powers w_i to have large cells.

For a system of hard balls, the Laguerre tessellation relative to the ball centres marked by the respective radii has convex cells, where each ball is contained in a cell. A Voronoi tessellation is sufficient to obtain a tessellation the cells of which contain the balls only in the special case in which the balls have identical radii. Figure 9.6 and the book cover, respectively, show systems of hard discs in the plane and hard balls in space and the corresponding Laguerre tessellations; the weights are the ball radii.

(b) *Additively weighted Voronoi tessellations*   The second definition considered here is

$$\text{dist}_{aw}([x_i; w_i], y) = \|x_i - y\| - w_i. \tag{9.6}$$

It leads to the so-called *additively weighted Voronoi tessellation*. Not all its cells are convex; in the planar case the boundaries are pieces of hyperbolic curves. An important related model is the Johnson–Mehl tessellation described below. In these cases the definition of tessellations has to be modified to include nonpolygonal cells, see Zähle (1988).

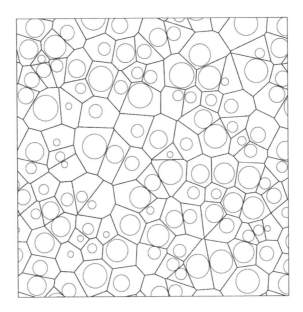

Figure 9.6 A Laguerre tessellation with respect to irregularly distributed discs. The Voronoi tessellation with respect to the disc centres does not have the property that all its cells contain the discs shown. Courtesy of C. Redenbach.

Voronoi S-tessellations

Of great practical interest is also the *Voronoi S-tessellation* in $\mathbb{R}^3$, where 'S' comes from 'distance to the surface'; see Medvedev (2000), Alinchenko *et al.* (2004) and Medvedev *et al.* (2006). As in the case of the Laguerre tessellation, this tessellation is defined relative to a system of hard balls, and each cell contains the respective ball. However, some of the cell boundaries are curved, hyperboloidal surfaces. Physicists consider this tessellation as more natural than the Laguerre tessellation, which they view only as a useful mathematical tool.

A definition which is more general than that for the case of hard balls is as follows. Let $K_1, K_2, \ldots$ be a sequence of non-overlapping convex bodies, for example balls with different radii. The cell containing K_i is defined by

$$C_i = \{x \in \mathbb{R}^d : d(x, K_i) \leq d(x, K_j) \text{ for all } j \neq i\}, \tag{9.7}$$

where $d(x, K)$ is the usual distance between the point x and the set K.

Clearly, if the K_i are identical balls, then the Voronoi S-tessellation coincides with the classical Voronoi tessellation.

The paper Medvedev *et al.* (2006) describes the construction of the Voronoi S-tessellation relative to given ensembles of balls of different radii.

Johnson–Mehl tessellations

The *Johnson–Mehl tessellation* is a structure resulting from a birth-and-growth process (see Section 6.6.4). This generalises the process leading to the Voronoi tessellation. Here 'nuclei' are generated by some birth process of intensity $\lambda(t)$, and each nucleus grows with speed $v(t)$ so that at time t after its birth it occupies all the previously vacant region within the ball of

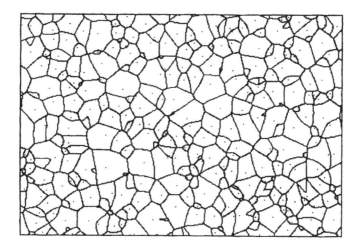

Figure 9.7   A realisation of a Johnson–Mehl model for a time-stationary birth process. This tessellation was drawn using a program which constructs the hyperbolic edges of the cells. Courtesy of L. Heinrich and E. Schüle.

radius $\int_0^t v(u)\mathrm{d}u$ centred on its original location. Consequently, if a nucleus is born at a point in the current cell of another nucleus then it vanishes immediately, while the survivors form cells that grow radially in each direction with age-dependent velocity $v(t)$, till cells of other nuclei are met in all directions. Figure 9.7 shows a part of a realisation of a planar Johnson–Mehl model. (A variant of the model considers that at time t after its birth, a nucleus born at time t_0 will occupy vacant region within a ball of radius $\int_{t_0}^{t+t_0} v(u)\mathrm{d}u$. However, it may lead to nuclei not lying in their own cells and/or disconnected cells and hence is less appealing.)

In the simplest Johnson–Mehl model, a homogeneous Poisson process of constant intensity in $\mathbb{R}^d \times [0, \infty)$ forms the nucleus birth locations and times. However, models with births that are stationary in space but not in time are also mathematically tractable (Møller, 1992). Section 9.9 contains some formulae for such time-inhomogeneous Johnson–Mehl models.

If all nuclei are born at the same instant then the resulting tessellation is Voronoi; but in the general model the cells of the tessellations are not even convex, and lens-shaped cells may appear. If the growth speed is a constant v then additively weighted Voronoi tessellations (generated by weighted nuclei $\{[x_i; -vt_i]\}$, where t_i is the birth time of the nucleus born at location x_i) are obtained.

The relevance of the model for crystal growth is discussed in Kolmogorov (1937), Johnson and Mehl (1939), Avrami (1939), Meijering (1953), Gilbert (1962) and Saltykov (1974). A nice application is given by Horálek (1988, 1990), who showed that the so-called ASTM model used in metallurgy can be well described by a particular Johnson–Mehl model.

The models of the DNA replication in Vanderbei and Shepp (1988) and Cowan *et al.* (1995) and the autoinhibited release of neurotransmitters at a synapse in Bennett and Robinson (1990) are also examples of applications of the Johnson–Mehl model.

Centroidal Voronoi tessellations

Voronoi tessellations in which the cell nuclei are also the centres of gravity of the cells are called *centroidal Voronoi tessellations*. Such tessellations exhibit a high degree of order.

Figure 9.8   A sample of a Gilbert tessellation. The ● are the starting points.

They can be computed iteratively using the Lloyd method (Du *et al.*, 1999) or a probabilistic algorithm (Ju *et al.*, 2002), and have applications in many different areas, such as image compression and meshless computing.

Crack and STIT tessellations

Many tessellations in nature are the result of fracture or crack processes. The tessellations are generated by edge growth instead of cell growth. Examples include crack systems on pottery surfaces, in rocks and in drying mud.

Typically, the edge systems of such tessellations show forms of hierarchical ordering, since the cracks appear step by step. In the planar case the vertices are typically T-shaped (π-vertices), and analogously in the spatial case. Such tessellations are not face-to-face.

A model of this class is the *Gilbert tessellation* (Gilbert, 1967), which is based on a marked point process in the plane, where each point is marked by the orientation of a line. The tessellation edges are generated by a growth process: each edge starts at one of the points and grows at constant rate in the two opposing directions specified by the mark of the point. Growth in a particular direction continues until further edges are hit. Figure 9.8 shows a part of a simulation of such a tessellation. Mathematically, the Gilbert tessellation is extremely difficult, even if the points form a homogeneous Poisson process and the orientations are uniform and independent; see Burridge *et al.* (2013).

A further crack tessellation extensively studied by mathematicians is the STIT tessellation, introduced by Nagel and Weiss (2003, 2004, 2005). The name comes from 'stability under iteration', a property which is related to a successive construction principle. Here the model construction (i.e. simulation) is described in the planar and isotropic case, following Mecke *et al.* (2008) and Calka (2010). For simplicity, the construction is described for the bounded planar case, that is, the tessellation is constructed in a bounded convex polygon W (e.g. $[0, 1]^2$). A definition for the unbounded and stationary case (for the whole $\mathbb{R}^d$) is possible; see Mecke *et al.* (2008) and Calka (2010). A further generalisation to the inhomogeneous case in $\mathbb{R}^2$ can be found in Mecke (2010).

The construction is dynamic and takes place over a time interval $[0, t]$. Intuitively, each polygon (including the window W itself) waits for an independent exponentially distributed time, with mean equal to its perimeter, and then divides into two polygons. The precise definition is based on an i.i.d. sequence $\{\tau_i, g_i\}$ of exponentially distributed random times τ_i and random lines g_i from a motion-invariant Poisson line process as in Section 8.2.2 that hit W (and in later steps the smaller polygons resulting from divisions). The parameter of the exponential distribution of the τ_i is $L(W)$. (According to Formula (8.4) this is the motion-invariant measure of the set of all lines hitting W.) The construction steps are as follows.

1. If $\tau_1 > t$, the process does not start at all and the tessellation is simply the entire window W.

2. If $\tau_1 \leq t$, W is cut at time τ_1 by the line g_1 into two polygons W_+ and W_-. These are then treated in the following way, separately and independently.

   (a) If $\tau_1 + \tau_2 > t$, W_+ remains as it is and will be a part of the final tessellation.

   (b) If $\tau_1 + \tau_2 \leq t$, W_+ is divided into two new polygons by g_2 if it intersects W_+. If g_2 does not hit, the next potential division of W_+ happens at time $\tau_1 + \tau_2 + \tau_3$, provided that it is less than t, and so on, so that either W_+ is eventually divided into two new polygons by g_{i_0} at time $\tau_1 + \cdots + \tau_{i_0} \leq t$ for some $i_0 \geq 2$, or W_+ remains as it is because none of $g_2, \ldots, g_{i_0-1}$ hits W_+ and $\tau_1 + \cdots + \tau_{i_0} > t$.

   (c) Repeat steps (a) and (b) above for W_-, with the same τ_1 but $\{(\tau_2, g_2), \ldots, (\tau_{i_0}, g_{i_0})\}$ is replaced by an independent sequence $(\tau_{i_0+1}, g_{i_0+1}), (\tau_{i_0+2}, g_{i_0+2}), \ldots$, until either W_- is divided into two new polygons or the sum of the exponential times is greater than t.

3. If both W_+ and W_- are not divided further in step 2, the process stops and the resultant structure is the tessellation. Otherwise, the same procedure is applied to each new polygon with independent exponential times and random lines iteratively until no more division is possible (i.e. after time t).

Because the polygons do not overlap, dividing whether W_+ or W_- first does not matter. Figure 9.9 shows two steps of tessellating a square by the STIT algorithm.

Although τ_i have the same parameter, in step 2(b) it is clear that a smaller polygon has to wait for a longer time until it is hit and divided by a random line. An equivalent formulation is that each polygon waits for an exponential time with parameter given by the length of the polygon perimeter, and then, unless the time is up, it will be divided by a random line that hits the polygon.

In the given form the model depends only on time t, which controls the density of edges. In the anisotropic case the rose of directions of the lines is a further parameter. In the stationary case the line density L_A and specific surface S_V, respectively, serve as parameters.

Section 9.6 presents some results for stationary STIT tessellations, and in Section 10.6 a remark on the stereology of such tessellations is given, that is, on the tessellations resulting from planar sections of spatial STIT tessellations.

Generalisations of the STIT model are considered in Cowan (2010), in which STIT tessellations are the special case that employs the 'perimeter-weighted' cell selection rule.

Horgan and Young (2000) present a model for the geometry of two-dimensional crack growth in soil, producing patterns which resemble that in Figure 9.2 on p. 345. Also here the cracks appear stepwise; similar is the fragmentation model in Hernandez et al. (2012).

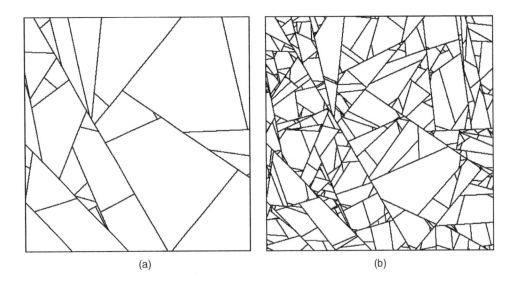

(a)   (b)

Figure 9.9   Two subsequent realisations of a planar isotropic STIT tessellation: (a) after 64 iterations; (b) after 1024 iterations. Courtesy of J. Ohser.

Further methods of tessellation construction

More complicated tessellations can be constructed by means of operations on simpler tessellations or by combining the methods given above. Here are two examples:

Superposition

If Θ_1 and Θ_2 are two tessellations with edge systems (regarded as fibre processes) E_{Θ_1} and E_{Θ_2}, respectively, then the union of E_{Θ_1} and E_{Θ_2} forms a further fibre process which defines the edge system of the *superposition tessellation*. For example, Mecke (1983, 1984b) and Santaló (1980, 1984) discuss this operation. Baccelli *et al.* (2000) study superpositions of planar Voronoi tessellations. Nagel and Weiss (2003) show that superpositions of suitably rescaled independent stationary planar tessellations converge to Poisson line tessellations.

Iterated division of cells

Each cell of a tessellation Θ may be divided into further cells by use of some rule of tessellation. There may be some form of hierarchy in the sense that the new cells tend to be an order smaller than the cells of Θ and are formed in close relation to them. For example, points may be scattered in the cells of Θ and then each cell itself is further tessellated by the Voronoi tessellation relative to the points in the cell, independent of the points in other cells, or as in Figure 9.10, each cell of a Voronoi tessellation is further subdivided by an independent line process. The process of division of cells may be continued in an iterative manner.

A general theory of iterated tessellations is presented in Maier and Schmidt (2003) and Maier *et al.* (2004). This includes formulae for intensities of face processes, for mean intrinsic volumes of typical faces and for the typical cell. Important components of these models are Poisson line and Poisson-Voronoi tessellations. A special construction principle leads to STIT tessellations.

Figure 9.10 A part of a Voronoi tessellation in which the cells are further subdivided by independent line processes. Courtesy of V. Schmidt.

Iterated or nested tessellations are used as models of road networks in telecommunication; see Gloaguen *et al.* (2006). The edges of a first tessellation stand for the main roads. In its cells, independent copies of a secondary tessellation with higher edge density are drawn to obtain the secondary roads that end up in the main roads. The types of both tessellations are Voronoi, Delaunay, Poisson line or regular tessellations.

9.3   General ideas and results for stationary planar tessellations

This section is still restricted to tessellations consisting of polygonal cells only. However, some notions such as the typical cell and tessellation faces are also applicable under a more general definition of tessellations allowing cells with curved boundaries, see Section 9.9.

9.3.1   Point processes related to tessellations

Tessellations are often accompanied by other stochastic structures. For example, a planar tessellation Θ automatically produces a fibre process, more precisely a segment process, namely the edge system E_Θ. A corresponding important tessellation characteristic is the line density L_A as defined in Section 8.3.1, the mean total edge length per unit area.

Together with E_Θ, three planar point processes can be constructed from Θ: the *set of vertices* $\alpha_0(\Theta)$, the *set of edge midpoints* $\alpha_1(\Theta)$, and the *set of cell centroids* $\alpha_2(\Theta)$.

If Θ is stationary then so are E_Θ and the $\alpha_k(\Theta)$. The intensities of the $\alpha_k(\Theta)$ are denoted by λ_k for $k = 0, 1$ and 2; for example, λ_2 is the mean number of cells per unit area. It is assumed that L_A and all three λ_k are positive and finite.

Further characteristics of a tessellation can be obtained by marking the point processes $\alpha_k(\Theta)$ in various geometrically natural ways. The ensuing mark distributions or their means provide further valuable tessellation characteristics. The following are some useful real-valued marks. Note that, because some important models for tessellations (such as STIT) are not face-to-face, it is necessary to distinguish between k-faces of tessellation and k-faces of polygons. In the notation above and below, the number subscripts '$_0$', '$_1$' and '$_2$' indicate the dimension(s) of the k-faces of the tessellation, namely, vertices, edges and cells, whilst the letter subscripts '$_C$' and '$_S$' are used to indicate polygonal k-faces, that is, corners and sides.

Marks for a vertex x in $\alpha_0(\Theta)$:

$n_{01}(x) = n_{02}(x) = \#\{C \in \Theta : x \in C\} = $ the number of edges emanating from the vertex x,

$n_{0S}(x) = $ the number of polygonal sides containing the vertex x,

$l_0(x) = $ the total length of the edges emanating from the vertex x.

Note that $n_{0S}(x)$ and $n_{02}(x)$, even when there is no π-vertex, are not the same. See Formula (9.11) on p. 360 and the explanation below it.

Mark for an edge midpoint x in $\alpha_1(\Theta)$:

$$l_1(x) = \text{the length of the edge through } x.$$

Marks for a cell centroid x in $\alpha_2(\Theta)$:

$$n_{20}(x) = \text{the number of vertices on the boundary of } C,$$
$$c_2(x) = \text{the number of corners of } C,$$
$$l_2(x) = L_2(C), \text{ the perimeter of } C,$$
$$a_2(x) = v_2(C), \text{ the area of } C,$$

where $x \in C$ and $C \in \Theta$. Note that $c_2(x)$ must not be confused with the number of polygon corners *adjacent to* polygon C (which would be naturally denoted by $n_{2C}(x)$). First, the π-vertices, if any, on the boundary of C are still corners of some other polygons and have to be counted in $n_{2C}(x)$ but not in $c_2(x)$. Second, a corner of C is also a corner of other polygons adjacent to C and hence, similar to the case in $n_{0S}(x)$, will be counted more than once in $n_{2C}(x)$ but only once in $c_2(x)$.

Each of these marks leads to a mark distribution. Denote by n_{02}, n_{0S}, l_0, l_1, n_{20}, c_2, l_2, and a_2, respectively, random variables following these mark distributions. The corresponding means are denoted by $\bar{n}_{02}$, $\bar{n}_{0S}$, $\bar{l}_0$, $\bar{l}_1$, $\bar{n}_{20}$, $\bar{c}_2$, $\bar{l}_2$ and $\bar{a}_2$. Each random variable and its mean are referred to as the appropriate random value and the mean value of the corresponding geometrical quantity. For example, l_1 is the random length of the typical edge with mean $\bar{l}_1$, n_{20} the random number of vertices on the boundary of the typical cell with mean $\bar{n}_{20}$ (for the definition of typical vertex, edge and cell; see Section 9.3.2 below). Since the number of vertices on the boundary of a cell is equal to the number of edges, n_{20} is sometimes also denoted as n_{21}.

9.3.2 Typical vertex, edge and cell

Another marking of the points of $\alpha_2(\Theta)$ uses the corresponding cells, taken as compact convex sets. If x_n is a cell centre and C_n the corresponding cell, then $\{[x_n; C_n]\}$ is a germ–grain process as in Section 6.5, which inherits stationarity from Θ. The typical grain of $\{[x_n; C_n]\}$ is called the *typical cell* of Θ in the tessellation context and is denoted by C^o. Analogously, for $\alpha_1(\Theta)$, the edge midpoints and the corresponding edges form another stationary germ–grain process, in which the typical grain is called the *typical edge*; for $\alpha_0(\Theta)$ with marks, the mark distribution is known as the distribution of the marks of the *typical vertex*.

Note that the typical cell should not be viewed as a particular cell selected from a given tessellation. Instead, it is a valuable distributional single-cell characteristic of a stationary tessellation. In a heuristic way the typical cell can be seen as the cell that is sampled from the population of all cells by random selection where all cells have the same chance to be chosen. The mean values for characteristics of the typical cell can be computed by using the Palm distribution of $\alpha_2(\Theta)$ or the ergodic theorem; see Section 9.3.4 below.

Simulating the typical cell of a stationary tessellation

The aim to simulate an i.i.d. sequence of the typical cell of a stationary tessellation can be achieved approximately by, first, simulating a sample of the tessellation in a large window, then enumerating the cells in some sequential order and finally selecting the cells uniformly at random. Clearly, this method is not very efficient; there may be stochastic dependence between the cells obtained from the same sample of the tessellation; and edge-effects may play a rôle.

In the case of Poisson-process-related tessellations considered in Sections 9.5 and 9.7 there are efficient special constructions. The Slivnyak–Mecke theorem is the basis for the simulations.

9.3.3 Zero cell

In addition to the typical cell, another single cell is associated with a tessellation, namely the *zero cell* C_0. This is that cell that contains the origin of the plane. Because of the different definitions, zero cell and typical cell are objects with different probability distributions. As will be shown in inequalities (9.30) and (9.31), the zero cell tends to be larger than the typical cell.

Another way to understand the distribution of the zero cell C_0 uses random sampling with points. Take a random test point and choose the cell containing the test point. By stationarity, the corresponding cell has the same distribution as the zero cell. It is plausible that the test point tends to fall with greater probability into larger cells. Thus the zero cell can be seen as a size-weighted version of a single cell.

Motivated by these heuristic considerations, Matheron (1975) used the terms *number law* for the distribution of the typical cell and *volume law* for that of the zero cell.

9.3.4 Mean-value relationships for stationary planar tessellations

There are a number of relationships holding between the above mean values in the case of stationary tessellations. They have been established by means of Palm distributions, as in Mecke (1980), and by means of limiting arguments involving ergodicity, as in Cowan (1978, 1980). As Mecke (1984b) showed, for almost surely face-to-face tessellations all these mean values can be expressed by the *three* parameters λ_0, λ_2 (or $\bar{n}_{02}$) and L_A. This follows from simple geometric reasoning, which has to be made precise in one of the two ways above. For general tessellations one more parameter is needed, namely, the proportion ϕ of π-vertices; see Weiss and Cowan (2011).

For example, the Euler formula for *finite* polyhedra in $\mathbb{R}^3$ states that

$$\#\text{corners} - \#\text{sides} + \#\text{facets} = 2.$$

Since a tessellation can be interpreted as the surface of an *infinite* polyhedron with the vertices and edges of the tessellation as the sides and corners of the polyhedron, limiting arguments show that

$$\lambda_0 - \lambda_1 + \lambda_2 = 0. \tag{9.8}$$

This basic formula exhibits a duality between vertices and cells; that is, between the indices '$_0$' and '$_2$'. Any tessellation is associated with a *dual* tessellation obtained by taking the original cell centroids as new vertex and connecting pairs of new vertices if they are centroids of neighbouring old cells. This association explains the duality between indices which underlies several of the formulae given below:

$$\bar{n}_{02} = 2 + \frac{2\lambda_2}{\lambda_0}, \tag{9.9}$$

$$\bar{n}_{20} = 2 + \frac{2\lambda_0}{\lambda_2}, \tag{9.10}$$

$$\bar{n}_{0S} = 2\bar{n}_{02} - \phi \tag{9.11}$$

$$\bar{c}_2 = \bar{n}_{20} - \frac{2\phi}{\bar{n}_{02} - 2}, \tag{9.12}$$

$$\bar{l}_0 = \frac{2L_A}{\lambda_0}, \tag{9.13}$$

$$\bar{l}_2 = \frac{2L_A}{\lambda_2}, \tag{9.14}$$

$$\bar{a}_2 = \frac{1}{\lambda_2}, \tag{9.15}$$

$$\bar{l}_1 = \frac{L_A}{\lambda_0 + \lambda_2} = \frac{L_A}{\lambda_1}. \tag{9.16}$$

Formula (9.11) can be understood easily from Figure 9.11, in which the three polygons meet at the π-vertex c; each of the polygons A and B has two sides adjacent to c, whilst polygon C has only one, which is the union of two edges. Now, first, note that for a non-π-vertex, each edge adjacent to it is a side (or part of a side) of two polygons adjacent to the vertex and will be counted twice, explaining the factor 2 in Formula (9.11). Then, consider the edges at a π-vertex, the two edges lying on the same polygonal side will together contribute not 4 but only 3 to the total count of sides adjacent to the π-vertex, leading to the term $-\phi$ in Formula (9.11).

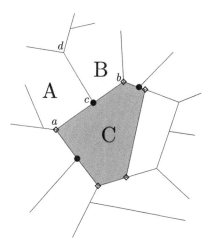

Figure 9.11   A tessellation with π-vertices. Polygon C has eight vertices, five of them are corners, represented by $\diamond$, and three are π-vertices, represented by $\bullet$. Vertex c is a corner of polygon A and polygon B. Polygon A has two sides adjacent to vertex c, namely (a, c) and (c, d); polygon B also has two sides adjacent to vertex c, namely (b, c) and (c, d); polygon C, however, has only one side adjacent to vertex c, namely (a, b). Thus, in total there are five polygonal sides adjacent to vertex c. On the other hand, three edges of tessellation are adjacent to vertex c, namely (a, c), (b, c) and (c, d).

From the above formulae others can be derived:

$$\frac{1}{\bar{n}_{02}} + \frac{1}{\bar{n}_{20}} = \frac{1}{2},$$ (9.17)

$$\bar{l}_0 = \bar{n}_{02}\bar{l}_1,$$ (9.18)

$$\bar{l}_2 = \bar{n}_{20}\bar{l}_1,$$ (9.19)

$$\bar{n}_{20} = \frac{2\bar{n}_{02}}{\bar{n}_{02} - 2},$$ (9.20)

$$\bar{c}_2 = \frac{2(\bar{n}_{02} - \phi)}{\bar{n}_{02} - 2},$$ (9.21)

and finally

$$\bar{n}_{02}, \bar{n}_{20}, \bar{c}_2 \geq 3,$$ (9.22)

$$\bar{n}_{20} \leq 6 - \phi.$$ (9.23)

More mean-value formulae for tessellations with π-vertices can be found in Weiss and Cowan (2011). Most formulae related to polygonal k-faces involve two more parameters, namely, the second moment of n_{02} and the mean number of edges emanating from the typical π-vertex.

Kendall and Mecke (1987) study the range

$$0 < \lambda_0 \leq 2\lambda_2, \qquad \lambda_2 \leq 2\lambda_0, \qquad L_A \geq 0$$

and show that each triplet satisfying these relations can be derived from a suitable tessellation. *The ordinary equilibrium* or *trivalent state* is an important special case. It holds when

$$n_{02}(x) \equiv 3 \qquad \text{for all } x \in \alpha_0(\Theta), \tag{9.24}$$

or

$$n_{02} \equiv 3 \qquad \text{almost surely,} \tag{9.25}$$

that is, when the number of edges emanating from each vertex is always three, as in Figures 9.1, 9.4 and 9.17. Tessellations with this property are also called *trivalent*. In this case simpler relationships hold:

$$\bar{n}_{20} = \bar{n}_{21} = 6, \qquad 2\lambda_2 = \lambda_0, \qquad \lambda_1 = 3\lambda_2. \tag{9.26}$$

Thus in the ordinary equilibrium state the mean number of edges (or vertices) of the typical cell is 6.

Formulae for the particular case of tessellations formed by line processes can be found in Section 9.5. In this case four edges emanate from each vertex and so the ordinary equilibrium state does not hold.

Proofs of some mean-value formulae

The following gives proofs for four examples of the general mean-value formulae. The proofs are taken from Mecke (1980) and are good examples of the use of the theorems concerning Palm distribution calculations. To begin with, some notation is introduced.

If Θ is a stationary planar tessellation with distribution P, then the $\alpha_k(\Theta)$ ($k = 0, 1, 2$) and also $\beta(\Theta) = \alpha_0(\Theta) \cup \alpha_1(\Theta) \cup \alpha_2(\Theta)$ are stationary point processes. The intensity of $\beta(\Theta)$ is $\lambda' = \lambda_0 + \lambda_1 + \lambda_2$, and under the general assumptions on tessellations λ' is positive and finite. The Palm distribution P_o of $\beta(\Theta)$ is

$$\lambda' P_o(Y) = \int_{\mathbb{T}} \sum_{x \in \beta(\theta) \cap [0,1]^2} \mathbf{1}_Y(\theta - x) P(d\theta) \qquad \text{for } Y \in \mathscr{T}. \tag{9.27}$$

This may be regarded intuitively as the probability distribution of a tessellation configuration given that a point of $\beta(\Theta)$ is at o. It is useful to further distinguish which of the $\alpha_k(\Theta)$ contains this point, called the typical point. If $T_k = \{\theta \in \mathbb{T} : o \in \alpha_k(\theta)\}$, then let

$$E_k\big(g(\Theta)\big) = \frac{\lambda'}{\lambda_k} \int_{T_k} g(\theta) P_o(d\theta) \qquad \text{for } k = 0, 1, 2,$$

where $g : T_k \to [0, \infty)$ is measurable. The quantity $E_k\big(g(\Theta)\big)$ can be interpreted as the mean of $g(\Theta)$ under the condition that o is a point of $\alpha_k(\Theta)$.

Proposition 9.1. *Let $C_0(\Theta) = C_0$ be the zero cell of the tessellation Θ. Then*

$$\mathbf{E}\big(f(\Theta)\big) = \lambda_2 E_2 \left(\int_{C_0(\Theta)} f(\Theta - x)\,dx \right) \tag{9.28}$$

for every measurable nonnegative function f on $\mathbb{T}$.

Proof. The right-hand side, using the definition of $E_2(\cdot)$ and change of coordinates $x \to -x$, becomes

$$\lambda_2 E_2 \left(\int_{C_0} f(\Theta - x)\,dx \right) = \lambda' \int_{\mathbb{T}} \int_{\mathbb{R}^2} \mathbf{1}_{T_2}(\theta)\mathbf{1}_{C_0(\theta)}(-x)f(\theta + x)\,dx\,P_o(d\theta).$$

A variation of Theorem 7.1(a) yields

$$\lambda' \int_{\mathbb{T}} \int_{\mathbb{R}^2} v(x, \theta)\,dx\,P_o(d\theta) = \int_{\mathbb{T}} \sum_{x \in \beta(\theta)} v(x, \theta - x)P(d\theta),$$

which can be applied to show that the right-hand side above equals

$$\int_{\mathbb{T}} \sum_{x \in \beta(\theta)} \mathbf{1}_{T_2}(\theta - x)\mathbf{1}_{C_0(\theta - x)}(-x)f(\theta)P(d\theta)$$

$$= \int_{\mathbb{T}} f(\theta)\cdot \#\{x \in \alpha_2(\theta) : -x \text{ belongs to the cell of } \theta - x \text{ containing } o\}\,P(d\theta).$$

If o is in the edge system E_θ of θ then there can be no cells containing it. Otherwise there can be only one such cell, and so only one corresponding cell centroid x in $\alpha_2(\theta)$. Also by stationarity

$$\mathbf{P}(o \in E_\Theta) = 0.$$

Hence the right-hand side of the equation in the proposition must equal

$$\int f(\theta)P(d\theta) = \mathbf{E}\big(f(\Theta)\big),$$

which proves the proposition.   □

Formula (9.28) can be explained intuitively as follows. The left-hand side is a kind of average of $f(\theta - x)$ over all x using an ergodic-theoretic interpretation of expectation and the stationary nature of Θ. The right-hand side is a kind of sum over all cell centroids (hence the factor λ_2) of integrals of $f(\theta - x)$ for x in the zero cell. It is natural to suppose that the two sides should be equal. The proof above uses Palm distributions to make this argument rigorous.

Taking $f \equiv 1$, Formula (9.28) yields

$$1 = \lambda_2 E_2 \left(\int_{C_0(\Theta)} dx \right) = \lambda_2 \bar{a}_2$$

which proves (9.15).

With $f(\Theta) = v_2(C_0(\Theta))^{-1}$, Formula (9.28) yields

$$\mathbf{E}(v_2(C_0)^{-1}) = \lambda_2 = \frac{1}{\bar{a}_2}. \tag{9.29}$$

An application of Jensen's inequality for $\mathbf{E}(f(X))$ with convex f for the special $f(x) = x^{-1}$ shows that

$$\bar{a}_2 \leq \mathbf{E}(v_2(C_0)), \tag{9.30}$$

which says that the mean area of the typical cell is smaller than that of the zero cell. This is one aspect of the property that the zero cell is 'larger' than the typical cell. Mecke (1999) shows more: The distribution functions of the areas of C_0 and C^o, denoted by $F_0(a)$ and $F^o(a)$ respectively, satisfy the inequality

$$F_0(a) \leq F^o(a) \qquad \text{for all } a \geq 0, \tag{9.31}$$

which implies that all moments of order greater than 1 of the area of the zero cell are greater than or equal to that of the typical cell, that is, the area of the zero cell is stochastically larger than that of the typical cell.

Mecke (1999) refines this inequality for Poisson line/plane tessellations and Poisson-Voronoi tessellations by showing that, by a suitable interpretation, the typical cell of each of two tessellations can be embedded as a subset in the zero cell of the tessellation, that is, there is $\subseteq_{st}$-order as defined on p. 212.

Møller (1989, Theorem 5.1) (see also Schneider and Weil, 2008, p. 493, Theorem 10.4.1) proves the following equation, which was originally given in Miles (1974a) for Poisson line/plane tessellation:

$$\mathbf{E}(f(C_0)) = \mathbf{E}(f(C^o)v_2(C^o))/\bar{a}_2 \tag{9.32}$$

for general stationary tessellations and any measurable, translation invariant function f that assigns to convex bodies nonnegative real numbers. An example is:

$$\mathbf{E}(N(C_0)) = \mathbf{E}(N(C^o)v_2(C^o))/\bar{a}_2, \tag{9.33}$$

where $N(C)$ is the number of neighbours of C. It can be used to determine the mean number of neighbours of the zero cell. Equation (9.32) shows in a clear way that the zero cell can be interpreted, up to translations, an area-weighted version of the typical cell.

Stoyan (1986) shows by counter-examples that mean perimeter and mean number of neighbours of the zero cell are not necessarily larger than their counterparts for the typical cell.

Proposition 9.2. *For a tessellation θ let $S(\theta)$ be the set of all cell centroids of cells containing o. Then for a stationary tessellation Θ*

$$\lambda_2 E_2 \left(\sum_{x \in \alpha_0(\Theta) \cap C_0(\Theta)} h(x, \Theta) \right) = \lambda_0 E_0 \left(\sum_{x \in S(\Theta)} h(-x, \Theta - x) \right) \tag{9.34}$$

for any nonnegative measurable function h on $\mathbb{R}^2 \times \mathbb{T}$.

Proof. A variation on Theorem 7.1(c) yields

$$\lambda' \int_{\mathbb{T}} \sum_{x \in \beta(\theta)} v(x, \theta) P_o(d\theta) = \lambda' \int_{\mathbb{T}} \sum_{x \in \beta(\theta)} v(-x, \theta - x) P_o(d\theta) \tag{9.35}$$

for nonnegative measurable functions v on $\mathbb{R}^2 \times \mathbb{T}$. Taking

$$v(x, \theta) = \mathbf{1}_{T_2(\theta)} \cdot \mathbf{1}_{\alpha_0(\theta) \cap C_0(\theta)}(x) \cdot h(x, \theta)$$

and applying (9.35) to the left-hand side of (9.34) give

$$\lambda_2 E_2 \left(\sum_{x \in \alpha_0(\Theta) \cap C_0(\Theta)} h(x, \Theta) \right) = \lambda' \int_{T_2} \sum_{x \in \alpha_0(\theta) \cap C_0(\theta)} h(x, \theta) P_o(d\theta)$$

$$= \lambda' \int_{\mathbb{T}} \sum_{x \in \beta(\theta)} v(x, \theta) P_o(d\theta)$$

$$= \lambda' \int_{\mathbb{T}} \sum_{x \in \beta(\theta)} v(-x, \theta - x) P_o(d\theta)$$

$$= \lambda' \int_{\mathbb{T}} \sum_{x \in \alpha_2(\theta)} h(-x, \theta - x) \mathbf{1}_{\alpha_0(\theta-x)}(-x) \mathbf{1}_{C_0(\theta-x)}(-x) P_o(d\theta).$$

If x is in $\alpha_2(\theta)$ then the summand above is zero unless o is a vertex, in which case $\theta \in T_0$, and x is the cell centroid of one of the cells in θ containing o as a vertex. Thus the last integrand above is equal to

$$\mathbf{1}_{T_0}(\theta) \sum_{x \in S(\theta)} h(-x, \theta - x)$$

from which (9.34) follows. □

As with (9.28), Formula (9.34) can be understood intuitively by an 'averaging' argument. The left-hand side averages over all cell centroids and counts the quantity $h(x, \theta)$ for x running through the vertices of the typical cell. The right-hand side averages over all vertices and counts the appropriate quantity for x running through the cell centroids of cells meeting at the typical vertex.

With $h \equiv 1$, Formula (9.34) yields

$$\lambda_2 \bar{n}_{20} = \lambda_0 \bar{n}_{02}.$$

If $h(x, \theta)$ is defined as the angle at the vertex x in $C_0(\theta)$ then the left-hand side of (9.34) involves the sum of all angles of the convex polygon $C_0(\theta)$. This sum equals

$$\big(\#\{\text{vertices of } C_0(\theta)\} - 2\big)\pi,$$

and so the left-hand side of (9.34) is $\lambda_2(\bar{n}_{20} - 2)\pi$. The right-hand side is $2\pi\lambda_0$ and so

$$\lambda_2 \bar{n}_{20} - 2\lambda_2 = 2\lambda_0.$$

Together these two derived equations yield the Formulae (9.9), (9.10) and (9.17).

9.3.5 The neighbourhood of the typical cell

The neighbourhood of the typical cell is of great importance, especially in materials science and physics. A well-known result is the so-called *Aboav's law* or *Aboav–Weaire's law* (Aboav, 1970, 1980; Weaire, 1974), which quantifies the relationship between n and $m(n)$, the mean number of edges of a randomly chosen neighbouring cell of the typical cell, under the condition that the typical cell has n edges. Lewis (1931) observed a remarkable tendency for few-edged cells to be in contact with many-edge cells and vice versa in epithelia. Aboav in a series of papers asserts empirically that

$$m(n) = 5 + \frac{8}{n}, \tag{9.36}$$

and Weaire suggested semi-empirically

$$m(n) = 5 + \frac{6 + \text{var}(n_{20})}{n}, \tag{9.37}$$

where $\text{var}(n_{20})$ denotes the variance of the number of edges (vertices) of the typical cell. Chiu (1994) showed mathematically that for trivalent tessellations,

$$m(n) = 5 + \frac{6}{n} + \frac{\text{cov}\big(k(n, n_{20}), n_{20}\big)}{n p_n} \qquad \text{for } n = 3, 4, \ldots, \tag{9.38}$$

where $p_n = \mathbf{P}(n_{20} = n)$ and $k(n, j)$ is the mean number of n-edged cells which belong to the complex of the typical cell, which has j edges. The complex of the typical cell is the union of the typical cell and its neighbouring cells, that is, its first shell. For an extensive review see Chiu (1995a) and Okabe *et al.* (2000, Section 5.5.3); for further work in this direction see Sahimi (2003, p.89) and Hilhorst (2006).

For unconditional mean-value formulae for the neighbourhood of the typical cell see Weiss (1995). She showed, in particular, that the mean area A_1 of the first shell of cells around the typical cell is given by

$$A_1 = \bar{a}_2 \mathbf{E}\big(N(C_0)\big), \tag{9.39}$$

where $\mathbf{E}\big(N(C_0)\big)$ is the mean number of edges of the zero cell. An analogous formula holds for tessellations in $\mathbb{R}^d$. Voloshin *et al.* (2010) study the volumes of the further shells by simulation. This is related to shell map analysis, see p. 393.

A heuristic proof of (9.39) goes as follows. Consider a large disc containing n cells. If edge-effects are ignored, the mean shell area A_1 is

$$A_1 = \frac{1}{n} \sum_i \sum_k a_{i,k},$$

where i goes over the cells and $a_{i,k}$ is the area of the kth cell in the shell around cell i. Since the area a_i of cell i appears in the double-sum as many times as the total number n_i of its neighbours, the mean shell area can be rewritten as

$$A_1 = \frac{1}{n} \sum_i n_i a_i = \frac{\sum_i n_i a_i}{\sum_i a_i} \cdot \frac{\sum_i a_i}{n} = \mathbf{E}\big(N(C_0)\big) \bar{a}_2.$$

Note that the last equality comes from the interpretation that the zero cell is the size-weighted version of a single cell.

The neighbourhoods of typical k-faces of the Poisson-Voronoi tessellation are studied in Baumstark and Last (2007).

Other characteristics of an n-edged cell

Another empirically suggested relationship for an n-edge cell is the so-called *Lewis' law*, which asserts a linear relation for the mean area $A(n)$ of the typical cell under the condition that it has n edges, namely

$$A(n) = \bar{a}_2 \left(\frac{n - n_0}{6 - n_0} \right), \tag{9.40}$$

for some constant n_0. The value 6 in the denominator comes from the mean number of edges of the typical cell in a trivalent tessellation. Quine and Watson (1984) conjecture that for the Poisson-Voronoi tessellation $n_0 = 3/2$, leading to

$$A(n) = \bar{a}_2 \left(\frac{2n - 3}{9} \right), \tag{9.41}$$

see Chiu (1995a) and Okabe *et al.* (2000, p. 315) for more empirically suggested forms.

Rivier and Lissowski (1982) develop a maximum entropy argument to explain heuristically that the linear relationship might result from statistical equilibrium, which might also give rise to Aboav's law (Peshkin *et al.*, 1991). However, Chiu (1995b) proves that their argument could only lead to tautological statements.

Conditional mean perimeter and mean edge length of an n-edged cell in a Poisson-Voronoi tessellation have been studied by simulation; see Okabe *et al.* (2000, p. 319).

9.4   Mean-value formulae for stationary spatial tessellations

Mean-value relationships hold for stationary spatial tessellations and tessellations of $\mathbb{R}^d$ just as for stationary planar tessellations. The relationships are due to Mecke (1980) and Stoyan and Mecke (1983a, Chapter 8), to Radecke (1980) for $\mathbb{R}^3$, and to Møller (1989) for general $\mathbb{R}^d$, and are displayed below for $d = 3$.

Again the terminologies in Weiss and Cowan (2011) are adopted. For the spatial case, the 0-, 1-, and 2-faces of a tessellation are called *vertices*, *edges* and *plates*, whilst those of the constituent polyhedra are called *apices*, *ridges* and *facets*. The polyhedra (3-faces) themselves are still called *cells*. A cell k-face may be a union of some k-faces of the tessellation, and lower dimensional faces may be in the relative interior of higher dimensional faces. When an edge whose interior lies in the interior of a facet, it is called a π-*edge*, and a vertex lying in the interior of a facet is a *hemi-vertex*; the same as in the planar case, a vertex in the interior of a ridge is a π-*vertex*. A tessellation without π-vertices, π-edges and hemi-vertices is *face-to-face*. When there are always four cells meeting at a vertex and three cells meeting at an edge, the tessellation is said in *ordinary equilibrium state*.

The definition of a face-to-face and ordinary equilibrium planar tessellation can be extended to a tessellation in $\mathbb{R}^d$. A tessellation is face-to-face, or *regular*, if the set of all k-faces, $k = 0, 1, \ldots, d$, of the tessellation and that of the constituent cells coincide. If each k-face of a tessellation of $\mathbb{R}^d$ lies on the boundaries of $d - k + 1$ cells for $k = 0, 1, \ldots, d - 1$, the tessellation is said to be in ordinary equilibrium state or *normal*. The latter term is used because many real-life tessellations possess this property. Some authors additionally require that a normal tessellation has to be face-to-face (see e.g. Lautensack and Zuyev, 2008), but this book treats the two properties as separate.

Let λ_k be the intensity of the point process of centroids of k-faces ($k = 0, 1, 2, 3$), let L_V be the intensity of the fibre process of edges (1-faces of Θ), and let S_V be the intensity of the surface process of plates (2-faces of Θ). If the vertices and edges are marked or weighted with the numbers of adjacent cells then the vertex point process and edge fibre process can be used to produce stationary random measures (multiplying by the marks or weights) whose corresponding intensities are denoted by T_V and Z_V.

The following notation is used.

$$\bar{n}_{kl} = \text{the mean number of } l\text{-faces of the tessellation adjacent to}$$
$$\text{the typical } k\text{-face of the tessellation, for } k \text{ and } l \text{ in } \{0, 1, 2, 3\},$$

$\bar{l}_1 = $ mean length of the typical edge (1-face of the tessellation),

$\bar{l}_2 = $ mean perimeter of the typical plate (2-face of the tessellation),

$\overline{A}_2 = $ mean area of the typical plate (2-face of the tessellation),

$\bar{l}_3 = $ mean total edge length of the typical cell,

$\overline{S}_3 = $ mean surface area of the typical cell,

$\overline{V}_3 = $ mean volume of the typical cell,

$\overline{\bar{b}}_3 = $ mean average breadth of the typical cell.

The quantity $\bar{n}_{30}$ is sometimes called *mean coordination number*.

Mecke (1984c) showed that, when restricted to tessellation k-faces, all these means can be expressed by the *seven* parameters

$$\lambda_0, \ \lambda_3, \ \lambda = \lambda_1 + \lambda_2, \ L_V, \ S_V, \ T_V, \ \text{and } Z_V,$$

whilst Weiss and Cowan (2011) found that, when cell k-faces are also of interest, four more parameters are needed, namely,

$\xi = $ proportion of π-edges,

$\kappa = $ proportion of hemi-vertices,

$\psi = $ mean number of ridge interiors adjacant to the typical vertex,

$\tau = $ mean number of plate-side interiors adjacent to the typical vertex.

The relevant formulae for k-faces of tessellation are as follows:

$$2\lambda_1 = \lambda + \lambda_0 - \lambda_3, \qquad\qquad 2\lambda_2 = \lambda + \lambda_3 - \lambda_0, \qquad\qquad (9.42)$$

$$\lambda_0 \bar{n}_{03} = T_V, \qquad\qquad \lambda_3 \bar{n}_{30} = T_V, \qquad\qquad (9.43)$$

$$\lambda_0 \bar{n}_{02} = T_V + \lambda - \lambda_0 - \lambda_3, \qquad\qquad \lambda_0 \bar{n}_{01} = \lambda + \lambda_0 - \lambda_3, \qquad\qquad (9.44)$$

$$\lambda_3 \bar{n}_{32} = \lambda + \lambda_3 - \lambda_0, \qquad\qquad \lambda_3 \bar{n}_{31} = T_V + \lambda - \lambda_0 - \lambda_3, \qquad\qquad (9.45)$$

$$\bar{n}_{12} = \bar{n}_{13} = 2\frac{T_V + \lambda - \lambda_0 - \lambda_3}{\lambda + \lambda_0 - \lambda_3}, \qquad \bar{n}_{21} = \bar{n}_{20} = 2\frac{T_V + \lambda - \lambda_0 - \lambda_3}{\lambda + \lambda_3 - \lambda_0}, \qquad (9.46)$$

$$\bar{l}_1(\lambda + \lambda_0 - \lambda_3) = 2L_V, \qquad\qquad \bar{l}_2(\lambda + \lambda_3 - \lambda_0) = 2Z_V, \qquad\qquad (9.47)$$

$$\bar{A}_2(\lambda + \lambda_3 - \lambda_0) = 2S_V, \qquad\qquad (9.48)$$

$$\lambda_3 \bar{l}_3 = Z_V, \qquad\qquad \lambda_3 \bar{S}_3 = 2S_V, \qquad\qquad (9.49)$$

$$\lambda_3 \bar{V}_3 = 1, \qquad\qquad 4\lambda_3 \bar{\bar{b}}_3 = Z_V - 2L_V. \qquad\qquad (9.50)$$

These formulae imply the relationships:

$$\lambda_0 - \lambda_1 + \lambda_2 - \lambda_3 = 0, \qquad\qquad (9.51)$$

$$\bar{n}_{01} - \bar{n}_{02} + \bar{n}_{03} = 2, \qquad \bar{n}_{30} - \bar{n}_{31} + \bar{n}_{32} = 2, \qquad\qquad (9.52)$$

$$\bar{n}_{12}\bar{n}_{01} = 2\bar{n}_{02}, \qquad \bar{n}_{21}\bar{n}_{32} = 2\bar{n}_{31}, \qquad \bar{n}_{02}\bar{n}_{30} = \bar{n}_{31}\bar{n}_{03}, \qquad (9.53)$$

$$\bar{S}_3 = \bar{n}_{32}\bar{A}_2. \qquad\qquad (9.54)$$

Some of these formulae display symmetries between the indices '0' and '3' and between the indices '1' and '2'.

For cell k-faces, denote by $\bar{m}_{kl}$ the mean number of cell l-faces of the typical k-face of the tessellation. The same as in the planar case, $\bar{m}_{kl}$ must not be confused the mean number of cell l-faces *adjacent to* the typical k-face, because when adjacency is of interest, each cell l-face will be counted as many times as the number of cells containing it. The relationships analogous to (9.21) are

$$\bar{m}_{30} = \bar{n}_{30} - \frac{2(\kappa + \psi)\bar{n}_{20}}{\bar{n}_{01}\bar{n}_{12} - \bar{n}_{20}(\bar{n}_{01} - 2)}, \qquad (9.55)$$

$$\bar{m}_{31} = \bar{n}_{31} - \frac{(\xi\bar{n}_{01} + 2\psi)\bar{n}_{20}}{\bar{n}_{01}\bar{n}_{12} - \bar{n}_{20}(\bar{n}_{01} - 2)}, \qquad (9.56)$$

$$\overline{m}_{32} = \overline{n}_{32} - \frac{(\xi \overline{n}_{01} - 2\kappa)\overline{n}_{20}}{\overline{n}_{01}\overline{n}_{12} - \overline{n}_{20}(\overline{n}_{01} - 2)}, \tag{9.57}$$

$$\overline{m}_{20} = \overline{m}_{21} = \overline{n}_{20}\left(1 - \frac{2\tau}{\overline{n}_{01}\overline{n}_{12}}\right). \tag{9.58}$$

More relationships involving cell k-faces can be found in Weiss and Cowan (2011).

Consider the special case of tessellations that are face-to-face. Then cell faces (apices, ridges and facets) and tessellation faces (vertices, edges and plates) coincide (and in this case the terms vertices, edges and cell facets are used in the literature). If in addition the tessellation is normal, that is, the random numbers n_{kl} of l-faces meeting at the typical k-face are

$$n_{03} \equiv 4 \text{ and } n_{13} \equiv n_{12} \equiv 3 \qquad \text{almost surely,} \tag{9.59}$$

then the intensities satisfy

$$\lambda_1 = 2\lambda_0, \qquad\qquad T_V = 4\lambda_0, \tag{9.60}$$
$$\lambda_2 = \lambda_0 + \lambda_3, \qquad\qquad Z_V = 3L_V, \tag{9.61}$$

and furthermore

$$\overline{n}_{01} = \overline{n}_{03} = 4, \qquad\qquad \overline{n}_{02} = 6. \tag{9.62}$$

Lautensack (2007) shows that each normal and face-to-face tessellation in $\mathbb{R}^d$ with $d > 2$ (the case $d = 2$ is excluded) can be considered as a Laguerre tessellation with some system of points and weights. Perhaps this striking fact will in due course find application in tessellation statistics and stereology.

For such tessellations the following important formula holds

$$\overline{n}_{32} = \frac{12}{6 - \overline{n}_{21}}. \tag{9.63}$$

It can be proved by means of Schneider and Weil (2008, Theorem 10.1.7).

Zero cells in the spatial case are also of interest. Their definition and properties are analogous to those in the planar case, and Inequality (9.31) remains true; see Mecke (1999).

It is not surprising that many of the mean-value relations are true under more general assumptions, since they have a topological character. That means that the convexity assumptions can be dropped. This is shown by Stoyan (1986), Zähle (1987b, 1988), Weiss and Zähle (1988) and Leistritz and Zähle (1992).

Formulae analogous to the Aboav law (9.37) and (9.38) for spatial tessellations have been suggested empirically in Aboav (1991, 1992) and established mathematically in Chiu (1994), respectively.

For planar sections of spatial tessellations see Section 10.6.

9.5   Poisson line and plane tessellations

Poisson line and plane processes produce tessellations in the plane and space respectively. Formulae for these tessellations are presented here; some results for higher-dimensional Poisson hyperplane tessellations can be found in Miles (1971a, 1974a) and Matheron (1975, Chapter 6); see also Schneider and Weil (2000, Section 6.3; 2008, Section 10.3).

9.5.1   Poisson line tessellations

Let Ψ be a planar motion-invariant Poisson line process of intensity L_A; the line intersections form the vertices and a segment of a line in Ψ with vertices at both endpoints but no vertices in the segment interior is an edge. Thus, the Ψ induces an edge system of a motion-invariant tessellation, known as the *Poisson line tessellation*. Such tessellations have no π-vertices and hence are face-to-face. However, there are always four edges emanating from each vertex and so they are not trivalent.

The quantity

$$\varrho = P_L = \frac{2L_A}{\pi}, \tag{9.64}$$

the mean number of lines intersected by a test line segment of unit length, serves as a convenient model parameter. The intensity of the point process of intersection points on a fixed line g in Ψ is also given by ϱ. Hence

$$\bar{l}_1 = \frac{1}{\varrho}. \tag{9.65}$$

With probability one the Poisson line process is bundle-free (has no triplets of lines meeting in a single point) and so each vertex is formed by the intersection of precisely two lines. Thus

$$n_{02} \equiv 4 \qquad \text{almost surely,} \tag{9.66}$$

and the formulae in Section 9.3.4 yield

$$\lambda_0 = \frac{\pi \varrho^2}{4}, \quad \lambda_1 = \frac{\pi \varrho^2}{2}, \quad \lambda_2 = \frac{\pi \varrho^2}{4}, \quad \text{and} \quad \bar{l}_0 = \frac{4}{\varrho}. \tag{9.67}$$

Suppose the cell centroid process is marked by the corresponding polygons. The mark distribution is then the distribution of the typical polygon of the tessellation process, also known as the *Poisson polygon*.

Several formulae are known for the distributions of characteristics of this polygon. First and second moments are given in Table 9.1. They are mostly due to Miles (1964a,b, 1973). Information about density functions of area, perimeter, and breadth can be found in Miles (1964a,b, 1973), Matheron (1975), Crain and Miles (1976), Deng and Dodson (1994) and Stoyan and Stoyan (1994).

The distribution $\{p_n\}$ of the number of edges is partially known and tabulated in Table 9.2; see Tanner (1983a,b), who also gave the values of the corresponding third and fourth moments, and Calka (2003b), who obtained the distribution of the number of edges for the zero cell.

Crain and Miles (1976) noted that the distribution $\{p_n\}$ can be approximated by a shifted Poisson distribution of parameter 1:

$$p_n \approx \frac{e^{-1}}{(n-3)!} \qquad \text{for } n = 3, 4, \ldots . \tag{9.68}$$

Table 9.1 Poisson polygon formulae.

		Area	Perimeter	Edge number
First moments:		A	L	N
		$4/(\pi\varrho^2) = \bar{a}_2$	$4/\varrho = \bar{l}_2$	$4 = \bar{n}_{20}$
Second moments:		A	L	N
	A	$8/\varrho^4$		
	l_2	$4\pi/\varrho^3$	$2(\pi^2+4)/\varrho^2$	
	N	$2\pi/\varrho^2$	$(\pi^2+8)/\varrho$	$(\pi^2+24)/2$

They found the values p_5, p_6, ... and much other distributional information by simulation of the Poisson polygon. The method was to generate the tessellation in a large circle and to determine the desired quantities by statistical estimation. George (1987) and Maier *et al.* (2004) described how to simulate single Poisson polygons. Michel and Paroux (2007) provided another fast simulation method.

The mean area and mean number of edges of the zero cell C_0, the cell containing the origin, are

$$\mathbf{E}\big(v_2(C_0)\big) = \frac{2\pi}{\varrho^2}, \quad \text{and} \quad \mathbf{E}\big(N(C_0)\big) = \frac{\pi^2}{2}, \tag{9.69}$$

which are larger than the corresponding parameters of the typical cell given in Table 9.1. Michel and Paroux (2007) estimate higher moments by simulation.

Note that mean-value formulae can also be derived in the *anisotropic* case. Let $\mathscr{R}$ be the rose of directions of the anisotropic Poisson line process generating the tessellation. Then

$$\bar{n}_{02} = 4 \tag{9.70}$$

Table 9.2 The distribution of the number of edges of the Poisson polygon and the zero cell.

Number of edges	Poisson polygon	zero cell
3	$2 - \pi^2/6 \approx 0.3551$	0.0767
4	$\pi^2 \ln 2 - 1/3 - 7\pi^2/36$ $-(7/2)(1 + 2^{-3} + 3^{-3} + \cdots) \approx 0.3815$	0.3013
5	0.1873	0.3415
6	0.0596	0.1905
7	0.0129	0.0682
8	0.0023	0.0155
⋮	⋮	⋮

and

$$\lambda_0 = \frac{L_A^2 \zeta}{2}, \tag{9.71}$$

where

$$\zeta = \int_{(0,\pi]} \int_{(0,\pi]} |\sin(\alpha - \beta)| \mathcal{R}(d\alpha) \mathcal{R}(d\beta)$$

(see Matheron, 1975; Thomas, 1984). From this one may conclude (using the mean-value formulae in Section 9.3.4) that

$$\bar{n}_{20} = 4, \tag{9.72}$$

$$\lambda_1 = L_A^2 \zeta, \qquad \lambda_0 = \lambda_2 = \frac{L_A^2 \zeta}{2}, \tag{9.73}$$

$$\bar{l}_1 = \frac{1}{L_A \zeta}, \qquad \bar{l}_0 = \bar{l}_2 = \frac{4}{L_A \zeta}, \tag{9.74}$$

$$\bar{a}_2 = \frac{2}{L_A^2 \zeta}. \tag{9.75}$$

The equation $\lambda_0 = \lambda_2$ is true for general stationary bundle-free line tessellations; for this the Poisson assumption is not necessary (Stoyan and Mecke, 1983a, Chapter 8).

9.5.2 Poisson plane tessellations

Let Ψ now be a motion-invariant Poisson plane process in $\mathbb{R}^3$ with intensity S_V. Similarly to the planar case, the quantity

$$\varrho = P_L = \frac{S_V}{2}, \tag{9.76}$$

the mean number of planes intersected by a test line segment of unit length, is used as the model parameter. With probability one there is no point of $\mathbb{R}^3$ in which more than three Poisson planes intersect and so

$$\bar{n}_{01} = 6, \qquad \bar{n}_{02} = 12, \tag{9.77}$$

$$\bar{n}_{03} = 8, \qquad \bar{n}_{12} = 4. \tag{9.78}$$

It is known that

$$\bar{n}_{30} = 8, \qquad \bar{n}_{31} = 12, \tag{9.79}$$

$$\bar{n}_{32} = 6, \qquad \bar{n}_{21} = 4, \tag{9.80}$$

(see Miles, 1974a; Mecke, 1984c). These are the same values as for a regular cubical tessellation. Basic parameters can be used, as in Section 9.4, to give

$$\lambda_0 = \lambda_3 = \frac{\pi \varrho^3}{6}, \tag{9.81}$$

$$\lambda = \pi \varrho^3, \qquad\qquad L_V = \frac{\pi \varrho^2}{2}, \tag{9.82}$$

$$S_V = 2\varrho, \qquad\qquad T_V = \frac{4\pi \varrho^3}{3}, \tag{9.83}$$

$$Z_V = 2\pi \varrho^2, \qquad\qquad \bar{l}_1 = \frac{1}{\varrho}. \tag{9.84}$$

Some formulae are known for the typical polyhedron (the *Poisson polyhedron* in the sense of the 'number law' of Matheron, 1975); these are mainly due to Miles (1972b). Table 9.3 gives first and second moments. Miles (1972b) and Matheron (1975) give further formulae. Also formulae are known for the zero polyhedron; see Matheron (1975) and Schneider and Weil (2008).

Several inequalities for Poisson line and plane tessellations are given in Mecke (1995) and Mecke *et al.* (1990, Section 3.12). Typically the isotropic case provides extremal values.

The Poisson polyhedron finds useful employment as a model component, for example as a typical grain for Boolean models, as in Serra (1982, p. 499) and Example 3.3 in this book.

In conclusion it should be noted that the planar section of a Poisson plane tessellation is a Poisson line tessellation defined on the plane of intersection with the same parameter ϱ as the plane tessellation. This allows application of the formulae of Section 9.5.1.

Mecke (1984c), see also Schneider (1987), showed that for general stationary bundle-free hyperplane tessellations in $\mathbb{R}^d$ the intensities satisfy

$$\lambda_k = \binom{d}{k} \cdot \lambda_0 \qquad \text{for } k = 1, 2, \ldots, d. \tag{9.85}$$

Table 9.3 Poisson polyhedron formulae; see Santaló (1976).

	Volume	Surface	Average breadth	Facet number
First moments:	V	S	$\bar{b}$	$N = n_{32}$
	$6/(\pi \varrho^3) = \bar{V}_3$	$24/(\pi \varrho^2) = \bar{S}_3$	$3/(2\varrho) = \bar{b}_3$	$6 = \bar{n}_{32}$
Second moments:	V	S	$\bar{b}$	N
V	$48/\varrho^6$			
S	$96/\varrho^5$	$240/\varrho^4$		
$\bar{b}$	$2(\pi^2 + 3)/(\pi \varrho^4)$	$(7\pi^2 + 12)/(\pi \varrho^3)$	$(13\pi^2 + 48)/(48\varrho^2)$	
N	$4(\pi^2 + 3)/(\pi \varrho^3)$	$2(7\pi^2 + 24)/(\pi \varrho^2)$	$(13\pi^2 + 120)/(12\varrho)$	$(13\pi^2 + 336)/12$

Note that a hyperplane process in $\mathbb{R}^d$ is said to be *bundle-free*, if each point of $\mathbb{R}^d$ is contained in at most d hyperplanes of the process.

Furthermore, the mean number $\bar{n}_{kj}$ of j-faces adjacent to the typical k-face satisfies

$$\bar{n}_{kj} = 2^{k-j} \binom{k}{j} \qquad \text{for } k = 1, 2, \ldots, d \text{ and } j = 0, 1, \ldots, k. \tag{9.86}$$

These formulae are the same as for a cubical tessellation generated by hyperplanes parallel to the coordinate hyperplanes.

The construction of the Poisson polyhedron, the typical cell of the Poisson hyperplane tessellation, is described in Calka (2010, pp. 149–50). He generalises a procedure of Miles (1973), which was originally made for the planar case.

For some applications the *twin flat model* may be of value, which can be found in Serra (1982). It is constructed from a Poisson plane process by duplicating each plane by a parallel plane separated from it by a random distance.

9.6 STIT tessellations

A stationary STIT tessellation is uniquely determined by its rose of directions and the densities L_A (planar case) and S_V (spatial case). Thus these appear in the formulae for tessellation characteristics.

Formulae for the typical cell are not given explicitly here since the interior of the typical cell of a stationary STIT tessellation has the same distribution as the interior of the typical cell of a Poisson line or plane tessellation with the same value of L_A and S_V, respectively; see Nagel and Weiss (2003). However, for the mean number of l-faces on the boundary of a cell, it is necessary to draw on the distinction between cell faces (corners and sides in the planar case; apices, ridges and facets in the spatial case) and tessellation faces (vertices and edges in the planar case; vertices, edges and plates in the spatial case). For example, in the planar case a π-vertex belongs to three polygons, but is not an apex of one of them, and in the spatial case all edges in STIT are π-edges. Weiss and Cowan (2011) worked out the following mean values for the spatial case:

$$\psi = 2, \qquad \kappa = \frac{2}{3}, \qquad \xi = 1, \qquad \tau = \frac{4}{3}, \tag{9.87}$$

$$\bar{n}_{30} = 24, \qquad \bar{n}_{31} = 36, \qquad \bar{n}_{32} = 14, \qquad \bar{n}_{21} = \bar{n}_{20} = \frac{36}{7}, \tag{9.88}$$

$$\bar{m}_{30} = 8, \qquad \bar{m}_{31} = 12, \qquad \bar{m}_{32} = 6, \qquad \bar{m}_{21} = \bar{n}_{20} = 4. \tag{9.89}$$

Recall that $\bar{n}_{kl}$ refer to tessellation k- and l-faces whilst $\bar{m}_{kl}$ to cell k- and l-faces. They also report mean values for the adjacency between a cell face and a tessellation face, for example the typical cell facet comprises, on average, 7/3 tessellation plates, 10 edges and 26/3 vertices, whilst the typical cell ridge contains two edges and three vertices.

Some further characteristics for isotropic STIT tessellations are, in the planar case,

$$\overline{l}_1 = \frac{\pi}{3L_A},$$

(9.90)

$$\lambda_0 = \frac{2L_A^2}{\pi}, \qquad \lambda_1 = \frac{3L_A^2}{\pi}, \qquad \lambda_2 = \frac{L_A^2}{\pi},$$

(9.91)

and, in the spatial case,

$$L_V = \frac{\pi S_V^2}{4},$$

(9.92)

$$\overline{l}_1 = \frac{1}{S_V}, \qquad \overline{l}_2 = \frac{36}{7S_V},$$

(9.93)

$$\overline{A}_2 = \frac{48}{7\pi S_V^2},$$

(9.94)

$$\lambda_0 = \frac{\pi S_V^3}{8}, \qquad \lambda_1 = \frac{\pi S_V^3}{4}, \qquad \lambda_2 = \frac{7\pi S_V^3}{48}, \qquad \lambda_3 = \frac{\pi S_V^3}{48}.$$

(9.95)

These formulae are proved in Nagel and Weiss (2006, 2008), in which the anisotropic case is also treated; see also Thäle and Weiss (2010).

9.7 Poisson-Voronoi and Delaunay tessellations

9.7.1 General

Voronoi and Delaunay tessellations generated by homogeneous Poisson processes have been studied extensively, both mathematically and by simulation. This section presents a brief overview of known results for the planar and spatial cases. Poisson-Voronoi tessellations in general $\mathbb{R}^d$ are studied in Møller (1989, 1994).

With probability one the Poisson-Voronoi and Poisson-Delaunay tessellations are normal and face-to-face. Thus, cell faces coincide with tessellation faces and in the literature one speaks only about vertices, edges and cell facets.

To the typical vertex of a Poisson-Voronoi tessellation there belong $d + 1$ nearest generating points of the same distance Δ. The distribution of Δ is given in Muche (1996a); see also Muche (2005). Mecke and Muche (1995) show that the distributional behaviour of the pattern of the generating points outside the ball of radius Δ centred at the typical vertex is analogous to that of a Poisson process. Furthermore, they prove that the distribution of any characteristic of the typical edge coincides with that of a randomly chosen edge adjacent to the typical vertex; see also Baumstark and Last (2007).

Błaszczyszyn and Schott (2003, 2005) consider the case of inhomogeneous Poisson processes with piecewise constant intensity function, which allows a decomposition approximation of the distributions by a mixture of distributions in the homogeneous case.

9.7.2 Planar Poisson-Voronoi tessellations (Poisson-Dirichlet tessellations)

A Poisson-Dirichlet tessellation is the Voronoi tessellation relative to a planar homogeneous Poisson process.

Denote the intensity of the generating point process by ϱ, which will serve as the model parameter. Then, following Meijering (1953) and Miles (1970), the mean-value parameters of Section 9.3.4 are given by

$$\lambda_0 = 2\varrho, \tag{9.96}$$

$$\lambda_1 = 3\varrho, \tag{9.97}$$

$$\lambda_2 = \varrho, \tag{9.98}$$

$$L_A = 2\sqrt{\varrho}. \tag{9.99}$$

All other mean values follow from these by the mean-value formulae on pp. 360 and 361. For example, $\bar{n}_{20} = 6$, so that the typical cell of the Poisson-Dirichlet tessellation has an average of six vertices. At the typical vertex the interior angle has a mean of $2\pi/3$ and variance $\pi^2/9 - 5/6$.

Gilbert (1962) calculated the second moment of the area of the typical polygon of the Poisson-Dirichlet tessellation, obtaining the value of $1.280\varrho^{-2}$. In principle all other characteristics of this random polygon can also be obtained by numerical integration. For example, Miles and Maillardet (1982) derived integral formulae for the probabilities p_n of the polygon having exactly n vertices. Table 9.4 gives the first values of p_n, based on Calka (2003a); see also Hayen and Quine (2000a,b). Hilhorst (2008) discussed the behaviour of the p_n for large n and Hilhorst (2007) suggested a special simulation method for their determination.

For many characteristics of the Poisson-Dirichlet tessellation it is difficult to derive formulae or known formulae are complicated integrals, and hence it is helpful to determine such characteristics by simulation. Hinde and Miles (1980) obtained a series of estimates by simulating the typical cell directly rather than by dealing with the whole tessellation. Some of these results are given in Table 9.5 together with variances and correlations for geometrical characteristics of the typical cell (see also Okabe et al., 2000, Table 5.5.1).

Hinde and Miles gave histograms for the distributions of perimeter, area, and inner angles (see also Quine and Watson, 1984, and Icke and van de Weygaert, 1987) and suggested approximations for perimeter and area distribution by the so-called generalised gamma densities, using the form

$$f(x) = \frac{\gamma \chi^{\nu/\gamma}}{\Gamma(\nu/\gamma)} x^{\nu-1} \exp(-\chi x^\gamma) \qquad \text{for } x \geq 0, \tag{9.100}$$

Table 9.4 The probability p_n that the typical Poisson-Dirichlet polygon has n vertices.

n	3	4	5	6	7	8	9	...
p_n	0.0112	0.1068	0.2595	0.2947	0.1988	0.0897	0.0295	...

Table 9.5 Second moments of characteristics of the typical cell of the Poisson-Dirichlet tessellation.

		Area	Perimeter	Edge number $= n_{21}$
Variance:				
		A	l_2	N
		$0.2802\varrho^{-2}$	$0.9455\varrho^{-1}$	1.7808
Correlation coefficient:		A	l_2	N
	A	1		
	l_2	0.953	1	
	N	0.568	0.5021	1

where γ, ν and χ are positive parameters. (Good choices for the area case, supposing $\varrho = 1$, are the values $\gamma = 1.08$, $\nu = 3.31$ and $\chi = 3.03$; see Hinde and Miles, 1980.)

Note in passing that the classical gamma distribution with integer shape parameter (the Erlang distribution) appears frequently in stochastic geometry in the context of Poisson processes as model elements. This is systematically investigated in various papers; see Miles (1971a), Møller and Zuyev (1996), Zuyev (1999), Cowan et al. (2003), Cowan (2006) and Baumstark and Last (2009). There measures of so-called stopping sets are often sums of i.i.d. exponentially distributed random variables.

Mean-value characteristics of the zero cell of the Poisson-Dirichlet tessellation can be derived using equation (9.32) and the numerical values in Table 9.5. The mean edge number is greater than 6 ($=$ the mean edge number of the typical cell), namely 6.40.

The point process of vertices of the Poisson-Dirichlet tessellation has also been considered. Its intensity is denoted by λ_0 as above. A study of its Palm distribution leads to the determination of the pair correlation function $g(r)$. It can be given in the form of a sum of four numerically tractable double-parameter integrals; see Heinrich and Muche (2008). The function $g(r)$ has a pole of order 1 at $r = 0$, that is, $g(r) = O(r^{-1})$ as $r \to 0$, and followed by a minimum and a subsequent maximum; see Figure 9.12(a). The pole results from very short edges.

The length of the typical edge was obtained by numerical integration in Brakke (1986a) (see also Muche, 1996b, and Schlather, 2000, for other integral expressions).

The semi-empirical Aboav law (9.37) for the mean number of edges of a random neighbour of a n-edged cell does not hold for the Poisson-Voronoi tessellation. This is discussed in detail in Hilhorst (2006).

Muche and Stoyan (1992) considered distributional characteristics of the random closed set of the edge system E_Θ and derived integral formulae for the linear and spherical contact distribution functions $H_l(r)$ and $H_s(r)$. Muche (2010) obtained formulae for the probability density functions of the linear contact distribution for the planar, as well as for the spatial case, see Figure 9.13, and also found formulae for the probability density functions of the spherical contact distributions and corresponding moments.

By Equation (6.66), an expression of $H_l(r)$ leads to one for the chord length distribution function $L(r)$, that is, the distribution function of the length of the chord generated by

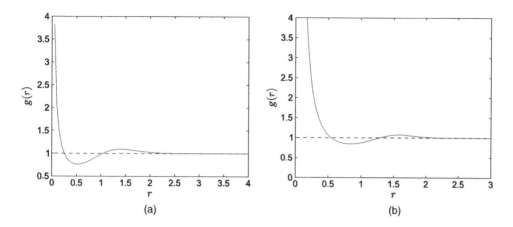

Figure 9.12   The pair correlation function of the point process of vertices of the Poisson-Voronoi tessellation in the (a) planar case and (b) spatial case, for $\varrho = 1$. Courtesy of L. Heinrich and L. Muche.

a random line with the typical Voronoi cell; see Figure 9.14. The corresponding first two moments are

$$\bar{l} = \frac{\pi}{4}\varrho^{-1/2} \quad \text{and} \quad \overline{l^2} \approx 0.806\varrho^{-1}. \tag{9.101}$$

Some more discussions will be found around (9.107) on p. 382.

Redenbach (2011) derived integral expressions for A_A and L_A of $E_\Theta \oplus B(0, r)$, the dilated set of the edge system of the Poisson-Voronoi tessellation.

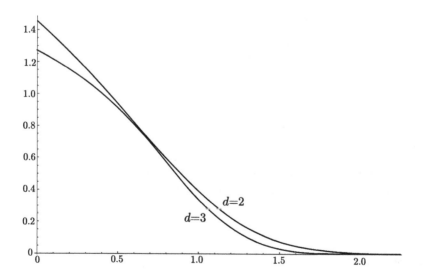

Figure 9.13   Probability densities of the linear contact distribution for the Poisson-Voronoi tessellation in $\mathbb{R}^2$ and $\mathbb{R}^3$ for $\varrho = 1$. Courtesy of L. Muche.

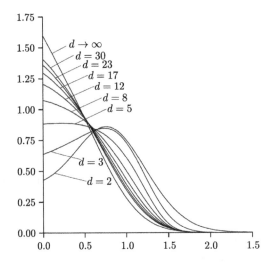

Figure 9.14 Probability density functions for the chord lengths through the typical Poisson-Voronoi polygon in $\mathbb{R}^d$ for $d = 2, 3, 5, 8, 12, 17, 23, 30, \infty$ with $\varrho = 1$. Reproduced from Muche (2010) with permission of the Applied Probability Trust.

9.7.3   Spatial Poisson-Voronoi tessellations

Here ϱ is the intensity of the underlying three-dimensional Poisson point process. The mean-value parameters of Section 9.4 take the values

$$\lambda_0 = \frac{24\pi^2 \varrho}{35} \approx 6.768\varrho, \qquad \lambda_3 = \varrho, \qquad \lambda = 3\lambda_0 + \varrho, \tag{9.102}$$

$$L_V = \frac{16}{15}\left(\frac{3}{4}\right)^{1/3} \pi^{5/3} \Gamma\left(\frac{4}{3}\right)\varrho^{2/3} \approx 5.832\varrho^{2/3}, \tag{9.103}$$

$$S_V = 4\left(\frac{\pi}{6}\right)^{1/3} \Gamma\left(\frac{5}{3}\right)\varrho^{1/3} \approx 2.910\varrho^{1/3}, \tag{9.104}$$

$$T_V = 4\lambda_0, \qquad Z_V = 3L_V, \tag{9.105}$$

see Meijering (1953), Miles (1972b) and Møller (1989, 1994). All other mean values follow from these results. For example, the mean volume of the typical cell is given by

$$\overline{V}_3 = \varrho^{-1},$$

the mean surface area by

$$\overline{S}_3 = \left(\frac{256\pi}{3}\right)^{1/3} \Gamma\left(\frac{5}{3}\right) \varrho^{-2/3} \approx 5.821\varrho^{-2/3},$$

and the mean average breadth by

$$\overline{\overline{b}}_3 = \frac{1}{5}\left(\frac{16\pi^5}{243}\right)^{1/3}\Gamma\!\left(\frac{1}{3}\right)\varrho^{-1/3} \approx 1.458\varrho^{-1/3}.$$

The mean total edge length and the means of numbers of apices (or vertices), ridges (or edges) and facets (or plates) of the typical cell are $17.496\,\varrho^{-1/3}$, 27.07, 40.61, and 15.54, respectively. For a table with numerical values and formulae of other means, see Okabe *et al.* (2000, pp. 316–7).

Gilbert (1962) uses numerical integration to estimate the second moment of the volume of the typical cell as $1.179\varrho^{-2}$. Muche and Ballani (2011) confirm this result (see also Brakke, 1986b) by analytical calculations, which leads to

$$\mathbf{E}(V^2) = \frac{1}{\varrho^2}\left(-\frac{4}{3^5}\left(16\mathscr{A} - 3\pi^2\right) + \frac{8}{27}\sqrt{3}\,\pi\right) \approx 1.179\varrho^{-2} \tag{9.106}$$

with the constant

$$\mathscr{A} \approx 3.4955.$$

Furthermore, they are able to give higher moments of the length of the typical edge of a k-dimensional section through a spatial Poisson-Voronoi tessellation.

The variances of surface area, total edge length and numbers of vertices, edges and facets of the typical cell have also been derived either by simulation or numerical integration, namely $2.19\varrho^{-4/3}$, $13.63\varrho^{-2/3}$, 43.98, 99.00, and 11.01, respectively.

Quine and Watson (1984) (see also Møller, 1994, Section 4.5) establish an important simulation method which produces i.i.d. realisations of the typical cell of the Poisson-Voronoi tessellation in $\mathbb{R}^d$ by means of the radial method, with an additional point at the origin o. For this point pattern the Voronoi tessellation is constructed, and the corresponding zero cell is a realisation of the typical cell in the Palm distribution sense. This allows a quick simulation of samples of typical cell realisations, which lead to precise estimates of distributional characteristics of the cells. Møller (1995) shows how to generalise this for Johnson–Mehl tessellations, and Lautensack (2007, Chapter 4) describes the procedure for Poisson-Laguerre tessellations.

The correlation coefficients in Table 9.6 are determined by simulation by Lorz and Hahn (1993). Their paper, as well as those by Quine and Watson (1984), van de Weygaert (1994) and Fleischer *et al.* (2009), also contains histograms of Voronoi cell (spatial as well as planar) characteristics. A gamma density function approximation such as (9.100) is possible; for the volume case good values are $\gamma = 1.41$, $\nu = 2.81$ and $\chi = 4.12$, for $\varrho = 1$, after Tanemura (1988); see also Andrade and Fortes (1988) and Kumar *et al.* (1992). A good survey can be found in Okabe *et al.* (2000, Section 5.5.4).

The mean-value characteristics of the zero cell of the Poisson-Voronoi tessellation can be derived using equation (9.32) and the numerical values in Table 9.6. The mean edge number is greater than 15.54 (the mean edge number of the typical cell), namely 16.58.

As mentioned in the previous section, Muche and Stoyan (1992) derive integral formulae for $H_l(r)$ and $H_s(r)$ of the system of cell facets; see Figure 9.13 for the former, which lead to

Table 9.6 Second moments of characteristics of the typical cell of the Poisson-Voronoi tessellation.

		Volume	Surface	Average breadth	Facet number
Variance:		V_3	S_3	$\bar{b}_3$	n_{32}
		$0.179\varrho^{-2}$	$2.19\varrho^{-4/3}$	$0.030\varrho^{-2/3}$	11.01
Correlation coefficients:		V_3	S_3	$\bar{b}_3$	n_{32}
	V_3	1			
	S_3	0.982	1		
	$\bar{b}_3$	0.945	0.987	1	
	n_{32}	0.736	0.711	0.671	1

the chord length distribution function $L(r)$, and the first two moments of it are

$$\bar{l} \approx 0.687\varrho^{-1/3} \quad \text{and} \quad \overline{l^2} \approx 0.631\varrho^{-2/3}. \tag{9.107}$$

Figure 9.14 shows the corresponding density functions for various dimensions d. The shape of these functions demonstrates what some statisticians have in mind when they say that the 'variability' of Poisson-Voronoi tessellations increases with increasing d. (The study of tessellations in $\mathbb{R}^d$ with $d > 3$ has practical applications: their intersections with three-dimensional subspaces are spatial tessellations; see Miles, 1986.)

Alishahi and Sharifitabar (2008) express the limits of $H_l(r)$ and $L(r)$ as $d \to \infty$ in terms of standard functions. The limiting value of the probability density function of $L(r)$ for $r = 0$ is given explicitly as 1.5968.

The probability density function of the spherical contact distribution is

$$h_s(r) = \frac{64\pi^2 \varrho^2 r^5}{3} \int_0^1 \left(2t^2 \left(3 + t^2 \right) \exp\left(-\frac{32\pi\varrho r^3}{3}(1 + t^2) \right) \right.$$
$$\left. + \frac{(1 + t)^3}{t^6} \exp\left(-\frac{4\pi\varrho r^3}{3} \frac{(1 + t)^4}{t^3} \right) \right) dt \quad \text{for } r \geq 0, \tag{9.108}$$

see Muche (2010), in which formulae for the densities of the discoidal and bilinear contact distributions (i.e. the spherical contact distribution in a planar and a linear section, respectively) are also given; see Figure 9.15.

The fundamental ideas for obtaining these results are as follows. For a convex compact structuring element B containing the origin the contact distribution function $H_B(r)$ satisfies

$$H_B(r) = 1 - \mathbf{P}(rB \subset C_0) \quad \text{for } r \geq 0,$$

where C_0 is the zero cell containing the origin. This cell is generated by the nearest neighbour of o in the Poisson point process of nuclei. The probability above is obtained by means of the void-probability of the nuclei process and use of geometrical properties of the cells.

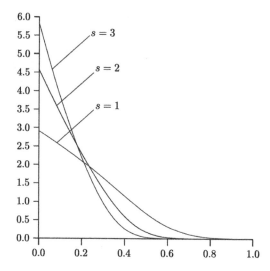

Figure 9.15 Probability density functions for the spherical contact distribution of an s-dimensional section of the spatial Poisson-Voronoi tessellation with $\varrho = 1$; $s = 3$, 2, and 1 correspond to the original spatial case, planar section and linear section, respectively. Reproduced from Muche (2010) with permission of the Applied Probability Trust.

Similar to the planar case, the distribution of the length of the typical edge is obtained by numerical integration in Brakke (1986b); see also Muche (1996b), Schlather (2000) and Muche (2005), who consider also higher dimensions. The corresponding density functions are similar but not equal to the density functions of Figure 9.14. Additionally, Muche (1996b, 1998, 2005) and Schlather (2000) give density functions for the angles between facets emanating in the same edge and between the typical edge and the line passing one of its endpoints and the nucleus of a neighbouring cell.

For the pair correlation function $g(r)$ of the vertices, the order of the pole of $g(r)$ is 2 (and in general is $d - 1$ in $\mathbb{R}^d$), followed by a minimum and then a maximum; see Figure 9.12(b). Thus the point process of vertices of the Poisson-Voronoi tessellation is not a suitable model for the galaxies in the universe since the pair correlation function for the latter follows the power law and has a pole of order around 1.8; see Snethlage *et al.* (2002) and Martínez *et al.* (2009).

Scheike (1994) gives mean-value formulae for scaled, that is, linearly transformed, Poisson-Voronoi tessellations. Such a tessellation can be interpreted as the result of a modified Voronoi construction with a metric which considers ellipses under the Euclidean metric as discs.

Intersections of spatial Poisson-Voronoi tessellations with planes are considered in Section 10.6.

9.7.4 Poisson-Delaunay tessellations

The dual graph of the Poisson-Voronoi tessellation, the *Poisson-Delaunay tessellation* has also been studied by many authors. With probability 1, it is face-to-face but not normal. Distributions and moments of characteristics of the typical Delaunay cell are more tractable than those of the typical Voronoi cell. For example, the kth moments of the volume of the typical cell of the Poisson-Delaunay tessellation in $\mathbb{R}^d$ are given by

$$\mathbf{E}(V_d^k) = \frac{\Gamma(\frac{d^2}{2})\Gamma(d+k)\Gamma(\frac{d^2+dk+k+1}{2})\Gamma^{d-k+1}(\frac{d+1}{2}) \prod_{i=2}^{d+1}\left(\Gamma(\frac{k+i}{2})\middle/\Gamma(\frac{i}{2})\right)}{\Gamma(d)\Gamma\left(\frac{d^2+1}{2}\right)\Gamma\left(\frac{d^2+dk}{2}\right)\Gamma^{d+1}\left(\frac{d+k+1}{2}\right)(2^d\pi^{(d-1)/2}\varrho)^k}, \tag{9.109}$$

for $k = 1, 2, \ldots$; see Miles (1974a) and Møller (1989, 1994). In particular, the mean area and volume are

$$\mathbf{E}(V_2) = \bar{a}_2 = \frac{1}{2\varrho}, \tag{9.110}$$

and

$$\mathbf{E}(V_3) = \overline{V}_3 = \frac{35}{24\pi^2\varrho}. \tag{9.111}$$

Size and shape of the Delaunay cells can also be characterised; see Miles (1974a) and D. G. Kendall (1983). In the case of a planar Poisson-Delaunay tessellation, each cell is a triangle, whose shape can be characterised by

$$\sqrt{\frac{4}{3}\left(\sin^2\vartheta_1 + \sin^2\vartheta_2 + \sin^2\vartheta_3\right)}, \tag{9.112}$$

where the ϑ_i are its interior angles (D.G. Kendall et al., 1999, Section 8.3). The joint density of the interior angles was derived by Miles (1970):

$$f(\vartheta_1, \vartheta_2) = \frac{8}{3\pi}\left(\sin\vartheta_1 \sin\vartheta_2 \sin(\vartheta_1 + \vartheta_2)\right) \qquad \text{for } \vartheta_1, \vartheta_2 \geq 0 \text{ and } \vartheta_1 + \vartheta_2 < \pi, \tag{9.113}$$

which gives $\mathbf{E}(\vartheta_1) = \pi/3$. The maximum density at $\vartheta_1 = \vartheta_2 = \pi/3$ indicates that an equilateral triangle is the most likely shape. See Okabe et al. (2000, pp. 398–400) for the densities of the minimum, middle and maximum interior angles.

Rathie (1992) obtained a general expression for the distribution function of the volume of the typical Poisson-Delaunay cell in $\mathbb{R}^d$. The corresponding density function for the area $f_A(a)$ in the planar case is given by

$$f_A(a) = \frac{8}{9}\pi\varrho^2 a K_{1/6}^2\left(\frac{2\pi\varrho a}{3\sqrt{3}}\right) \qquad \text{for } a \geq 0, \tag{9.114}$$

where $K_{1/6}$ is the modified Bessel function of order $1/6$,

$$K_{1/6}(x) = \frac{\sqrt{\pi}\left(\frac{x}{2}\right)^{1/6}}{\Gamma(\frac{2}{3})} \int_1^\infty \exp(-xt)(t^2 - 1)^{-1/3}\, dt.$$

However, for the spatial case Rathie's expression involves a sequence of complicated analytical functions. Muche (1996a) found numerically tractable forms for the area A in the planar case and the volume V in the spatial case,

$$f_A(a) = \frac{\pi \varrho^2 a}{6} \int_0^{2\pi} \int_0^{2\pi - \theta_1} g(\theta_1, \theta_2) \exp\left(\frac{-\pi \varrho a g(\theta_1, \theta_2)}{2}\right) d\theta_2 \, d\theta_1, \qquad \text{for } a \geq 0,$$

(9.115)

and

$$f_V(v) = \frac{35\varrho^2 v}{2} \int_0^{2\pi} \int_0^{2\pi - \theta_1} \int_0^{\pi} \sin \theta_3 \exp\left(\frac{-2\pi \varrho v g(\theta_1, \theta_2)}{(1 + \cos \theta_3) \sin^2 \theta_3}\right) d\theta_3 \, d\theta_2 \, d\theta_1, \qquad \text{for } v \geq 0,$$

(9.116)

in both of which

$$g(\theta_1, \theta_2) = \left(\sin \frac{\theta_1}{2} \sin \frac{\theta_2}{2} \sin \frac{\theta_1 + \theta_2}{2}\right)^{-1}.$$

Collins (1968) (see also Sibson, 1980) gave the probability density function $f_{l_1}(r)$ of the length l_1 of the typical Poisson-Delaunay edge (equal to the distance between two 'neighbouring' points of the generating Poisson process) for the planar case, and Muche (1998) (see also Muche, 1996a) derived it for the spatial case:

$$f_{l_1}(r) = \begin{cases} \dfrac{\varrho \pi r}{3}\left(\sqrt{\varrho r}\exp\left(-\dfrac{\varrho \pi r^2}{4}\right) + \mathrm{erfc}\left(\dfrac{\sqrt{\varrho \pi} r}{2}\right)\right), & \text{for } d = 2, \\[2em] \dfrac{35\pi \varrho r^2}{28}\left(\left(\dfrac{\pi \varrho r^3}{4} + \dfrac{5}{2}\right)\exp\left(-\dfrac{\pi \varrho r^3}{6}\right)\right. \\[1em] \qquad \left. - \dfrac{\pi \varrho r^2}{3}\int_{r/2}^{\infty}\left(1 + \dfrac{\pi \varrho r^2 t}{6}\right)\exp\left(-\dfrac{4\pi \varrho t^3}{3}\right)dt\right), & \text{for } d = 3, \end{cases}$$

(9.117)

for $r \geq 0$, where $\mathrm{erfc}(x)$ is the complementary error function, which is related to the standard Gaussian distribution function $\Phi(x)$ by

$$2\Phi(x\sqrt{2}) + \mathrm{erfc}(x) = 2.$$

(9.118)

Consequently

$$\mathbf{E}(l_1) = \bar{l}_1 = \begin{cases} \dfrac{32}{9\pi \sqrt{\varrho}} \approx 1.132 \varrho^{-1/2}, & \text{for } d = 2, \\[2em] \dfrac{1715}{2304}\left(\dfrac{3}{4\pi \varrho}\right)^{1/3} \Gamma\left(\dfrac{1}{3}\right) \approx 1.237 \varrho^{-1/3}, & \text{for } d = 3. \end{cases}$$

(9.119)

The distributions of some other characteristics, such as the perimeter in the planar case and the surface area and the average breadth in the spatial case, are also known; see Okabe *et al.* (2000, Section 5.11) for more details. Some important mean values are:

for $d = 2$:

$$\bar{l}_2 = \frac{32}{3\pi\sqrt{\varrho}} \approx 3.395\varrho^{-1/2}, \tag{9.120}$$

$$\bar{n}_{02} = \bar{n}_{01} = 6; \tag{9.121}$$

for $d = 3$:

$$\bar{S}_3 = \frac{3500(3/4)^{2/3}\Gamma(2/3)}{243\pi^{5/3}\varrho^{2/3}} \approx 2.389\varrho^{-2/3}, \tag{9.122}$$

$$\bar{b}_3 \approx 1.118\varrho^{-1/3}, \tag{9.123}$$

$$\bar{n}_{03} = \frac{96\pi^2}{35} \approx 27.071, \quad \bar{n}_{02} = \frac{144\pi^2}{35} \approx 40.606, \quad \bar{n}_{01} = \frac{48\pi^2}{35} + 2 \approx 15.536, \tag{9.124}$$

$$\bar{n}_{12} = \bar{n}_{32} = \frac{144\pi^2}{24\pi^2 + 35} \approx 5.228. \tag{9.125}$$

9.8 Laguerre tessellations

Lautensack (2007) and Lautensack and Zuyev (2008) show that for the rather complicated Laguerre tessellations numerical results can be obtained not only by simulation. For the case of the Poisson-Laguerre tessellation they derived formulae for mean-value characteristics, which are, though containing complicated multiple integrals, useful in numerical calculations. A Poisson-Laguerre tessellation in $\mathbb{R}^d$ is the Laguerre tessellation with respect to an independently marked Poisson process in $\mathbb{R}^d$. The intensity of the Poisson process is denoted by ϱ and the distribution function of the (positive) marks is denoted by $F(w)$. The resulting tessellation is normal and face-to-face. The following presents some of the results from Lautensack and Zuyev (2008) for the planar case, just to show the reader some formulae and properties of planar Laguerre tessellations.

The characteristics of interest are λ_0, the mean number of vertices per area unit, and L_A, the mean edge length per area unit. It holds

$$\lambda_0 = \frac{\varrho^3}{12} \int_0^\infty \int_0^\infty \int_0^\infty \int_{-\min_i w_i^2}^\infty \exp\left(-\varrho\pi \int_0^\infty [t + w^2]^+ dF(w)\right)$$

$$\times V_{2,0}\left(\sqrt{t + w_0^2}, \sqrt{t + w_1^2}, \sqrt{t + w_2^2}\right) dt\, dF(w_0)\, dF(w_1)\, dF(w_2), \tag{9.126}$$

where

$$[x]^+ = \max\{0, x\},$$

$$V_{2,0}(w_0, w_1, w_2) = 2 \int_{S^1} \cdots \int_{S^1} \Delta(w_0 u_0, \ldots, w_2 v_2) \sigma(du_0) \cdots \sigma(du_2),$$

in which $\Delta(x_1, x_2, x_3)$ is the area of the triangle formed by the points x_1, x_2 and x_3, and σ is the length measure on the unit circle S^1. Furthermore,

$$L_A = \varrho^2 \pi \int_0^\infty \int_0^\infty \int_{-\min_i w_i^2}^\infty \frac{2t + w_0^2 + w_1^2}{\sqrt{(t + w_0^2)(t + w_1^2)}} \int_0^\infty \exp\left(-\varrho\pi \int_0^\infty [t + s + w^2]^+ dF(w)\right)$$

$$\times \sqrt{s}\, ds\, dt\, dF(w_0)\, dF(w_1). \qquad (9.127)$$

These two formulae provide the parameters that are required for determining all mean values of the tessellation characteristics using Mecke's mean-value formulae in Section 9.3.4. (Only two are sufficient because of the face-to-face and normality property of Poisson-Laguerre tessellations.) In particular, the probability p_0 that the typical point of the underlying Poisson process of intensity ϱ generates a non-empty cell is

$$p_0 = \frac{\lambda_0}{\varrho}. \qquad (9.128)$$

The spherical and linear contact distributions and the chord length distribution have also been derived, which leads to the mean chord length:

$$\bar{l} = \frac{\pi}{2L_A} \qquad (9.129)$$

(Lautensack, 2007, p. 63).

Example 9.1. *A planar Poisson-Laguerre tessellation with a two-atom mark distribution (Lautensack and Zuyev, 2008)*
The intensity of the Poisson process is $\varrho = 100$, the mark distribution is a two-atom distribution with the atoms $a = 0.01$ and $b = 0.01, 0.05, 0.10, 0.15, 0.20, 0.25$ and 0.30, with probability 0.5. The case $a = b$ corresponds to the Poisson-Voronoi tessellation.

The numerical values are summarised in Table 9.7. For comparison, the values for Poisson-Voronoi tessellations with intensity $\varrho = 100$ (this is the case of $a = b = 0.01$) and $\varrho = 50$ are included.

The values for $b = 0.30$ are close to those for the Poisson-Voronoi tessellation with $\varrho = 50$. This observation illustrates a limit theorem for Poisson-Laguerre tessellations: when there is an atom at the maximum weight b and when b is large, nearly all of the cells generated by the points with smaller weights will become empty and vanish, leaving the Voronoi tessellation generated by the points with the larger weight b.

Table 9.7 Mean values of cell characteristics for a two-dimensional Laguerre tessellation generated by a homogeneous Poisson process of intensity $\varrho = 100$ with weights independently drawn from a two-atom distribution taking the values 0.01 and b with probability 0.5 each. The columns PV_{100} and PV_{50} contain the values for Poisson-Voronoi tessellations of intensity $\lambda = 100$ and $\lambda = 50$, respectively. Courtesy of C. Redenbach and S. Zuyev.

				b				
	PV_{100}	0.05	0.10	0.15	0.20	0.25	0.30	PV_{50}
λ_0	200.000	192.406	148.398	110.968	101.050	100.043	100.001	100.000
λ_1	300.000	288.609	222.597	166.452	151.574	150.065	150.001	150.000
λ_2	100.000	96.203	74.199	55.484	50.525	50.022	50.001	50.000
L_A	20.000	19.203	16.283	14.529	14.173	14.143	14.142	14.142
$\overline{l_1}$	0.06667	0.06654	0.07315	0.08729	0.09351	0.09425	0.09428	0.09428
$\overline{a_2}$	0.0100	0.0104	0.0135	0.0180	0.0198	0.0200	0.0200	0.0200
$\overline{l_2}$	0.4000	0.3992	0.4389	0.5237	0.5610	0.5655	0.5657	0.5657

9.9 Johnson–Mehl tessellations

Johnson–Mehl tessellations, which are defined on p. 352, are more complicated than Voronoi tessellations, depend on more model parameters and do not necessarily have convex cells. General formulae are correspondingly more intricate. This section considers the birth locations and birth times of nuclei form a Poisson process in $\mathbb{R}^d \times [0, \infty)$ with time-dependent intensity function $\lambda(t)$, meaning that the births are homogeneous in space but not necessarily in time. The mean number of births per unit volume in the time interval $[s, t]$ is $\int_s^t \lambda(u)du$, but this is not the mean number of cells because some nuclei are born in occupied regions and then disappear without a trace.

An important characteristic is the volume fraction $V_V(t)$ of the space occupied by the Johnson–Mehl cells at time t. In the particular case of constant growth rate and time-homogeneous births

$$v(t) = v \qquad \text{and} \qquad \lambda(t) = \alpha,$$

it is given by

$$V_V(t) = 1 - \exp\left(-\frac{b_d \alpha v^d t^{d+1}}{d+1}\right) \qquad \text{for } t \geq 0. \tag{9.130}$$

A frequently considered time-inhomogeneous case is the one corresponding to the Weibull model of the underlying birth-and-growth process in Section 6.6.4:

$$\lambda(t) = ct^{m-1}, \qquad \text{where } c, m > 0,$$

considered in Horálek (1988, 1990) and Møller (1992).

In the general case with time-dependent birth rate $\lambda(t)$ and growth rate $v(t)$ the quantity $V_V(t)$ is given by

$$V_V(t) = 1 - \exp\left(-b_d \int_0^t \lambda(s)\left(\int_0^{t-s} v(u)\mathrm{d}u\right)^d \mathrm{d}s\right) \qquad \text{for } t \geq 0, \qquad (9.131)$$

which differs from Formula (6.188) because, as explained on p. 353, in the Johnson–Mehl model the growth rate of a cell depends on the age of the cell (i.e. for a cell born at time s its growth rate at time t is equal to $v(t-s)$, see the age-dependent growth model (6.175)), whilst in Formula (6.188) the growth rate is a function of the time t (i.e. at time t the growth rate of all existing cells is $v(t)$).

These results can be obtained by means of the formulae for the Boolean model and were first given by Kolmogorov (1937), Johnson and Mehl (1939) and Avrami (1939). Formulae (9.130) and (9.131) are used for estimating $\lambda(t)$ statistically when $V_V(t)$ and $v(t)$ are observable.

The mean number of cells per volume unit of the resultant tessellation is

$$\lambda_d = \int_0^\infty V_V(t)\lambda(t)\,\mathrm{d}t. \qquad (9.132)$$

Clearly, the mean volume of the typical cell is λ_d^{-1}.

Further mean-value formulae for fundamental characteristics of the cells and intersections are given in Møller (1992), who distinguishes between faces and *interfaces*. By faces he means the intersections of cells, the same as the tessellation faces introduced on p. 367, so that a k-face is the intersection of $d - k + 1$ cells. However, in contrast to the convex case, a face may consist of several (connected) components, which Møller calls k-*interfaces*. A 0-interface is a vertex, while a $(d-1)$-interface is a facet between two cells.

Let λ_k be the intensity of k-interfaces, and let $\bar{n}_{kl}$ be the same mean as on p. 368 but with 'interface' instead of 'face'. Then in the particular case of $d = 2$ some of Møller's formulae are

$$\lambda_0 = 2\lambda_2, \qquad \lambda_1 = 3\lambda_2 \qquad (9.133)$$

(generalising Formulae (9.96)–(9.98) from the Poisson-Dirichlet tessellation to this non-convex case), and

$$\bar{n}_{20} = \bar{n}_{21} = 6 \qquad (9.134)$$

(again the same as in the Poisson-Dirichlet tessellation case).

The edges of planar Johnson–Mehl tessellations are hyperbolic arcs; this can be used for constructing such tessellations, as described in Møller (1995).

In the spatial case the situation may be more complicated. As Møller (1992) explained, it is possible there that 'a cell C_1 is surrounded by only two other cells C_2 and C_3. Then C_1 contains two facets and one edge but no vertices ... the edge is a closed curve and the 2-interface $C_2 \cap C_3$ is not simply connected, since C_1 causes a "gap" in the relative interior of $C_2 \cap C_3$. Such gaps represent the possibility of forming lenses'. This implies deviations from those mean-value relations which apply for tessellations with convex cells in the ordinary equilibrium state. While in the convex ordinary equilibrium case

$$\lambda_2 = \lambda_0 + \lambda_3 \qquad \text{and} \qquad \lambda_1 = 2\lambda_0,$$

for a Johnson–Mehl tessellation the corresponding values satisfy

$$\lambda_2 \geq \lambda_0 + \lambda_3 \qquad \text{and} \qquad \lambda_1 \geq 2\lambda_0,$$

with equality holds only in the Voronoi tessellation case. The fragments are pieces of rotation hyperboloids. Møller (1995) showed how typical Johnson–Mehl cells can be simulated. Heinrich and Schüle (1995) discussed the simulation of the typical cell for some non-Poisson cases.

Mahin *et al.* (1980) studied planar sections of Johnson–Mehl tessellations by simulation.

Several asymptotic results for this tessellation observed in $[0, L]^d$, as $L \to \infty$, are mathematically tractable. Chiu (1995c) and Chiu and Yin (2000) derived the limiting distribution of the time of complete tessellation, which is the process duration for underlying birth-and-growth processes (see Section 6.6.4). Chiu (1997), Chiu and Quine (1997, 2001) and Chiu and Lee (2002) established central limit theorems for the total number of cells.

9.10 Statistics for stationary tessellations

9.10.1 Reconstruction

Frequently, the first step of statistics for tessellation data is their (re)construction. Empirical tessellations are often given only in a rough form, without clearly visible cells. It is a demanding task to transform such data to real tessellations. There are two established methods, which are applied to planar and spatial tessellations.

The first uses a chain of procedures of image processing and finally yields a unique tessellation. Typical references are Schwertel and Stamm (1997), Coster *et al.* (2005), Dillard *et al.* (2005), Lautensack and Sych (2006), Lautensack (2008) and Redenbach (2009b). The given image is binarised, then the Euclidean distance transform (see Ohser and Schladitz, 2009, p. 114ff.) is applied to yield for each pixel its distance to the system of cell boundaries, which leads to cell centres at the local maxima. Finally, the watershed algorithm divides the inverted distance image into cells; see Figure 9.16. As Lautensack (2008) reported, some manual polishing of the results obtained may be necessary.

The second method uses ideas of Bayesian statistics and MCMC algorithms. Typical references are Blackwell and Møller (2003), Skare *et al.* (2007) and Møller and Stoyan (2014), where one starts with patterns of points scattered around the invisible cell edges. The prior is a deformed Voronoi tessellation obtained by random perturbations of the vertices, the nuclei and the unknown parameters, whilst the posterior is product of the likelihood of the data modelled by the deformed tessellation and the prior. The MCMC output consists of many random reconstructions, which give the user some impression of the uncertainty of the result.

9.10.2 Summary characteristics

Given a sample of a tessellation, corresponding summary characteristics can be estimated. The methods for point processes, fibre and surface processes and random sets can be applied to the various related structures.

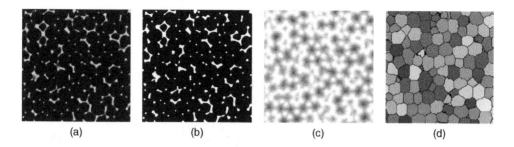

(a)　　　　　　　(b)　　　　　　　(c)　　　　　　　(d)

Figure 9.16   Reconstruction of a tessellation: (a) a section of the original three-dimensional image; (b) binarised image; (c) inverted distance image; (d) reconstructed cells. Reproduced from Redenbach (2009b) with permission of Società Editrice Esculapio Srl.

For example, the system of vertices is a point process and its intensity λ_0 can be easily estimated by the standard intensity estimator. The estimation of the intensity of the point process of cell centroids is more complicated; see the discussion below for the planar case. There the tessellations are viewed as particular germ–grain models.

The parameters L_A and L_V belong to the fibre process of edges and can be estimated by the methods described in Sections 8.3 and 8.4. Finally, the linear and spherical contact distributions $H_l(r)$ and $H_s(r)$ for the system of edges (planar case) and plates (spatial case) can be estimated by the methods of Section 6.4.5.

9.10.3   Statistics for planar tessellations

Some statistics for a tessellation are possible by considering the tessellation as a germ–grain model and applying the random-set methods of Chapter 6. This leads to information on the cells. Tessellation-specific methods start with the estimation of the four fundamental parameters: edge density L_A, intensity of vertices λ_0, the proportion ϕ of π-vertices, and mean number of cells per unit area (intensity of cell centroids) λ_2.

The characteristic L_A can be estimated by applying the methods for planar fibre processes as in Section 8.3, thus direct length measurement can be made or intersection methods may be applied. In the case of isotropy one can use Formula (8.38) in Section 8.3.2:

$$L_A = \frac{\pi}{2} P_L, \qquad (9.135)$$

where P_L is the intensity of the intersection point processes on the test lines.

Estimation of the mean number λ_0 of vertices per unit area is a simple task: one has simply to estimate the intensity of a planar stationary point process, that of the vertices. Given a window of observation W of area $A(W)$ an unbiased estimator is clearly

$$\hat{\lambda}_0 = \frac{\#\{\text{vertices observed in } W\}}{A(W)}. \qquad (9.136)$$

The proportion ϕ of π-vertices can be estimated analogously.

Unbiased estimation of λ_2, the mean number of cells per unit area, is a bit more complicated because of the influence of edge-effects. In the case of a trivalent tessellation, that is, $n_{02} \equiv 3$, in each vertex precisely three edges emanate, Formula (9.26) can be used to obtain

$$\hat{\lambda}_2^{(\mathrm{tri})} = \frac{\hat{\lambda}_0}{2}, \tag{9.137}$$

and so the problem is reduced to estimation of λ_0.

If trivalence is not given, then in the case of isotropy,

$$\hat{\lambda}_2^{(\mathrm{iso})} = \frac{N(W) - 1 - \frac{1}{2}N_{\mathrm{e}}(W)}{A(W)} \tag{9.138}$$

for a convex window of observation W is an unbiased estimator. Here $N(W)$ is the number of cells intersecting the window W while $N_{\mathrm{e}}(W)$ is the number of edges intersecting the boundary of W.

Proof of unbiasedness of $\hat{\lambda}_2^{(\mathrm{iso})}$. Formula (6.123) yields

$$\lambda_2 A(W) = \mathbf{E}\big(N(W)\big) - 1 - \frac{L_A^* \, L_2(W)}{2\pi},$$

where L_A^* is the boundary length density of the germ–grain model formed by the cells. Clearly, $A_A = 1$ and $L_A^* = 2L_A$. Formula (9.135) can be used to replace L_A by $(\pi/2)P_L$, which is equal to $(\pi/2)\mathbf{E}\big(N_{\mathrm{e}}(W)\big)/L_2(W)$. This yields

$$\lambda_2 A(W) = \mathbf{E}\big(N(W)\big) - 1 - \frac{\mathbf{E}\big(N_{\mathrm{e}}(W)\big)}{2}. \qquad \square$$

If W is a rectangle and if the cells are so small that each intersecting cell produces at most two intersections of edges with the boundary of W then Formula (9.138) gives a third estimator, which is suggested by Saltykov (1974):

$$\hat{\lambda}_2^{(\mathrm{rect})} = \frac{z(W) + \frac{1}{2}w(W) + \frac{1}{4}u(W)}{A(W)}, \tag{9.139}$$

where

$z(W) =$ number of cells completely in W,

$w(W) =$ number of cells that intersect the sides of W but *not* its corners,

$u(W) =$ number of cells containing the corners of $W (= 4)$.

In the trivalent case all three estimators $\hat{\lambda}_2^{(\mathrm{tri})}$, $\hat{\lambda}_2^{(\mathrm{iso})}$ and $\hat{\lambda}_2^{(\mathrm{rect})}$ yield the same result, and no isotropy assumption is necessary; see Hahn (1995).

Because of Formula (9.15) estimates of λ_2 also lead to estimates of mean cell area $\bar{a}_2$. Unbiased direct estimation of $\bar{a}_2$ is via the methods in Section 6.4. However, the following estimator $\hat{\bar{a}}_2$ for the mean cell area is biased, known to yield estimates too large:

$$\hat{\bar{a}}_2 = \frac{\sum\{\text{cell areas}\}}{\#\{\text{cells}\}}, \tag{9.140}$$

where the cells are a randomly chosen cell and the cells in some shells around; see Voloshin *et al.* (2010). Nevertheless, 'shell map analysis' as described by Aste *et al.* (1996) and Aste (1999) is a valuable tool in tessellation statistics.

Furthermore, it is a natural idea in the statistical analysis of an irregular system of hard objects, in particular balls, to construct the corresponding Voronoi, Voronoi S or Laguerre tessellation and to analyse these tessellations statistically. This idea is systematically treated in Medvedev (2000) and Alinchenko *et al.* (2004), under the name Voronoi–Delaunay method. Of particular interest are analyses of:

- Delaunay tetrahedra (shape and size);
- void geometry, using interstitial balls centred at the tessellation vertices;
- the whole network of edges of the tessellation.

If three-dimensional tessellation data are given, then at least in principle one can apply methods similar to those discussed above in the planar case. Until now such tessellations have been analysed mainly by means of particular models, as described in the next section.

9.10.4   Statistics for Voronoi, Laguerre and Johnson–Mehl tessellations

Planar Poisson-Voronoi tessellations

It would be convenient if one could conclude that each tessellation in the ordinary equilibrium state is interpretable as a Voronoi tessellation derivable from some point pattern. Unfortunately a simple degrees-of-freedom argument shows that this cannot be the case. So the Poisson-Voronoi tessellation is not more than a kind of 'benchmark'; the mean-value quantities extracted there provide values which (i) may serve as rough approximative estimates for the tessellation of interest or (ii) may give some impression about the deviations from a Poisson-Voronoi tessellation.

If it is known that a given tessellation is really a sample of a Poisson-Voronoi tessellation, then there are at least four possible procedures for estimation of the single model parameter ϱ, the mean number of cells per unit area in the planar case, based on determination of estimates of:

- λ_0 (intensity of vertices),
- λ_1 (intensity of edges),
- λ_2 (intensity of cells), and
- L_A (intensity of the fibre process of edges).

In the first and last case no edge-correction is necessary, while in the second and third case the ideas of Section 6.5.7 can be used, where the 'grains' are either the edges or the cells, respectively. The estimation variances for all four intensities above are smaller than the variance of estimation of the intensity of a Poisson process with parameter ϱ; see Hahn (1995). However, with increasing window size the difference vanishes. (The variances behave analogously if a Voronoi tessellation relative to another stationary point process is considered.)

If $\bar{n}_{02} = 3$ then the question may arise as to whether the analysed tessellation is Poisson-Dirichlet. One could first ask whether the tessellation is one that could be a Dirichlet

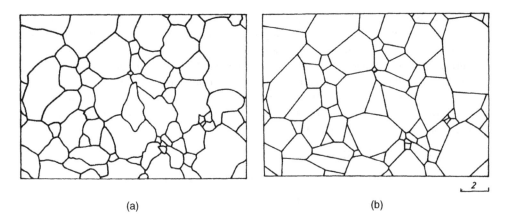

(a)   (b)

Figure 9.17   Planar section through a specimen of steel. (a) The lines correspond to smoothed boundaries of Austenite grains. (b) A tessellation with convex cells of the same topological form.

tessellation for any point process; if so then one might attempt to reconstruct the points of the generating pattern and test whether this pattern could be considered to be a Poisson process pattern. A quicker approach, and one that at any rate assesses the tessellation against the Poisson-Voronoi tessellation *qua* benchmark, uses estimates λ_0, λ_2 and L_A, and looks how these values satisfy the Poisson-Voronoi tessellation relationships

$$\lambda_0 = 2\lambda_2 \quad \text{and} \quad L_A = 2\sqrt{\lambda_2}.$$

Example 9.2. *Austenite grain boundaries in steel*
Figure 9.17 shows a planar tessellation which results from a naïve 'linearisation' of an image obtained by planar section through a sample of steel. (A better analysis of such tessellations should start with a reconstruction of the tessellation by the methods described in Section 9.10.1.) The edges correspond to the Austenite grain boundaries. From the tessellation the following values are obtained:

total edge length $= 172.2$,   number of vertices $= 124$, $N(W) = 78$, $N_e(W) = 30$.

These give

$$\hat{L}_A = 0.96 \quad \text{and} \quad \hat{\lambda}_0 = 0.69.$$

Since $\bar{n}_{02} = 3$, the mean number of cells per unit area λ_2 can be estimated in different ways. Formula (9.137) yields $\hat{\lambda}_2 = 0.34$, while (9.138) leads to $\hat{\lambda}_2 = 0.35$. Formula (9.139) is inappropriate here, as there are two cells hitting two sides of W each.

These estimates differ considerably from those expected for a Poisson-Dirichlet tessellation: from L_A the estimate $\hat{\lambda}_2 = 0.23$ is derived and this is considerably smaller than the direct estimates. But of course the sample is small, arguably too small for model tests to be realistic.

Note that it is methodologically doubtful to use a Poisson-Dirichlet tessellation in this application: probably there is no spatial tessellation of which the planar section is a

Poisson-Dirichlet tessellation. It is known that the planar section of a spatial Poisson-Voronoi tessellation is not a Dirichlet tessellation with respect to any point process (Chiu *et al.*, 1996); see also Section 10.6.

Further discussion of this example can be found in Example 10.4 on p. 442.

In the spatial case, stereological formulae given in Section 10.2 can be used.

Spatial Laguerre tessellations

As mentioned on p. 370, each normal and face-to-face spatial tessellation can be considered as a Laguerre tessellation with some system of points and weights. Many planar tessellations may also be well approximated by Laguerre tessellations. However, a marked Poisson process will produce Laguerre cells that are much more irregular in shape than those found in most empirical tessellations in the nature. It is more reasonable to consider Laguerre tessellations relative to centres of hard (non-overlapping) balls marked by the respective radii; see Figure 9.6 and the book cover. The balls are contained in their cells, which are more regular in shape. This approach is well demonstrated by Lautensack (2008), Lautensack *et al.* (2008) and Redenbach (2009a,b) for open and closed polymer, ceramic and metallic foams. Redenbach *et al.* (2012) consider modelling structures of strongly varying cell sizes.

The fitting involves two steps. The first step explores the type of the hard ball packing and that of the ball diameter distribution. Packings resulting from RSA or force-biased algorithms are successful examples. The ball volume distribution is heuristically assumed to be of the same type as that of the cell volumes, which is estimated from the given sample. Typical distributions for this purpose are gamma and lognormal. Their parameters can be estimated by classical methods of statistics.

In the second step the parameters of the ball volume distribution and the volume fraction of the corresponding hard ball packing are chosen to minimise the contrast d:

$$d = \sqrt{\sum_{i=1}^{8} \left(\frac{\hat{c}_i - c_i}{\hat{c}_i} \right)^2}, \tag{9.141}$$

with

c_1 (c_2) = mean (variance) of cell volume,

c_3 (c_4) = mean (variance) of cell surface,

c_5 (c_6) = mean (variance) of number of facets per cell,

c_7 (c_8) = mean (variance) of average cell breadth.

The c_i are model characteristics under the respective parameters, obtained by simulation, while the $\hat{c}_i$ are the empirical counterparts. The Nelder–Mead algorithm may be used in this optimisation problem with simulated data. Instead of L^2-norm, it is also possible to use L^1- or L^∞-norm in (9.141).

Example 9.3. *Cell boundaries in thermal insulation materials (Lautensack, 2008)*
A three-dimensional grey value image of the cell boundaries in a material for the thermal insulation of buildings was taken. Then, the image processing procedures mentioned in

Section 9.10.1 were applied to yield the tessellation structure, from which the empirical model characteristics were

$$\hat{c}_1 = 0.0055532, \qquad \hat{c}_3 = 0.17586, \qquad \hat{c}_5 = 14.637, \qquad \hat{c}_7 = 0.24698,$$
$$\hat{c}_2 = 0.0000065422, \qquad \hat{c}_4 = 0.00280, \qquad \hat{c}_6 = 10.378, \qquad \hat{c}_8 = 0.00143.$$

Since $\hat{c}_5 = 14.637$, which is smaller than that (15.535) of the typical Poisson-Voronoi cell, but greater than that (13.0–14.2) in the typical Laguerre cell generated by dense packing of hard balls, systems of hard balls generated by RSA were considered.

By comparing the empirical cell volume distribution with the lognormal and gamma distributions, with parameters estimated by the maximum likelihood, the gamma distribution was chosen for the ball volumes. Finally, the volume fraction of the RSA and the parameters of the gamma distribution were chosen, by using large scale simulation to obtain the c_i, to minimise d in (9.141). The fitted model gave

$$c_1 = 0.0055532, \qquad c_3 = 0.17507, \qquad c_5 = 15.111, \qquad c_7 = 0.24920,$$
$$c_2 = 0.0000072912, \qquad c_4 = 0.00251, \qquad c_6 = 11.079, \qquad c_8 = 0.00115.$$

With the exception of the estimate for c_6, all these values are closer to the empirical $\hat{c}_i$ than the corresponding values of the Poisson-Voronoi tessellation of the same intensity.

Example 9.4. *Polyurethane foam (Lautensack et al., 2008)*
A three-dimensional grey value image of polyurethane foam was studied. In the reconstructed tessellations, the mean number of facets per cell was $\hat{c}_5 = 13.722$, suggesting that, unlike the sample in Example 9.3, a Laguerre tessellation relative to a system of random dense packing of hard balls could be tried. However, 2.9% of the cells were very small and irregular in shape, meaning that the cell sizes came from a mixture of two distributions. Thus, a system of two classes of balls, one having gamma distributed sizes and one having fixed small size, was used. First the force-biased algorithm was tried for both classes of balls, but it turned out that in the corresponding Laguerre tessellations the small cells were too regularly distributed. A better result was obtained by generating a dense packing of the gamma distributed balls by means of the force-biased algorithm, followed by distributing the small fixed size balls according to the RSA principle. A model-fitting procedure by minimising the contrast to estimate the gamma distribution parameters, the radius of the 2.9% small balls, and the volume fraction of balls was then applied.

Johnson–Mehl tessellations

For a Johnson–Mehl tessellation, even if a spatially homogeneous Poisson process for the birth locations is assumed, a sample of only the resultant tessellation has not enough information for the estimation of the time-dependent birth rate and the growth rate. In order to apply the estimation techniques given in last part of Section 6.6.4, temporal information, especially the birth times of nuclei, is essential.

Example 9.5. *Application to neurobiology (Quine and Robinson, 1992; Chiu et al., 2000; Molchanov and Chiu, 2000; Chiu et al., 2003)*
The terminal of a neuronal axon has branches consisting of strands. At a synapse an action potential triggers the release of neurotransmitter at randomly scattered sites on these strands.

Each quantum released would cause release of an inhibitory substance that diffuses along the terminal bi-directionally at a constant rate preventing further releases in the inhibited region. Thus, it can be modelled by the birth-and-growth process of a Johnson–Mehl tessellation on the line. The data are measurement of times of actual release of transmitters (birth times of nuclei) and amplitudes of releases that, after the inverse power transformation, could be served as surrogates for birth locations. Estimation methods described in Section 6.6.4 could then be applied to these 'tessellation' data.

9.11 Random geometrical networks

9.11.1 Introduction

This section considers random geometrical structures constructed by vertices, in this context often called 'nodes', and edges, which can be curves. The nodes are points randomly placed in $\mathbb{R}^d$.

Examples of geometrical networks

- *Edge systems of tessellations* are geometrical networks. A geometrical network is also obtained when some edges of the tessellation are removed, but so that isolated vertices are avoided. An example is the (planar) Arak polygonal Markov field, which is based on the Poisson line process; see Arak and Surgailis (1989), Arak *et al.* (1993) and van Lieshout (2012). It finds application in the context of image segmentation.

- *Crack or fracture networks* as in Figures 8.1 and 9.2 are geometrical networks, if the endpoints of free cracks (crack tips) are considered as vertices. A simple planar model is that the vertices are considered as primary elements and the straight line segment edges then are stepwise constructed.

- *Polymer networks* are an intensively studied class of geometrical networks. There the edges or 'chains' stand for chemical junctions. These networks exist in $\mathbb{R}^3$ and occupy there space. Edges connect preferably vertices close together. Physicists and chemists developed models of reaction kinetics that explain the generation of such networks, for which real images are not available. The classical theory of Flory (1953, 1976) uses an f-degree Cayley tree or Bethe lattice (a connected graph without cycles, see p. 403, where each vertex is connected with f other vertices, usually with $f = 3$ or 4; see Matoušek and Nešetřil, 2008), which is made incomplete and random, in the sense that only a fraction p of edges is realised. This theory has been generalised and extended; see Macosko and Miller (1991). This book has not the space to give even a sketch of these theories, which even use ideas from knot theory; see Michalke *et al.* (2001).

- *Communication networks* are of course an important class of networks, where also geometrical aspects play an important rôle; see Baccelli *et al.* (1997), Dousse *et al.* (2006), Gloaguen *et al.* (2006), Franceschetti and Meester (2008), Baccelli and Błaszczyszyn (2009a,b) and Zuyev (2010). A geometrical network of interest in this context is, for example, the 'connectivity network' or transmission range network, which connects all

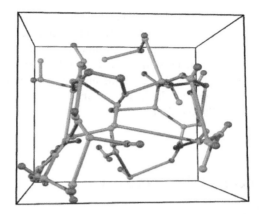

Figure 9.18   A small part of a three-dimensional geometrical network, showing the topology of a human forearm (radius) trabecular bone structure. It results from skeletonisation. Reproduced from Tscheschel and Stoyan (2003) with permission of Wiley.

antennas the SINR cells (see Section 6.5.4) of which each other overlap. However, in such networks many effects play rôles that are beyond the aims of the present section, which are only geometrical–topological.

- *Porous-media networks* result from geometrical constructions aiming at reducing porous media to networks that characterise their topological structure (perhaps to those of systems of balls and throats). Many geometrical networks could be considered as porous-media networks. The networks stand either for the system of pores or for its complement; in the latter case one speaks about 'skeletonisation'. Examples of such constructions can be found in Lowry and Miller (1995), Pothuaud *et al.* (2000), Sok *et al.* (2002), van Dalen *et al.* (2007), Thiedmann *et al.* (2009), Kadashevich and Stoyan (2010) and Stroeven *et al.* (2012). See Figure 9.18.

- *Traffic networks* (railway, highway) and *drainage networks*. See for example Gloaguen *et al.* (2006).

9.11.2   Formal definition of random geometrical networks

A *geometrical network* is a union δ of smooth curves, called edges, in $\mathbb{R}^d$ of finite length with two endpoints, which may coincide to allow self-loops. The system of curves is locally finite, that is, each compact set intersects δ in only a finite number of curves. Two arbitrary curves contact one another only in one common endpoint. Isolated points (segments of zero length) do not exist, but there may exist endpoints that are not connected with other curves ('dead ends'). The joint endpoints at the contacts of curves, the dead ends and the coincided endpoints in loops are all called vertices. Figure 9.18 shows a piece of a three-dimensional network.

Two edges are connected if they have a common vertex; two vertices are connected if there is a sequence of connected edges between them; a network is connected if every two vertices are connected.

A geometrical network is a closed subset of $\mathbb{R}^d$ and, if the curves are finite chains of straight line segments, polyconvex. Let $\mathbb{N}_{GN}$ be the set of all geometrical networks of $\mathbb{R}^d$. The corresponding σ-algebra $\mathcal{N}_{GN}$ is the trace of $\mathcal{F}$ on $\mathbb{N}_{GN}$, where $\mathcal{F}$ is the hitting σ-algebra as introduced in Section 6.1.2. A *random geometrical network* Δ is a random element of $[\mathbb{N}_{GN}, \mathcal{N}_{GN}]$ or an $(\mathcal{A}, \mathcal{N}_{GN})$-measurable mapping from a probability space $[\Omega, \mathcal{A}, \mathbf{P}]$ into $\mathbb{N}_{GN}$. In short, Δ is a special random closed set the realisations of which are geometrical networks.

Stationarity and isotropy of random geometrical networks are defined analogously as for tessellations in Section 9.1. This section considers mainly stationary random geometrical networks. These have necessarily infinitely many edges. For random finite networks, it would be more natural to study them as random graphs; see Section 9.12.2.

Geometrical networks belong to the class of so-called *cell complexes* as studied by Zähle (1988). These consist not only of curves (edges), but also of lower-dimensional manifolds. For them generalisations of the formulae below hold.

9.11.3 Summary characteristics of stationary random geometrical networks

Characteristics for geometrical networks are often topological, except perhaps only those originated from fibre processes, such as L_A and L_V. The definitions and notations below are given for the spatial case, and those for the planar case are analogous.

As in the case of tessellations, the following characteristics of a random geometrical network Δ are natural:

- λ_0 = vertex intensity, the mean number of vertices per unit volume.

- λ_1 = edge intensity, the mean number of edge midpoints per unit volume.

- L_V = mean total length of edges per unit volume (and so the mean length of the typical edge is equal to L_V/λ_1).

- N_V = specific connectivity or specific Euler–Poincaré characteristic, which is defined as either

$$N_V = \lim_{K \uparrow \mathbb{R}^3} \frac{\mathbf{E}\big(\chi(\Delta \cap K)\big)}{v_3(K)}, \tag{9.142}$$

where χ is the connectivity or Euler–Poincaré characteristic, or

$$N_V = \text{intensity of the 0-curvature measure corresponding to } \Delta.$$

- f = mean degree (also known as 'mean functionality' and 'mean coordination number') of the typical vertex, where the degree of a vertex is the number of edges emanating in it. This is the analogue to $\bar{n}_{01}$ for tessellations, and is defined as

$$f = \sum_{k=1}^{\infty} k p_k, \tag{9.143}$$

where p_k is the probability that the typical vertex has degree k. This probability is given by

$$p_k = v_k / \lambda_0, \qquad (9.144)$$

in which v_k is the mean number per unit volume of k-degree vertices. Because each edge connects two vertices, it follows

$$\sum_{k=1}^{\infty} v_k = 2\lambda_1. \qquad (9.145)$$

The characteristic f is considered dominant in the physical context (Chubynsky and Thorpe, 2001).

That some of the characteristics introduced above are closely related is shown by the formula

$$N_V = \lambda_0 - \lambda_1, \qquad (9.146)$$

see Mecke and Stoyan (2001) for a proof. It is based on the fact that the 0-curvature measure is purely atomic and concentrated at the vertices, and assigns the mass $1 - k/2$ to every k-degree vertex.

Finally, it is the case that

$$N_V = (1 - f/2)\lambda_0. \qquad (9.147)$$

Formula (9.147) shows that for connected networks the connectivity number is always nonpositive because $f \geq 2$, and becomes 'more negative' with increasing mean degree f. However, positive values are also possible. In the case of a rather strange 'network' that consists only of non-intersecting curves it is $N_V = \lambda_1$.

9.11.4   Statistics for networks

Tscheschel and Stoyan (2003) study a statistical problem for spatial networks, namely the estimation of their connectivity number N_V. By Equation (9.147) it can be reduced to a problem of point process statistics. While the network studied in this paper is a constructed porous-media network, the network in Lück *et al.* (2010, 2013) is a real filament network, the graph structure of which is reconstructed by means of image-analytical tools. Both papers characterise the network variability by pair correlation functions of associated point processes; see also Car and Parrinello (1988).

An example for network modelling and goodness-of fit testing is Thiedmann *et al.* (2009). The authors compare empirical and model minimum spanning trees, tortuosity characteristics and degree distributions.

Okabe and Sugihara (2012) study various statistical analyses of events (such as traffic accidents) on and alongside networks. Aldous and Shun (2010) consider statistics for planar networks which measure the trade-off between total graph length and efficiency of transportation.

9.11.5   Models of random geometrical networks

All tessellation models are of course models for geometrical networks. However, the converse is not true. In the following models, there is no guarantee that the construction procedures will lead to tessellations.

Nearest neighbour embracing graphs

Chiu and Molchanov (2003) introduced a model, now known as the *nearest neighbour embracing graph*, to incorporate not only distances but also directions. Each point of a given (finite or infinite) point process Φ in $\mathbb{R}^d$ is connected by a directed edge to its nearest neighbour, then to the second nearest, etc. From each point directed edges are continued to emanate until either the point is contained in the interior of the convex hull of the neighbours it connects to or the point is connected to all other points. If Φ is planar homogeneous Poisson, the mean out-degree (the number of directed edges going out) of the typical vertex is 5 (cf. the mean degree in the Poisson-Delaunay tessellation is 6). A realisation is shown in Figure 9.19(a). An optimal construction algorithm was proposed in Chan *et al.* (2004, 2006). The model has been used to study for example visibility and illumination in cameras and robot vision systems (Abellanas *et al.*, 2006, 2007a,b, 2009), limited range coverage of antennas (Matos, 2009), coverage boundary detection for wireless networks (Zhang *et al.*, 2006, 2009) and data depth in multivariate statistics (Cascos, 2007).

Poisson graphs with i.i.d. degrees

Deijfen *et al.* (2012) considered the following model, called the *polygamous Poisson process*. The starting point is a homogeneous Poisson point process in $\mathbb{R}^d$. Its points serve as vertices of the random geometrical network. The degrees of the vertices follow independently a prescribed distribution P. The first step is to attach an independent P-distributed random number of stubs

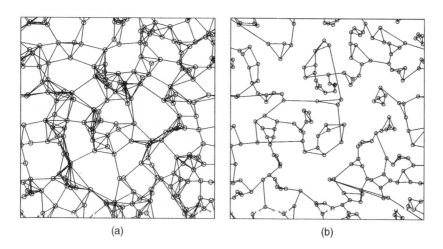

(a)   (b)

Figure 9.19   Realisations of network models for a sample of 200 i.i.d. uniform points in the unit square with periodic boundary conditions: (a) a nearest neighbour embracing graph; (b) a Poisson graph with degree 3 in all vertices. Note that some vertices in the graph seem as if they had degree 2; in fact these are pairs of vertices. Courtesy of T. Rajala.

to each vertex. Then a stable matching algorithm is applied to connect stubs. It has been shown that if the vertices form a homogeneous Poisson process, the matching scheme will be perfect and will lead to a stationary random network with the degree distribution P. Figure 9.19(b) shows a simulated sample of such a graph for the particular case that the degree of all vertices is 3, as it is in many tessellations. Clearly, the pattern is not a tessellation; it could be interpreted as a segment process. Deijfen (2009) studied the edge length distribution of such geometrical networks.

9.12   Random graphs

9.12.1   Introduction

This last section of the chapter presents some ideas of the theory of random graphs. Clearly, also tessellations and geometrical networks can be interpreted as graphs, but now only their topological nature matters, while locations and lengths are ignored. In many contexts such an approach is sufficient.

A *graph* consists of a nonvoid set of vertices connected by edges. Denote by $\mathbf{N}$ the set of all natural numbers. When vertices are labelled by natural numbers, a graph of n points can be represented by its $n \times n$ adjacency matrix, the (i, j)-entry of which indicates the number of edges connecting vertices i and j. Thus, $(\mathbf{N} \cup \{0\})^{\mathbf{N} \times \mathbf{N}}$ is the space of all graphs. A random element of this space endowed with the Borel σ-algebra with respect to the discrete topology (i.e. all subsets are open) is called a *random graph*.

Each tessellation and network can be interpreted as a graph by considering only the connectivity properties of vertices and edges.

A wide range of systems in nature and society can be described and hence modelled by random graphs. See Albert and Barabási (2002), Dorogovtsev and Mendes (2002) and Newman (2003) for details.

Classical random graph theory, initiated by Erdős and Rényi (1959), considers randomly chosen graphs among all possible N-edged graphs with n vertices. In later development, random graphs are considered as results of stochastic construction processes. A simple model of this type, $\mathbb{G}(n, p)$, proposed independently by Solomonoff and Rapoport (1951) and Gilbert (1959), is constructed as follows: there are given n vertices, and each pair of the vertices is connected independently with probability p. This model and its variants are (collectively) known as the *Erdős–Rényi model*, which is, however, too simple to model real-life situations.

In the following discussion, more complicated models and their properties are considered. For statistical analysis of empirical graphs, see for example Fienberg (2010a,b), in which the term 'network' is used instead of 'graph'.

The *typical vertex* of a random graph is just a randomly (uniformly) chosen vertex among the existing n vertices. Denote by p_k the probability that the typical vertex is of *degree k*, where the degree of a vertex is the number of edges connected to it. The distribution $\{p_k\}$ is called the *degree distribution*.

A *path* is a sequence of distinct vertices and corresponding edges, each of which connects two consecutive vertices in the sequence, and the length of a path is measured by the number of its edges. For a graph $\mathbb{G}_n$ with n vertices, let the *typical distance* H_n be the length of the shortest path between the members of a randomly and uniformly chosen pair of connected vertices. The *characteristic path length* L_n (also known as the *average path length* or simply the *average distance*) is the mean of H_n, while the *diameter* $\mathrm{diam}(\mathbb{G}_n)$ is the maximum of the shortest path lengths between any pair of connected vertices.

A *cycle* is a 'closed path', that is, a path plus an edge connecting the two vertices with degree 1 (before adding the edge) in the path, and the cycle length is equal to its number of edges. A connected graph without any cycle is a *tree*.

9.12.2   Random graph models and their properties

Degree distributions

A *scale-free random graph* is one whose degree distribution satisfies the power law,

$$p_k \simeq Ck^{-\tau}, \tag{9.148}$$

for large k, where C and τ are some constants. Power-law degree distributions have been observed statistically in many natural graphs with τ around 2 to 3; see the examples tabulated in Albert and Barabási (2002, Table II) and Dorogovtsev and Mendes (2002, Table I).

The Erdős-Rényi model is not scale-free. For $\mathbb{G}(n, p)$ the degree distribution is binomial. As $n \to \infty$ such that $np \to \lambda$, its degree distribution converges to Poisson.

An important scale-free model is the Barabási–Albert model, also called *preferential attachment model*; see Barabási and Albert (1999). It starts with m_0 vertices without any edges and then grows by adding vertices, one by one sequentially, such that at each step the new vertex has m ($\leq m_0$) edges that link the new vertex to m different already existing vertices in such a way that an existing vertex is linked with probability proportional to its current degree. Thus, large degrees tend to become even larger and hence this model is also known as the *rich-get-richer model*.

The description given above is incomplete because it does not explain how the first edge is connected when all existing vertices are of zero degree, does not describe the dependencies between edges added at each step, and does not mention whether self-loops are allowed and whether the degrees should be intermediately updated. Since Bollobás *et al.* (2001) formulated the model rigorously many variants have been developed; see van der Hofstad (2010b, Chapter 8).

However, each of the different ways to make the model precise leads to '*the same asymptotic behaviour*' (Durrett, 2007, p. 90) and '*the results, in particular the occurrence of power laws and the power-law exponent, do not depend sensitively on the respective choices*' (van der Hofstad, 2010b, p. 166).

The limit of the degree distribution in the incomplete Barabási–Albert model above, as the number n of vertices goes to infinity, satisfies the power law

$$p_k \simeq \frac{2m^2}{k^3}, \tag{9.149}$$

for large k, giving $\tau = 3$ in Formula (9.148), which turns out to be true also in precise models; see Bollobás *et al.* (2001, Theorem 1) and Durrett (2007, Formula (4.1.1)).

To alter the value of the exponent τ, consider the *affine preferential attachment model* (van der Hofstad, 2010a,b) in which a third model parameter δ is introduced, which is larger than $-m$. The model starts with a single vertex with m self-loops. If $m = 1$, the new edge attached to the new vertex will be linked to the new vertex itself to form a self-loop with probability proportional to $1 + \delta$ and linked to an existing vertex with degree k with probability proportional to $k + \delta$. If $m > 1$, the m new edges attached to a new vertex are linked to vertices sequentially with intermediate updating of their degrees. The ith new edge is linked to an existing vertex with probability proportional to its current degree after linking the $(i-1)^{\text{st}}$ edge plus δ, and linked to the new vertex itself with probability proportional to its intermediately updated degree (with the convention that its degree is 1 when $i = 1$) plus $i\delta/m$. The special case $\delta = 0$ gives the preferential attachment model.

The power law given in (9.148) still holds with

$$\tau = 3 + \frac{\delta}{m} > 2, \qquad (9.150)$$

see van der Hofstad (2010a, Theorem 6.16).

Whilst the preferential attachment model turned out to have a power-law degree distribution, Molloy and Reed (1995) constructed random graphs with other, specified degree distributions. Their model is described as follows.

Take a sequence $\{k_i\}_{i=1}^{n}$ of independent degrees from a given degree distribution $\{p_k\}$. Then k_i stubs, or half-edges, are attached to vertex i and pairs of stubs are chosen without replacement at random to be connected together. This model is called the *configuration model*, in which self-loops are possible. See Durrett (2007, Chapter 3) and van der Hofstad (2010b, Chapters 7 and 10) for details and references.

A generalisation of the Erdős–Rényi model $\mathbb{G}(n, p)$ is the *inhomogeneous random graph*, proposed by Bollobás *et al.* (2007) and further developed in van der Hofstad (2010b, Chapters 6 and 9). Instead of the same probability p for all pairs of vertices, the model has p_{ij} for the probability that the vertices i and j are connected. As an example, consider such a model with double randomness. Suppose $\{w_i\}$ are i.i.d. random weights assigned to the vertices. Conditionally on the weights $\{w_i\}$, letting

$$p_{ij} = 1 - \exp\left(-\frac{w_i w_j}{\sum_{k=1}^{n} w_k}\right)$$

gives the model in Norros and Reittu (2006). If the tail of the probability density of the w_i follows a power law with exponent $-\tau$, then the power law (9.148) holds (van der Hofstad, 2010a, Section 6.3.2.1).

Path lengths

The 'small-world phenomenon' is a famous, even notorious, term. It is popularly known as the *six degrees of separation* phenomenon, corresponding to the assertion that any two persons on this planet can be connected by a chain of at most six persons knowing each other on a first-name basis; see Newman *et al.* (2006). This happens when there is a relatively short path connecting any two vertices in a graph.

Many social networks of ten thousands to a million vertices have average distances of around 3 to 8 (Newman, 2003, Table 3.1), which loosely agrees with the six degrees of separation, while the average distance of the World Wide Web, with over 800 million vertices, was in the year 1999 around 19 (Albert *et al.*, 1999).

The *Watts–Strogatz model* is a one-parameter small-world model proposed by Watts and Strogatz (1998). Start from a one-dimensional lattice with periodic boundary conditions, that is, a ring lattice, with n vertices. Assume that there are k edges per vertex, and then each edge may, with probability p independent of the others, be rewired by connecting one end of it to another vertex chosen at random.

In order to obtain a sparse but connected graph, one has to ensure that

$$n \gg k \gg \log n \gg 1,$$

in which the condition $k \gg \log n$ guarantees that during the rewiring process the random graph is connected at all times. Watts and Strogatz (1998) showed that in this model, the average distance L_n of a graph of n vertices has the following asymptotic properties:

a large world: $L_n \simeq \dfrac{n}{2k}$,   as $p \to 0$ (regular lattice),

a small world: $L_n \simeq \dfrac{\log n}{\log k}$,   as $p \to 1$ (entirely random).

The crossover (or phase transition) between the large world, in which L_n grows linearly with n, and the small world, in which L_n grows only logarithmically with n, is of some interest. For the Watts–Strogatz model and many of its variants, including models starting with a d-dimensional lattice, numerical simulations and analytical arguments led to the widely accepted general form

$$L_n \simeq \frac{n^{1/d}}{k} f(p \cdot k \cdot n), \qquad (9.151)$$

where

$$f(u) = \begin{cases} \text{constant}, & \text{if } u \ll 1, \\[2mm] \dfrac{\log u}{u}, & \text{if } u \gg 1; \end{cases}$$

see Albert and Barabási (2002, Section VI) and Newman (2003, Section 6.3). Relation (9.151) shows that switching from a large world to a small world and vice versa can be achieved by changing just the single scaling variable, namely, the product $p \cdot k \cdot n$, which is the average number of rewiring of edges. However, the Watts–Strogatz model is not scale-free because the degree distribution decays exponentially (Albert and Barabási, 2002, Formula (77)).

For the affine preferential attachment model above, the random typical distance H_n, with probability tending to 1 as $n \to \infty$, satisfies

$$C_1 \log n \leq H_n \leq C_2 \log n, \qquad \text{if } \tau > 3,$$

$$C_3 \log \log n \leq H_n \leq C_4 \log \log n, \qquad \text{if } 2 < \tau < 3,$$

$$(9.152)$$

and for the configuration model and inhomogeneous random graphs,

$$H_n \simeq C_5 \log n, \qquad \text{if } \tau > 3,$$

$$H_n \simeq C_6 \log \log n, \qquad \text{if } 2 < \tau < 3,$$

$$(9.153)$$

for some positive constants C_i; the asymptotic equalities are in the sense of convergence in probability as $n \to \infty$ (van der Hofstad, 2010a, Theorems 6.18 and 6.19).

Note that the relations (9.152) and (9.153) do not include the case that $\tau = 3$. For the inhomogeneous random graph above where the p_{ij} are proportional to (deterministic) weight products $w_i w_j$, Chung and Lu (2003, Theorem 2.6) prove that

$$L_n \simeq C \frac{\log n}{\log \log n}, \qquad \text{for } \tau = 3, \tag{9.154}$$

for some positive constant C. However, it is not clear to what extent (9.154) is true also for other models.

Bollobás and Riordan (2004, Theorem 1) show that the preferential attachment model with a fixed $m \geq 2$ is also a small world in the sense that for any positive ϵ the probability that the following inequalities hold tends to 1 as $n \to \infty$:

$$(1 - \epsilon)\frac{\log n}{\log \log n} \leq \text{diam}(\mathbb{G}_n) \leq (1 + \epsilon)\frac{\log n}{\log \log n}. \tag{9.155}$$

The lower bound in (9.155) also holds for $m = 1$, but the upper bound does not.

However, in general, the diameter is less robust and informative than the average and typical distance. Albert and Barabási (2002, p. 58) argue that the diameters of graphs of the same size n do not vary much and are concentrated around some value of order $O(\log n)$. This statement has been made more precise and proved for some special models (Chung and Lu, 2001 and Bollobás *et al.*, 2007, Theorem 3.16).

Clustering

In a small world, clustering is typical: it is likely that a friend of your friend is also your friend. A random graph is said to be *highly clustered* when two randomly chosen nearest neighbours of an arbitrary vertex are more likely to be connected to each other than an arbitrary pair of vertices. The *Barrat–Weigt clustering* (or *transitivity*) *coefficient* $C_{BW}^{\mathbb{G}}$ of a random graph $\mathbb{G}$ is defined as

$$C_{BW}^{\mathbb{G}} = \frac{3 \times \text{number of triangles in } \mathbb{G}}{\text{number of connected triples}} = \frac{6 \times \text{number of triangles in } \mathbb{G}}{\text{number of paths of length two}}, \tag{9.156}$$

where the factors 3 and 6 arise in the numerators because each triangle has been counted three and six times, respectively, in the denominators (Barrat and Weigt, 2000). It tells the proportion of connected triples that form triangles and hence $0 \leq C_{BW}^{\mathbb{G}} \leq 1$. A graph $\mathbb{G}$ containing n vertices and N edges is highly clustered if

$$C_{BW}^{\mathbb{G}} \gg \frac{2N}{n(n-1)}. \tag{9.157}$$

It is also useful to define a local value to measure the strength of clustering of a vertex:

$$c_i^{\mathbb{G}} = \frac{\text{number of triangles in } \mathbb{G} \text{ connected to vertex } i}{\text{number of triples centred on vertex } i}, \tag{9.158}$$

with the convention that $0/0 = 0$ (Watts and Strogatz, 1998). This value reflects the extent to which neighbours of vertex i are also neighbours of each other. Define the *Watts–Strogatz*

Table 9.8 Barrat–Weigt clustering coefficients of random graph models.

Model of $\mathbb{G}$	Asymptotic $\mathbb{E}(C_{BW}^{\mathbb{G}})$	Source
Erdős–Rényi	$\dfrac{\lambda}{n}$ (where $np \to \lambda$)	Watts and Strogatz (1998)
Watts–Strogatz	$\dfrac{3(k-1)}{2(2k-1)}(1-p)^3$	Barrat and Weigt (2000)
Barabási–Albert	$\dfrac{m-1}{8}\dfrac{(\log n)^2}{n}$ *	Bollobás and Riordan (2003)
Affine preferential attachment with $\delta > 0$	$C_{m,\delta}\dfrac{\log n}{n}$ †	Eggemann and Noble[‡] (2011)

* This result is very different from the experimental value $O(n^{-0.75})$ in Albert and Barabási (2002).

† $C_{m,\delta}$ is an explicitly known value depending solely on m and δ.

‡ Their precise model is slightly different from the one given here.

clustering coefficient as

$$C_{WS}^{\mathbb{G}} = \frac{\sum_{i \in \mathbb{V}} C_i^{\mathbb{G}}}{n}. \tag{9.159}$$

Both clustering coefficients above are of order n^{-1} for the Erdős–Rényi model. In contrast, for an empirical small-world network, it is suspected that either coefficient will tend to a non-zero limit as $n \to \infty$. Newman (2003, Table 3.1) reports that the mathematics coauthorship network of 253,339 vertices and 496,489 edges has $C_{BW}^{\mathbb{G}} = 0.15$ and $C_{WS}^{\mathbb{G}} = 0.34$. Table 9.8 summarises the known asymptotic formulae for the mean Barrat–Weigt clustering coefficients of various models.

Phase transition, percolation and resilience

One of the main questions in random graph theory is to determine critical model parameters ω_c for phase transition. The term *phase transition* (Janson *et al.*, 2000, Chapter 5) refers to the phenomenon that for some property, such as connectivity, graphs with model parameter $\omega < \omega_c$ are very unlikely (and almost surely not, as $n \to \infty$) to possess the property, whilst graphs with model parameter $\omega > \omega_c$ are very likely (and almost surely, as $n \to \infty$) to possess it.

Erdős and Rényi (1960) show that for the model $\mathbb{G}(n, p)$, the critical probability p for the existence of a giant component is $1/n$ and the critical probability p for connectivity is $\log n/n$. More precisely, with probability increasing to 1 as $n \to \infty$, when $p < 1/n$, none of the connected components has more than $O(\log n)$ vertices, whilst when $p > 1/n$, the number of vertices of the largest component is asymptotic to rn, for some $0 < r < 1$. In particular, at the critical probability $p = 1/n$, the largest component has a size of order $n^{2/3}$; for connectivity, if $p < \log n/n$, then $\mathbb{G}(n, p)$ contains isolated vertices and hence be disconnected, whilst if $p > \log n/n$, then $\mathbb{G}(n, p)$ is connected.

There is an analogous result for the existence of a giant component in scale-free models. The model parameter for such a phase transition is

$$
\omega = \begin{cases}
m & \text{for affine preferential attachment models,} \\[2em]
\dfrac{\mathbf{E}\big(K(K-1)\big)}{\mathbf{E}(K)} & \text{for the configuration model,} \\[2em]
\dfrac{\mathbf{E}(W^2)}{\mathbf{E}(W)} & \text{for inhomogeneous random graphs,}
\end{cases} \tag{9.160}
$$

where K denotes the random degree of the typical vertex, following the distribution $\{p_k\}$, and W denotes a random variable having the same distribution as the random weights w_i in inhomogeneous random graphs. For these three model classes, the critical values are the same and equal to $\omega_c = 1$ (see van der Hofstad 2010a,Theorem 6.17 for a precise statement).

One of the most striking features of natural networks is their high tolerance for errors. Albert *et al.* (2000) studied the effects of random errors (random removals of vertices, including edges attached to them) and targeted attack (targeted removal of high degree vertices) on the Internet and the World Wide Web, as well as the Erdős–Rényi model and the Barabási–Albert model. In the Erdős–Rényi model, the average path length increases monotonically with the percentage of removals, no matter random or targeted. In contrast, real networks and the Barabási–Albert model are robust to random errors but vulnerable to targeted attacks. The reader is referred to Albert and Barabási (2002, Section IX), Dorogovtsev and Mendes (2002, Section XI), Durrett (2007, Section 4.7) and Newman (2003, Sections 3.4 and 8.1) for detailed discussions and further references. Another way to study network vulnerability is to consider a stochastic model, such as a random walk, for the traffic of signals on a geometrical network; a vertex breaks down when the number of signals, such as random walkers, arrived exceeds a prescribed threshold; see for example Kishore *et al.* (2011, 2012).

The problem of connectivity of an infinite graph or network with random removals of vertices or edges is known as site percolation or bond percolation, respectively, which corresponds to the problem of the existence of a giant component in a finite graph. For a comprehensive review and further references; see van der Hofstad (2010a).

Some models for finite geometrical networks

While stationary random tessellations and geometrical networks are infinite graphs, finite geometrical networks are also important. In the following some popular models are briefly described.

Since physical measurements for these graphs are possible, the length of a path can be measured either topologically (i.e. still in terms of the number of edges) or physically (i.e. in terms of the sum of the Euclidean distances between consecutive vertices in the path). Of course properties of the topological length and of the physical length can be substantially different.

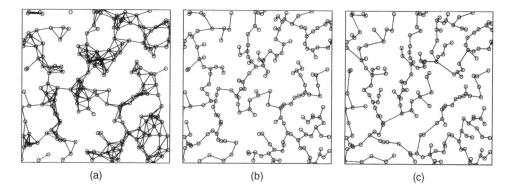

(a)   (b)   (c)

Figure 9.20   Realisations of finite random geometrical network models in the unit square for the same 200 points used in Figure 9.19: (a) a random geometric graph with $r = 0.1$; (b) a minimum spanning tree with the Euclidean distances as weights; (c) a radial spanning tree with respect to the origin, the lower left corner.

Random geometric graphs

The simplest model for a finite geometrical network is probably the *random geometric graph* $\mathbb{G}(\Phi_n, r)$, where $\Phi_n = \{x_1, \ldots, x_n\}$ denotes a finite point process in $\mathbb{R}^d$ and x_i and x_j are connected if the distance between them is not more than r; see Figure 9.20(a). Its infinite counterpart for the Poisson process case is discussed in Section 3.3.3. Probabilistic properties of $\mathbb{G}(\Phi_n, r)$ have been studied extensively in Penrose (2003). The existence of a giant component in $\mathbb{G}(\Phi_n, r)$ is related to the problem of continuum percolation, discussed in detail in Section 3.3.4 and Meester and Roy (1996), which plays an important rôle in studying actual percolation processes that occur in real heterogeneous materials (see e.g. Torquato, 2002, Section 9.2 and Chapter 10).

Special geometrical networks can be built based on $\mathbb{G}(\Phi_n, r)$. Examples include the restricted Delaunay triangulation (Chen, 2008), which contains the common edges in $\mathbb{G}(\Phi_n, r)$ and the Delaunay tessellation generated by Φ_n, and the geometric preferential attachment model (Flaxman *et al.*, 2006, 2007), where each new vertex in the preferential attachment model can only be connected to vertices within distance r. These models can be extended to the stationary case, using a homogeneous Poisson process instead of Φ_n.

Minimum spanning tree

A spanning tree of a finite set of vertices is a tree connecting all vertices. When each edge is given a weight, often but not necessarily its length, a *minimum spanning tree* is a spanning tree with the total weight not more than the total weight of any other spanning tree. Figure 9.20(b) shows a minimum spanning tree with Euclidean distance between vertices as weights.

Minimal directed spanning forests

Often edges in natural systems are directed. Such edges appear in the *minimal directed spanning forests* (Penrose and Wade, 2010), motivated by modelling communication and drainage networks. Suppose that the vertices have a partial order, such as the coordinatewise partial order. The distance between two vertices is measured by for example the L^p-norm. Each non-minimal vertex under the partial order is connected to the nearest neighbouring vertex of lower order, or a randomly chosen one if there are more than one. Each connected component is a tree and the number of trees in this forest is equal to the number of minimal vertices. Different partial orders and distance functions lead to a rich variety of models. For example, the *radial spanning tree* is obtained if the partial order is the distance from the origin and the weights are the Euclidean distances; see Figure 9.20(c).

The minimal directed spanning forest must not be confused with the minimum spanning tree. The latter does not involve a partial order and its vertices are connected in such a way that the total weight is minimised; they do not necessarily connect to their nearest neighbours.

10

Stereology

10.1 Introduction

The objective of stereology is to draw inferences about the geometrical properties of three-dimensional structures when information is only available in lower-dimensional form via planar sections, linear probes, or projections of thick slices. Its application arises in the study of geometrical structure of constituents in opaque or partially opaque bodies such as metals, minerals, synthetic materials, food or biological tissues. Often these structures appear at the microscale, but sometimes also the macroscale is of interest, for example in studies of rock or soil structures. In all these cases an elegant way to acquire spatial information is via planar sections or linear probes.

Fortunately, today there exist various techniques that yield spatial information in a direct way, for example computerised tomography (CT) based on absorption of X-rays or synchrotron radiation and reconstruction of three-dimensional (3D) data from stacks of two-dimensional (2D) images obtained by confocal laser scanning microscopy. In the first case the inverse Radon transform is applied, in the second methods of differential calculus. Furthermore, 3D images can be obtained by serial sections combined with classical microscopic imaging such as focused ion beam thinning combined with scanning electron microscopy. The data obtained can be analysed with the methods described, for example, in Chapter 6. The 3D-data-yielding methods can be seen as using many sections close together, while in stereology as understood in this chapter the number of sections is small and these are placed in large distances such that the measurement results can be considered as independent.

Some stereological methods that were too complicated and instable are now obsolete. However, there are many very simple and elegant stereological techniques which make the use of 3D techniques unnecessary, as for example, the methods for determining the so-called 'global parameters' (Rhines, 1986) such as volume fraction V_V, specific surface area S_V and specific line length L_V; see the discussion in Exner (2004). Additionally, planar sections are sometimes observable 'with much higher precision than is available in 3D and hence the only way to gain information about the 3D objects is to study 2D sections of them', in particular if

Stochastic Geometry and its Applications, Third Edition.
Sung Nok Chiu, Dietrich Stoyan, Wilfrid S. Kendall and Joseph Mecke.
© 2013 John Wiley & Sons, Ltd. Published 2013 by John Wiley & Sons, Ltd.

one is interested in nano-scale objects (Nagel, 2010). The wording in Torquato (2002, p. 294) describing stereology as a 'poor man's' tomography or in Baddeley and Jensen (2005, p. 81) as 'cheap technique' should not be considered ruling out the use of stereology. Stereology should be applied when other techniques are too expensive or cannot be successfully applied. (Wise rich men use their resources economically.)

Traditionally, stereology represents an important part of the applications of stochastic geometry and spatial statistics. This chapter introduces many of the results and ideas of stereology, limited mainly to cases where statistically homogeneous structures are of interest. Such applications are typical in materials science, geological science and food science (Baddeley and Jensen, 2005) and in some cases in medicine or biology. There are various textbooks on the topic starting with the classical books by Underwood (1970) and Saltykov (1974), namely Weibel (1980), Saxl (1989), Ohser and Mücklich (2000), Hilliard and Lawson (2003), Baddeley and Jensen (2005), Howard and Reed (2005) and Mouton (2011). And the fine survey of Nagel (2010) written for mathematicians has to be mentioned.

The methods and formulae of stereology relate characteristics of three-dimensional structures to quantities arising from measurements of planar sections (such as slices through metallic or mineral material), projections of thick sections (such as foils or sheets of biological material studied by light or electron microscopy), or linear probes (such as arising in biopsy or geological drilling).

The step from spatial structures to planar or linear sections or from planar structures to linear probes involves much loss of information and so stereological methods commonly can yield only 'global' information of a statistical character. A further consequence of this loss of information is that some of the applications require the solution of *ill-posed* mathematical problems, that is, numerical problems in which small deviations of data (perhaps due to measurement errors and statistical fluctuations) can lead to large discrepancies in the final solution. For example, this is the case for the Wicksell corpuscle problem, the unfolding problem for ball diameters in Section 10.4.

Stereological formulae depend on some assumption of 'randomness'. This can come in one of two forms: either the structure itself is assumed to be random or the section process is taken to involve a randomising element. In the first case stereologists speak about the

$$\boxed{\textit{model-based approach}}$$

and in the second case about the

$$\boxed{\textit{design-based approach}}.$$

For scientists using stereology in materials science and technology, mineralogy, geology, and food science it is usually natural to follow the model-based approach and to assume that the structures are motion-invariant, that is, stationary (statistically homogeneous) and isotropic. In many cases motion-invariance is the only mathematical assumption, and it alone already leads to valuable results. (However, finer model assumptions such as that the observed structure is a system of balls are often of interest and hence are also discussed as special cases.) In the mathematical theory it is then possible to assume that the section plane or line is fixed, for example the section plane is taken to be the (x_1, x_2)-plane. This book mainly follows the model-based approach. The alternative approach typically assumes that the sectioning planes or lines are members of motion-invariant plane or line processes, which is more natural in biological applications. The design-based approach is described in detail in the book by Baddeley and

Jensen (2005). A reader of that book may read the term 'modern stereology' as 'modern biological stereology'. In actual practice, many stereological formulae are generally similar for both approaches, differing only in the statistical theory and underlying interpretation; see Jensen (1984), Stoyan (1990b) and Baddeley and Jensen (2005).

A special instance of design-based techniques is the area of *local stereology*. 'The target is here quantitative parameters of structures which can be regarded as neighbourhoods of points, called reference points. An important example is the case where the structure is a biological cell and the reference point is the cell nucleus or some identifiable part of the nucleus such as the nucleolus' (Jensen, 1998, p. 27). Inferences are drawn using random lines or planes through the reference point, based on physical or optical sections. The Horvitz–Thompson estimator plays an important rôle.

Design-based stereology as in Baddeley and Jensen (2005) is substantially based on ideas of sampling theory and opens the way to characterise the precision of stereological estimators. It applies various random sampling designs which are adapted to the case of inhomogeneous or even finite structures.

Historically, Delesse (1847) and Rosiwal (1898) are landmark papers in the development of stereology, recognising that under certain conditions the volume fraction of a component in a body equals the area or linear fraction in planar or linear section; see Sections 6.4.2 and 10.2.2. Wicksell (1925) solved the problem of inferring the size distribution for systems of balls given planar sections; see Section 10.4.2. Beginning with the 1940s the stereological theory has been developed by researchers such as Bach, Cruz-Orive, DeHoff, Giger, Hilliard, Moran, and Saltykov. The mathematical basis of modern stereology is due to Ambartzumian, Baddeley, Coleman, Cruz-Orive, Davy, Gundersen, Jensen, Miles, Saxl and others, with important impulses for its development coming repeatedly from the needs of practical applications. The modern mathematical–statistical theory is a synthesis of integral geometry, stochastic geometry and sampling theory. For example, the proof on p. 428 demonstrates the use of point processes; random measure theory occurs in Section 8.3 in the proofs of section-formulae for fibre processes.

10.2   The fundamental mean-value formulae of stereology

10.2.1   Notation

Stereology is built on a small number of mean-value formulae concerning intensities of random measures, which belong to the *global parameters*. These quantities, often also called *densities* or *specific values*, arise here as intensities of stationary (often isotropic) random measures. These are listed systematically in Table 10.1, using a system described originally in Underwood (1976) and Weibel (1980). It is used throughout this book and is here explained systematically.

Each mean-value quantity is denoted by one symbol subscripted by another. The main symbol is the quantity measured while the subscripted symbol indicates the dimensionality of the section employed in the measurement.

For example, the symbol A_A denotes the mean area of some constituent of the structure being considered per unit area. It is therefore the *area density* or *area fraction* of the constituent. The notation S_V stands for the mean surface area per unit volume, the *surface density* or *specific surface (area)*; it is measured in units of m^2/m^3.

In the following, Ξ_V is the 3D constituent of interest and Ξ_A its planar section.

Table 10.1 Symbols in common use in stereology.

	Interpretation of main symbol
V	Volume (in m^3)
S	Surface area (in m^2)
M	Integral of mean curvature (in m)
A	Planar area (of profiles or section structures) (in m^2)
L	Length (of linear elements in the space or plane) (in m)
N	Number of grains, particles or section profiles (for non-overlapping germ–grain models) or more generally the connectivity number
P	Number of points (in a section)
	Interpretation of symbol in subscript
V	per unit volume
A	per unit area
L	per unit length
P	per point

10.2.2 Planar and linear sections

The following presents the general stereological formulae that arise when the structure is sectioned by a plane or line; see Figure 10.1. Each formula will be stated and followed by either a brief indication of its proof or a reference.

Volume fraction

The most important stereological formula is very simple. The volume fraction V_V of a stationary structure satisfies

$$V_V = A_A = L_L = P_P. \qquad (10.1)$$

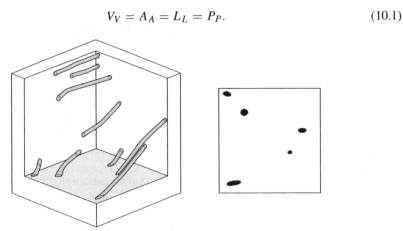

(a)   (b)

Figure 10.1   (a) A three-dimensional sample of fibres and (b) a planar section. The figure is based on a CT image kindly provided by Fraunhofer ITWM, showing a concrete sample with steel fibres obtained from the civil engineering department of TU Kaiserslautern.

That means, V_V can be obtained by planar or linear sections, or by measurement with a point lattice — without any assumption of isotropy. In mathematical terminology, the 'structure' is a stationary random set. The section plane or line can be arbitrary, but must be independent of the structure.

Equation-chain (10.1) can be interpreted as follows. The volume fraction V_V is equal to the quantity obtained by intersecting Ξ_V by an arbitrary test plane, line, or point system and measuring respectively the area fraction A_A, linear fraction L_L, or fraction P_P of test points in Ξ_V, respectively. Section 6.4.2 considers this further and discusses the estimation of A_A (or p in the notation there).

For example, let Ξ_A be the planar random closed set resulting from intersection of the stationary Ξ_V with a test plane, which is considered as the (x_1, x_2)-plane. Then Ξ_A is also stationary and so,

$$A_A = \mathbf{E}\big(\nu_2(\Xi_A \cap [0, 1]^2)\big)$$
$$= \int_{[0,1]^2} \mathbf{P}(x \in \Xi_A)\,dx = \mathbf{P}(o \in \Xi_A) = \mathbf{P}(o \in \Xi_V)$$
$$= p = V_V,$$

as discussed in Section 6.3.1.

Specific surface area

For a *motion-invariant* surface process, typically the surface of a three-dimensional motion-invariant random closed set, the following formulae were found by Saltykov (1945):

$$S_V = \frac{4}{\pi} L_A, \tag{10.2}$$

and

$$S_V = 2P_L. \tag{10.3}$$

This means that S_V can be estimated from planar sections by measuring line lengths and determining the intensity L_A, or from linear sections by counting the points formed by intersection of the linear probe with the surface of interest and calculating the mean number of intersection points per unit length; see Section 8.5.1.

In the anisotropic case there also exist elegant stereological methods for the determination of S_V. The following two cases are of particular interest.

(a) The surface process of interest is the result of a *homogeneous deformation* of an originally motion-invariant structure. Such a deformation is a mapping

$$(x_1, x_2, x_3) \mapsto (\eta_1 x_1, \eta_2 x_2, \eta_3 x_3)$$

with

$$\eta_1 \eta_2 \eta_3 = 1,$$

that is, it is volume-conserving. Samples of rolled steel can be described in this way, where $\eta_1 \le \eta_2 \le \eta_3$ represent the amounts of working. In this case lineal analysis measurements in the directions of x_1-, x_2- and x_3-axes lead to S_V as follows.

One determines the mean number $P_L^{(i)}$ of intersections of a test line in the direction of x_i-axis with the surface process of interest per length unit and so obtains estimators for the η_i via

$$\eta_1 P_L^{(1)} = \eta_2 P_L^{(2)} = \eta_3 P_L^{(3)} \tag{10.4}$$

and

$$\eta_1^3 = \frac{P_L^{(2)} P_L^{(3)}}{(P_L^{(1)})^2}, \qquad \eta_2^3 = \frac{P_L^{(1)} P_L^{(3)}}{(P_L^{(2)})^2}, \qquad \eta_3^3 = \frac{P_L^{(1)} P_L^{(2)}}{(P_L^{(3)})^2}. \tag{10.5}$$

The specific surface area S_V satisfies

$$S_V = 4 P_L^{(1)} \frac{S(\eta_1, \eta_2, \eta_3)}{\pi \eta_2 \eta_3}, \tag{10.6}$$

$$S_V = 4 P_L^{(2)} \frac{S(\eta_1, \eta_2, \eta_3)}{\pi \eta_1 \eta_3}, \tag{10.7}$$

$$S_V = 4 P_L^{(3)} \frac{S(\eta_1, \eta_2, \eta_3)}{\pi \eta_1 \eta_2}, \tag{10.8}$$

see Ohser and Mücklich (2000, p. 90), where $S(a, b, c)$ is the surface area of an ellipsoid of semiaxes a, b and c. They also present an algorithm for the estimation of S_V for given estimates of the $P_L^{(i)}$, $i = 1, 2, 3$. Furthermore, there is a well-documented example with drawn wire, where $\eta_1 < \eta_2 = \eta_3$; the x_1-direction is the drawing direction.

(b) The surface process of interest is stationary and, in addition, its distribution is invariant with respect to rotations about axes parallel to a given direction, in the following considered as the x_3-axis. (This x_3-rotation invariance corresponds to the equality $\eta_1 = \eta_2$ in case (a) above.) In this situation the specific surface area can be estimated from only one section plane, by *vertical section*.

A vertical plane is a plane parallel to the vertical axis. It yields a line of intersection with the (x_1, x_2)-plane, which uniquely determines the section plane.

The intersection of the vertical plane with the surface process yields a planar fibre process. This fibre process is intersected with a system of test curves as shown in Figure 10.2 and the number of fibre–curve intersection points is counted. The mean number P_L of intersection points per unit curve length leads, via Formula (10.3), to S_V. Note that the line density L_A of the fibre process in general does not satisfy Formula (10.2).

The curves in Figure 10.2 are so-called cycloids given by the parametric form

$$x_1 = \theta - \sin \theta \quad \text{and} \quad x_2 = 1 - \cos \theta$$

for $0 \leq \theta \leq \pi$. Such a curve has vertical height 2, horizontal width π and length 4. The tangent directions have the direction probability density function $f(\alpha) = \frac{1}{2} \sin \alpha$, which appears in Section 8.4.

Practical vertical S_V-measurement consists in taking n random vertical planes (constructed from random lines in the (x_1, x_2)-plane as in Section 8.2.2) and in using on each of them a

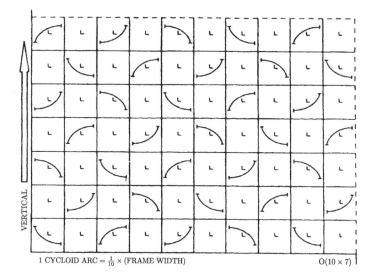

VERTICAL

1 CYCLOID ARC = $\frac{1}{10}$ × (FRAME WIDTH)   O(10 × 7)

Figure 10.2   A test system of cycloids for estimating S_V from vertical sections. Reproduced from Baddeley *et al.* (1986) with permission of Wiley.

system of cycloids as shown in Figure 10.2. For the ith plane, where the total length of all cycloids is l_i, the number P_i of intersection points of the surface process with the cycloids is determined, $i = 1, 2, \ldots, n$. The sum of the P_i divided by the sum of the total lengths l_i leads to the estimator

$$\hat{S}_V = 2 \sum_{i=1}^{n} P_i \Big/ \sum_{i=1}^{n} l_i. \tag{10.9}$$

This method was developed by Spektor (1950) and rediscovered by Hilliard (1967). In a very general context it is described and discussed in Baddeley and Jensen (2005, Section 8.1).

Specific curve length

For a stationary and isotropic fibre process the specific curve length satisfies the following relation found by Smith and Guttman (1953):

$$L_V = 2P_A. \tag{10.10}$$

This means that L_V can be determined from a planar section by counting the number of points formed by intersection of a plane with the fibre process. Remarks on the proof are to be found in Section 8.4.

Specific integral of mean curvature

For a stationary and isotropic surface process the specific integral of mean curvature or integral of mean curvature density M_V is defined as the intensity of the mean curvature measure associated with the process; see Section 7.3.4. It satisfies

$$M_V = C_A, \tag{10.11}$$

where C_A denotes the density of curvature of fibres in the planar section. In the terminology of Section 7.3.4 it is the intensity of the planar curvature measure. Formulae (7.56) and (10.11) yield

$$M_V = 2\pi N_A. \tag{10.12}$$

In the case of non-overlapping grains this formula is due to Bodziony (1965), where N_A is the mean number of section profiles per area unit. In the general case N_A is the specific planar connectivity number as introduced on p. 294.

Functional summary characteristics

In the motion-invariant case there are some random-set functional summary characteristics that can be easily determined from planar sections. As shown in Section 6.3, they are of great value in exploratory analyses. These characteristics are:

- covariance $C(r)$,
- linear contact distribution function $H_l(r)$,
- chord length distribution function $L(r)$,
- discoidal contact distribution function $H_d(r)$,
- all characteristics related to the capacity functional $T_{\Xi_A}(K)$ for planar K.

By the way, the first three characteristics can be obtained even from linear probes.

Stereological reconstruction

A powerful but computation-intensive method is stereological reconstruction for constituents in random structures that can be modelled by motion-invariant random sets. This method is a particular application of the general reconstruction method described in Section 6.7. Here the data come from a section plane, which is in the following identified with the (x_1, x_2)-plane, as a black–white pixel set. This pattern is analysed with the statistical methods for planar random sets as in Section 6.4 to yield estimates of $V_V = A_A$, $C(r)$, $H_l(r)$ and $H_d(r)$. Then a three-dimensional random set is simulated in the 'positive half-space' $\mathbb{R}^3_+ = \{(x_1, x_2, x_3) : x_3 \geq 0\}$ with the reconstruction method described in Section 6.7. The energy function used is an expression like

$$E = \alpha_C \int_0^{R_C} \left(\hat{C}(r) - C^s(r)\right)^2 dr + \alpha_l \int_0^{R_l} \left(\hat{H}_l(r) - H_l^s(r)\right)^2 dr + \alpha_d \int_0^{R_d} \left(\hat{H}_d(r) - H_d^s(r)\right)^2 dr \tag{10.13}$$

with suitable weight parameters α_C, α_l and α_d and upper limits R_C, R_l and R_d. The aim is to obtain a pixel structure in $\mathbb{R}^3_+$ whose characteristics are as close as possible (in the sense that E is minimised) to the estimated characteristics $\hat{V}_V$, $\hat{C}(r)$, $\hat{H}_l(r)$ and $\hat{H}_d(r)$ of the given pattern. In (10.13), $C^s(r)$, $H_l^s(r)$ and $H_d^s(r)$ are the corresponding characteristics of the simulated structure. It is easy to guarantee that $V_V^s = \hat{V}_V$ in the simulated structure, but for the other characteristics only approximation is possible, usually with simulated annealing.

This simulation is carried out *conditionally*, that is, on the (x_1, x_2)-plane the data structure is fixed during the whole simulation. This implies that for points with small x_3-coordinates the colours of the pixels are similar to the values at their neighbours in the section plane.

The three-dimensional pixel structure obtained by this simulation can be statistically analysed by the methods for three-dimensional random sets as in Section 6.4, and this leads to estimated characteristics that usually cannot by obtained by planar sections, such as the spherical contact distribution function $H_s(r)$, connectivity parameters beginning with the specific Euler–Poincaré characteristic N_V, or information on permeability and percolation.

The two papers of Talukdar *et al.* (2002,b) present interesting applications of this method. In both cases the two-dimensional data, which come from a pack of glass balls and North Sea chalk respectively, result from electron microscopic images. It is remarkable that these authors could demonstrate that their method produced good results though they only worked with $C(r)$ and $H_l(r)$ or $L(r)$. However, in general, using only $C(r)$ and $H_l(r)$ would result in reconstructed structures having poor pore space connectivity; see for example Øren and Bakke (2002, 2003) and Čapek *et al.* (2009). Li *et al.* (2012) compares reconstructions of three-dimensional structures using different summary characteristics.

10.2.3   Thick sections

To understand what a thick section is, see Figure 10.3 on p.435. The random closed set Ξ_V under study is intersected by two parallel planes separated by a distance $t > 0$. For the sake of definiteness one of these planes is taken to be the (x_1, x_2)-plane. The 'slice' that is to be examined is the set

$$T^t = \{x = (x_1, x_2, x_3) : 0 \le x_3 \le t\}$$

lying between these two planes. The experimental data are the planar image Ξ^t of the orthogonal projection of $\Xi_V \cap T^t$ onto the (x_1, x_2)-plane, which is again a random set. The projected thick section Ξ^t inherits the properties of stationarity and isotropy from Ξ_V.

Note that the term 'thin sections' is also often used in the literature. The term 'thick sections' makes the contrast to 'planar sections', which have no thickness, whilst the term 'thin' refers to the fact that the thickness of such a section is very small; both terms can be used for the same problem.

Application of the classical stereological Formulae (10.1) and (10.2) to the planar data $A_A(t)$ and $L_A(t)$ explained below would yield biased estimates of V_V and S_V. Therefore refined formulae are necessary.

Volume fraction and specific surface area of porous media

There seem to be no exact general formulae relating V_V and S_V to intensities that are measurable on Ξ^t. However, formulae are known for particular cases. Section 10.3.2 gives examples. In the motion-invariant case an approximate formula is given by

$$V_V \approx A_A(t) - \frac{t L_A(t)}{\pi}, \tag{10.14}$$

where $A_A(t)$ is the area fraction and $L_A(t)$ the mean boundary length per unit area of Ξ^t. This formula was given by Cahn and Nutting (1959) for systems of convex non-overlapping particles and then a heuristic argument based on the theories of random closed sets and surface processes was given for more general structures; see Stoyan (1984a).

An approximation of a similar nature for S_V is

$$S_V \approx \frac{4}{\pi} L_A(t) - 4t N_A(t), \tag{10.15}$$

where $N_A(t)$ is the specific number of section profiles; see Ohser and Mücklich (2000, p. 79)

An alternative to (10.14) is the formula

$$A_A(t) = 1 - (1 - V_V) \exp\left(-\frac{S_V t}{4(1 - V_V)}\right), \tag{10.16}$$

derived by Overby and Johnson (2005). For S_V the approximation formula (10.15) can be plugged in.

Note that with $t \to 0$ the classical stereological formulae are obtained.

Length density of fibre processes

Let now Ξ_V be a spatial motion-invariant fibre process. The fibre pieces between the two section planes are orthogonally projected and Ξ^t is a planar motion-invariant fibre process. The length density L_V of Ξ_V satisfies

$$L_V = \frac{4}{\pi t} L_A(t), \tag{10.17}$$

where $L_A(t)$ is the specific curve length of Ξ^t. Nagel (1983) gave a proof using the theory of fibre processes. Together with Formula (8.38) this yields

$$L_V = \frac{2}{t} P_L(t), \tag{10.18}$$

where $P_L(t)$ is the mean number of intersection points of the x_1-axis with the projected fibres in the slice. Note that here the limit $t \to 0$ does not make sense.

Formula (10.18) is the base of estimation of L_V in the anisotropic case by means of vertical sections as on p. 416, as suggested by Gokhale (1990, 1992, 1993). The quantity $P_L(t)$ is then the mean number of intersections with projected fibres per unit test line length. The test lines are cycloids with minor axis *perpendicular* to the vertical axis.

Stoyan and Gerlach (1987) show how to determine the spatial curvature distribution function (the distribution function of curvature in the typical fibre point) from the planar curvature distribution function.

10.2.4   Stereology for excursion sets

Stereology for excursion sets as defined in Section 6.6.3 is extremely simple. The planar section of a spatial random field yields a planar random field. If the former is Gaussian then so is the latter. And both fields have the same mean and correlation function $k(r)$. Finally, for given level u the excursion sets of the spatial field and that of the planar field coincide on the section plane. Consequently, it suffices to estimate u and $k(r)$ for the planar excursion set by means of the method sketched in Section 6.6.3 and all necessary spatial information is obtained.

10.2.5 On the precision of stereological estimators

The precision of stereological estimators, usually expressed by estimation variances, depends on the size of samples taken and on the spatial variability of the structure under investigation. The latter can be described by second-order characteristics. These are usually not known and must be estimated, often from the same samples as those for estimating the mean-value parameters.

The estimation of volume fraction is discussed in Section 6.4.2. The second-order characteristic needed there is the covariance $C(r)$. Hall and Ziegel (2011) developed bootstrap methods for a consistent estimation of the distribution of $\hat{V}_V$ and hence constructed, in various ways, confidence intervals for V_V.

In the case of stereological estimators based on counts of randomly distributed intersection points, it can be useful to employ the Poisson approximation. Then one can use the asymptotic confidence interval for the parameter of a Poisson distribution, as in Section 2.6.2, and so obtain confidence intervals for S_V, L_V and M_V. Thus, for example, an approximate $(1 - \alpha)100\%$ confidence interval for S_V is given by

$$\frac{2}{l} \left(\sqrt{n} - \frac{z_{\alpha/2}}{2} \right)^2 \leq S_V \leq \frac{2}{l} \left(\sqrt{n+1} + \frac{z_{\alpha/2}}{2} \right)^2 \tag{10.19}$$

where l is the total length of the test lines employed, n the number of points of intersection of the surface with the test system, and $z_{\alpha/2}$ the $(1 - \alpha/2)100^{\text{th}}$ percentile of the standard normal distribution.

More on statistics in stereology can be found in Ohser and Mücklich (2000), Beneš and Rataj (2004, Chapter 4) and Baddeley and Jensen (2005), as well as in the references therein.

10.3 Stereological mean-value formulae for germ–grain models

10.3.1 Planar sections

Suppose that Ξ_V is a spatial stationary germ–grain model and let $\{[x_n; \Xi_n]\}$ be the corresponding stationary germ–grain process, following the notation of Section 6.5. The distribution of the typical grain $\Xi_{V,0}$ is assumed to be rotation-invariant, but Ξ_V itself need not be isotropic. Consider the intersection Ξ_A of Ξ_V with an arbitrary test plane T,

$$\Xi_A = \Xi_V \cap T.$$

The position and normal direction of T can be chosen for convenience (because of the assumptions on Ξ_V), and so T is considered to be the (x_1, x_2)-plane. The intersection Ξ_A, if non-empty, is itself a planar germ–grain model.

The aim is to relate mean-value characteristics, such as the mean grain volume $\overline{V}$ and the germ-process intensity λ_V, of Ξ_V to the mean-value characteristics, such as the mean grain area $\overline{A}$ and germ-process intensity λ_A, of Ξ_A. (In the case of non-overlapping grains N_V is the same as λ_V: the mean number of grains per unit volume. Analogous relationships hold for λ_A.)

The next two sections consider the problem of estimating shape and size distributions for the grains of Ξ_V from planar information, a so-called *unfolding problem*.

Throughout this section the grains of Ξ_V are assumed to be convex. This assumption is often made; it seems to be realistic in many cases and measurement difficulties arise when it does not hold, for example, when a nonconvex grain is intersected by the plane and the resultant planar grain is not connected. However, the basic Formula (6.122) has a generalised version holding for nonconvex particles, and so consideration of the nonconvex case is possible; see the discussion in Miles (1983, 1984). For the application of his results it is necessary to be able to decide whether different components of a section profile actually arise from the same grain.

Under the assumption of convex grains it is clear that the intersection Ξ_A is also a stationary germ–grain model with convex grains. Because the typical grain $\Xi_{V,0}$ of Ξ_V has rotation-invariant distribution, so has the typical grain $\Xi_{A,0}$ of the planar section structure. If Ξ_V is a Boolean model then so is Ξ_A; see Section 3.2.5.

Theorem 10.1. *If the grains are convex and the components of overlapping grains can be identified, the mean-value characteristics of Ξ_V and Ξ_A are connected by*

$$\lambda_V \overline{V} = \lambda_A \overline{A}, \tag{10.20}$$

$$\lambda_V \overline{S} = \frac{4\lambda_A \overline{L}}{\pi}, \tag{10.21}$$

$$\lambda_V \overline{\overline{b}} = \frac{\lambda_V \overline{M}}{2\pi} = \lambda_A, \tag{10.22}$$

where $\overline{\overline{b}}$ is the mean average breadth of the typical grain $\Xi_{V,0}$.

Proof. (Stoyan, 1979c)   Consider the disc in the (x_1, x_2)-plane

$$c_r = \{x = (x_1, x_2, x_3) : x_1^2 + x_2^2 \le r^2, x_3 = 0\}.$$

By construction the numbers of grains hitting c_r must be the same for Ξ_V and Ξ_A, for all r. The corresponding mean values are given by Formula (6.124) as

$$\lambda_V \mathbf{E}\left(\nu_3(\check{\Xi}_{V,0} \oplus c_r)\right) = \lambda_V \left(\overline{V} + \frac{\pi \overline{S} r}{4} + \pi \overline{\overline{b}} r^2\right)$$

and

$$\lambda_A \mathbf{E}\left(\nu_2(\check{\Xi}_{A,0} \oplus c_r)\right) = \lambda_A(\overline{A} + \overline{L}r + \pi r^2)$$

for all $r \ge 0$. The ν_2-term in the second formula was calculated by means of the Steiner formula (1.23) and the ν_3-term in the first by means of the generalisation (6.29). Comparison of the coefficients of powers of r leads to the Formulae (10.20)–(10.22).   □

There are more general formulae for d-dimensional germ–grain models intersected by k-dimensional flats: Formulae (3.66) and (3.67), given in Section 3.2.5 for the Boolean model, hold generally for stationary germ–grain models. Indeed, the formulae do not depend on assumptions such as independence of the grains or particular distributions for the germ process such as being a Poisson process.

The spatial characteristics λ_V, $\overline{V}$, $\overline{S}$ and $\overline{\overline{b}}$ cannot be determined directly from the formulae in the theorem. There are only three equations available for four unknowns. How to overcome this difficulty is a fundamental problem of classical stereology. Stereologists have tried to solve it by several tricks. One of them is to make use of *form factors* such as Blaschke's coefficients

$$f = \frac{4\pi \overline{S}}{\overline{M}^2}, \tag{10.23}$$

$$g = \frac{48\pi^2 \overline{V}}{\overline{M}^3}. \tag{10.24}$$

The idea is to assume that f (or g) is known and fixed and so to reduce the number of unknowns to three.

The situation is not much easier even in the case of a Boolean model.

Example 10.1. *Planar section of a Boolean model*
The mean characteristics of a spatial stationary Boolean model Ξ_V with convex isotropic grains are

$$\lambda_V, \quad \overline{V}, \quad \overline{S} \quad \text{and} \quad \overline{\overline{M}} = 2\pi \overline{\overline{b}}.$$

The planar section Ξ_A is again a Boolean model with mean characteristics

$$\lambda_A, \quad \overline{A} \quad \text{and} \quad \overline{L}.$$

These three mean values are connected with area fraction A_A, specific boundary length L_A and specific convexity number N_A^+ of Ξ_A by the formulae in Section 3.2.2:

$$A_A = 1 - \exp(-\lambda_A \overline{A}) = 1 - \exp(-\lambda_V \overline{V}), \tag{10.25}$$

$$L_A = \lambda_A \overline{L} \exp(-\lambda_A \overline{A}) = \lambda_V \frac{\pi}{4} \overline{S} \exp(-\lambda_V \overline{V}), \tag{10.26}$$

$$N_A^+ = \lambda_A \exp(-\lambda_A \overline{A}) = \lambda_V \overline{\overline{b}} \exp(-\lambda_V \overline{V}). \tag{10.27}$$

Again there are given three equations for the four unknowns λ_V, $\overline{V}$, $\overline{S}$ and $\overline{\overline{b}}$. The situation does not change if the specific connectivity number N_A is used.

If the grain distribution is given by a two-parameter distribution (e.g. spherical grains with lognormal distribution) then the equations above can be used to estimate these two parameters and λ_V; see the discussion in Section 3.4.3.

Another way to avoid this situation may be to use the disector. For example, the specific convexity number N_V^+ can be estimated by using this technique, as described in Bindrich and Stoyan (1991). Formula (3.49) gives N_V^+ as

$$N_V^+ = \lambda_V \exp(-\lambda_V \overline{V}). \tag{10.28}$$

10.3.2   Thick sections of spatial germ–grain models

The mean-value characteristics of Ξ_V can be related to those of the orthogonal projection Ξ^t of a thick section when Ξ_V is a motion-invariant germ–grain model. The formulae are

$$\lambda_V \left(\overline{V} + \frac{\overline{S}t}{4} \right) = \lambda_A \overline{A}, \tag{10.29}$$

$$\lambda_V \left(\frac{\pi \overline{S}}{4} + \pi \overline{\overline{b}} t \right) = \lambda_A \overline{L}, \tag{10.30}$$

$$\lambda_V (\overline{\overline{b}} + t) = \lambda_A. \tag{10.31}$$

The practical value of these formulae is limited since projections of non-intersecting grains may overlap and the determination of the parameters λ_A, $\overline{A}$ and $\overline{L}$ may be difficult.

Voss and Stoyan (1985) give a proof of these formulae which follows that of Theorem 10.1. The disc c_r there is replaced by the cylinder of height t

$$C_r = \{ x = (x_1, x_2, x_3) : x_1^2 + x_2^2 \leq r^2, \quad 0 \leq x_3 \leq t \}.$$

Formula (10.29) is due to Cahn and Nutting (1959) and Formula (10.30) to DeHoff (1968).
The mean-value characteristics of the typical projected grain of Ξ_t are

$$\overline{A} = \frac{\overline{V} + \overline{S}t/4}{\overline{\overline{b}} + t} = \frac{4\pi \overline{V} + \pi \overline{S}t}{2\overline{M} + 4\pi t}, \tag{10.32}$$

$$\overline{L} = \frac{\pi \overline{S}/4 + \pi \overline{\overline{b}}t}{\overline{\overline{b}} + t} = \frac{\pi^2 \overline{S} + 2\pi \overline{M}t}{2\overline{M} + 4\pi t}, \tag{10.33}$$

see Voss (1980).

The formulae for the Boolean model in Section 3.2.2 give formulae for characteristics of Ξ^t if Ξ_V is a Boolean model. The formulae concern the area fraction $A_A(t)$, specific boundary length $L_A(t)$ and specific connectivity number $N_A(t)$:

$$A_A(t) = 1 - \exp\left(-\lambda_V\left(\overline{V} + \frac{\overline{S}t}{4}\right)\right), \tag{10.34}$$

$$L_A(t) = \pi\lambda_V\left(\frac{\overline{S}}{4} + \overline{\overline{b}}t\right)\exp\left(-\lambda_V\left(\overline{V} + \frac{\overline{S}t}{4}\right)\right), \tag{10.35}$$

$$N_A(t) = \left(\lambda_V\left(\overline{\overline{b}} + t\right) - \frac{\pi}{4}\lambda_V^2\left(\overline{\overline{b}}t + \frac{\overline{S}}{4}\right)^2\right)\exp\left(-\lambda_V\left(\overline{V} + \frac{\overline{S}t}{4}\right)\right), \tag{10.36}$$

as in Miles (1976).

The Equation system (10.29)–(10.31), as well as (10.34)–(10.36) and (10.25)–(10.27), has properties similar to those of the system (10.20)–(10.22). Again there are three equations and four unknown variables. One way of resolving this problem is to use two probes of different thicknesses t_1 and t_2 as proposed by Giger (1967) and Miles (1976); see also Weibel (1980). This approach has a disadvantage in that the resulting equations are ill-conditioned and sensitive with respect to measurement errors. Another possibility is to use a whole sequence of different probes with thicknesses $t_1, t_2, \ldots, t_n$ to measure the corresponding planar quantities and to determine the parameters of the spatial model by regression.

10.3.3 Tubular structures and membranes

It should be noted that there are also stereological methods which are related to spatial fibre and surface processes just as germ–grain models are related to point processes. These fibre- and surface-based models are systems of tubes and thin surfaces, for example tissue membranes. Stereological techniques yield V_V, S_V, L_V and so forth; see Gundersen (1979), Weibel (1980) and Baddeley and Averback (1983). Methods for the determination of the distribution of the thickness of membranes are presented in Ohser and Mücklich (2000).

10.4 Stereological methods for spatial systems of balls

10.4.1 Introduction

This section presents methods for a particular case of a stationary germ–grain model, namely, a random system of balls in $\mathbb{R}^3$. The case of planar sections is studied in detail while thick and linear sections are touched on briefly. The mathematical theory can apply to both the case of systems of non-overlapping balls and the case of systems of penetrable balls, under the condition that all balls, even in the case of overlapping, can be identified.

10.4.2   Planar sections and the Wicksell corpuscle problem

Theoretical solution

The following quantities can be determined from a planar section by means of the methods discussed in Section 6.5.7:

- N_A = the mean number of section discs per unit area,
- $D_A(r)$ = the distribution function of diameters of section discs.

The distribution function $D_A(r)$ always has a density function, denoted here by $\delta_A(r)$; the corresponding mean is denoted by d_A. The aim of stereology is to determine the following three-dimensional quantities:

- N_V = the mean number of balls per unit volume,
- $D_V(r)$ = the distribution function of the diameters of balls,
- d_V = the mean ball diameter.

Procedures for solving the important problem of determining $D_V(r)$ from knowledge of $D_A(r)$ are called *unfolding* procedures. The unfolding problem was first posed and theoretically solved by Wicksell (1925, 1926). However, as will become clear, there are serious statistical and numerical difficulties in implementing the solution.

The following formulae relate these quantities to each other:

$$d_V = \frac{\pi}{2}\left(\int_0^\infty \frac{\delta_A(r)}{r}\,dr\right)^{-1}, \tag{10.37}$$

$$N_V = \frac{N_A}{d_V}, \tag{10.38}$$

$$D_V(r) = 1 - \frac{2d_V}{\pi}\int_r^\infty \frac{\delta_A(x)}{\sqrt{x^2 - r^2}}\,dx \qquad \text{for } r \geq 0, \tag{10.39}$$

$$D_A(r) = 1 - \frac{1}{d_V}\int_0^\infty \left(1 - D_V\left(\sqrt{r^2 + x^2}\right)\right)dx \qquad \text{for } r \geq 0, \tag{10.40}$$

$$N_A\left(1 - D_A(r)\right) = N_V \int_r^\infty \sqrt{x^2 - r^2}\,dD_V(x) \qquad \text{for } r \geq 0. \tag{10.41}$$

Formula (10.38) is a particular case of (10.22) since $\overline{\overline{b}} = d_V$, and Formula (10.39) implies (10.37) with $r = 0$. If $D_V(r)$ possesses a density function $\delta_V(r)$ then

$$\delta_A(r) = \frac{r}{d_V}\int_r^\infty \frac{\delta_V(x)}{\sqrt{x^2 - r^2}}\,dx \qquad \text{for } r \geq 0. \tag{10.42}$$

The solution of this Abel integral equation is

$$\delta_V(r) = -\frac{2d_V r}{\pi}\int_r^\infty \frac{1}{\sqrt{x^2 - r^2}}\frac{d}{dx}\left(\frac{\delta_A(x)}{x}\right)dx \qquad \text{for } r \geq 0. \tag{10.43}$$

There are nice relations between the *moments* of $D_A(r)$ and $D_V(r)$ given by (10.67), presented in the context of thick sections below. When $t = 0$ these formulae specialise to the case considered here.

Drees and Reiss (1992) study the tail behaviour of $D_A(r)$ and $D_V(r)$. For example, they show that $D_A(r)$ and $D_V(r)$ belong to the same type of extreme value distributions (Fréchet, Weibull and Gumbel).

Example 10.2. *Particular forms of $D_A(r)$*
(a) Let the diameters of the balls be constant. Then Formula (10.40) implies

$$D_A(r) = 1 - \frac{\sqrt{d_V^2 - r^2}}{d_V}, \tag{10.44}$$

$$\delta_A(r) = \frac{r}{d_V \sqrt{d_V^2 - r^2}} \qquad \text{for } 0 \le r \le d_V.$$

The mean and variance of the section disc diameters are

$$d_A = \frac{\pi d_V}{4} \qquad \text{and} \qquad \sigma_A^2 = (32 - 3\pi^2)\frac{d_V^2}{48}. \tag{10.45}$$

These equations can be used to test whether the diameters are constant and if so, then to estimate d_V.

(b) If the diameters of the balls have a Rayleigh distribution, that is, if

$$\delta_V(r) = \frac{r}{a^2} \exp\left(\frac{-r^2}{2a^2}\right) \qquad \text{for } r \ge 0, \quad \text{with mean } d_V = a\sqrt{\frac{\pi}{2}}, \tag{10.46}$$

then the distribution reproduces itself:

$$\delta_V(r) = \delta_A(r) \qquad \text{for } r \ge 0. \tag{10.47}$$

This is the only distribution with this property; see Drees and Reiss (1992). It makes the use of the Rayleigh distribution attractive for stereology. The paper by Rysz and Wiencek (1980) gives an application of this idea.

(c) If the diameters have a uniform distribution on $[0, a]$, then

$$\delta_A(r) = \frac{2r}{a^2} \cosh^{-1}\left(\frac{a}{r}\right) \qquad \text{for } 0 \le r \le a, \tag{10.48}$$

see Ohser and Mücklich (2000, p. 170).

Before turning to the statistical problems in the determination of N_V, d_V and $D_V(r)$ by the above formulae a proof of Formulae (10.37)–(10.40) is presented, following Mecke and Stoyan (1980b).

Proof of Formulae (10.37)–(10.40). The system of balls can be described by a motion-invariant marked point process

$$\Psi_V = \{[x_n, y_n, z_n; d_n]\}.$$

Here (x_n, y_n, z_n) is the centre of the nth ball and the mark d_n stands for its diameter. The corresponding intensity measure Λ_V (including the marks) has the form

$$\Lambda_V(B \times M) = N_V \nu_3(B) \mathscr{D}_V(M) \qquad \text{for Borel } B \times M \subset \mathbb{R}^3 \times [0, \infty),$$

where $\mathscr{D}_V$ denotes the mark distribution and ν_3 the volume measure. Thus $\mathscr{D}_V$ is the distribution of the diameter of the typical ball

$$\mathscr{D}_V([0, r]) = D_V(r) \qquad \text{for } r \geq 0.$$

All that is observed of a ball $[x, y, z; d]$ is its intersection with the (x, y)-plane. The condition that a section is not empty is precisely

$$4z^2 \leq d^2$$

and the resulting section diameter s is given by

$$s = \sqrt{[d^2 - 4z^2]^+},$$

where $[x]^+ = \max\{0, x\}$. Consider the system of all section discs. This can be described by another marked point process, which this time lies in the (x, y)-plane:

$$\Psi_A = \{[x_m, y_m; s_m]\}.$$

Its intensity measure Λ_A is given by

$$\Lambda_A(B' \times M') = N_A \nu_2(B') \mathscr{D}_A(M'), \qquad \text{for Borel } B' \times M' \subset \mathbb{R}^2 \times [0, \infty),$$

where $\mathscr{D}_A$ is the diameter distribution of the typical section disc (ignoring those of null section) and ν_2 the area measure, so

$$D_A(r) = \mathscr{D}_A([0, r]) \qquad \text{for } r \geq 0.$$

The value of $\Lambda_A(C)$ for the set $C = [0, 1) \times [0, 1) \times (r, \infty)$ can be expressed in two ways. First, the formula for Λ_A yields directly

$$\Lambda_A(C) = N_A \big(1 - D_A(r)\big).$$

Second, since the points of Ψ_A result from points of Ψ_V whose balls hit the (x, y)-plane, such points correspond to points $[x, y, z; d]$ of Ψ_V with

$$\left(x, y, \sqrt{[d^2 - 4z^2]^+}\right) \in C.$$

Therefore the expected number of Ψ_A points in C must equal the expected number of Ψ_V points for which the above is true:

$$\Lambda_A(C) = \mathbf{E}\left(\sum_{[x, y, z; d] \in \Psi_V} 1_C(x, y, w)\right),$$

where $w = \sqrt{[d^2 - 4z^2]^+}$. The formula for Λ_V and the Campbell theorem for marked point processes (4.20) yield

$$\Lambda_A(C) = \int_{\mathbb{R}^3 \times [0,\infty)} \mathbf{1}_C(x, y, w) \, \Lambda_V\big(d(x, y, z; d)\big)$$
$$= N_V \int_0^\infty \mathscr{D}_V\Big(\big(\sqrt{r^2 + t^2}, \infty\big)\Big) dt.$$

Formula (10.38) is obtained by equating both expressions for $\Lambda_A(C)$, via

$$\mathscr{D}_V\big((u, \infty)\big) = 1 - D_V(u),$$

setting r equal to zero, and using

$$d_V = \int_0^\infty \big(1 - D_V(r)\big) \, dr.$$

This yields Formula (10.40) immediately. Integration shows that (10.40) has as solution the result summarised in (10.39). Setting $r = 0$ in (10.39) gives (10.37).   □

Determination of section disc diameters

The statistics begins with determination of the section disc diameters. Usually, these diameters are measured from profiles that are not exactly circular. Thus methods of image analysis are employed; see for example Balslev *et al.* (2000). A first step may be application of operations of mathematical morphology, to close gaps (closing) in the profiles and to separate aggregates of objects (opening). By the way, these 'objects' do not necessarily stand for single discs, since overlappings of balls in space and thus of section discs are possible. Thus aggregates of discs must be divided into single objects. For the resulting objects then 'centres' are determined (usually as centres of gravity) and finally discs with these centres are constructed such that the disc areas are equal to the areas of the former objects. This method is demonstrated in Kadashevich *et al.* (2005) for the case where the balls approximate partially overlapping pores. The edge-correction used here is minus-sampling.

The end result is a sample of n diameters $d_1, \ldots, d_n$ from a window W.

Simple mean-value estimators

A simple *ad hoc* estimator for d_V is based on Formula (10.37) and serves to deliver an initial example:

$$\hat{d}_V = \frac{n\pi}{2} \Big/ \sum_{i=1}^n \frac{1}{d_i}. \tag{10.49}$$

This estimator has been suggested by Saltykov (1950) and Fullman (1953). It has been studied by Watson (1971) and Franklin (1977, 1981). The latter constructed a confidence interval for d_V. It has been criticised on the grounds that its variance is infinite; see for example Ripley (1981, p. 210). Estimators for other moments of $D_V(r)$ have been studied in Nicholson (1970, 1976, 1978), Watson (1971) and Jakeman and Schaeffer (1978).

A counterpart to $\hat{d}_V$ is

$$\hat{N}_V = \frac{2}{n v_2(W)} \sum_{i=1}^{n} \frac{1}{d_i}. \tag{10.50}$$

Approaches to solve the unfolding problem

The determination of the distribution function $D_V(r)$ is an ill-posed problem of a well-known nature, which is typical for the inversion of integral equations of the first kind. Many statistical procedures have been developed to handle the ill-posed nature of the problem, which are all biased but lead to acceptable approximations. The work until 1995 is documented in SKM95; it seems that since then interest in the problem has faded out. A simple solution recommended in SKM95 seems to be the best solution; see below. For historical interest and to prevent rediscoveries the list of SKM95 is given here in abridged form:

(a) Discretisation of Formula (10.40) or (10.41) by discretising $D_V(r)$, which corresponds to grouping the sample. The integral is then approximated by a finite sum, and a finite system of linear equations is produced. Methods of this kind have been proposed by Goldsmith, Saltykov, Scheil, Schwartz, and Wicksell; see Lewis et al. (1973), Saltykov (1974) and Weibel (1980). Differences between the methods lie in choosing the representatives of the histogram classes: midpoints, upper or lower limits. A particular case is the following.

The Scheil–Schwartz–Saltykov method
As above, $\{d_1, \ldots, d_n\}$ is a sample of diameters of section discs in a planar section. An interval $(0, d_u)$ is chosen, with suitable upper limit d_u exceeding all the d_j. This interval is divided into k cells of width $\Delta = d_u/k$. The empirical number $\hat{n}_i$ of discs per unit area with diameters in $C_i = ((i-1)\Delta, i\Delta]$ is recorded, for each $i = 1, \ldots, k$. The aim is to provide from $\{\hat{n}_1, \ldots, \hat{n}_k\}$ good estimates $\hat{N}_j$ for the expected number of balls per unit volume having diameters in C_j for $j = 1, \ldots, k$. The density of the distribution $D_V(r)$ can then be estimated as the histogram value $\hat{N}_j/\sum_{i=1}^{k} \hat{N}_i$ in the cell C_j.

If N_j is the expected number of balls per unit volume with diameters in C_j then in a plane section these balls contribute to the expected numbers n_i of section disc per unit area with diameters in C_i approximately as follows:

$$n_i = \Delta \sum_{j=i}^{k} b_{ij} N_j \qquad \text{for } i = 1, \ldots, k, \tag{10.51}$$

where the coefficients are

$$b_{ij} = \sqrt{j^2 - (i-1)^2} - \sqrt{j^2 - i^2}, \qquad \text{for } i \leq j \text{ and } j = 1, \ldots, k, \tag{10.52}$$

Here the approximation is made that all balls contributing to n_i actually have diameter $j\Delta$. Note that other approximations, such as that the diameters are $j\Delta + \Delta/2$, or that they are uniformly distributed in $((j-1)\Delta, j\Delta)$, do not give better results.

Estimates $\hat{N}_j$ of the N_j can now be obtained by solving for N_j in these equations, using $\hat{n}_i$. The solutions have the form

$$\hat{N}_j = \frac{1}{\Delta} \sum_{i=1}^{k} c_{ji} \hat{n}_i, \qquad (10.53)$$

where the c_{ij} may be obtained by back-substitution in the linear equations for n_i above.

As a by-product estimators for N_V and d_V are obtained:

$$\hat{N}_V = \sum_{j=1}^{k} \hat{N}_j \qquad \text{and} \qquad \hat{d}_V = \frac{n}{\hat{N}_V v_2(W)}. \qquad (10.54)$$

(b) Application of Formula (10.39), replacing $D_A(r)$ by an empirical distribution function; see Blödner et $al.$ (1984).

(c) Application of Formula (10.43) and use of spectral differentiation; see Anderssen and Jakeman (1975a,b).

(d) Parametric methods. These involve making a parametric assumption about the form of $D_V(r)$ and then estimating the parameters; see Likeš (1963), DeHoff (1965), Giger and Ried-wyl (1970), Keiding et $al.$ (1972) and Suwa et $al.$ (1976).

The idea is simple: A parametric form of $\delta_V(r)$ is taken, whose parameters θ have to be estimated. The formula is plugged in (10.42), which leads to an expression for $\delta_A(r)$, where the same parameters θ as in $\delta_V(r)$ appear.

Assuming that the section disc diameters in the sample $\{d_1, \ldots, d_n\}$ can be considered as coming from independent random variables, one can maximise the likelihood function

$$L(d_1, \ldots, d_n; \theta) = \prod_{i=1}^{n} \delta_A(d_i; \theta) \qquad (10.55)$$

with respect to θ, usually by means of numerical methods, to yield estimates of θ.

(e) Trial and error methods, which consist of guessing a diameter distribution $D_V^{(1)}(r)$, computing the corresponding $D_A^{(1)}(r)$, comparing this with the empirical $D_A(r)$, and then making a second guess at the diameter distribution, and so forth; see Voss and Schubert (1975).

(f) Regularisation. This popular and powerful method has been also applied to the particular case of the Wicksell problem. In the case of estimation of the probability density function, let $\hat{\delta}_A(r)$ be an empirical density function of the section disc diameters, and let $\delta_A^*(r)$ be the function obtained by means of Formula (10.42) with a $\delta_V(r)$ which depends on some parameters. In a naïve solution method the parameters are determined by minimising

$$S = \int_0^\infty \left(\hat{\delta}_A(r) - \delta_A^*(r) \right)^2 dr.$$

The idea of regularisation is to introduce the term

$$d = \int_0^\infty \left(\delta_V''(r) \right)^2 dr$$

and to minimise

$$S + \varepsilon d$$

for a suitable positive ε. The additional term helps to find smooth solutions, which can be easily interpreted. Typical references for the application of regularisation in the Wicksell problem are Nychka (1983b,a), Nychka et al. (1984) and Weese (1995).

(g) EM algorithm. Wilson (1987) and Silverman et al. (1990) used modified forms of the EM algorithm; see also Reiss (1993, pp. 163–7) and Ohser and Mücklich (2000, pp. 185–6). The standard EM algorithm consists of two steps, an Expectation-step and a Maximisation-step; see McLachlan and Krishnan (2008). The latter belongs to maximum-likelihood estimation. The modification consists in adding a smoothing-step. To justify the maximisation-step these authors assume that the ball centres form a homogeneous Poisson point process. Instead it is better to interpret the M-step as a Moment-method-step. There seems to be no convergence proof as yet, but practical experience shows that the method works even when the Poisson process assumption is inappropriate.

(h) Kernel methods. Taylor (1983) used a kernel estimator with variable bandwidth for the density $\delta_A(r)$ and then determines $\delta_V(r)$ by means of Formula (10.43); see also Möller (1989). Hall and Smith (1988) used a kernel estimator for the density function of the squared disc diameters in order to obtain the density function of the squared ball diameters. For the squared diameters the integral equation is easier. Van Es and Hoogendoorn (1990) postponed the kernel smoothing until after a transformation step. They determined first an estimator $D_V(r)$ by means of Formula (10.39) and then derived an estimator of $\delta_V(r)$ by kernel smoothing. Weese (1995) discussed the relationship of kernel methods and regularisation techniques, and Weese et al. (1997) combined the advantages of both.

(i) Wavelet method. Antoniadis et al. (2001) demonstrated the application of wavelet methods. The authors first converted the integral equation to a form suitable for the application of thresholding wavelet methods. Then they derived the asymptotic properties of their estimators and compared with other numerical methods.

(j) Nonparametric maximum likelihood; see van Es (1991) and Jongbloed (1991).

(k) Particular integration methods. Mase (1995) suggested rewriting the integral in Formula (10.39) as

$$\int_{-1}^{1} \frac{|x|}{\sqrt{1 - x^2}} \delta_A(x)\, dx$$

and then using the Gauss–Chebyshev quadrature method and a smoothing technique for the empirical $D_A(r)$.

No method has clear advantages over all others. Statistical and numerical difficulties interact in a delicate fashion.

The two-step method

The authors recommend a *two-step method* as also in other places of this book.

- First step: Apply one of the numerical methods above as a pilot study, in order to generate a plausible assumption on the type of the density function $\delta_V(r)$. Since only a qualitative problem is solved, high precision is not necessary. Indeed, even biased estimators may be useful here.

- Second step: Use a parametric method, preferably the maximum likelihood method, for the estimation of the parameters of $\delta_A(r)$ obtained via Equation (10.42). It is important to note that statistics should be based only on the diameters d_i of the section discs, as these produce the only (relatively) safe data.

In view of Ohser and Mücklich (2000, Tables 6.6–6.7), it is suggested to use for the first step the classical Scheil–Schwartz–Saltykov method; see (a) above, perhaps in the improved form of EM algorithm. Note that in that book there are source codes for the numerical calculation.

The application of this method is demonstrated in various papers; see for example Groos and Kopp-Schneider (2006) and Kadashevich *et al.* (2005). In the latter paper the spatial diameter distribution is a mixture of an exponential distribution (for very small pores) and normal distribution (for larger pores).

10.4.3   Linear sections

The following quantities are available from a linear section:

- N_L = mean number of chords per unit length,
- $D_L(r)$ = distribution function of chord lengths.

The distribution function $D_L(r)$ always has a density, denoted by $\delta_L(r)$.

The aims of statistical analysis in the case of linear unfolding are the same as in the case of planar unfolding. The mathematical problem of establishing relationships between N_L and $D_L(r)$ on the one hand and N_V and $D_V(r)$ on the other was solved by Spektor (1950) and Lord and Willis (1951). The following formulae can be proved in a way similar to that of the planar case:

$$N_V = \frac{4N_L}{\pi d_V^{(2)}}, \tag{10.56}$$

$$D_V(r) = 1 - \frac{d_V^{(2)}}{2} \frac{\delta_L(r)}{r}, \tag{10.57}$$

$$D_L(r) = \frac{2}{d_V^{(2)}} \int_0^r x\big(1 - D_V(x)\big)\, dx, \tag{10.58}$$

$$N_L\big(1 - D_L(r)\big) = N_V \frac{\pi}{4} \int_r^\infty (x^2 - r^2)\, dD_V(x) \qquad \text{for } r \geq 0, \tag{10.59}$$

where $d_V^{(2)}$ denotes the second moment of $D_V(r)$. Formula (10.56) can be interpreted as a particular case of Formula (10.3) since the mean ball surface is $4\pi d_V^{(2)}$.

Example 10.3. *Particular forms of $D_L(r)$*
(a) If the ball diameters are constant then by (10.58)

$$D_L(r) = \frac{r^2}{d_V^2},$$
(10.60)

$$\delta_L(r) = \frac{2r}{d_V^2} \qquad \text{for } 0 \le r \le d_V.$$
(10.61)

The mean chord length satisfies

$$d_L = \frac{2d_V}{3}.$$
(10.62)

(b) As in the case of planar sections, the Rayleigh distribution has a reproduction property:

$$D_V(r) = D_L(r) \qquad \text{for } r \ge 0.$$

(c) Let the ball diameters have an exponential distribution with parameter μ. Then

$$D_L(r) = 1 - (1 + \mu r)\exp(-\mu r),$$

$$\delta_L(r) = \mu^2 r \exp(-\mu r) \qquad \text{for } r \ge 0.$$

Relation (10.57) implies that

$$d_V^{(2)} = 2 \lim_{r \downarrow 0} \left(\frac{r}{\delta_L(r)} \right).$$

Consequently a density function $f(r)$ is a chord length density function only if

$$0 < \lim_{r \downarrow 0} \left(\frac{f(r)}{r} \right) = f'(0) < \infty.$$
(10.63)

Among the family of gamma (or Schulz) distributions with densities

$$f(r) = \frac{\mu^\alpha r^{\alpha-1}}{\Gamma(\alpha)} \exp(-\mu r) \qquad \text{for } r \ge 0 \text{ and } \alpha \ge 0,$$

the condition given in (10.63) is fulfilled only when $\alpha = 2$, that is, when the ball diameters have an exponential distribution.

The linear section method is more unstable than the planar section method. Tests suggest that 'it should not be used whenever planar sectioning is possible'; see Ripley (1981, p. 212). Measurement errors and statistical errors play a still greater rôle. Ohser and Mücklich (2000, p. 175) present methods for numerical solution.

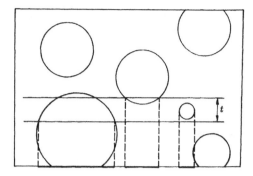

Figure 10.3 A thick section through a system of opaque balls lying in a transparent medium. The projections are shown at the bottom of the window.

10.4.4 Thick sections

In this section a system of opaque balls embedded in a transparent medium is considered. The system is intersected by two parallel planes separated by (a small) distance t. Only the projection of the part of the system lying between the two planes is observed, as in Section 10.2.3; see Figure 10.3. The contrary case, studied by Coleman (1981, 1982, 1983), is when the balls are transparent and the medium is opaque, as would be the case for holes in Swiss cheese or closed foams.

Just as in the case of planar sections one obtains relationships between $N_A(t)$ and $D_{A,t}(r)$ on the one hand and N_V and $D_V(r)$ on the other. However, the planar quantities must be interpreted more carefully: $D_{A,t}(r)$ is the distribution function of the *projection* disc diameters, where all projections are counted, also such which are completely overlapped ('overprojection') by larger projections; $N_A(t)$ is the corresponding number density.

Bach (1965, 1967, 1976), Crompton *et al.* (1966) and Mecke and Stoyan (1980b) studied this problem. Bach showed that the solution for $D_V(r)$, which is surely only of theoretical value, is

$$D_V(r) = 1 - \frac{t\left(1 - D_{A,t}(r)\right) - \int_r^\infty p\left(\frac{\sqrt{\pi}}{\sqrt{2t}}\sqrt{x^2 - r^2}\right)\frac{1 - D_{A,t}(x)}{\sqrt{x^2 - r^2}}x\,dx}{t - \int_0^\infty p\left(\frac{\sqrt{\pi}x}{\sqrt{2t}}\right)\left(1 - D_{A,t}(x)\right)dx} \qquad \text{for } r \geq 0,$$

(10.64)

where $p(x)$ is given by

$$p(x) = 1 - \exp(x^2/2)\int_x^\infty \exp(-t^2/2)\,dt \qquad \text{for } x \geq 0.$$
(10.65)

The intensity N_V satisfies

$$N_V = \frac{N_A(t)}{(d_V + t)}.$$
(10.66)

Bach (1959, 1967) found some important relationships between the moments of $D_V(r)$ and $D_{A,t}(r)$. (In the case $t = 0$ they reduce, of course, to those for planar sections.) Let

$$M_1 = d_V,$$

$$M_k = \int_0^\infty x^k \, dD_V(x) \quad \text{and} \quad m_k = \int_0^\infty x^k \, dD_{A,t}(x) \qquad \text{for } k = 0, 1, 2, \ldots.$$

Then the moment relationships are

$$(t + d_V) m_k = J_{k+1} M_{k+1} + t M_k, \tag{10.67}$$

where

$$J_r = \frac{\sqrt{\pi}}{2} \cdot \frac{\Gamma(r/2 + 1/2)}{\Gamma(r/2 + 1)} \qquad \text{for } r = 1, 2, \ldots.$$

The numerics of the unfolding problem is discussed in Ohser and Mücklich (2000, pp. 193–8).

10.4.5 Sieving distributions for balls

Sieving distributions for balls are a particular case of the volume-weighted distributions discussed in Section 6.5.5. Such distributions can often be more interesting than the number-weighted distributions such as $D_V(r)$. This is the case when it is important to know the fraction of volume corresponding to a certain class of diameters; a few very big balls may be more important than many tiny balls.

Let $D_V^S(r)$ be the volume-weighted or *sieving diameter distribution*. The quantity $D_V^S(r)$ is the total volume of balls of diameter less than or equal to r (per unit volume) divided by the total volume of all balls (again per unit volume). This is related to $D_V(r)$ itself in a manner familiar from Section 6.5.5; the corresponding densities satisfy

$$\delta_V^S(r) = r^3 \delta_V(r) \bigg/ \int_0^\infty x^3 \delta_V(x) \, dx \qquad \text{for } r \geq 0. \tag{10.68}$$

More on sieving (and more general, on weighted) distributions can be found in Ohser and Mücklich (2000, Section 7.5). For example, there (p. 225) is an integral equation connecting the volume-weighted ball diameter distribution with the area-weighted section disc diameter distribution. The numerics is here more stable than for the nonweighted distributions.

10.5 Stereological problems for nonspherical grains (shape-and-size problems)

10.5.1 General remarks

Section 10.4 demonstrates that stereology gives rise to hard problems if more is required than mean values; and these difficulties arise already in the simple case of spherical grains. The case of nonspherical grains is even more complicated. Here the section figures or *profiles* have varying shapes and it is often difficult to make even inferences about the true three-dimensional shape. The size distribution cannot be investigated at all without some information about the shape of the grains. But if such information is given, it helps to introduce an additional parameter (such as the number of vertices of a polygonal section profile), which may stabilise the unfolding procedure.

A popular procedure is to approximate the grains by spherical grains matched in some way so as to be 'equivalent'. For example, the areas of the section profiles can be measured and thus diameters calculated for section discs with the same areas. This method may sometimes produce fairly accurate results but can also yield nonsense. Wicksell (1926) first suggested this method.

In this section two important special cases are considered which are often considered in the literature: those of ellipsoidal and polyhedral grains. Under shape assumptions such as these it is formally possible to write down integral equations for the size distributions; see for example Cruz-Orive (1976, 1978), Ohser and Mücklich (2000) and Nagel (2010).

It is generally assumed that the structure considered is a motion-invariant germ–grain model. Note that this can be more general than a system of isolated and independent grains, for example, the cells of a tessellation also produce a germ–grain model. So long as only shape and size of grains is of interest, the following theorem shows how to proceed.

Theorem 10.2. *(Stoyan, 1982) Let Ξ be a motion-invariant germ–grain model with typical grain Ξ_0. Consider the intersection of Ξ with an arbitrary plane or line independent of Ξ. Let Ξ_0^S be the typical grain of the planar or linear germ–grain model resulting from the section. Then Ξ_0^S has a distribution that is independent of any assumptions concerning the distribution of the point process of germs and the correlations between germs and grains.*

Consequently, in order to obtain a stereological integral equation between spatial and sectional characteristics of the individual grains there is no loss of generality in assuming that the germs form a homogeneous Poisson process, that the grains are independent, and that sections of different grains are distinguishable when they overlap. Furthermore, for obtaining the distribution of Ξ_0^S it is sufficient to determine the distribution of the section profile of a single typical grain intersected by a Poisson line or plane process.

Planar sections

For the case where all grains are convex and have the same shape but different sizes, Ohser and Mücklich (2000, Equation (7.1)) gave a general integral equation, which is presented as Formula (10.69) below. They used a size characteristic, measured in length, for the grains, such as their average or maximum breadth. For brevity in this section the chosen size characteristic will simply be called size, and its distribution function is denoted by $F_V(a)$. The mean number of grains per volume unit is N_V.

The corresponding characteristics for a planar section are the profile-size distribution function $F_A(u)$ and the grain-profile intensity N_A.

The standard grain has size 1 and (deterministic) average breadth $\bar{b}_s$. (If the standard grain is a ball and size characteristic is the maximum breadth, then $\bar{b}_s = 1$.)

The integral equation, which connects $F_V(a)$ and $F_A(a)$, is

$$N_A\big(1 - F_A(a)\big) = N_V \bar{b}_s \int_a^\infty u\big(1 - G_u(a)\big)\mathrm{d}F_V(u) \qquad \text{for } a \geq 0. \qquad (10.69)$$

where $G_u(a)$ is the conditional size distribution of section profiles of a grain of size u. The kernel function $p(u, a) = \overline{b}_s u \big(1 - G_u(a)\big)$ in (10.69) depends on grain shape. In the case of spherical grains it is explicitly given by Formula (10.40).

Formula (10.69) implies Formula (10.22) with $a = 0$.

The analytical determination of the kernel is usually difficult. Thus simulation is used: the typical grain is intersected by random planes (so-called IUR-planes: isotropic uniform random planes) and the section profiles are evaluated; see Ohser and Mücklich (1995; 2000, p. 204).

For the numerical determination of $F_V(u)$ a similar procedure as in the case of spherical grains is recommended: the two-step method, where in the first step ideas of the EM algorithm should be used; see Ohser and Mücklich (2000, pp. 208 and 218).

Linear sections

Linear sections are sometimes of interest when there is no other way to get spatial information. In particular, scattering investigations lead to chord length distributions; see p. 23. In the following the same setting as for planar sections is assumed. All grains have the same shape, but their sizes are random, represented by some size characteristic measured in length with probability density function $f_V(a)$. The second moment of the size is $\overline{a^2}$. The chord length distribution function of a grain with size a is denoted by $L_a(l)$. It is assumed that it is known; see the references in Section 1.7.3. The following integral gives the distribution function $L(l)$ of the lengths of chords through the germ–grain model considered:

$$L(l) = \int_0^\infty L_a(l)\, a^2 f_V(a)\, da \Big/ \overline{a^2}. \qquad (10.70)$$

The term $a^2 / \overline{a^2}$ yields the necessary weighting: larger grains are intersected more frequently than smaller ones, and so area weighting should be used, since Cauchy's formula says that the mean area $\mathbf{E}(F)$ of a random orthogonal projection of a convex body K with surface area $S(K)$ satisfies

$$\mathbf{E}(F) = \frac{1}{4} S(K);$$

see Santaló (1976, p. 218). Typically, there is a maximum length $l(a)$ with the property $L_a(l) = 1$ for $l > l(a)$. This makes the explicit form of the integral in (10.70) a bit difficult. For the particular case of spherical grains Formula (10.70) yields (10.58). There the size characteristic is the diameter and $\overline{a^2}$ corresponds to $d_V^{(2)}$ and $f_V(a)$ to $\delta_V(r)$. The maximum length is $l(a) = a$ and

$$L_a(l) = \begin{cases} \dfrac{l^2}{a^2} & \text{for } 0 \le l \le a, \\[2mm] 1 & \text{otherwise.} \end{cases}$$

10.5.2   Two particular grain shapes

Ellipsoids

Ellipsoids form a frequently studied case of nonspherical grains. Here the stereologist wants to determine the joint distribution of size and shape. Cruz-Orive (1976, 1978) gives an excellent introduction to work done up to 1976 (dating back to Wicksell, 1926). As Cruz-Orive points out, it is not possible to solve completely the general problem of reconstructing from planar sections the shape and size distribution of ellipsoids. One must assume either restrictions on the form of the ellipsoids (for example, that the ellipsoids are all spheroids, that is, ellipsoids of revolution) or restrictions on the orientation of the ellipsoids relative to the plane of intersection. Cruz-Orive (1976) presents an example employing on the one hand oblate and on the other hand prolate spheroids, that is, ellipsoids of revolution generated by ellipses revolving around their major and minor axes respectively; in either case the same collection of planar elliptical section figures is produced. This settles a conjecture of Moran (1972) that the general ellipsoidal problem is undetermined.

In the case of spheroids Cruz-Orive (1976) derives and solves an integral equation connecting the bivariate densities of principal semi-axes and eccentricity parameters of ellipses. Flynn *et al.* (2009) use a parametric approach (normal distribution) and maximum likelihood estimation in the context of spatial analysis of mitochondrial networks.

Cruz-Orive (1978) suggests estimation of the size–shape bivariate histogram of spheroids from the corresponding histogram of elliptical parameters measured in the plane of intersection. Using numerical experiments Cruz-Orive concludes that the spherical approximation is acceptable when the ratio of minor axis to major axis is greater than 0.7. The book by Weibel (1980) is still an excellent reference to stereology of ellipsoids.

Beneš and Krejčíř (1997), Beneš *et al.* (1997) and Beneš and Rataj (2004, Chapter 6) consider stereological unfolding of the trivariate size–shape–orientation distribution for the case of spheroids.

Convex polyhedra

The case of polyhedral grains is rather complicated. This will be appreciated immediately if the reader considers that the shape of an intersection profile varies considerably depending on the plane of intersection. For example, for a parallelepiped the intersection profiles can be polygons with 3, 4, 5, or 6 sides; see Figure 8.4 on p. 300.

The literature is devoted to the shape and the size problem. There is a general theory for an important aspect of the shape problem: determination of the probabilities p_n that the typical intersection profile has n sides. This has been developed by Ambartzumian (1974b, 1982), Sukiasian (1982) and Voss (1982). Ambartzumian's method (also used by Sukiasian) is based on his solution of the Buffon–Sylvester problem in $\mathbb{R}^3$, and it studies random plane sections through systems of needles, where the needles correspond to the edges of the polyhedron. His 1973 paper gives the probabilities for the cube. Voss and Sukiasian tabulate p_n for various polyhedra. Table 10.2 displays some of their results. Sukiasian (1982) finds some equations and inequalities that the p_n must satisfy.

Solutions of the size problem assume that shape is known. Then chord length or section area distributions may help determine the size distribution. For chord lengths see the references in Section 1.7.3. The corresponding distributions are rather complicated; see for example

Table 10.2 The probability p_n that the number of sides of a random section of a polyhedron is n.

Polyhedron	Height to side length ratio	Number of sides of section profile					
		3	4	5	6	7	8
Regular tetrahedron		0.711	0.289	—	—	—	—
Regular triangular pyramid with	1	0.715	0.285	—	—	—	—
height to side length ratio of:	2	0.774	0.226	—	—	—	—
Cube		0.280	0.487	0.187	0.046	—	—
Parallelepiped with square base and	2	0.248	0.536	0.183	0.033	—	—
height to side length ratio of:	4	0.180	0.653	0.152	0.014	—	—
Regular hexagonal straight prism	1	0.145	0.525	0.235	0.088	0.005	0.001
with height to side length ratio of:	8	0.109	0.206	0.193	0.421	0.066	0.005

Figure 10.4. Ohser and Mücklich (2000, Sections 7.4 and 8.1.2) study in detail the cases of cubic and prismatic grains, starting from planar sections.

Other candidates for polyhedral grains are the typical cells of the Poisson plane ('Poisson polyhedron') and Poisson-Voronoi tessellations. Note that the typical planar profile of the Poisson polyhedron is the Poisson polygon. Consequently the formulae for this polygon, such as the p_n tabulated in Table 9.2 on p. 372, can be employed. The typical planar profile of the Poisson-Voronoi tessellation is much more complicated; see the next section.

10.6 Stereology for spatial tessellations

The intersection of a motion-invariant spatial tessellation by a plane is a planar tessellation. Its main parameters can be estimated and by the standard stereological formulae transformed in characteristics of the spatial tessellation:

$$N_V \overline{\overline{b}} = N_A(\text{cells}), \tag{10.71}$$

$$S_V = \frac{4}{\pi} L_A(\text{edges}), \tag{10.72}$$

$$L_V = 2 P_A(\text{vertices}). \tag{10.73}$$

If the mean number of plates emanating from the typical edge is 3 for the spatial tessellation then the mean number of edges emanating in a typical vertex is also 3 for the planar tessellation.

The disector described in Section 6.5.7 can be applied for tessellations to estimate N_V, the mean number of cells per volume unit. This is described in Liu *et al.* (1994).

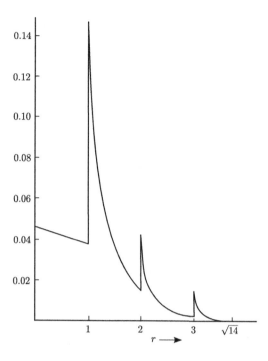

Figure 10.4 Probability density of length of random chords for a parallelepiped with side lengths 1, 2 and 3. The curve was computed by a program written by Gille (1988).

If the spatial tessellation is a *Poisson-Voronoi tessellation* with parameter λ_3 (the mean number N_V of cells per unit volume, in Chapter 9 denoted by ϱ) then for the resulting planar tessellation

$$n_{02} \equiv 3 \text{ almost surely,} \qquad \bar{n}_{20} = 6,$$

where n_{02} is the number of cells adjacent at the typical vertex and $\bar{n}_{20}$ the mean number of vertices (edges) of the typical cell. The planar densities above are

$$N_A(\text{cells}) = 1.46\lambda_3^{2/3}, \tag{10.74}$$

$$L_A(\text{edges}) = 2.29\lambda_3^{1/3}, \tag{10.75}$$

$$P_A(\text{vertices}) = 2.92\lambda_3^{2/3}. \tag{10.76}$$

Because of (9.26), which means $2N_A = P_A$ in the given case, (10.74) and (10.76) say the same. Further characteristics of the planar section tessellation are given by

$$\lambda_1^{(2)} = 4.37\lambda_3^{2/3}, \qquad\qquad \bar{l}_1 = 0.52\lambda_3^{-1/3}, \tag{10.77}$$

$$\bar{l}_0 = 1.57\lambda_3^{-1/3}, \qquad\qquad \overline{a_2^2} = 0.70\lambda_3^{-4/3}, \tag{10.78}$$

where $\lambda_1^{(2)}$ is the mean number of edges per area unit, $\bar{l}_1$ the mean length of the typical edge, $\bar{l}_0$ the mean length of edges emanating at the typical vertex, and $\overline{a_2^2}$ the second moment of area of the typical cell.

Formulae (10.74)–(10.76) offer methods for estimating $\lambda_3 = N_V$, for example

$$\hat{N}_V = 0.20\hat{P}_A^{3/2}, \tag{10.79}$$

by counting planar vertices. Hahn and Lorz (1994a) and Lorz (1995) investigate biases and estimation variances of these estimators. They show that the biases are small and the variances are practically the same for all three estimators. The coefficient of variation ($= \sqrt{\text{variance}}/\text{mean}$) is approximately $1/\sqrt{n}$, where n is the number of cells observed in the planar sampling window. Consequently, the estimator based on counting vertices given by Formula (10.79) is preferable; its application does not require an edge-correction.

Example 10.4. *Austenite grain boundaries in steel*
This is a continuation of Example 9.2 on p. 394. Figure 9.17(a) shows the boundaries of Austenite grains in a section through some steel. The cells in the figure are not convex but it seems appropriate to approximate them by convex polygons as in Figure 9.17(b). For this, the following values were obtained using the methods discussed in Section 9.10.3

$$\hat{N}_A = 0.34, \qquad \hat{L}_A = 0.96, \qquad \hat{P}_A = 0.69,$$

with length unit as in Figure 9.17(b). Under the assumption that the planar tessellation is a planar section of a spatial Poisson-Voronoi tessellation the corresponding parameter λ_3 can be estimated using N_A, L_A and P_A. The formulae above give the values 0.112, 0.074 and 0.115 (note that $2N_A = P_A$, without '$\wedge$') as estimates for λ_3. The discrepancy between these values throws doubt on whether the pattern really corresponds to a section of a Poisson-Voronoi tessellation. Conceivably a parametric bootstrap test could be conducted to decide whether the fluctuations are significant or whether they are consistent with the spatial Poisson-Voronoi tessellation assumption.

Comparison between Figures 9.7 and 9.17(a) suggests that a variant of the Johnson–Mehl model might be appropriate; note in particular the curved boundaries, their concavity for large grains and the clusters of small grains. This might be the result of two stages of generation of Austenite-grain germs. Saltykov (1974, Chapter 2) gives an interesting and relevant discussion.

Note that the planar section of a Poisson-Voronoi tessellation is not a planar Poisson-Voronoi tessellation. This is obvious since the formulae above give $L_A^2/P_A = 1.80$ for the section tessellation while $L_A^2/P_A = 2$ for the planar Poisson-Voronoi tessellation. Figure 10.5 shows a planar section through a simulated Poisson-Voronoi tessellation. It cannot be a Dirichlet tessellation generated by some point pattern because of the occurrence of very small cells as neighbours of large cells in opposite positions. The existence of such configurations can be used to prove that planar sections of Poisson-Voronoi tessellations cannot be Dirichlet tessellations; see Chiu *et al.* (1996).

However, Voronoi tessellations belong to the class of Laguerre tessellations. A planar section of a Laguerre tessellation is a Laguerre tessellation, and thus the planar section of a Voronoi tessellation is a Laguerre tessellation (Nagel, 2010).

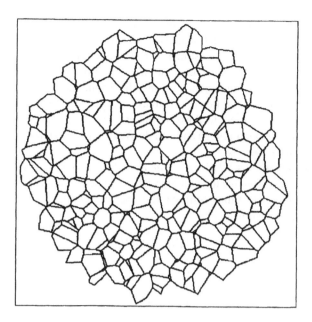

Figure 10.5 Planar section through a Poisson-Voronoi tessellation. Inspection shows that this tessellation cannot be produced by a Dirichlet tessellation of a point-pattern because of the occurrence of very small cells as neighbours of large cells in opposite positions.

If the real structure deviates from the Poisson-Voronoi model then of course the methods above become inaccurate. Perhaps the estimator using N_A is then still the best.

Tests of the hypothesis that a given tessellation is a Poisson-Voronoi tessellation are suggested and discussed in Hahn and Lorz (1994b), based only on planar sections. If the alternative hypotheses are Voronoi tessellations relative to other point processes then tests based on the variability of section cell area seem to be most powerful.

Perhaps surprisingly, mean-value characteristics of planar sections of Voronoi tessellations relative to non-Poisson processes may not differ very much from those corresponding to Poisson-Voronoi tessellations. This and practical experience suggest that good estimates of N_V for arbitrary spatial tessellations can be obtained by the following method: Obtain the empirical N_A from planar sections, plug it into Formula (10.74) to get the corresponding λ_3, and use the latter as an estimate of N_V, under the working assumption that the given tessellation is in some aspects not 'far from' a Poisson-Voronoi tessellation.

Schwertel and Stamm (1997) applied this method (and others) to samples of austenitic steel. They found that N_V can reliably estimated in this way, but it 'is not sufficient in predict the higher moments reliably'. Coster *et al.* (2005) made a similar study for cerine, a sintered ceramic, CeO_2, which led them to a particular Johnson–Mehl model.

An analogous approach is possible for linear sections: the mean number of intersections of cell faces with a test line per unit length is

$$P_L = 1.456\lambda_3^{1/3}. \tag{10.80}$$

It is easy to see that sections of stationary STIT tessellations inherit the STIT property. In particular, a linear section yields a one-dimensional Poisson point process. Furthermore, the distribution of any stationary STIT tessellation is uniquely determined by its intensity (S_V, L_A or P_L, respectively) and the directional distribution of the cell boundaries. Thus the stereological problems are reduced to the estimation of these characteristics and are quite similar to those in the case of a Poisson plane tessellation. For motion-invariant STIT tessellations it suffices to use just one of the fundamental mean-value formulae.

Lorz (1990) studies the distribution of cell areas of the planar section tessellations of spatial Voronoi tessellations with respect to Poisson, Matérn cluster and hard-core processes. This is continued by Lorz and Hahn (1993) and van de Weygaert (1994), the latter also considers form factors, boundaries and edges.

If the spatial tessellation is a *Poisson plane tessellation* with parameter $\varrho = S_V/2$, then the planar section is a Poisson line tessellation with the same parameter ϱ, which in the planar context is the mean number of lines hitting a test line segment per unit length. The quantity N_A(cells) can be used to estimate ϱ by

$$\varrho = 2\sqrt{N_A(\text{cells})/\pi}. \tag{10.81}$$

Note that

$$N_V = \frac{\pi \varrho^3}{6}. \tag{10.82}$$

Ohser and Mücklich (2000, Section 8.2 and Chapter 10) describe stereological analysis of general tessellations based on the fact that the cells are polyhedra, using stereological methods for polyhedra as in Section 10.5.2.

A further problem for tessellations in $\mathbb{R}^3$ concerns the determination of the distribution of the angles between the plates or facets at the typical edge. Under 'equilibrium conditions' one would expect an angle of $120°$; see Saltykov (1974) and Harker and Parker (1945). If the angles are constant at $120°$ then the distribution of the angle at the typical vertex in the planar section tessellation can be calculated under the assumption of stationarity and isotropy; the mean is again $120°$ and the standard deviation is $22°$. Duvalian (1971) calculates the standard deviations for other spatial angles, both larger and smaller than $120°$. The inverse problem of estimating the spatial angle distribution given the angle distribution in a planar section is an interesting problem of stereology; see Schwandtke (1985) and Miles (1987).

10.7 Second-order characteristics and directional distributions

10.7.1 Introduction

Previous sections describe stereological methods that yield mean values or distributional information on particles. However, more is possible: the spatial variability of three-dimensional structures can also be explored stereologically. In one case this was already explained: As Section 10.2.2 says, the covariance $C(r)$, a second-order characteristic of random sets, can be easily determined by planar or linear sections. This is possible even in the anisotropic case.

Even more is possible: For germ–grain models with spherical grains the pair correlation function of the point process of ball centres can be determined from thick or planar sections. Of course, in this case high quality images of sections are needed and the grains must be spherical in very good approximation. Furthermore, pair correlation functions can be estimated from planar sections in the case of fibre processes as well. These methods yield at least qualitative information that may be of value for model building.

10.7.2 Stereological determination of the pair correlation function of a system of ball centres

Let Φ_V be a motion-invariant point process of balls centres with pair correlation function $g_V(r)$. Taking a thick section of thickness t, parallel to the (x_1, x_2)-plane, and projecting onto the (x_1, x_2)-plane give the planar point process Φ_A of the centres of projected section discs. (If $t = 0$, then the thick section becomes a planar section.) Let Φ_A have pair correlation function $g_A(r)$. The function $g_A(r)$ can be estimated statistically using the methods of Section 4.7.4. Under some simplifying assumptions on the diameter distribution, $g_A(r)$ can be related to $g_V(r)$ by a stereological integral equation.

One example is that if the ball diameters are constant, equal to d_V, then

$$g_A(r) = \frac{2}{(d_V + t)^2} \int_0^{d_V+t} (d_V + t - x)\, g_V(\sqrt{r^2 + x^2})\, dx \qquad \text{for } r \geq 0. \qquad (10.83)$$

Figure 10.6 shows $g_A(r)$ for a random dense packing of identical balls, obtained by simulation and statistical estimation based on 2683 section discs. The graph of $g_A(r)$, for the smaller r, looks like that of the pair correlation function of a planar point process with a soft-core behaviour, but for the larger r it is quite similar to the pair correlation function $g(r)$ for the ball centres shown in Figure 4.7 on p. 142. Note that indeed in a planar section, inter-point distances smaller than the ball diameter are possible for the section disc centres.

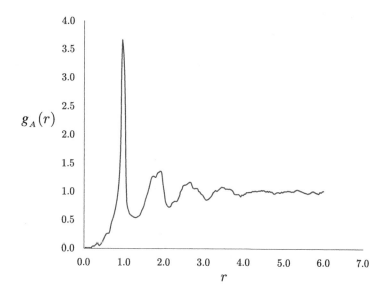

Figure 10.6 Pair correlation function $g_A(r)$ for a system of identical hard balls in $\mathbb{R}^3$. The function was determined statistically from a simulation of the hard ball system with the force-biased algorithm, see Section 6.5.3, based on 2683 section discs in a square window.

If the balls have independent random diameters with distribution function $D_V(r)$ and if individual section discs, even in case of overlapping, can be identified, then $g_A(r)$ and $g_V(r)$ are related by

$$g_A(r) = \int_0^\infty f_V(x, t) g_V(\sqrt{x^2 + r^2}) \, dx \qquad \text{for } r \geq 0, \qquad (10.84)$$

where

$$f_V(x, t) = \frac{2}{(d_V + t)^2} \int_x^\infty \left(1 - D_V([\|2x - u\| - t]^+)\right)\left(1 - D_V([u - t]^+)\right) du \quad \text{for } x \geq 0, \qquad (10.85)$$

see Hanisch (1983), where $[x]^+ = \max\{x, 0\}$. Formula (10.84) was already given for the special case $t = 0$ of planar sections in Hanisch and Stoyan (1980).

A very simple but rough nonstereological approximation is

$$g_V(r) \approx g_A(r). \qquad (10.86)$$

In fact, this is exact in the case of a Poisson process, and in general it is not bad for medium and large r; compare, for example, Figures 10.6 and 4.7. Note also that the covariances satisfy

$$C_A(r) = C_V(r), \qquad (10.87)$$

and that Formula (6.100) holds.

Difficulties arise with the above formulae in general because $D_V(r)$ and d_V are unknown and, as seen in Section 10.4, their determination is not easy. However, there is a helpful recommendation for the integral equation (10.88), found by Hanisch (1984a), which expresses $g_V(r)$ in terms of quantities that can be estimated directly from planar sections.

Remark on the proof of Formula (10.83)
The second-order factorial moment measure $\alpha_A^{(2)}$ of Φ_A can be expressed in terms of that of Φ_V, denoted by $\alpha_V^{(2)}$, by

$$\alpha_A^{(2)}(B_1 \times B_2) = \alpha_V^{(2)}(B_1 \times [-\Delta, \Delta] \times B_2 \times [-\Delta, \Delta])$$

$$= \int_{B_1} \int_{-\Delta}^{\Delta} \int_{B_2} \int_{-\Delta}^{\Delta} \alpha_V^{(2)} \, d\big((x_1, x_2, x_3), (y_1, y_2, y_3)\big),$$

where $\Delta = (d_V + t)/2$ and B_1 and B_2 are both planar Borel sets. This equation holds because all points of Φ_A in B_1 and B_2 result from points of Φ_V in $B_1 \times [-\Delta, \Delta]$ and $B_2 \times [-\Delta, \Delta]$, and is true for all planar Borel sets B_1 and B_2. The pair correlation functions are densities of the corresponding second-order factorial moment measures (up to constant factors $N_A^2 = N_V^2(d_V + t)^2$ and N_V^2). Therefore a similar equation holds for the pair correlation functions,

$$g_A(r) = \frac{1}{(d_V + t)^2} \int_{-\Delta}^{\Delta} \int_{-\Delta}^{\Delta} g_V\left(\sqrt{r^2 + (x_3 - y_3)^2}\right) dx_3 \, dy_3. \qquad (10.88)$$

This relationship can be transformed into (10.83). The proofs of the other formulae are similar. □

In the case of Formula (10.83) there exists a theoretical solution, $g_V(r)$ can be obtained if $g_A(r)$ is given, as shown by Kalmykov and Shepilov (2000).

If there is an r_1 (the smallest one) such that $g_V(r) = 1$ for $r \geq r_1$, the solution is

$$g_V(r) = \frac{g^*(r)}{\pi} - \frac{d_V}{\pi r} \int_r^{r_1} \frac{x g^{*'}(x)}{\sqrt{x^2 - r^2}} \, dx + \frac{1}{\pi} \int_r^{r_1} g^*(x)$$

$$\times \left(\frac{2x}{\pi d_V^2} \exp\left(\frac{x^2 - r^2}{\pi d_V^2} \right) \left(1 + \text{erf}\left(\sqrt{\frac{x^2 - r^2}{\pi d_V^2}} \right) \right) \right)$$

$$+ \frac{2x}{\pi d_V \sqrt{x^2 - r^2}} - \frac{d_V}{(x + r)\sqrt{x^2 - r^2}} \right) dx \qquad \text{for } \sqrt{r_1^2 - d_V^2} \leq r \leq r_1,$$

$$(10.89)$$

where

$$g^*(r) = g_A(r) - 1 - \frac{r_1^2 - r^2}{d_V^2} + \frac{2}{d_V} \sqrt{r_1^2 - r^2},$$

and $\text{erf}(x)$ is the error function, which is related to the complementary error function in (9.118) on p. 385 by

$$\text{erf}(x) + \text{erfc}(x) = 1.$$

Shepilov et al. (2006) consider an example from glass research. The starting point is a sample of electron microscopic images, representing planar sections. The authors show ordering effects in the arrangement of particles in phase-separated sodium borosilicate glasses.

As with the other integral equations of stereology, (10.83) and (10.84) are difficult to solve numerically, and the solution formula (10.89) is only of limited value, since usually only estimates are known for $g_A(r)$. It may be useful to consult Hanisch (1984a), where $g_A(r)$ is calculated for some spatial models with known $g_V(r)$. Also here a two-step method can be recommended, where in the first step the form of $g_V(r)$ is explored and in the second the corresponding parameters are estimated, using (10.84).

Example 10.5. *Sinter metal balls (Hanisch, 1984a)*
Figure 10.7(a) shows a planar section through a specimen of sinter metal, and Figure 10.7(b) displays the corresponding estimated pair correlation function $g_A(r)$ of the point process of section disc centres. Hanisch assumed that the ball diameters were constant at 1.4; an assumption supported by (10.45) and the measurements for mean and standard deviation of diameters of section discs,

$$\bar{x} = 0.99 \qquad \text{and} \qquad s = 0.32.$$

Numerical solution of the integral equation (10.83) yielded the graph of $g_V(r)$ displayed in Figure 10.7(b).

The differences between $g_A(r)$ and $g_V(r)$ are typical for the case of independent (constant) diameters: the hard-core distance corresponding to g_V is larger than that for $g_A(r)$ while the peaks of $g_V(r)$ are higher and appear at larger values of r. Comparison of $g_V(r)$ with Figure 4.7 on p. 142 suggests the possibility that the balls in the probe form a random dense packing.

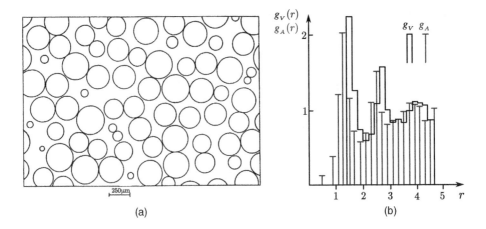

Figure 10.7   (a) Planar section through a specimen of sinter metal. The length unit is approximately 250 μm. (b) The pair correlation functions $g_A(r)$ and $g_V(r)$ for the sinter metal probe. The form of $g_V(r)$ suggests the assumption that the balls form a random dense packing; compare with Figures 4.7 and 10.6.

If the diameters are dependent (as in random dense packings with different ball diameters or in the Stienen model) then different things may happen. This case is studied in a general second-order theory for ball systems given by stationary and isotropic marked point processes as on p. 428; see Stoyan *et al.* (1990). There appears an integral equation which generalises (10.40), involving the planar and spatial pair correlation functions and the family of so-called two-point mark distributions $\mathscr{D}_{V,r}$, which are the joint distributions of the diameters of two balls with given inter-centre distance r under the condition that at two given points of distance r there are ball centres. Stoyan (1990a) uses this integral equation to calculate the pair correlation function $g_A(r)$ of the point process of section disc centres corresponding to the Stienen model; see Figure 10.8, which also contains pair correlation function graphs for the generalised Stienen model, where all diameters are multiplied by a constant factor $\alpha < 1$. (This produces a further system of non-overlapping balls in space, but now without contact between balls and with volume fraction less than 0.125.)

A statistician who is familiar with pair correlation function statistics as described in Illian *et al.* (2008) would classify the pair correlation functions $g_A(r)$ shown in Figures 10.7(b) and 10.8 as pair correlation functions of point processes with a weak form of short-range order of the inhibition type. And indeed, there is a form of inhibition: a point pair of very close section disc centres is only possible if the corresponding balls have a rather extraordinary position. This may explain the form of $g_A(r)$. But of course, the pair correlation function $g_V(r)$ of the Stienen model ball centres is that of a Poisson process and in this case $g_V(r) \equiv 1$. Thus $g_A(r)$ reflects weak inner order of the Stienen model, where the diameters are not independent.

10.7.3   Second-order analysis for spatial fibre systems

Stoyan (1984b, 1985a,b) suggested approximate stereological methods for the determination of the pair correlation function for spatial fibre systems. The method applies both to planar and to thick sections. Under the simplifying assumption that one can ignore the spatial correlations

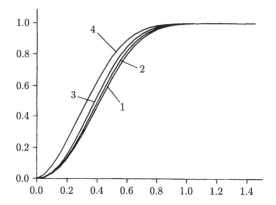

Figure 10.8 The pair correlation function of the point process of section disc centres of the Stienen model with intensity $\lambda = N_V = 1$. Furthermore, the pair correlation functions for generalised Stienen models with $\alpha < 1$ are shown. See text for explanation. (1 : $\alpha = 0.25$, 2 : $\alpha = 0.5$, 3 : $\alpha = 0.7$, 4 : $\alpha = 1$.)

of fibre directions at the fibre section points, an integral equation similar to (10.83) can be derived for the case of thick sections. It connects the pair correlation functions $g_V(r)$ and $g_A(r)$ of the spatial and the projected fibre processes. Figure 10.9(a) shows a thick section of a spatial fibre system, representing a network of dislocation lines induced by diffusion in a

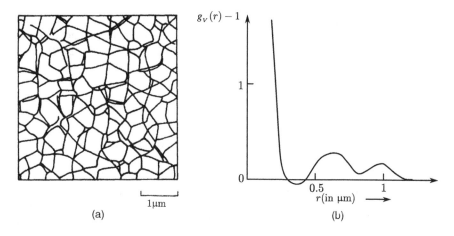

(a)   (b)

Figure 10.9 (a) Network of dislocation lines induced by diffusion in a silicon crystal, taken from a photograph of a transmission electromicroscopic image in Hanisch *et al.* (1985). The thickness of the foil was about 0.2μm. (b) Estimated spatial pair correlation function $g_V(r)$ of the dislocation lines. The limiting value $g_V(0+)$ is $+\infty$. This is typical for fibre processes as discussed in Section 8.3.1 for the planar case.

single silicon crystal. Figure 10.9(b) gives the corresponding spatial pair correlation function; see Hanisch *et al.* (1985).

Further simplification, as described in Krasnoperov and Stoyan (2006), leads in the case of planar sections to the simple approximation

$$g_V(r) \approx g_A(r), \qquad (10.90)$$

where $g_A(r)$ is the pair correlation function of the planar point process of fibre section points, which in applications are usually the centres of section profiles of thick fibres. (Recall that the same approximation has been suggested in (10.86) for systems of ball centres.) Krasnoperov and Stoyan (2006) apply these ideas to the case of blood capillaries.

The case of straight fibres makes some difficulties in all second-order stereological approaches; in Stoyan's papers it is usually excluded or handled in a particular way. Two simple models are the spatial Poisson line process and the Boolean model with segments as grains. In both cases the section with a plane yields a Poisson point process, and thus the pair correlation function $g_A(r)$ does not give interesting spatial information: it is

$$g_A(r) \equiv 1.$$

However, such fibre processes will hardly ever appear in practical applications. They could be called fibre processes 'without structure', since their only spatial correlation comes from the inner connectedness of the segments or lines; between different segments or lines there are no correlations, and each object produces at most only one point of intersection with the section plane.

The following discusses the quality of the approximation (10.90) for a simple fibre process model 'with structure'.

Example 10.6. *The Boolean circle model*
Consider a Boolean model, with germ intensity λ_V. The grains are fibres, closed circular lines of constant radius R of random uniform orientation. Thus the grains are not convex. The union of all circles is the fibre process of interest. Its fibre density is $L_V = 2\pi R \lambda_V$ and the pair correlation function is

$$g_V(r) = 1 + \begin{cases} \dfrac{1}{\pi r^2 L_V} \dfrac{R}{\sqrt{4R^2 - r^2}} & \text{for } r \leq 2R, \\[2ex] 0 & \text{otherwise.} \end{cases}$$

This can be proved as Formula (8.27) in Example 8.4; see also Stoyan and Stoyan (1986).

A planar section produces a Poisson cluster process, where each cluster consists of exactly two points, which belong to the same intersected circle. The intensity of the Poisson process of parent points is $\lambda_A = \pi R^2 \lambda_V/2$, using Formula (3.66). Thus the intensity of the point process of all fibre section points is $P_A = \pi R^2 \lambda_V$. Its pair correlation function is

$$g_A(r) = 1 + \begin{cases} \dfrac{1}{2\pi r P_A} \dfrac{r}{2R\sqrt{4R^2 - r^2}} & \text{for } r \leq 2R, \\[2ex] 0 & \text{otherwise.} \end{cases}$$

This example shows what the approximation (10.90) can yield: (a) $g_A(r)$ usually does not have a pole at $r = 0$, whilst $g_V(r)$ always has, but (b) other structural properties are indicated in a qualitatively correct form. In the given example $g_V(r)$ as well as $g_A(r)$ have a pole at $r = 2R$, which is related to the circle diameter.

10.7.4   Determination of directional distributions

Stereological problems increase in complexity when the structures under investigation are anisotropic. Anisotropy occurs, for example, for capillaries in skeletal muscles or for grains in rolled steel. In such cases it is common to make sections of different orientations, generally in transverse and longitudinal directions to the main axis of anisotropy if this is known. If the structure can be interpreted as a 'pressed' structure, then formulae as on p. 324 may be employed.

In this section a brief discussion will be given of the case of directional distributions of anisotropic fibre systems; similar considerations for surface systems can be found in Weibel (1980) and Beneš and Rataj (2004).

Several authors suggest parametric approaches; see Weibel (1980), Mathieu *et al.* (1983), Mattfeldt and Mall (1984), Cruz-Orive *et al.* (1985), Jensen *et al.* (1985) and Poole *et al.* (1992). Each fibre direction corresponds to a direction on a hemisphere (or more properly the projective plane) which allows the use of standard distributions on the hemisphere as considered in the theory of directional data; see Fisher *et al.* (1987). In particular the *Fisher* or the *Watson* distribution (Fisher *et al.*, 1987, pp. 86 and 89) may be employed. The latter depends on a concentration parameter κ and this can be determined together with the intensity L_V by using a nonlinear regression procedure. The procedure uses as data the mean number of intersection points of fibres per unit area in planes of various known directions. This approach also enables tests of the goodness-of-fit of the distributional assumption. Given κ it is sufficient to consider just one plane of intersection. Cruz-Orive *et al.* (1985) applied the method to an example concerning the study of orientations of capillaries in the skeletal muscle of cats.

Commencing from the observation that fibres are often tubes, a nonparametric method has been developed; see Stoyan (1984b, 1985b) and also papers on the stereology of tubes such as Baddeley and Averback (1983). If the fibres are of circular cross-section with slowly varying diameter, the angles at which the fibres intersect the plane may be determined by observation of the intersection figures in planar or thick sections. Figure 10.10 illustrates this point.

Figure 10.10 shows that the intersection angle β is given by

$$\cos \beta = \frac{bl + t/\sqrt{l^2 + t^2 - b^2}}{l^2 + t^2}. \tag{10.91}$$

It is possible to measure β automatically by exploiting this formula. An alternative means of measuring β uses the possibility of moving the focal plane up and down when observing thick sections of biological tissue by microscope. The line density L_V can be estimated by

$$L_V = \sum_{i=1}^{n} \frac{1}{v_2(W) \cos \beta_i}, \tag{10.92}$$

where β_i is the angle corresponding to the ith intersection figure, and n is the number of intersection figures in the window of observation W.

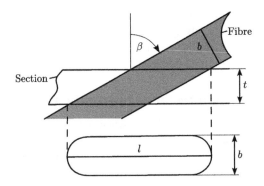

Figure 10.10 Intersection figure for a cylindrical fibre in a thick section. The same figure would occur if the intersection angle was still β and the orientation of the fibre was reversed (so that it would come from left above to right below).

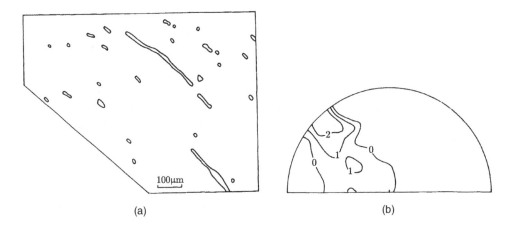

(a)   (b)

Figure 10.11 (a) Intersections of capillaries with a section from the medullary zone of the human brain, and (b) the corresponding directional distribution, using Schmidt's net representation; the density is given up to a factor of proportionality by contour lines.

Stoyan (1985b) demonstrates how the directional distribution can be determined. Figure 10.11(b) shows the directional distribution, displayed in Schmidt's net, for the specimen of capillaries in the medullary zone of the human brain in Figure 10.11(a) (a description of Schmidt's net can be found in for example Fisher *et al.*, 1987, p. 38). The intersection figures only give fibre directions up to reflection in the normal to the plane of intersection (see Figure 10.10) and so only a half-circle is used.

Beneš and Rataj (2004, section 5.3) discuss a convex-geometry approach.

References

Abbe, E. (1879). Über Blutkörperzählung. *Jena Z. Med. Naturwiss.* **13** (Neue Serie), 98–105.

Abellanas, M., Bajuelos, A., Hernández, G., Hurtado, F., Matos, I., and Palop, B. (2006). Good illumination of minimum range. arXiv:cs/0606013v1 [cs.CG].

Abellanas, M., Bajuelos, A., and Matos, I. (2007a). Some problems related to good illumination. In Gervasi, O. and Gavrilova, M., eds, *Computational Science and its Application — ICCSA 2007*, Lecture Notes in Computer Science **4705**, pp. 1–14. Springer-Verlag, Berlin.

Abellanas, M., Bajuelos, A., and Matos, I. (2007b). Variations of good illumination. In *Proceedings of XII Spanish Workshop on Computational Geometry*, pp. 265–72, Valladolid, Spain.

Abellanas, M., Bajuelos, A. L., Hernádez, G., Matos, I., and Palop, B. (2009). The embracing Voronoi diagram and closest embracing number. *J. Math. Sci.* **161**, 909–18.

Aboav, D. A. (1970). The arrangement of grains in a polycrystal. *Metallography* **3**, 383–90.

Aboav, D. A. (1980). The arrangement of cells in a net. *Metallography* **13**, 43–58.

Aboav, D. A. (1991). The stereology of the intergranular surface of a metal. *Acta Stereol.* **10**, 43–54.

Aboav, D. A. (1992). The topology of a polycrystal in three dimensions. *Mater. Sci. Forum* **94–6**, 275–80.

Adler, F. R. (1996). A model of self-thinning through local competition. *Proc. Natl. Acad. Sci. USA* **93**, 9980–4.

Adler, P. M., Mourzenko, V. V., Thovert, J.-F., and Bogdanov, I. (2005). Study of single and multiphase flow in fractured porous media, using a percolation approach. In Faybishenko, B., Witherspoon, P. A., and Gale, J., eds, *Dynamics of Fluids and Transport in Fractured Rock*, Geophysical Monograph Series 162, pp. 33–41. Amer Geophysical Union, Washington, D. C.

Adler, P. M. and Thovert, J.-F. (1999). *Fractures and Fracture Networks*. Kluwer Academic Publishers, Dordrecht.

Adler, R. J. (1981). *The Geometry of Random Fields*. John Wiley & Sons, Ltd, Chichester.

Adler, R. J. (2000). On excursion sets, tube formulas and maxima of random fields. *Ann. Appl. Probab.* **10**, 1–74.

Adler, R. J., Samorodnitsky, G., and Taylor, J. E. (2010). Excursion sets of three classes of stable random fields. *Adv. Appl. Prob.* **42**, 293–318.

Adler, R. J. and Taylor, J. E. (2007). *Random Fields and Geometry*. Springer-Verlag, New York.

Adler, R. J. and Taylor, J. E. (2011). *Topological Complexity of Smooth Random Functions. École d'Été de Probabilités de Saint-Flour XXXIX-2009*. Lecture Notes in Mathematics **2019**. Springer-Verlag, New York.

Albert, R. and Barabási, A.-L. (2002). Statistical mechanics of complex networks. *Rev. Mod. Phys.* **74**, 47–97.

Albert, R., Jeong, H., and Barabási, A.-L. (1999). Diameter of the World-Wide Web. *Nature* **401**, 130–1.

Albert, R., Jeong, H., and Barabási, A.-L. (2000). Error and attack tolerance of complex networks. *Nature* **406,** 378–82.

Aldous, D. (1989). *Probability Approximations via the Poisson Clumping Heuristic.* Springer-Verlag, New York.

Aldous, D. J. and Kendall, W. S. (2008). Short-length routes in low-cost networks via Poisson line patterns. *Adv. Appl. Prob.* **40,** 1–21.

Aldous, D. J. and Shun, J. (2010). Connected spatial networks over random points and a route-length statistic. *Statist. Sci.* **25,** 275–88.

Aletti, G., Bongiorno, E. G., and Capasso, V. (2009). Statistical aspects of fuzzy monotone set-valued stochastic processes. Application to birth-and-growth processes. *Fuzzy Sets and Systems* **160,** 3140–51.

Aletti, G., Bongiorno, E. G., and Capasso, V. (2011). Integration in a dynamical stochastic geometric framework. *ESAIM Probab. Stat.* **15,** 402–16.

Alinchenko, M. G., Anikeenko, A. V., Medvedev, N. N., Voloshin, V. P., Mezei, M., and Jedlovszky, P. (2004). Morphology of voids in molecular systems. A Voronoi-Delaunay analysis of a simulated DMPC membrane. *J. Phys. Chem. B* **108,** 19056–67.

Alishahi, K. and Sharifitabar, M. (2008). Volume degeneracy of the typical cell and the chord length distribution for Poisson–Voronoi tessellations in high dimensions. *Adv. Appl. Prob.* **40,** 919–38.

Allard, D. and Fraley, C. (1997). Nonparametric maximum likelihood estimation of features in spatial point processes using Voronoï tessellation. *J. Amer. Statist. Assoc.* **92,** 1485–93.

Altendorf, H. and Jeulin, D. (2009). 3D directional mathematical morphology for analysis of fiber orientations. *Image Anal. Stereol.* **28,** 143–53.

Ambartzumian, R. V. (1966). On an equation for stationary point processes (in Russian). *Dokl. Akad. Nauk Armjanskoi SSR* **42,** 141–7.

Ambartzumian, R. V. (1973). On random fields of segments and random mosaics in the plane (in Russian). *Teor. Veroyatn. Prim.* **18,** 515–26. Corrections: **19,** 600.

Ambartzumian, R. V. (1974a). Convex polygons and random tessellations. In Harding, E. F. and Kendall, D. G., eds, *Stochastic Geometry*, pp. 176–91. John Wiley & Sons, Ltd, London.

Ambartzumian, R. V. (1974b). The solution to the Buffon–Sylvester problem in $\mathbb{R}^3$. *Z. Wahrscheinlichkeitsth. verw. Geb.* **27,** 53–74.

Ambartzumian, R. V. (1977). Stochastic geometry from the standpoint of integral geometry. *Adv. Appl. Prob.* **9,** 792–823.

Ambartzumian, R. V. (1981). Stereology of random planar segment processes. *Rend. Sem. Mat. Torino* **39,** 147–59.

Ambartzumian, R. V. (1982). *Combinatorial Integral Geometry.* John Wiley & Sons, Ltd, Chichester.

Ambartzumian, R. V. (1990). *Factorization Calculus and Geometric Probability.* Cambridge University Press, Cambridge.

Ambartzumian, R. V. and Ohanian, V. K. (1975). Homogeneous and isotropic fibre fields in the plane (in Russian). *Izv. AN Armen. SSR Ser. Math.* **10,** 509–28.

Ambrosio, L., Capasso, V., and Villa, E. (2009). On the approximation of mean densities of random closed sets. *Bernoulli* **15,** 1222–42.

Andersen, P. K., Borgan, Ø., Gill, R. D., and Keiding, N. (1993). *Statistical Models Based on Counting Processes.* Springer-Verlag, New York.

Anderssen, R. S. and Jakeman, A. J. (1975a). Abel-type integral equations in stereology. *J. Microsc.* **105,** 121–53.

Anderssen, R. S. and Jakeman, A. J. (1975b). Product integration for functionals of particle size distributions. *Utilitas Math.* **8,** 111–26.

Andersson, J., Häggström, O., and Månsson, M. (2006). The volume fraction of a non-overlapping germ–grain model. *Electron. Commun. Probab.* **11**, 78–88.

Andrade, P. N. and Fortes, M. A. (1988). Distribution of cell volumes in a Voronoi partition. *Phil. Mag.* **58**, 671–4.

Ang, Q. W., Baddeley, A., and Nair, G. (2012). Geometrically corrected second order analysis of events on a linear network, with applications to ecology and criminology. *Scand. J. Statist.* **39**, 591–617.

Antoniadis, A., Fan, J., and Gijbels, I. (2001). A wavelet method for unfolding sphere size distributions. *Can. J. Statist.* **29**, 251–68.

Arak, T., Clifford, P., and Surgailis, D. (1993). Point-based polygonal models for random graphs. *Adv. Appl. Prob.* **25**, 348–72.

Arak, T. and Surgailis, D. (1989). Markov fields with polygonal realizations. *Probab. Theory Related Fields* **80**, 543–79.

Arbeiter, E. and Zähle, M. (1994). Geometric measures for random mosaics in spherical spaces. *Stoch. Stoch. Rep.* **46**, 63–77.

Armitage, P. (1949). An overlapp problem arising in particle counting. *Biometrika* **36**, 257–66.

Armitage, P., Berry, G., and Matthews, J. N. S. (2002). *Statistical Methods in Medical Research*. Blackwell, Malden, MA., 4th edition.

Arns, C. H., Knackstedt, M. A., and Mecke, K. R. (2002). Characterising the morphology of disordered materials. In Mecke, K. R. and Stoyan, D., eds, *Morphology of Condensed Matter. Physics and Geometry of Spatially Complex Systems*, Lecture Notes in Physics **600**, pp. 37–74. Springer-Verlag, Berlin.

Arns, C. H., Mecke, J., Mecke, K., and Stoyan, D. (2005). Second-order analysis by variograms for curvature measures of two-phase structures. *Eur. Phys. J. B* **47**, 397–409.

Artstein, Z. (1984a). Convergence of sums of random sets. In Ambartzumian, R. V. and Weil, W., eds, *Stochastic Geometry, Geometric Statistics, Stereology*, Teubner-Texte zur Mathematik **65**, pp. 34–42. B. G. Teubner Verlagsgesellschaft, Leipzig.

Artstein, Z. (1984b). Limit laws for multifunctions applied to an optimization problem. In Salinetti, G., ed., *Multifunctions and Integrands, Stochastic Analysis, Approximation and Optimization*, Lecture Notes in Math. **1091**, pp. 66–79. Springer-Verlag, Berlin.

Artstein, Z. and Vitale, R. A. (1975). A strong law of large numbers for random compact sets. *Ann. Prob.* **3**, 879–82.

Asano, A., Miyagawa, M., and Fujio, M. (2003). Morphological texture analysis using optimization of structuring elements. In Asano, T., Klette, R., and Ronse, C., eds, *Geometry, Morphology, and Computational Imaging*, Lecture Notes in Computer Science **2616**, pp. 141–52. Springer-Verlag, Berlin.

Aste, T. (1999). The shell map: the structure of froths through a dynamical map. In Sadoc, J. F. and Rivier, N., eds, *Foams and Emulsions*, pp. 497–510. Kluwer Academic Publishers, Dordrecht.

Aste, T., Saadatfar, M., Sakellariou, A., and Senden, T. J. (2004). Investigating the geometrical structure of disordered sphere packings. *Physica A* **339**, 16–23.

Aste, T., Saadatfar, M., and Senden, T. J. (2005). Geometrical structure of disordered sphere packings. *Phys. Rev. E* **71**, 061302.

Aste, T., Szeto, K. Y., and Tam, W. Y. (1996). Statistical properties and shell analysis in random cellular structures. *Phys. Rev. E* **54**, 5482–92.

Aste, T. and Weaire, D. (2008). *The Pursuit of Perfect Packing*. Taylor & Francis, New York, 2nd edition.

Åström, J. A., Mäkinen, J. P., Alava, M. J., and Timonen, J. (2000). Elasticity of Poissonian fiber networks. *Phys. Rev. E* **61**, 5550–6.

Athreya, K. B. and Lahiri, S. N. (2010). *Measure Theory and Probability Theory*. Springer-Verlag, New York.

Aubin, J.-P. (1999). *Mutational and Morphological Analysis. Tools for Shape Evolution and Morphogenesis*. Birkhäuser, Boston.

Aubin, J.-P. and Frankowska, H. (1990). *Set-Valued Analysis*. Birkhäuser-Verlag, Basel, Boston.

Aumann, R. J. (1965). Integrals of set-valued functions. *J. Math. Anal. Appl.* **12**, 1–12.

Aurenhammer, F. (1991). Voronoi diagrams–a fundamental geometric data structure. *ACM Computing Surveys* **23**, 345–405.

Averkov, G. and Bianchi, G. (2009). Confirmation of Matheron's conjecture on the covariogram of a planar convex body. *J. Eur. Math. Soc.* **11**, 1187–1202.

Avrami, M. (1939). Kinetics of phase change I. *J. Chem. Phys* **7**, 1103–12.

Ayala, G., Ferrandiz, J., and Montes, F. (1991). Random set and coverage measure. *Adv. Appl. Prob.* **23**, 972–4.

Baccelli, F. and Błaszczyszyn, B. (2009a). Stochastic geometry and wireless networks I: Theory. *Foundations and Trends in Networking* **3**, 249–449.

Baccelli, F. and Błaszczyszyn, B. (2009b). Stochastic geometry and wireless networks II: Applications. *Foundations and Trends in Networking* **4**, 1–312.

Baccelli, F. and Brémaud, P. (2003). *Elements of Queueing Theory: Palm Martingale Calculus and Stochastic Recurrences*. Springer-Verlag, Berlin.

Baccelli, F., Gloaguen, C., and Zuyev, S. (2000). Superposition of planar Voronoi tessellations. *Stochastic Models* **16**, 69–98.

Baccelli, F., Klein, M., Lebourges, M., and Zuyev, S. (1997). Stochastic geometry and architecture of communication networks. *Telecommunication Systems* **7**, 209–27.

Baccelli, F. and Zuyev, S. (1997). Stochastic geometry models of mobile communication networks. In Dshalaow, J. H., ed., *Frontiers of Queuing Theory: Models and Applications in Science and Engineering*, pp. 227–43. CRC Press, Boca Raton.

Bach, G. (1959). Über die Größenverteilung von Kugelschnitten in durchsichtigen Schnitten endlicher Dicke. *Z. wiss. Mikrosk.* **64**, 265–86.

Bach, G. (1965). Über die Bestimmung von charakteristischen Größen einer Kugelverteilung aus der unvollständigen Verteilung der Schnittkreise. *Metrika* **9**, 228–33.

Bach, G. (1967). Kugelgrößenverteilung und Verteilung der Schnittkreise; ihre wechselseitigen Beziehungen und Verfahren zur Bestimmung der einen aus den anderen. In Weibel, E. R. and Elias, H., eds, *Quantitative Methods in Morphology*, pp. 23–45. Springer-Verlag, Berlin.

Bach, G. (1976). Über die Auswertung von Schnittflächenverteilungen. *Biometrical J.* **18**, 407–12.

Baddeley, A. J. (1980). Absolute curvatures in integral geometry. *Proc. Camb. Phil. Soc.* **88**, 45–58.

Baddeley, A. J. (1984). Stochastic geometry and image analysis. *CWI Newsletter* (Centrum voor Wiskunde en Informatica, Amsterdam) September 1984, pp. 2–20.

Baddeley, A. J. (2010). Modeling strategies. In Gelfand, A. E., Diggle, P. J., Fuentes, M., and Guttorp, P., eds, *Handbook of Spatial Statistics*, pp. 339–69. CRC Press, Boca Raton.

Baddeley, A. J. and Averback, P. (1983). Stereology of tubular structures. *J. Microsc.* **131**, 323–40.

Baddeley, A. J. and Cruz-Orive, L. M. (1995). The Rao–Blackwell theorem in stereology and some counterexamples. *Adv. Appl. Prob* **27**, 2–19.

Baddeley, A. J. and Dereudre, D. (2013). Variational estimators for the parameters of Gibbs point processes models. *Bernoulli*. Forthcoming.

Baddeley, A. J. and Gill, R. D. (1997). Kaplan–Meier estimators of distance distributions for spatial point processes. *Ann. Statist.* **25**, 263–92.

Baddeley, A. J., Gundersen, H. J. G., and Cruz-Orive, L. M. (1986). Estimation of surface area from vertical sections. *J. Microsc.* **142,** 259–76.

Baddeley, A. J. and Jensen, E. B. V. (2005). *Stereology for Statisticians.* Chapman & Hall, Boca Raton.

Baddeley, A. J. and Møller, J. (1989). Nearest-neighbour Markov point processes and random sets. *Int. Statist. Rev.* **57,** 89–121.

Baddeley, A. J. and Nair, G. (2012). Fast approximation of the intensity of Gibbs point processes. *Electron. J. Statist.* **6,** 1155–69.

Baddeley, A. J. and Silverman, B. W. (1984). A cautionary example for the use of second order methods for analyzing point patterns. *Biometrics* **40,** 1089–94.

Baddeley, A. J., van Lieshout, M. N. M., and Møller, J. (1996). Markov properties of cluster processes. *Adv. Appl. Prob.* **28,** 346–55.

Baecher, G. B. (1983). Statistical analysis of rock mass fracturing. *Math. Geol.* **15,** 329–48.

Balberg, I., Anderson, C. H., Alexander, S., and Wagner, N. (1984). Excluded volume and its relation to the onset of percolation. *Phys. Rev. B* **30,** 3933–43.

Ballani, F. (2006a). On modelling of refractory castables by marked Gibbs and Gibbsian-like processes. In Baddeley, A., Gregori, P., Mahiques, J. M., and Stoyan, D., eds, *Case Studies in Spatial Point Process Modeling*, Lecture Notes in Statistics **185,** pp. 153–67. Springer-Verlag, New York.

Ballani, F. (2006b). On second-order characteristics of germ–grain models with convex grains. *Mathematika* **53,** 255–85.

Ballani, F. (2011). Multiple-point hit distribution functions and vague convergence of related measures. *Math. Nachr.* **284,** 938–47.

Ballani, F., Daley, D. J., and Stoyan, D. (2006). Modelling the microstructure of concrete with spherical grains. *Comp. Mat. Sci.* **35,** 399–407.

Ballani, F., Kabluchko, Z., and Schlather, M. (2012). Random marked sets. *Adv. Appl. Prob.* **44,** 603–16.

Ballani, F. and van den Boogaart, K. G. (2013). Weighted Poisson cells as models for random convex polytopes. *Methodol. Comput. Appl. Probab.* Forthcoming.

Balslev, I., Døring, K., and Eriksen, R. D. (2000). Weighted central moments in pattern recognition. *Pattern Recogn. Lett.* **21,** 381–4.

Bandemer, H. and Näther, W. (1992). *Fuzzy Data Analysis.* Kluwer Academic Publishers, Dordrecht.

Barabási, A.-L. and Albert, R. (1999). Emergence of scaling in random networks. *Science* **286,** 509–12.

Barber, P. R., Vojnovic, B., Ameer-Beg, S. M., Hodgkiss, R. J., Tozer, G. M., and Wilson, J. (2003). Semi-automated software for the three-dimensional delineation of complex vascular networks. *J. Microsc.* **211,** 54–62.

Bargieł, M. (2008). Geometrical properties of simulated packings of spherocylinders. In Bubak, M., van Albada, G. D., Dongarra, J., and Sloot, P. M. A., eds, *Computational Science – ICCS 2008. Part II*, Lecture Notes in Computer Science **5102,** pp. 126–35. Springer-Verlag, Berlin.

Barrat, A. and Weigt, M. (2000). On the properties of small-world network models. *Eur. Phys. J. B* **13,** 547–60.

Bartlett, M. S. (1954). Processus stochastiques ponctuels. *Ann. Inst. H. Poincaré* **14,** 35–60.

Bartlett, M. S. (1975). *The Statistical Analysis of Spatial Pattern.* Chapman & Hall, London.

Baumstark, V. and Last, G. (2007). Some distributional results for Poisson–Voronoi tessellations. *Adv. Appl. Prob.* **39,** 16–40.

Baumstark, V. and Last, G. (2009). Gamma distributions for stationary Poisson flat processes. *Adv. Appl. Prob.* **41,** 911–39.

Beer, G. (1993). *Topologies on Closed and Closed Convex Sets.* Kluwer Academic Publishers, Dordrecht.

Beneš, V. (1994). On second-order formulas in anisotropic stereology. *Adv. Appl. Prob.* **27,** 326–43.

Beneš, V., Chadœuf, J., and Ohser, J. (1994). Estimation of intensity in anisotropic fibre processes. *Math. Nachr.* **169**, 5–17.

Beneš, V., Jiruše, M., and Slámová, M. (1997). Stereological unfolding of the trivariate size-shape-orientation distribution of spheroidal particles with application. *Acta Materialia* **45**, 1105–13.

Beneš, V. and Krejčíř, P. (1997). Decomposition in stereological unfolding problems. *Kybernetika* **33**, 245–58.

Beneš, V. and Rataj, J. (2004). *Stochastic Geometry: Selected Topics*. Kluwer Academic Publishers, Boston.

Beneš, V. and Gokhale, A. M. (2000). Planar anisotropy revisited. *Kybernetika* **36**, 149–64.

Bennett, M. R. and Robinson, J. (1990). Probabilistic secretion of quanta from nerve terminals at synaptic sites on muscle cells: nonuniformity, autoinhibition and the binomial hypothesis. *Proc. Roy. Soc. Lond. Ser. B* **239**, 329–58.

Beresteanu, A., Molchanov, I., and Molinari, F. (2011). Partial identification using random set theory. *J. Econometrics* **166**, 17–32.

Berg, C. (1969). Corps convexes et potentials sphériques. *Mat.-Fys. Medd.* **37**, 6.

Berg, C., Christensen, J. P. R., and Ressel, P. (1984). *Harmonic Analysis on Semigroups*. Springer-Verlag, Berlin.

Berkowitz, B. and Adler, P. M. (1998). Stereological analysis of fracture network structure in geological formations. *J. Geophys. Res.* **103**, 15339–60.

Bernal, J. D. (1960). Geometry of the structure of monatomic liquids. *Nature* **185**, 68–70.

Bernardeau, F. and van de Weygaert, R. (1996). A new method for accurate estimation of velocity field statistics. *Mon. Not. R. Astron. Soc.* **279**, 693–711.

Berryman, J. G. (1987). Relationship between specific surface area and spatial correlation functions for anisotropic porous media. *J. Math. Phys.* **28**, 244–5.

Besag, J. (1975). Statistical analysis of non-lattice data. *The Statistician* **24**, 179–95.

Besag, J. (1978). Some methods of statistical analysis for spatial data. *Bull. Int. Statist. Inst.* **47**, 77–92.

Besag, J. and Diggle, P. J. (1977). Simple Monte Carlo tests for spatial pattern. *Appl. Statist.* **26**, 327–33.

Bezrukov, A., Bargieł, M., and Stoyan, D. (2002). Statistical analysis of simulated random packings of spheres. *Part. Part. Syst. Charact.* **19**, 111–18.

Bezrukov, A. and Stoyan, D. (2006). Simulation and statistical analysis of random packings of ellipsoids. *Part. Part. Syst. Charact.* **23**, 388–98.

Bianchi, G. (2005). Matheron's conjecture for the covariogram problem. *J. London Math. Soc.* **71**, 203–20.

Bianchi, G. (2009). The covariogram determines three-dimensional convex polytopes. *Adv. Math.* **220**, 1771–1808.

Biggins, J. D. (1977). Martingale convergence in the branching random walk. *J. Appl. Prob.* **14**, 25–37.

Billera, L. J. and Diaconis, P. (2001). A geometric interpretation of the Metropolis–Hastings algorithm. *Statist. Sci.* **16**, 335–9.

Billingsley, P. (1995). *Probability and Measure*. John Wiley & Sons, Inc., New York, 3rd edition.

Bindrich, U. and Stoyan, D. (1991). Stereology for pores in wheat bread: statistical analyses for the Boolean model by serial sections. *J. Microcs.* **162**, 231–9.

Biswal, B., Øren, P.-E., Held, R. J., Bakke, S., and Hilfer, R. (2009). Modeling of multiscale porous media. *Image Anal. Stereol.* **28**, 23–34.

Blackwell, P. G. and Møller, J. (2003). Bayesian analysis of deformed tessellation models. *Adv. Appl. Prob.* **35,** 4–26.

Błaszczyszyn, B., Rau, C., and Schmidt, V. (1999). Bounds for clump size characteristics in the Boolean model. *Adv. Appl. Prob.* **31,** 910–28.

Błaszczyszyn, B. and Schott, R. (2003). Approximate decomposition of some modulated-Poisson Voronoi tessellations. *Adv. Appl. Prob.* **35,** 847–62.

Błaszczyszyn, B. and Schott, R. (2005). Approximations of functionals of some modulated-Poisson Voronoi tessellations with applications to modeling of communication networks. *Japan J. Indust. Appl. Math.* **22,** 179–204.

Blödner, R., Mühlig, P., and Nagel, W. (1984). The comparison by simulation of solutions of Wicksell's corpuscle problem. *J. Microsc.* **135,** 61–74.

Bodziony, J. (1965). On certain indices characterizing the geometric structure of rocks. *Bull. Acad. Polon. Science Ser. Science. Technol.* **13,** 469–75.

Bogachev, V. I. (2007). *Measure Theory, Volume I* and *II*. Springer-Verlag, New York.

Böhm, S. and Schmidt, V. (2003). Palm representation and approximation of the covariance of random closed sets. *Adv. Appl. Prob.* **35,** 295–302.

Boissonnat, J.-D. and Yvinec, M. (1998). *Algorithmic Geometry*. Cambridge University Press, Cambridge.

Bollobás, B., Janson, S., and Riordan, O. (2007). The phase transition in inhomogeneous random graphs. *Random Structures Algorithms* **31,** 3–122.

Bollobás, B. and Riordan, O. (2004). The diameter of a scale-free random graph. *Combinatorica* **24,** 5–34.

Bollobás, B. and Riordan, O. (2006). *Percolation*. Cambridge University Press, New York.

Bollobás, B., Riordan, O., Spencer, J., and Tusnády, G. (2001). The degree sequence of a scale-free random graph process. *Random Structures Algorithms* **18,** 279–90.

Bollobás, B. and Riordan, O. M. (2003). Mathematical results on scale-free random graphs. In Bornholdt, S. and Schuster, H. G., eds, *Handbook of Graphs and Networks: From the Genome to the Internet*, pp. 1–34. Wiley-VCH, Berlin.

Boots, B. N. (1982). The arrangements of cells in "random" networks. *Metallography* **15,** 53–62.

Boots, B. N. (1984). Comments on "Aboav's rule" for the arrangement of cells in a network. *Metallography* **17,** 411–18.

Boots, B. N. (1987). *Voronoi (Thiessen) Polygons*. Geo Books, Norwich.

Bosan, S., Kareco, T., Ruehlmann, D., Chen, K. Y. M., and Walley, K. R. (2003). Three-dimensional capillary geometry in gut tissue. *Microsc. Res. Tech.* **61,** 428–37.

Brakke, K. A. (1986a). Statistics of random plane Voronoi tessellations. Technical report, Department of Mathematical Sciences, Susquehanna University, Selinsgrove.

Brakke, K. A. (1986b). Statistics of three dimensional random Voronoi tessellations. Technical report, Department of Mathematical Sciences, Susquehanna University, Selinsgrove.

Brémaud, P. (1981). *Point Processes and Queues: Martingale Dynamics*. Springer-Verlag, New York.

Bretheau, T. and Jeulin, D. (1989). Caractéristiques morphologiques des constituants et comportement à la limite élastique d'un matériau biphasé fe/ag. *Revue Phys. Appl.* **24,** 861–9.

Brillinger, D. R. (1975). Statistical inferences for stationary point processes. In Puri, M., ed., *Stochastic Processes and Related Topics*, pp. 55–9. Academic Press, New York.

Brillinger, D. R. (1978). Comparative aspects to the study of ordinary time series and point processes. In Krishnaiah, P. R., ed., *Developments in Statistics, Volume I*, pp. 227–320. Academic Press, New York.

Brix, A. (1999). Generalized gamma measures and shot-noise Cox processes. *Adv. Appl. Prob.* **31,** 929–53.

Brix, A. and Kendall, W. S. (2002). Simulation of cluster point processes without edge effects. *Adv. Appl. Prob.* **34,** 267–80.

Brodatzki, U. and Mecke, K. R. (2001). Morphological model for colloidal suspensions. arXiv:cond-mat/0112009v1 [cond-mat.soft].

Brooks, S., Gelman, A., Jones, G. L., and Meng, X.-L., eds (2011). *Handbook of Markov Chain Monte Carlo.* CRC, Boca Raton.

Brumberger, H. and Goodisman, J. (1983). Voronoi cells: An interesting and potentially useful cell model for interpreting the small-angle scattering of catalysts. *J. Appl. Cryst.* **16,** 83–8.

Burger, C. and Ruland, W. (2001). Analysis of chord-length distributions. *Acta Cryst. A* **57,** 482–91.

Burridge, J., Cowan, R., and Ma, I. (2013). Full-and-half Gilbert tessellations with rectangular cells. *Adv. Appl. Prob.* **45,** 1–19.

Buryachenko, V. A. (2007). *Micromechanics of Heterogeneous Materials.* Springer-Verlag, New York.

Cahn, J. W. (1956). The kinetics of grain boundary nucleated reactions. *Acta Metallurgica* **4,** 449–59.

Cahn, J. W. (1967). The significance of average mean curvature and its determination by quantitative metallography. *Trans. AIME* **239,** 610–6.

Cahn, J. W. and Nutting, J. (1959). Transmission quantitative metallography. *Trans. AIME* **215,** 526–8.

Cai, Y. and Kendall, W. S. (2002). Perfect simulation for correlated Poisson random variables conditioned to be positive. *Statist. Comput.* **12,** 229–43.

Calka, P. (2003a). An explicit expression for the distribution of the number of sides of the typical Poisson–Voronoi cell. *Adv. Appl. Prob.* **35,** 863–70.

Calka, P. (2003b). Precise formulae for the distributions of the principal geometric characteristics of the typical cells of a two-dimensional Poisson–Voronoi tessellation and a Poisson line process. *Adv. Appl. Prob.* **35,** 551–62.

Calka, P. (2010). Tessellations. In Kendall, W. S. and Molchanov, I., eds, *New Perspectives in Stochastic Geometry,* pp. 145–69. Oxford University Press, Oxford.

Calka, P., Michel, J., and Paroux, K. (2009). Refined convergence for the Boolean model. *Adv. Appl. Prob.* **41,** 940–57.

Campbell, N. R. (1909). The study of discontinuous phenomena. *Proc. Camb. Phil. Soc.* **15,** 117–36.

Capasso, V., ed. (2003). *Mathematical Modelling for Polymer Processing: Polymerization, Crystallization, Manufacturing.* Mathematics in Industry **2.** Springer-Verlag, Berlin.

Capasso, V. and Micheletti, A. (2006). Stochastic geometry and related statistical problems in biomedicine. In Quarteroni, A., Formaggia, L., and Veneziani, A., eds, *Complex Systems in Biomedicine,* pp. 35–69. Springer-Verlag, Milan.

Capasso, V., Micheletti, A., and Morale, D. (2008). Stochastic geometric models, and related statistical issues in tumour-induced angiogenesis. *Math. Biosci.* **214,** 20–31.

Čapek, P., Hejtmánek, V., Brabec, L., Zikánová, A., and Kočiřík, M. (2009). Stochastic reconstruction of particulate media using simulated annealing: improving pore connectivity. *Trans. Porous Med.* **76,** 179–98.

Capiński, M. and Kopp, E. (2004). *Measure, Integral and Probability.* Springer-Verlag, London, 2nd edition.

Car, P. and Parrinello, M. (1988). Structural, dynamical and electronic properties of amorphous silicon: an *ab initio* molecular-dynamic study. *Phys. Rev. Lett.* **60,** 204–7.

Caravenna, F., den Hollander, F., and Pétrélis, N. (2011). Lectures on random polymers. Report 2011-07, Mathematisch Instituut, Universiteit Leiden.

Cascos, I. (2007). Depth functions based on a number of observations of a random vector. Statistic and Econometric Series 2007, Departamento de Estadística, Universidad Carlos III De Madrid. Working paper 07-29.

Cascos, I. (2010). Data depth: Multivariate statistics and geometry. In Kendall, W. S. and Molchanov, I., eds, *New Perspectives in Stochastic Geometry*, pp. 398–432. Oxford University Press, Oxford.

Cascos, I. and Molchanov, I. (2003). A stochastic order for random vectors and random sets based on the Aumann expectation. *Statist. Probab. Lett.* **63**, 295–305.

Čebašek, V., Eržen, I., Vyhnal, A., Janáček, J., Ribarič, S., and Kubínová, L. (2010). The estimation error of skeletal muscle capillary supply is significantly reduced by 3D method. *Microvascular Res.* **79**, 40–6.

Chadœuf, J., Bacro, J. N., Thébaud, G., and Labonne, G. (2008). Testing the boolean hypothesis in the non-convex case when a bounded grain can be assumed. *Environmetrics* **19**, 123–36.

Chadœuf, J., Goulard, M., and Garcia-Sanchez, L. (1996). Modeling soil surface roughness by Boolean random functions. *Microsc. Mircoanal. Microstruct.* **7**, 557–63.

Chadœuf, J., Senoussi, R., and Yao, J. F. (2000). Parametric estimation of a Boolean segment process with stochastic restoration estimation. *J. Comput. Graph. Statist.* **9**, 390–402.

Chan, M. Y., Chen, D., Chin, F. Y. L., and Wang, C. A. (2004). Construction of the nearest neighbor embracing graph of a point set. In Hagerup, T. and Katajainen, J., eds, *Algorithm Theory — SWAT 2004*, Lecture Notes in Computer Science **3111**, pp. 150–60. Springer-Verlag, Berlin.

Chan, M. Y., Chen, D. Z., Chin, F. Y. L., and Wang, C. A. (2006). Construction of the nearest neighbor embracing graph of a point set. *J. Comb. Optim.* **11**, 435–43.

Chen, A. (2008). Fast and efficient restricted Delaunay triangulation in random geometric graphs. *Internet Math.* **5**, 195–210.

Chen, F. and Kelly, P. (1992). Algorithms for generating and segmenting morphologically smooth binary images. In *Proceedings of the 26th Conference on Information Sciences and Systems*, pp. 902–6. Princeton University, Princeton.

Chiu, S. N. (1994). Mean-value formulae for the neighbourhood of the typical cell of a random tessellation. *Adv. Appl. Prob.* **26**, 565–76.

Chiu, S. N. (1995a). Aboav–Weaire's and Lewis–Rivier's law—a review. *Mater. Char.* **34**, 149–65.

Chiu, S. N. (1995b). A comment on Rivier's maximum entropy method of statistical crystallography. *J. Phys. A: Math. Gen.* **28**, 607–15.

Chiu, S. N. (1995c). Limit theorems for the time of completion of Johnson–Mehl tessellations. *Adv. Appl. Prob.* **27**, 889–910.

Chiu, S. N. (1997). A central limit theorem for linear Kolmogorov's birth–growth models. *Stoch. Process. Appl.* **66**, 97–106.

Chiu, S. N. (2003). Spatial point pattern analysis by using Voronoi diagrams and Delaunay tessellations — a comparative study. *Biometrical J.* **45**, 367–76.

Chiu, S. N. (2007). Correction to Koen's critical values in testing spatial randomness. *J. Statist. Comput. Simul.* **77**, 1001–4.

Chiu, S. N. (2010). Approximate and parametric bootstrap tests for two Poisson variates. *J. Statist. Comput. Simul,* **80**, 263–71

Chiu, S. N. and Lee, H. Y. (2002). A regularity condition and strong limit theorems for linear birth–growth processes. *Math. Nachr.* **241**, 21–7.

Chiu, S. N. and Liu, K. I. (2013). Stationarity tests for spatial point processes using discrepancies *Biometrics*. Forthcoming.

Chiu, S. N. and Molchanov, I. S. (2003). A new graph related to the directions of nearest neighbours in a point process. *Adv. Appl. Prob.* **35**, 47–55.

Chiu, S. N., Molchanov, I. S., and Quine, M. P. (2003). Maximum likelihood estimation for germination–growth processes with application to neurotransmitters data. *J. Statist. Comput. Simul.* **73**, 725–32.

Chiu, S. N. and Quine, M. P. (1997). Central limit theory for the number of seeds in a growth model in $\mathbb{R}^d$ with inhomogeneous Poisson arrivals. *Ann. Appl. Probab.* **7**, 802–14.

Chiu, S. N. and Quine, M. P. (2001). Central limit theorem for germination–growth models in $\mathbb{R}^d$ with non-Poisson locations. *Adv. Appl. Prob.* **33**, 751–5.

Chiu, S. N., Quine, M. P., and Stewart, M. (2000). Nonparametric and parametric estimation for a linear germination–growth model. *Biometrics* **56**, 755–60.

Chiu, S. N. and Stoyan, D. (1998). Estimators of distance distributions for spatial patterns. *Stat. Neerl.* **52**, 239–46.

Chiu, S. N., van de Weygaert, R., and Stoyan, D. (1996). The sectional Poisson–Voronoi tessellation is not a Voronoi tessellation. *Adv. Appl. Prob* **28**, 356–76.

Chiu, S. N. and Wang, L. (2009). Homogeneity tests for several Poisson populations. *Comput. Statist. Data Anal.* **53**, 4266–78.

Chiu, S. N. and Yin, C. C. (2000). The time of completion of a linear birth–growth model. *Adv. Appl. Prob.* **32**, 620–7.

Chiu, S. S. and Larson, R. C. (2009). Bertrand's paradox revisited: more lessons about the ambiguous word, random. *J. Ind. Syst. Eng.* **3**, 1–26.

Choquet, G. (1953/1954). Theory of capacities. *Ann. Inst. Fourier* **V**, 131–295.

Chubynsky, M. V. and Thorpe, M. F. (2001). Self-organization and rigidity in network glasses. *Curr. Opin. Solid State Mater. Sci.* **5**, 525–32.

Chung, F. and Lu, L. (2001). The diameter of sparse random graphs. *Adv. Appl. Math.* **26**, 257–79.

Chung, F. and Lu, L. (2003). The average distance in a random graph with given expected degrees. *Internet Math.* **1**, 91–114.

Ciccariello, S. (1995). Integral expressions of the derivatives of the small-angle scattering correlation function. *J. Math. Phys.* **36**, 219–46.

Ciccariello, S. (2009). The correlation functions of plane polygons. *J. Math. Phys.* **50**, 103527.

Ciccariello, S. (2010). The isotropic correlation function of planar figures: the triangle case. *J. Phys.: Conf. Ser.* **247**, 012014.

Clausius, R. (1857). Über Art der Bewegung, welche wir Wärme nennen. *Ann. Phys. (Berlin)* **100**, 353–80.

Clausius, R. (1858). Über die mittlere Länge der Wege, welche bei der Molecularbewegung gasförmiger Körper von den einzelnen Molecülen zurückgelegt werden; nebst einigen anderen Bemerkungen über die mechanischen Wärmetheorie. *Ann. Phys. (Berlin)* **181**, 239–58.

Coeurjolly, J.-F., Dereudre, D., Drouilhet, R., and Lavancier, F. (2012). Takacs–Fiksel method for stationary marked Gibbs point processes. *Scand. J. Statist.* **39**, 416–43.

Coleman, R. (1981). Size determination of transparent spheres in an opaque specimen from a slice. *J. Microsc.* **123**, 343–5.

Coleman, R. (1982). The sizes of spheres from profiles in a thin slice. I: Opaque spheres. *Biometrical J.* **24**, 273–86.

Coleman, R. (1983). The sizes of spheres from profiles in a thin slice. II: Transparent spheres. *Biometrical J.* **25**, 745–56.

Coleman, R. (1989). Random sections of a sphere. *Canad. J. Statist.* **17**, 27–39.

Coles, P. and Jones, B. (1991). A lognormal model for the cosmological mass distribution. *Mon. Not. R. Astr. Soc.* **248**, 1–13.

Collins, R. (1968). A geometrical sum rule for two dimensional fluid correlation functions. *J. Phys. C* **1,** 1461–72.

Comas, C., Mehtätalo, L., and Miina, J. (2013). Analysing space-time tree interdependence based on individual tree growth functions. *Stoch. Env. Res. Risk A.* Forthcoming. DOI: 10.1007/s00477-013-0704-3.

Corte, H. and Kallmes, O. J. (1962). Statistical geometry of a fibrous network. In Bolam, F., ed., *The Formation and Structure of Paper*, MCI Accession Number 28440, pp. 13–46. Technical Section of the British Paper and Board Makers' Association, London.

Coster, M., Arnould, X., Chermant, J. L., Moataz, A. E., and Chartier, T. (2005). A microstructural model by space tessellation for a sintered ceramic: Cerine. *Image Anal. Stereol.* **24,** 105–16.

Cowan, R. (1978). The use of the ergodic theorems in random geometry. *Suppl. Adv. Appl. Prob.* **10,** 47–57.

Cowan, R. (1979). Homogeneous line-segment processes. *Math. Proc. Camb. Phil. Soc.* **86,** 481–9.

Cowan, R. (1980). Properties of ergodic random mosaic processes. *Math. Nachr.* **97,** 89–102.

Cowan, R. (2006). A more comprehensive complementary theorem for the analysis of Poisson point processes. *Adv. Appl. Prob.* **38,** 581–601.

Cowan, R. (2010). New classes of random tessellations arising from iterative division of cells. *Adv. Appl. Prob.* **42,** 26–47.

Cowan, R., Chiu, S. N., and Holst, L. (1995). A limit theorem for the replication time of a DNA molecule. *J. Appl. Prob.* **32,** 296–303.

Cowan, R., Quine, M., and Zuyev, S. (2003). Decomposition of gamma-distributed domains constructed from Poisson point processes. *Adv. Appl. Prob.* **35,** 56–69.

Cox, D. R. (1955). Some statistical models connected with series of events. *J. Roy. Statist. Soc. B* **17,** 129–64.

Cox, D. R. and Isham, V. (1980). *Point Processes.* Chapman & Hall, London.

Cox, D. R. and Isham, V. (1988). A simple spatial-temporal model of rainfall. *Proc. Roy. Soc. London A* **415,** 317–28.

Cox, D. R. and Lewis, P. A. W. (1966). *The Statistical Analysis of Series of Events.* Methuen, London and John Wiley & Sons, Inc., New York.

Crain, I. K. (1976). Statistical analysis of geotectonics. In Merriam, D. F., ed., *Random Processes in Geology*, pp. 3–15. Springer-Verlag, Berlin.

Crain, I. K. and Miles, R. E. (1976). Monte Carlo estimates of the distribution of the random polygons determined by random lines in a plane. *J. Statist. Comput. Simul.* **4,** 293–325.

Cressie, N. (1993). *Statistics for Spatial Data.* John Wiley & Sons, Inc., New York, revised edition.

Cressie, N. and Hulting, F. L. (1992). A spatial statistical analysis of tumor growth. *J. Amer. Statist. Assoc.* **87,** 272–83.

Cressie, N. and Laslett, G. M. (1987). Random set theory and problems of modelling. *SIAM Rev.* **29,** 557–74.

Cressie, N. and Wikle, C. K. (2011). *Statistics for Spatio-Temporal Data.* John Wiley & Sons, Inc., Hoboken, New Jersey.

Crofton, M. W. (1885). Probability. In *Encyclopaedia Britannica*. IX edition.

Crompton, J. N. G., Waghorne, R. M., and Brook, G. B. (1966). The estimation of size distribution and density of precipitates from electron micrographs of thin foils. *Brit. J. Appl. Phys* **17,** 1301–5.

Cruz-Orive, L. M. (1976). Particle size-shape distributions: the general spheroid problem. I. *J. Microsc.* **107,** 235–53.

Cruz-Orive, L. M. (1978). Particle size-shape distributions: the general spheroid problem. II. *J. Microsc.* **112**, 153–67.

Cruz-Orive, L. M., Hoppeler, H., Mathieu, O., and Weibel, E. R. (1985). Stereological analysis of anisotropic structures using directional statistics. *Appl. Statist.* **34**, 14–32.

Daley, D. J. and Last, G. (2005). Descending chains, the lilypond model, and the mutual-nearest-neighbour matching. *Adv. Appl. Prob.* **37**, 604–28.

Daley, D. J., Mallows, C. L., and Shepp, L. A. (2000). A one-dimensional Poisson growth model with non-overlapping intervals. *Stoch. Process. Appl.* **90**, 223–41.

Daley, D. J., Stoyan, H., and Stoyan, D. (1999). The volume fraction of a Poisson germ model with maximally non-overlapping spherical grains. *Adv. Appl. Prob.* **31**, 610–24.

Daley, D. J. and Vere-Jones, D. (1988). *An Introduction to the Theory of Point Processes.* Springer-Verlag, New York.

Daley, D. J. and Vere-Jones, D. (2003). *An Introduction to the Theory of Point Processes. Volume I: Elementary Theory and Methods.* Springer-Verlag, New York, 2nd edition.

Daley, D. J. and Vere-Jones, D. (2008). *An Introduction to the Theory of Point Processes. Volume II: General Theory and Structure.* Springer-Verlag, New York, 2nd edition.

Davidson, R. (1974a). Construction of line-processes: second-order properties. In Harding, E. F. and Kendall, D. G., eds, *Stochastic Geometry*, pp. 55–75. John Wiley & Sons, Ltd, London.

Davidson, R. (1974b). Line-processes, roads and fibres. In Harding, E. F. and Kendall, D. G., eds, *Stochastic Geometry*, pp. 248–51. John Wiley & Sons, Ltd, London.

Davidson, R. (1974c). Stochastic processes of flats and exchangeability. In Harding, E. F. and Kendall, D. G., eds, *Stochastic Geometry*, pp. 13–45. John Wiley & Sons, Ltd, London.

Davison, A. C. and Hinkley, D. V. (1997). *Bootstrap Methods and their Application.* Cambridge University Press, Cambridge.

de Berg, M., Cheong, O., van Kreveld, M., and Overmars, M. (2008). *Computational Geometry: Algorithms and Applications.* Springer-Verlag, New York, 3rd edition.

de Gennes, P.-G. (1979). *Scaling Concepts in Polymer Physics.* Cornell University Press, Ithaca.

Debye, P., Anderson, H. R., and Brumberger, H. (1957). Scattering by an inhomogeneous solid. II. The correlation function and its application. *J. Appl. Phys.* **28**, 679–83.

DeHoff, R. T. (1965). The estimation of particle distributions from simple counting measurements made on random plane sections. *Trans. AIME* **233**, 25–9.

DeHoff, R. T. (1967). The quantitative estimation of mean surface curvature. *Trans. AIME* **239**, 617–21.

DeHoff, R. T. (1968). Curvature and topological properties of interconnected phases. In DeHoff, R. T. and Rhines, F. N., eds, *Quantitative Microscopy*, pp. 291–324. McGraw-Hill, New York.

Deijfen, M. (2003). Asymptotic shape in a continuum growth model. *Adv. Appl. Prob.* **35**, 303–18.

Deijfen, M. (2009). Stationary random graphs with prescribed iid degrees on a spatial Poisson process. *Electron. Commun. Probab.* **14**, 81–9.

Deijfen, M., Häggström, O., and Holroyd, A. E. (2012). Percolation in invariant Poisson graphs with i.i.d. degrees. *Ark. Mat.* **50**, 41–58.

Delaney, G. W., Di Matteo, T., and Aste, T. (2010). Combining tomographic imaging and DEM simulations to investigate the structure of experimental sphere packings. *Soft Matter* **6**, 2992–3006.

Delesse, M. A. (1847). Procede mecanique pour determiner la composition des roches. *C. R. Acad. Sci. (Paris)* **25**, 544–5.

Delfiner, P. (1972). A generalization of the concept of size. *J. Microsc.* **95**, 203–16.

Demichel, Y., Estrade, A., Kratz, M., and Samorodnitsky, G. (2011). How fast can the chord length distribution function decay? *Adv. Appl. Prob.* **43**, 504–23.

den Hollander, F. (2009). *Random Polymers*. Lecture Notes in Mathematics **1974**. Springer-Verlag, Berlin.

Deng, M. and Dodson, C. T. J. (1994). *Paper: an Engineered Stochastic Structure*. Tappi Press, Atlanta.

Dereudre, D., Drouilhet, R., and Georgii, H.-O. (2012). Existence of Gibbsian point processes with geometry-dependent interactions. *Probab. Theory Related Fields* **153**, 643–70.

Dereudre, D. and Lavancier, F. (2009). Campbell equilibrium equation and pseudo-likelihood estimation for non-hereditary Gibbs point processes. *Bernoulli* **15**, 1368–96.

Dereudre, D. and Lavancier, F. (2011). Practical simulation and estimation for Gibbs Delaunay–Voronoi tessellations with geometric hardcore interaction. *Comput. Statist. Data Anal.* **55**, 498–519.

Diggle, P. J. (1979). On parameter estimation and goodness-of-fit testing for spatial point-patterns. *Biometrics* **35**, 87–101.

Diggle, P. J. (1981). Binary mosaics and the spatial pattern of heather. *Biometrics* **37**, 531–9.

Diggle, P. J. (1983). *Statistical Analysis of Spatial Point Patterns*. Academic Press, London.

Diggle, P. J. (2003). *Statistical Analysis of Spatial Point Patterns*. Edward Arnold, London, 2nd edition.

Diggle, P. J. (2007). Spatio-temporal point processes: methods and applications. In Finkenstädt, B., Held, L., and Isham, V., eds, *Statistical Methods for Spatio-Temporal Systems*, pp. 1–45. Chapman & Hall/CRC, Boca Rota.

Diggle, P. J., Fiksel, T., Grabarnik, P., Ogata, Y., Stoyan, D., and Tanemura, M. (1994). On parameter estimation for pairwise interaction processes. *Int. Statist. Rev.* **62**, 99–117.

Diggle, P. J., Gates, D. J., and Stibbard, A. (1987). A non-parametric estimator for pairwise-interaction point processes. *Biometrika* **74**, 763–70.

Diggle, P. J. and Gratton, R. J. (1984). Monte Carlo methods of inference for implicit statistical models (with discussion). *J. Roy. Statist. Soc. B* **46**, 193–227.

Diggle, P. J. and Milne, R. K. (1983). Bivariate Cox processes: Some models for bivariate spatial point-patterns. *J. Roy. Statist. Soc. B* **45**, 11–21.

Diggle, P. J. and ter Braak, C. J. F. (1982). Point sampling of binary mosaics in ecology. In Ranneby, B., ed., *Statistics in Theory and Practice. Essays in Honour of Bertil Matérn*, pp. 107–22. Swedish University of Agricultural Sciences, Umea.

Dillard, T., N'guyen, F., Maire, E., Salvo, L., Forest, S., Bienvenu, Y., Bartout, J.-D., Croset, M., Dendievel, R., and Cloetens, P. (2005). 3D quantitative image analysis of open-cell nickel foams under tension and compression loading using X-ray microtomography. *Philos. Mag.* **85**, 2147–75.

Dirichlet, G. L. (1850). Über die Reduction der positiven quadratischen Formen mit drei unbestimmten ganzen Zahlen. *J. Reine und Angew. Math.* **40**, 209–27.

Dodson, C. T. J. and Sampson, W. W. (1999). Spatial statistics of stochastic fiber networks. *J. Stat. Phys.* **96**, 447–58.

Döge, G. (2001). Perfect simulation for random sequential adsorption of d-dimensional spheres with random radii. *J. Statist. Comput. Simul.* **69**, 141–56.

Doležal, F. (1982). The use of stereology for the evaluation of crack patterns in agricultural soils. In Saxl, I., ed., *Proceedings of the Colloquium on Mathematical Morphology, Stereology and Image Analysis, 14–16 September 1982, Prague*, pp. 305–7. Akademia, Prague.

Dominguez, M. and Torres, L. (1997). Analysis and synthesis of textures through the inference of Boolean functions. *Signal Processing* **59**, 1–16.

Donev, A., Torquato, S., and Stillinger, F. H. (2005). Neighbor list collision-driven molecular dynamics simulation for nonspherical hard particles. I. Algorithmic details. *J. Comput. Phys.* **202**, 737–64.

Dorogovtsev, S. N. and Mendes, J. F. F. (2002). Evolution of networks. *Adv. in Phys.* **51**, 1079–187.

Dousse, O., Franceschetti, M., Macris, N., Meester, R., and Thiran, P. (2006). Percolation in the signal to interference ratio graph. *J. Appl. Prob.* **43**, 552–62.

Douven, I., Decock, L., Dietz, R., and Égré, P. (2013). Vagueness: a conceptual spaces approach. *J. Philos. Logic* **42,** 137–60.

Dozzi, M., Merzbach, E., and Schmidt, V. (2001). Limit theorems for sums of random fuzzy sets. *J. Math. Anal. Appl.* **259,** 554–65.

Drees, H. and Reiss, R.-D. (1992). Tail behaviour in Wicksell's corpuscle problem. In Galambos, J. and Katai, I., eds, *Probability and Applications*, pp. 205–20. Kluwer Academic Publishers, Dortrecht.

Du, Q., Faber, V., and Gunzburger, M. (1999). Centroidal Voronoi tessellations: Applications and algorithms. *SIAM Rev.* **41,** 637–76.

Dubois, D. and Prade, H., eds (2000). *Fundamentals of Fuzzy Sets*. Handbooks of Fuzzy Sets Series **7.** Kluwer Academic Publishers, Dordrecht.

Durrett, R. (2007). *Random Graph Dynamics*. Cambridge University Press, Cambridge.

Durrett, R. (2010). *Probability. Theory and Examples*. Cambridge University Press, Cambridge, 4th edition.

Duvalian, A. V. (1971). A method for the approximate determination of the variance of dihedral angles in alloys (in Russian). *Zavod. Lab.* **37,** 939–41.

Edelsbrunner, H. (1987). *Algorithms in Combinatorial Geometry*. Springer-Verlag, Berlin.

Eggemann, N. and Noble, S. D. (2011). The clustering coefficient of a scale-free random graph. *Discrete Appl. Math.* **159,** 953–65.

El-Azab, A., Deng, J., and Tang, M. (2007). Statistical characterization of dislocation ensembles. *Philos. Mag.* **87,** 1201–23.

Elsner, A., Wagner, A., Aste, T., Hermann, H., and Stoyan, D. (2009). Specific surface area and volume fraction of the cherry-pit model with packed pits. *J. Phys. Chem. B* **113,** 7780–4.

Emery, X., Kracht, W., Egaña, Á., and Garrido, F. (2012). Using two-point set statistics to estimate the diameter distribution in Boolean models with circular grains. *Math. Geosci.* **44,** 805–22.

Erdős, P. and Rényi, A. (1959). On random graphs I. *Publ. Math. Debrecen* **6,** 290–7.

Erdős, P. and Rényi, A. (1960). On the evolution of random graphs. *Publ. Math. Inst. Hungar. Acad. Sci.* **5,** 17–61.

Estrade, A., Iribarren, I., and Kratz, M. (2012). Chord-length distribution functions and Rice formulae. Application to random media. *Extremes* **15,** 333–52.

Evans, J. W. (1993). Random and cooperative sequential adsorption. *Rev. Modern Phys.* **65,** 1281–1329.

Exner, H. E. (2004). Stereology and 3D microscopy: Useful alternatives or competitors in the quantitative analysis of microstructures? *Image Anal. Stereol.* **23,** 73–82.

Fairclough, A. R. N. and Davies, G. A. (1990). Poisson line processes in 2 space to simulate the structure of porous media: methods of generation, statistics and applications. *Chem. Eng. Comm.* **2,** 23–48.

Falconer, K. J. (1990). *Fractal Geometry. Mathematical Foundations and Applications*. John Wiley & Sons, Inc., New York.

Federer, H. (1959). Curvature measures. *Trans. Amer. Math. Soc.* **93,** 418–91.

Fellous, A., Granara, J., and Krickeberg, K. (1978). Statistics of stationary oriented line Poisson processes in the plane. In Miles, R. E. and Serra, J., eds, *Geometrical Probability and Biological Structures: Buffon's 200th Anniversary*, Lecture Notes in Biomathematics **23,** pp. 295–9. Springe-Verlag, Berlin.

Fienberg, S. E. (2010a). Introduction to papers on the modeling and analysis of network data. *Ann. Appl. Statist.* **4,** 1–4.

Fienberg, S. E. (2010b). Introduction to papers on the modeling and analysis of network data II. *Ann. Appl. Statist.* **4,** 533–4.

Finney, J. L. (1975). Volume occupation, environment and accessibility in proteins. The problem of the protein surface. *J. Mol. Biol.* **96,** 721–32.

Finney, J. L. (1979). A procedure for the construction of Voronoi polyhedra. *J. Comput. Phys.* **32,** 137–43.

Fischer, R. A. and Miles, R. E. (1973). The role of spatial pattern in the competition between crop plants and weeds: A theoretical analysis. *Math. Biosci.* **18,** 335–50.

Fisher, N. I., Lewis, T., and Embleton, B. J. J. (1987). *Statistical Analysis of Spherical Data.* Cambridge University Press, Cambridge.

Flaxman, A. D., Frieze, A. M., and Vera, J. (2006). A geometric preferential attachment model of networks. *Internet Math.* **3,** 187–205.

Flaxman, A. D., Frieze, A. M., and Vera, J. (2007). A geometric preferential attachment model of networks II. *Internet Math.* **4,** 87–112.

Fleischer, F., Eckel, S., Schmidt, I., Schmidt, V., and Kazda, M. (2006). Point process modelling of root distribution in pure stands of *Fagus sylvatica* and *Picea abies. Can. J. Forest Res.* **36,** 227–37.

Fleischer, F., Gloaguen, C., Schmidt, V., and Voss, F. (2009). Simulation of the typical Poisson–Voronoi–Cox–Voronoi cell. *J. Statist. Comput. Simul.* **79,** 939–57.

Fletcher, N. D. and Evans, A. N. (2005). Texture segmentation using area morphology local granulometries. In Ronse, C., Najman, L., and Decencière, E., eds, *Mathematical Morphology: 40 Years On,* Computational Imaging and Vision **30,** pp. 367–76. Springer-Verlag, Dordrecht.

Flory, P. J. (1953). *Principles of Polymer Chemistry.* Cornell University Press, Ithaca.

Flory, P. J. (1969). *Statistical Mechanics of Chain Molecules.* John Wiley & Sons, Inc., New York.

Flory, P. J. (1976). Statistical thermodynamics of random networks. *Proc. R. Soc. London A* **351,** 351–80.

Flynn, D. C., Nulton, J. D., Salamon, P., Frey, T. G., Rabinovitch, A., Sun, M. G., and Baljon, A. R. C. (2009). A stereological unfolding method for the study of the mitochondrial network. *Image Anal. Stereol.* **28,** 11–22.

Fortune, S. (1992). Voronoi diagrams and Delaunay triangulations. In Du, D.-Z. and Wang, F. K., eds, *Computing in Euclidean Geometry,* pp. 193–233. World Scientific, Singapore.

Foxall, R. and Baddeley, A. (2002). Nonparametric measures of association between a spatial point process and a random set, with geological applications. *J. Roy. Statist. Soc. C* **51,** 165–82.

Franceschetti, M. and Meester, R. (2008). *Random Networks for Communication.* Cambridge University Press, Cambridge.

Franken, P., König, D., Arndt, U., and Schmidt, V. (1981). *Queues and Point Processes.* Akademie-Verlag, Berlin and John Wiley & Sons, Ltd, Chichester.

Franklin, J. N. (1977). Some stereological principles in morphometric cytology. *SIAM J. Appl. Math.* **33,** 267–78.

Franklin, J. N. (1981). Confidence intervals for stereological estimators with infinite variance. *SIAM J. Appl. Math.* **40,** 179–90.

Frenkel, D. and Smit, B. (2002). *Understanding Molecular Simulation: From Algorithms to Applications.* Academic Press, San Diego, 2nd edition.

Freudenthal, A. (1950). *The Inelastic Behavior of Engineering Materials and Structures.* John Wiley & Sons, Inc., New York.

Frieden, B. R. (1983). *Probability, Statistical Optics and Data testing.* Springer-Verlag, New York.

Frisch, H. L. and Stillinger, F. H. (1963). Contribution to the statistical geometric basis of radiation scattering. *J. Chem. Phys.* **38,** 2200–7.

Fu, K.-a. and Zhang, L.-x. (2008). Strong limit theorems for random sets and fuzzy random sets with slowly varying weights. *Inform. Sci.* **178,** 2648–60.

Fullman, R. L. (1953). Measurement of particle sizes in opaque bodies. *J. Metals* **5,** 447–52.

Galbraith, R. F. (2005). *Statistics for Fission Track Analysis*. Chapman & Hall/CRC, Boca Rota.

Gärdenfors, P. (2000). *Conceptual Spaces: The Geometry of Thought*. MIT Press, Cambridge, MA.

Gardner, R. J., Kiderlen, M., and Milanfar, P. (2006). Convergence of algorithms for reconstructing convex bodies and directional measures. *Ann. Statist.* **34**, 1331–74.

Gates, D. J. and Westcott, M. (1986). Clustering estimates in spatial point processes with stable potentials. *Ann. Inst. Statist. Math.* **38**, 123–35.

Gavrikov, V. L., Grabarnik, P. Y., and Stoyan, D. (1993). Trunk-top relations in a Siberian pine forest. *Biometrical J.* **35**, 487–98.

Gavrilova, M. L., ed. (2008). *Generalized Voronoi Diagram: A Geometry-Based Approach to Computational Intelligence*. Studies in Computational Intelligence **158**. Springer-Verlag, Berlin.

Gelfand, A. E., Diggle, P. J., Fuentes, M., and Guttorp, P., eds (2010). *Handbook of Spatial Statistics*. CRC, Boca Raton.

Gentle, J. E. (2003). *Random Number Generation and Monte Carlo Methods*. Springer-Verlag, New York, 2nd edition.

George, E. I. (1987). Sampling random polygons. *J. Appl. Prob.* **24**, 557–73.

Georgii, H.-O. (1976). Canonical and grand canonical Gibbs states for continuum systems. *Comm. Math. Phys.* **48**, 31–51.

Georgii, H.-O. (1988). *Gibbs Measures and Phase Transition*. de Gruyter, Berlin.

Georgii, H.-O. (2000). Phase transition and percolation in Gibbsian particle models. In Mecke, K. R. and Stoyan, D., eds, *Statistical Physics and Spatial Statistics: the Art of Analyzing and Modeling Spatial Structures and Pattern Formation*, Lecture Notes in Physics **554**, pp. 267–94. Springer-Verlag, Berlin.

Georgii, H.-O., Häggström, O., and Maes, C. (2001). The random geometry of equilibrium phases. In Domb, C. and Lebowitz, J., eds, *Phase Transitions and Critical Phenomena*, pp. 1–142. Academic Press, London.

Geyer, C. J. (1992). Practical Markov chain Monte Carlo (with discussion). *Statist. Sci.* **7**, 473–511.

Ghorbani, M. (2012). Cauchy cluster process. *Metrika*. Forthcoming. DOI: 10.1007/s00184-012-0411-y.

Giger, H. (1967). Ermittlung der mittleren Maßzahlen von Partikeln eines Körpersystems durch Messungen auf dem Rand eines Schnittbereiches. *Z. angew. Math. Phys.* **18**, 883–8.

Giger, H. and Riedwyl, H. (1970). Bestimmung der Größenverteilung von Kugeln aus Schnittkreisradien. *Biometrisch Z.* **12**, 156–68.

Gilbert, E. N. (1959). Random graphs. *Ann. Math. Statist.* **30**, 1141–4.

Gilbert, E. N. (1961). Random plane networks. *J. SIAM* **9**, 533–43.

Gilbert, E. N. (1962). Random subdivisions of space into crystals. *Ann. Math. Statist.* **33**, 958–72.

Gilbert, E. N. (1967). Random plane networks and needle-shaped crystals. In Noble, B., ed., *Applications of Undergraduate Mathematics in Engineering*, chapter 16. Macmillan, New York.

Gill, R. D. (1994). Lectures on survival analysis. In Bernard, P., ed., *Lectures on Probability Theory: Ecole d'Eté de Probabilités de Saint-Flour XXII-1992*, Lecture Notes in Mathematics **1581**, pp. 115–241. Springer-Verlag, Berlin.

Gille, W. (1988). The chord length distribution on parallelepipeds with their limiting cases. *Exp. Technik Phys.* **36**, 197–208.

Gille, W. (2002). The set covariance of a dead leaves model. *Adv. Appl. Prob.* **34**, 11–20.

Gille, W. (2011). Scattering properties and structure functions of Boolean models. *Comput. Struct.* **89**, 2309–15.

Gille, W. (2014). *Particles, Puzzles and Scattering Patterns – Mysteries of Small-Angle Scattering*. Forthcoming.

Girling, A. J. (1982). Approximate variances associated with random configurations of hard spheres. *J. Appl. Prob.* **19**, 588–96.

Glagolev, A. A. (1933). On the geometrical methods of quantitative mineralogic analysis of rocks. *Trans. Inst. Econ. Min. Moscow* **59**, 1–47.

Glatter, O. (1979). The interpretation of real-space information from small-angle scattering experiments. *J. Appl. Cryst.* **12**, 166–75.

Gloaguen, C., Fleischer, F., Schmidt, H., and Schmidt, V. (2006). Fitting of stochastic telecommunication network models, via distance measures and Monte Carlo tests. *Telecommunication Systems* **31**, 353–77.

Glötzl, E. (1980). Lokale Energien und Potentiale für Punktprozesse. *Math. Nachr.* **96**, 196–206.

Glötzl, E. (1981). Time-reversible and Gibbsian point processes I. Markovian spatial birth and death processes on a general phase space. *Math. Nachr.* **102**, 217–22.

Gneiting, T. and Guttorp, P. (2010). Continuous parameter stochastic process theory. In Gelfand, A. E., Diggle, P. J., Fuentes, M., and Guttorp, P., eds, *Handbook of Spatial Statistics*, pp. 17–28. CRC Press, Boca Raton.

Gokhale, A. M. (1990). Unbiased estimation of curve length in 3-D using vertical slices. *J. Microsc.* **159**, 133–41.

Gokhale, A. M. (1992). Estimation of length density L_v from vertical slices of unknown thickness. *J. Microsc.* **167**, 1–8.

Gokhale, A. M. (1993). Utility of the horizontal slice for stereological characterization of lineal features. *J. Microsc.* **170**, 3–8.

Gokhale, A. M., Tewari, A., and Garmestani, H. (2005). Constraints on microstructural two-point correlation functions. *Scripta Materialia* **53**, 989–93.

Goulard, M., Chadœuf, J., and Bertuzzi, P. (1994). Random Boolean functions: non-parametric estimation of the intensity. Application to soil surface roughnessy. *Statistics* **25**, 123–36.

Goutsias, J. and Batman, S. (2000). Morphological methods for biomedical image analysis. In Sonka, M. and Fitzpatrick, J. M., eds, *Handbook of Medical Imaging. Volume 2: Medical Image Processing and Analysis*, pp. 175–272. SPIE Press, Bellingham, Washington.

Goutsias, J. and Heijmans, H. J. A. M., eds (2000). *Mathematical Morphology*. IOS Press, Amsterdam.

Goutsias, J., Mahler, R. P. S., and Nguyen, H. T. (1997). *Random Sets: Theory and Applications*. Springer-Verlag, New York.

Goutsias, J. and Sivakumar, K. (1998). A multiresolution morphological approach to stochastic image modeling. *CWI Quarterly* **11**, 347–69.

Grabarnik, P., Pagès, L., and Bengough, A. G. (1998). Geometrical properties of simulated maize root systems: consequences for length density and intersection density. *Plant Soil* **200**, 157–67.

Grandell, J. (1976). *Doubly Stochastic Poisson Processes*. Lecture Notes in Mathematics **529**. Springer-Verlag, Berlin.

Grandell, J. (1981). Some recent developments of theory and application of Cox processes. In Bereanu, B., Grigorescu, S., Iosifescu, M., and Postelnicu, T., eds, *Sixth Conf. Prob. Theory, Brasov 1979*, pp. 288–9. Bucuresti.

Gray, N. H., Anderson, J. B., Devine, J. D., and Kwasnik, J. M. (1976). Topological properties of random crack networks. *Math. Geol* **8**, 617–26.

Greco, A., Jeulin, D., and Serra, J. (1979). The use of the texture analyser to study sinter structure: application to the morphology of calcium ferrites encountered in basic sinters of rich iron ores. *J. Microsc.* **116**, 199–211.

Greeley, R. (1987). *Planetary Landscapes*. Allen & Unwin, Boston.

Greig-Smith, P. (1952). The use of random and contiguous quadrats in the study of the structure of plant communities. *Ann. Bot.* **16,** 293–316.

Greig-Smith, P. (1983). *Quantitative Plant Ecology.* Blackwell, Oxford, 3rd edition.

Grimmett, G. (1999). *Percolation.* Grundlehren der mathematischen Wissenschaften **321**. Springer-Verlag, Berlin, 2nd edition.

Groos, J. and Kopp-Schneider, A. (2006). Application of a color-shift model with heterogeneous growth to a rat hepatocarcinogenesis experiment. *Math. Biosci.* **202,** 248–68.

Groos, J. and Kopp-Schneider, A. (2010). Application of a two-phenotype color-shift model with heterogeneous growth to a rat hepatocarcinogenesis experiment. *Math. Biosci.* **224,** 95–100.

Gruber, P. M. and Wills, J. M. (1993). *Handbook of Convex Geometry.* Elsevier Science, Amsterdam.

Gu, M. G. and Zhu, H.-T. (2001). Maximum likelihood estimation for spatial models by Markov chain Monte Carlo stochastic approximation. *J. Roy. Statist. Soc. B* **63,** 339–55.

Guan, Y. (2008). A KPSS test for stationarity for spatial point processes. *Biometrics* **64,** 800–6.

Guderlei, R., Klenk, S., Mayer, J., Schmidt, V., and Spodarev, E. (2007). Algorithms for the computation of the Minkowski functionals of deterministic and random polyconvex sets. *Image Vision Comput.* **25,** 464–74.

Guibas, L. J. and Stolfi, J. (1988). Ruler, compass and computer: the design and analysis of geometric algorithms. In Earnshaw, R., ed., *Theoretical Foundation of Computer Graphics and CAD,* pp. 111–65. Springer-Verlag, Berlin.

Guinier, A. and Fournet, G. (1995). *Small-angle Scattering of X-rays.* John Wiley & Sons, Inc., New York.

Gundersen, H. J. G. (1979). Estimation of tubule or cylinder L_V, S_V, and V_V on thick sections. *J. Microsc.* **117,** 333–46.

Gut, A. (2013). *Probability: A Graduate Course.* Springer Science + Business Media, New York, 2nd edition.

Guttorp, P. (2007). Discussion of 'modern statistics for spatial point processes' by J. Møller and R. P. Waagepetersen. *Scand. J. Statist.* **34,** 692–3.

Guttorp, P. and Thorarinsdottir, T. L. (2012). What happened to discrete chaos, the Quenouille process, and the sharp Markov property? Some history of stochastic point processes. *Int. Stat. Rev.* **80,** 253–68.

Haas, A., Matheron, G., and Serra, J. (1967). Morphologie mathematique et granulometries en place I, II. *Ann. Mines* **11,** 736–53; **12,** 767–82.

Hadwiger, H. (1957). *Vorlesungen über Inhalt, Oberfläche und Isoperimetrie.* Springer-Verlag, Berlin.

Hadwiger, H. and Giger, H. (1968). Über Treffzahlwahrscheinlichkeiten im Eikörperfeld. *Z. Wahrscheinlichkeitsth. verw. Geb.* **10,** 329–34.

Häggström, O. and Meester, R. (1996). Nearest neighbor and hard sphere models in continuum percolation. *Random Structures Algorithms* **9,** 295–315.

Hahn, U. (1995). On the precision of some estimators of the number of cells per unit area in planar tessellations. Unpublished manuscript.

Hahn, U. and Lorz, U. (1994a). On the precision of some stereological estimators of the spatial Poisson–Voronoi tessellation. *Acta Stereol.* **13,** 245–50.

Hahn, U. and Lorz, U. (1994b). Stereological analysis of the spatial Poisson–Voronoi tessellation. *J. Microsc.* **175,** 176–85.

Hahn, U., Micheletti, A., Pohlink, R., Stoyan, D., and Wendrock, H. (1999). Stereological analysis and modelling of gradient structures. *J. Microsc.* **195,** 113–24.

Hall, P. (1985a). Distribution of size, structure and number of vacant regions in a high-intensity mosaic. *Z. Wahrscheinlichkeitsth. verw. Geb.* **70,** 237–61.

Hall, P. (1985b). On continuum percolation. *Ann. Probab.* **13**, 1250–66.

Hall, P. (1988). *Introduction to the Theory of Coverage Processes.* John Wiley & Sons, Inc., New York.

Hall, P. and Smith, R. L. (1988). The kernel method for unfolding sphere distributions. *J. Comput. Phys.* **74**, 409–21.

Hall, P. and Ziegel, J. (2011). Distribution estimators and confidence intervals for stereological volumes. *Biometrika* **98**, 417–31.

Hamilton, W. D. (1971). Geometry for the selfish herd. *J. Theor. Biol.* **31**, 295–311.

Hanisch, K.-H. (1981). On classes of random sets and point processes. *Serdica* **7**, 160–7.

Hanisch, K.-H. (1982). On inversion formulae for n-fold Palm distributions of point processes in LCS-spaces. *Math. Nachr.* **106**, 171–9.

Hanisch, K.-H. (1983). On stereological estimation of second-order characteristics and of hard-core distances of systems of sphere-centres. *Biometrical J.* **25**, 731–43.

Hanisch, K.-H. (1984a). On Palm and second-order quantities of point processes and germ–grain models. Technical Report Wissenschaftl. Sitzungen Stochastik WSS-01/84, Akademie der Wissenschaften der DDR, Berlin.

Hanisch, K.-H. (1984b). Scattering analysis of point processes and random measures. *Math. Nachr.* **117**, 235–45.

Hanisch, K.-H. (1984c). Some remarks on estimators of the distribution function of nearest-neighbour distance in stationary spatial point-patterns. *Math. Operationsf. Statist., ser. Statistics* **15**, 409–12.

Hanisch, K.-H. (1985). On the second order analysis of stationary and isotropic fibre processes by line-intercept methods. In Nagel, W., ed., *Geobild '85: Workshop on Geometrical Problems of Image Processing, Georgenthal (GDR), January 14–18, 1985: Proceedings*, pp. 141–6. Wissenschaftliche Beiträge der Friedrich-Schiller-Universität Jena.

Hanisch, K.-H., Klimanek, P., and Stoyan, D. (1985). Stereological analysis of dislocation arrangements in crystals from TEM images. *Cryst. Res. Technol.* **20**, 921–30.

Hanisch, K.-H. and Stoyan, D. (1980). Stereological estimation of the radial distribution function of centres of spheres. *J. Microsc.* **122**, 131–41.

Hansen, J.-P. and McDonald, I. R. (2006). *Theory of Simple Liquids.* Academic Press, London, 3rd edition.

Hansen, M. B., Baddeley, A. J., and Gill, R. D. (1999). First contact distribution for spatial patterns: regularity and estimation. *Adv. Appl. Prob.* **31**, 15–33.

Hansen, M. B., Gill, R. D., and Baddeley, A. (1996). Kaplan–Meier type estimators for linear contact distributions. *Scand. J. Statist* **23**, 129–55.

Harker, D. and Parker, E. R. (1945). Grain shape and grain growth. *Trans. Amer. Soc. Metals* **34**, 156–73.

Hasegawa, M. and Tanemura, M. (1976). On the pattern of space division by territories. *Ann. Inst. Math.* **28**, 509–19.

Hasegawa, M. and Tanemura, M. (1980). Spatial patterns of territories. In *Recent Developments in Statistical Inference and Data Analysis*, pp. 73–8. North-Holland, Amsterdam.

Hasegawa, M., Tanemura, M., and Takiguchi, S. (1981). Spatial patterns in ecology. In *Int. Roundtable Congress 50th Anniversary Jap. Statist. Soc. 1981*, pp. 146–61. Japan Statist. Soc.

Hayen, A. and Quine, M. P. (2000a). Calculating the proportion of triangles in a Poisson–Voronoi tessellation of the plane. *J. Statist. Comput. Simul.* **32**, 67–74.

Hayen, A. and Quine, M. P. (2000b). The proportion of triangles in a Poisson–Voronoi tessellation of the plane. *Adv. Appl. Prob.* **32**, 67–74.

Hazlett, R. D. (1997). Statistical characterization and stochastic modeling of pore networks in relation to fluid flow. *Math. Geol.* **29**, 801–22.

Heijmans, H. J. A. M. (1994). *Morphological Image Operators*. Academic Press, Boston, MA.

Heinrich, L. (1992a). *Mixing properties of Gibbsian point processes and asymptotic normality of Takacs–Fiksel estimates*. Preprint 92-051, Universität Bielefeld.

Heinrich, L. (1992b). On existence and mixing properties of germ–grain models. *Statistics* **23**, 271–86.

Heinrich, L. (1993). Asymptotic properties of minimum contrast estimators for parameters of Boolean models. *Metrika* **31**, 349–60.

Heinrich, L. (1998). Contact and chord length distribution of a stationary Voronoi tessellation. *Adv. Appl. Prob.* **30**, 603–18.

Heinrich, L. (2005). Large deviations of the empirical volume fraction for stationary Poisson grain models. *Ann. Appl. Probab.* **15**, 392–420.

Heinrich, L. and Molchanov, I. (1994). Some limit theorems for extremal and union shot noise processes. *Math. Nachr* **168**, 139–59.

Heinrich, L. and Muche, L. (2008). Second-order properties of the point process of nodes in a stationary Voronoi tessellation. *Math. Nachr.* **281**, 350–75. (Erratum: 2010, vol. 283, pp. 1674–6).

Heinrich, L. and Schmidt, V. (1985). Normal convergence of multidimensional shot noise and rates of their convergence. *Adv. Appl. Prob.* **17**, 709–30.

Heinrich, L. and Schüle, E. (1995). Generation of the typical cell of a non-Poissonian Johnson–Mehl tessellation. *Stochastic Models* **11**, 541–60.

Heinrich, L. and Spiess, M. (2009). Berry–Esseen bounds and Cramér-type large deviations for the volume distribution of Poisson cylinder processes. *Lith. Math. J.* **49**, 381–98.

Heinrich, L. and Werner, M. (2000). Kernel estimation of the diameter distribution in Boolean models with spherical grains. *J. Nonparametr. Stat.* **12**, 147–76.

Heinrich, P., Stoica, R. S., and Tran, V. C. (2012). Level sets estimation and Vorob'ev expectation of random compact sets. *Spatial Statistics* **2**, 47–61.

Heinzer, S., Krucker, T., Stampanoni, M., Abela, R., Meyer, E. P., Schuler, A., Schneider, P., and Müller, R. (2006). Hierarchical microimaging for multiscale analysis of large vascular networks. *NeuroImage* **32**, 626–36.

Hermann, H. (1991). *Stochastic Models of Heterogeneous Materials*. Materials Science Forum **78**. Trans Tech Publications Inc., Zürich.

Hermann, H. (1998). Transformation kinetics in partially crystallized amorphous alloys – a theoretical approach. *Europhys. Lett.* **41**, 245–50.

Hermann, H., Elsner, A., Hecker, M., and Stoyan, D. (2005). Computer simulated dense-random packing models as approach to the structure of porous low-k dielectrics. *Microelec. Eng.* **81**, 535–43.

Hermann, H., Elsner, A., and Stoyan, D. (2013). Surface area and volume fraction of random open-pore systems. *Submitted*.

Hermann, H., Wendrock, H., and Stoyan, D. (1989). Cell-area distributions of planar Voronoi mosaics. *Metallography* **23**, 189–200.

Hernandez, G., Leon, R., Salinas, L., and Dimnet, E. (2012). A fragmentation model with neighborhood interaction. *Appl. Math. Model.* **36**, 1694–702.

Hestroffer, D., Berthier, J., Descamps, P., Tanga, P., Cellino, A., Lattanzi, M., Di Martino, M., and Zappala, V. (2002). Asteroid (216) Kleopatra. Test of the radar-derived shape model. *Astron. Astrophys.* **392**, 729–33.

Heveling, M. and Last, G. (2005). Characterization of Palm measures via bijective point-shifts. *Ann. Probab.* **33**, 1698–1715.

Heveling, M. and Last, G. (2006). Existence, uniqueness, and algorithmic computation of general lily-pond systems. *Random Structures Algorithms* **29**, 338–50.

Heveling, M. and Reitzner, M. (2009). Poisson–Voronoi approximation. *Ann. Appl. Probab.* **19**, 719–36.

Hilfer, R. (1991). Geometric and dielectric characterization of porous media. *Phys. Rev. B* **44**, 60–75.

Hilfer, R. (2000). Local porosity theory and stochastic reconstruction for porous media. In Mecke, K. R. and Stoyan, D., eds, *Statistical Physics and Spatial Statistics: the Art of Analyzing and Modeling Spatial Structures and Pattern Formation*, Lecture Notes in Physics **554**, pp. 203–41. Springer-Verlag, Berlin.

Hilhorst, H. J. (2006). Planar Voronoi cells: the violation of Aboav's law explained. *J. Phys. A: Math. Gen.* **39**, 7227–43.

Hilhorst, H. J. (2007). New Monte Carlo method for planar Poisson–Voronoi cells. *J. Phys. A: Math. Gen.* **40**, 2615–38.

Hilhorst, H. J. (2008). Statistical properties of planar Voronoi tessellations. *Eur. Phys. J. B* **64**, 437–41.

Hill, B. J., Kendall, W. S., and Thönnes, E. (2012). Fibre-generated point processes and fields of orientations. *Ann. Appl. Statist.* **6**, 994–1020.

Hilliard, J. E. (1962). Specification and measurement of microstructural anisotropy. *Trans. Metall. Soc. Amer. Inst. Met. Eng.* **224**, 1201–11.

Hilliard, J. E. (1967). Determination of structural anisotropy. In Elias, H., ed., *Stereology: Proceedings of the Second International Congress for Stereology, Chicago, 1967*, pp. 219–27. Springer-Verlag, Berlin.

Hilliard, J. E. and Lawson, L. R. (2003). *Stereology and Stochastic Geometry*. Kluwer Academic Publishers, Dordrecht.

Hinde, A. L. and Miles, R. E. (1980). Monte Carlo estimates of the distributions of the random polygons of the Voronoi tessellation with respect to a Poisson process. *J. Statist. Comput. Simul.* **10**, 205–23.

Hjelle, Ø. and Dæhlen, M. (2006). *Triangulations and Applications*. Springer-Verlag, Berlin.

Ho, L. P. and Chiu, S. N. (2006). Testing the complete spatial randomness by Diggle's test without an arbitrary upper limit. *J. Statist. Comput. Simul.* **76**, 585–91.

Ho, L. P. and Chiu, S. N. (2009). Using weight functions in spatial point pattern analysis with application to plant ecology data. *Comm. Statist. Simulation Comput* **38**, 269–87.

Hodder, I. and Orton, C. (1976). *Spatial Analysis in Archaeology*. Cambridge University Press, Cambridge.

Hoffmann, L. M. (2007). Intersection densities of nonstationary Poisson processes of hypersurfaces. *Adv. Appl. Prob.* **39**, 307–17.

Holroyd, A. E. and Peres, Y. (2005). Extra heads and invariant allocations. *Ann. Probab.* **33**, 31–52.

Honda, H. (1983). Geometric models for cells in tissues. *Int. Rev. Cytology* **81**, 191–248.

Horálek, V. (1988). A note on the time-non-homogeneous Johnson–Mehl tessellation. *Adv. Appl. Prob.* **20**, 684–5.

Horálek, V. (1990). ASTM grain-size model and related random tessellation models. *Mater. Char.* **25**, 263–84.

Horgan, G. W. and Young, I. M. (2000). An empirical stochastic model for the geometry of two-dimensional crack growth in soil (with Discussion). *Geoderma* **96**, 263–76.

Hörig, M. and Redenbach, C. (2012). The maximum volume hard subset model for Poisson processes: simulation aspects. *J. Statist. Comput. Simul.* **82**, 107–21.

Hosemann, R. and Bagchi, S. N. (1962). *Direct Analysis of Diffraction by Matter*. North Holland, Amsterdam.

Howard, V. and Reed, M. G. (2005). *Unbiased Stereology: Three-dimensional Measurement in Microscopy*. Garland Science/BIOS Scientific Publishers, New York, 2nd edition.

Huang, F. and Ogata, Y. (2001). Comparison of two methods for calculating the partition functions of various spatial statistical models. *Aust. N.Z. J. Stat.* **43,** 47–65.

Huber, M. (2011). Spatial point processes. In Brooks, S., Gelman, A., Jones, G. L., and Meng, X.-L., eds, *Handbook of Markov Chain Monte Carlo*, pp. 227–52. CRC Press, Boca Raton.

Hug, D., Last, G., and Weil, W. (2002a). Generalized contact distributions of inhomogeneous Boolean models. *Adv. Appl. Prob.* **34,** 21–47.

Hug, D., Last, G., and Weil, W. (2002b). A survey on contact distributions. In Mecke, K. R. and Stoyan, D., eds, *Morphology of Condensed Matter. Physics and Geometry of Spatially Complex Systems*, Lecture Notes in Physics **600**, pp. 317–57. Springer-Verlag, Berlin.

Hug, D., Last, G., and Weil, W. (2006). Polynomial parallel volume, convexity and contact distributions of random sets. *Probab. Theory Related Fields* **135,** 169–200.

Hug, D. and Schneider, R. (2010). Large faces in Poisson hyperplane mosaics. *Ann. Probab.* **38,** 1320–44.

Hughes, B. D. (1996). *Random Walks and Random Environments. Volume 2: Random Environments.* Oxford University Press, New York.

Hunt, A. G. (2005). *Percolation Theory for Flow in Porous Media.* Lecture Notes in Physics **674**. Springer-Verlag, Berlin.

Icke, V. and van de Weygaert, R. (1987). Fragmenting the universe I. *Astron. Astrophys.* **184,** 16–32.

Illian, J., Penttinen, A., Stoyan, H., and Stoyan, D. (2008). *Statistical Analysis and Modelling of Spatial Point Patterns.* John Wiley & Sons, Ltd, Chichester.

Isokawa, Y. (2000). Poisson–Voronoi tessellations in three-dimensional hyperbolic space. *Adv. Appl. Prob.* **32,** 648–62.

Ivanoff, G. (1982). Central limit theorems for point processes. *Stoch. Process. Appl.* **12,** 171–86.

Jacobsen, M. (2006). *Point Process Theory and Applications: Marked Point and Piecewise Deterministic Processes.* Birkhäuser, Boston.

Jacod, J. and Joathon, P. (1971). Use of random genetic models in the study of sedimentary processes. *J. Assoc. Math. Geol.* **3,** 265–79.

Jakeman, A. J. and Schaeffer, R. L. (1978). On the properties of product integration estimators for linear functionals of particle size distributions. *Utilitas Math.* **14,** 117–28.

Jalilian, A., Guan, Y., and Waagepetersen, R. (2013). Decomposition of variance for spatial Cox processes. *Scand. J. Statist.* **40,** 119–37.

Jammalamadaka, S. R. and SenGupta, A. (2001). *Topics in Circular Statistics.* World Scientific, Singapore.

Janáček, J. and Kubínová, L. (2010). Variances of length and surface area estimates by spatial grids: preliminary study. *Image Anal. Stereol.* **29,** 45–52.

Jankowski, H. and Stanberry, L. (2010). Expectations of random sets and their boundaries using oriented distance functions. *J. Math. Imaging Vis.* **36,** 291–303.

Jankowski, H. and Stanberry, L. (2012). Confidence regions for means of random sets using oriented distance functions. *Scand. J. Statist.* **39,** 340–57.

Janson, S. (1986). Random coverings in several dimensions. *Acta Math.* **156,** 83–118.

Janson, S. and Kallenberg, O. (1981). Maximizing the intersection density of fibre processes. *J. Appl. Prob.* **18,** 820–8.

Janson, S., Łuczak, T., and Ruciński, A. (2000). *Random Graphs.* John Wiley & Sons, Inc., New York.

Jensen, E. B. (1984). A design-based proof of Wicksell's integral equation. *J. Microsc.* **136,** 345–8.

Jensen, E. B., Baddeley, A. J., Gundersen, H. J. G., and Sundberg, R. (1985). Recent trends in stereology. *Int. Statist. Rev.* **53,** 99–108.

Jensen, E. B., Kieû, K., and Gundersen, H. J. G. (1990a). On the stereological estimation of reduced moment measures. *Ann. Inst. Statist. Math.* **42,** 445–61.

Jensen, E. B., Kieû, K., and Gundersen, H. J. G. (1990b). Second-order stereology. *Acta Stereol.* **9,** 15–35.

Jensen, E. B. V. (1998). *Local Stereology.* World Scientific, Singapore.

Jeulin, D. (1987). Anisotropic rough surface modelling by random morphological functions. *Acta Stereol.* **6,** 183–9.

Jeulin, D. (1994). Random structure models for composite media and fracture statistics. In Markov, K. Z., ed., *Advances in Mathematical Modelling of Composite Materials*, pp. 239–89. World Scientific, Singapore.

Jeulin, D., ed. (1997). *Advances in Theory and Applications of Random Sets.* World Scientific, Singapore.

Jeulin, D. (2000). Random texture models for material structures. *Statist. Comput.* **10,** 121–32.

Jeulin, D. (2002). Modelling random media. *Image Anal. Stereol.* **21** (Suppl 1)**,** S31–40.

Jeulin, D. (2012). Multi scale random sets: from morphology to effective behaviour. In Günther, M., Bartel, A., Brunk, M., Schöps, S., and Striebel, M., eds, *Progress in Industrial Mathematics at ECMI 2010*, Mathemataics in Industry **17,** pp. 381–93. Springer-Verlag, Berlin.

Jeulin, D., Monnaie, P., and Péronnet, F. (2001). Gypsum morphological analysis and modeling. *Cement Concrete Comp.* **23,** 299–311.

Jeulin, D. and Moreaud, M. (2007). Percolation of random cylinder aggregates. *Image Anal. Stereol.* **26,** 121–7.

Jiao, Y., Stillinger, F. H., and Torquato, S. (2007). Modeling heterogeneous materials via two-point correlation functions: Basic principles. *Phys. Rev. E* **76,** 031110.

Jiao, Y., Stillinger, F. H., and Torquato, S. (2009). A superior descriptor of random textures and its predictive capacity. *Proc. Natl. Acad. Sci.* **106,** 17634–9.

Jiao, Y., Stillinger, F. H., and Torquato, S. (2010). Geometrical ambiguity of pair statistics. II. Heterogeneous media. *Phys. Rev. E* **82,** 011106.

Jiao, Y. and Torquato, S. (2011). Maximally random jammed packings of Platonic solids: Hyperuniform long-range correlations and isostaticity. *Phys. Rev. E* **84,** 041309.

Jodrey, W. S. and Tory, E. M. (1979). Simulation of random packing of spheres. *Simulation* **32,** 1–12.

Johnson, W. A. and Mehl, R. F. (1939). Reaction kinetics in processes of nucleation and growth. *Trans. AIME* **135,** 416–58.

Jolivet, E. (1986). Parametric estimation of the covariance density for a stationary point process on $\mathbb{R}^d$. *Stoch. Process. Appl.* **22,** 111–19.

Jongbloed, G. (1991). Non-parametric approach to Wicksell's corpuscle problem. Master's thesis, Faculty of Mathematics and Computer Science, Delft University of Technology.

Jónsdóttir, K. Ý., Schmiegel, J., and Jensen, E. B. V. (2008). Lévy-based growth models. *Bernoulli* **14,** 62–90.

Ju, L., Du, Q., and Gunzburger, M. (2002). Probabilistic methods for centroidal Voronoi tessellations and their parallel implementations. *Parallel Comput.* **28,** 1477–500.

Kadashevich, I., Schneider, H.-J., and Stoyan, D. (2005). Statistical modeling of the geometrical structure of the system of artificial air pores in autoclaved aerated concrete. *Cement Concrete Res.* **35,** 1495–502.

Kadashevich, I. and Stoyan, D. (2010). Simulation of brittle fracture of autoclaved aerated concrete. *Comput. Concrete* **7,** 39–51.

Kallenberg, O. (1976a). On the structure of stationary flat processes I. *Z. Wahrscheinlichkeitsth. verw. Geb.* **37,** 157–74.

Kallenberg, O. (1976b). *Random Measures*. Akademie-Verlag, Berlin.

Kallenberg, O. (1977). A counterexample to R. Davidson's conjecture on line processes. *Math. Proc. Camb. Phil. Soc.* **82,** 301–7.

Kallenberg, O. (1980). On the structure of stationary flat processes II. *Z. Wahrscheinlichkeitsth. verw. Geb.* **52,** 127–47.

Kallenberg, O. (1981). On the structure of stationary flat processes III. *Z. Wahrscheinlichkeitsth. verw. Geb.* **56,** 239–53.

Kallenberg, O. (1983a). *Random Measures*. Akademie-Verlag, Berlin and Academic Press, London, 3rd edition.

Kallenberg, O. (1983b). The invariance problem for stationary line and flat processes. In Jensen, E. B. and Gundersen, H. J. G., eds, *Second International Workshop on Stereology and Stochastic Geometry*, pp. 105–14. Memoirs No. 6. Institute of Mathematics, Department of Theoretical Statistics, University of Aarhus.

Kallenberg, O. (1986). *Random Measures*. Akademie-Verlag, Berlin and Academic Press, London, 4th edition.

Kallenberg, O. (2002). *Foundations of Modern Probability*. Springer-Verlag, New York, 2nd edition.

Kallmes, O. and Corte, H. (1960). The structure of paper. I: The statistical geometry of an ideal two dimensional fiber network. *Tappi J.* **43,** 737–52. (Erratum: 1961, vol. 44, p. 448).

Kalmykov, A. E. and Shepilov, M. P. (2000). Analytical solution to the equation for pair correlation function of particles formed in the course of phase separation in a glass. *Glass Phys. Chem.* **26,** 143–7.

Kanatani, K. I. (1984). Stereological determination of structural anisotropy. *Int. J. Eng. Sci.* **22,** 531–46.

Kanaun, S. K. and Levin, V. M. (1994). Effective field method in mechanics of matrix composite materials. In Markov, K. Z., ed., *Advances in Mathematical Modelling of Composite Materials*, pp. 1–58. World Scientific, Singapore.

Kanit, T., Forest, S., Galliet, I., Mounoury, V., and Jeulin, D. (2003). Determination of the size of the representative volume element for random composites: statistical and numerical approach. *Int. J. Solids. Structures* **40,** 3647–79.

Kärkkäinen, S., Miettinen, A., Turpeinen, T., Nyblom, J., Pötschke, P., and Timonen, J. (2012). A stochastic shape and orientation model for fibres with an application to carbon nanotubes. *Image Anal. Stereol.* **31,** 17–26.

Kärkkäinen, S., Penttinen, A., Ushakov, N. G., and Ushakova, A. P. (2001). Estimation of orientation characteristic of fibrous material. *Adv. Appl. Prob.* **33,** 559–75.

Karr, A. F. (1978). Derived random measures. *Stoch. Process. Appl.* **8,** 159–69.

Karr, A. F. (1984). Combined non-parametric inference and state estimation for mixed Poisson processes. *Z. Wahrscheinlichkeitsth. verw. Geb.* **66,** 81–96.

Karr, A. F. (1986). *Point Processes and Their Statistical Inference*. Marcel Dekker, New York.

Karr, A. F. (1991). *Point Processes and Their Statistical Inference*. Marcel Dekker, New York, 2nd edition.

Kautz, M., Berger, U., Stoyan, D., Vogt, J., Khan, N. I., Diele, K., Saint-Paul, U., Triet, T., and Nam, V. N. (2011). Desynchronizing effects of lightning strike disturbances on cyclic forest dynamics in mangrove plantations. *Aquat. Bot.* **95,** 173–81.

Keiding, N., Jensen, S. T., and Ranek, L. (1972). Maximum-likelihood estimation of the size distribution of linear cell nuclei from the observed distribution in a plane section. *Biometrics* **28,** 813–30.

Kellerer, A. M. (1983). On the number of clumps resulting from the overlap of randomly placed figures in a plane. *J. Appl. Prob.* **20,** 126–35.

Kendall, D. G. (1974). Foundations of a theory of random sets. In Harding, E. F. and Kendall, D. G., eds, *Stochastic Geometry*, pp. 322–76. John Wiley & Sons, Ltd, London.

Kendall, D. G. (1983). The shape of Poisson Delaunay triangles. In Demetrescu, M. C. and Iosifescu, M., eds, *Studies in Probability and Related Topics in Honour of Octav Onicescu*, pp. 321–30. Nagard, Montreal.

Kendall, D. G., Barden, D., Carne, T. K., and Le, H. (1999). *Shape and Shape Theory*. John Wiley & Sons, Ltd, Chichester.

Kendall, M. G. and Moran, P. A. P. (1963). *Geometrical Probability*. Griffin, London.

Kendall, W. S. (2011). Geodesics and flows in a Poissonian city. *Ann. Appl. Probab.* **21**, 801–42.

Kendall, W. S. and Le, H. (2010). Statistical shape theory. In Kendall, W. S. and Molchanov, I., eds, *New Perspectives in Stochastic Geometry*, pp. 348–73. Oxford University Press, Oxford.

Kendall, W. S. and Mecke, J. (1987). The range of the mean-value quantities of planar tessellations. *J. Appl. Prob.* **24**, 411–21.

Kendall, W. S. and Molchanov, I., eds (2010). *New Perspectives in Stochastic Geometry*. Oxford University Press, Oxford.

Kendall, W. S. and Møller, J. (2000). Perfect simulation using dominating processes on ordered spaces, with application to locally stable point processes. *Adv. Appl. Prob.* **32**, 844–65.

Kendall, W. S. and Thönnes, E. (1999). Perfect simulation in stochastic geometry. *Pattern Recognition* **32**, 1569–86.

Kendall, W. S., van Lieshout, M. N. M., and Baddeley, A. J. (1999). Quermass-interaction processes: conditions for stability. *Adv. Appl. Prob.* **31**, 315–42.

Kessler, M. A. and Werner, B. T. (2003). Self-organisation of sorted patterned ground. *Science* **299**, 380–3.

Khintchin, A. Y. (1955). *Mathematical Methods in the Theory of Queueing* (in Russian). Trudy Matematicheskogo Instituta imeni V. A. Steklova 49, Akad. Nauk, U.S.S.R. (English translation by D. M. Andrews and M. H. Quenouille in 1969, published by Griffin).

Khmaladze, E. and Toronjadze, N. (2001). On the almost sure coverage property of Voronoi tessellation: the $\mathbb{R}^1$ case. *Adv. Appl. Prob.* **33**, 756–64.

Kiang, T. (1966). Random fragmentation in 2 and 3 dimensions. *Z. Astrophys.* **64**, 433–9.

Kiderlen, M. (2001). Non-parametric estimation of the directional distribution of stationary line and fibre processes. *Adv. Appl. Prob.* **33**, 6–24.

Kiderlen, M. (2008). Estimation of the mean normal measure from flat sections. *Adv. Appl. Prob.* **40**, 31–48.

Kiderlen, M. and Jensen, E. B. V. (2003). Estimation of the directional measure of planar random sets by digitization. *Adv. Appl. Prob.* **35**, 583–602.

Kiderlen, M. and Pfrang, A. (2005). Algorithms to estimate the rose of directions of a spatial fibre system. *J. Microsc.* **219**, 50–60.

Kiderlen, M. and Rataj, J. (2006). On infinitesimal increase of volumes of morphological transforms. *Mathematika* **53**, 103–27.

Kingman, J. F. C. (1969). Random secants of a convex body. *J. Appl. Prob.* **6**, 660–72.

Kingman, J. F. C. (1977). Remarks on the spatial distribution of a reproducing population. *J. Appl. Prob.* **14**, 577–83.

Kingman, J. F. C. (1993). *Poisson Processes*. Oxford Studies in Probability **3**. Oxford University Press, Oxford.

Kingman, J. F. C. (2006). Poisson processes revisited. *Probab. Math. Statist.* **26**, 77–95.

Kishore, V., Santhanam, M. S., and Amritkar, R. E. (2011). Extreme events on complex networks. *Phys. Rev. Lett.* **106**, 188701.

Kishore, V., Santhanam, M. S., and Amritkar, R. E. (2012). Extreme events and event size fluctuations in biased random walks on networks. *Phys. Rev. E* **85**, 056120.

Klain, D. A. (1995). A short proof of Hadwiger's characterization theorem. *Mathematika* **42**, 329–39.

Klain, D. A. and Rota, G.-C. (1997). *Introduction to Geometric Probability*. Cambridge University Press, Cambridge.

Klenk, S., Schmidt, V., and Spodarev, E. (2006). A new algorithmic approach to the computation of Minkowski functionals of germ–grain models. *Comput. Geom.* **34**, 127–48.

Klette, R. and Rosenfeld, A. (2004). *Digital Geometry: Geometrical Methods for Digital Picture Analysis*. Morgan Kaufmann, San Francisco.

Klier, G. (1969). Mathematisch-statistische Untersuchungen zur Verteilung der Bäume im Bestand. *Wiss. Z. TU Dresden* **18**, 1061–5.

Kloeden, P. E. and Lorenz, T. (2011). Stochastic morphological evolution equations. *J. Differential Equations* **251**, 2950–79.

Koch, R. A., Pfeiffer, L., and Stammler, L. (1983). *Der Basalt von Stolpen in der Lausitz*. Deutscher Verlag für Grundstoffindustrie, Leipzig.

Kohutek, I. and Saxl, I. (1993). Properties of the Voronoi tessellation corresponding to the generalized planar Gauss-Poisson process. *Acta Stereol.* **12**, 155–60.

Kolmogorov, A. N. (1933). *Grundbegriffe der Wahrscheinlichkeitsrechnung Ergebnisse der Mathematik*. Springer-Verlag, Berlin.

Kolmogorov, A. N. (1937). Statistical theory of crystallization of metals. *Bull. Acad. Sci. USSR Mat. Ser.* **1**, 355–9.

Kopp-Schneider, A., Portier, C., and Bannasch, P. (1998). A model for hepatocarcinogenesis treating phenotypical changes in focal hepatocellular lesions as epigenetic events. *Math. Biosci.* **148**, 181–204.

Koschitzki, S. (1980). Some stereological problems for random discs in $\mathbb{R}^3$. *Math. Operationsf. Statist., Ser. Statistics* **11**, 75–83.

Kovalenko, I. N. (1997). A proof of a conjecture of D.G. Kendall concerning shapes of random polygons of large area. *Cybernet. Systems Anal.* **33**, 461–7.

Kovalenko, I. N. (1998). An extension of a conjecture of D.G. Kendall concerning shapes of random large polygons to Poisson Voronoï cells. In Engel, P. and Syta, H., eds, *Voronoï Impact on Modern Science, Book I*, Volume 21 of *Proceedings of the Institute of Mathematics of the National Academy of Sciences of Ukraine*, pp. 266–74. Institute of Mathematics, Kyiv.

Kovalenko, I. N. (1999). A simplified proof of a conjecture of D.G. Kendall concerning shapes of random polygons. *J. Appl. Math. Stochastic Anal.* **12**, 301–10.

Krasnoperov, R. A. and Gerasimov, A. N. (2003). Determination of size distribution of elliptical microvessels from size distribution measurement of their section profiles. *Exp. Biol. Med.* **228**, 84–92.

Krasnoperov, R. A. and Stoyan, D. (2006). Spatial correlation analysis of isotropic microvessels: methodology and application to thyroid capillaries. *Ann. Biomed. Eng.* **34**, 810–22.

Krätschmer, V. (2006). Integrals of random fuzzy sets. *Test* **15**, 433–69.

Kraynik, A. M., Reinelt, D. A., and van Swol, F. (2003). Structure of random monodisperse foam. *Phys. Rev. E* **67**, 031403.

Krebs, C. J. (1999). *Ecological Methodology*. Addison Wesley Longman, Menlo Park, CA.

Krickeberg, K. (1972). The Cox process. *Instituto Nazionale di Alta Matematicam Symposia Matematica* **9**, 151–67.

Krickeberg, K. (1974). Moments of point processes. In Harding, E. F. and Kendall, D. G., eds, *Stochastic Geometry*, pp. 89–113. John Wiley & Sons, Ltd, London.

Krickeberg, K. (1982). Processus Ponctuels en Statistique. In Hennequin, P. L., ed., *Ecole d'Ete de Probabilities de Saint-Flour X 1980*, Lecture Notes in Mathematics **929**, pp. 205–313. Springer-Verlag, Berlin.

Kroese, D. P., Taimre, T., and Botev, Z. I. (2011). *Handbook of Monte Carlo Methods*. John Wiley & Sons, Inc., Hoboken, New Jersey.

Kruse, R. and Meyer, K. D. (1987). *Statistics with Vague Data*. D. Reidel Publishing, Dortrecht.

Kumar, S., Kurtz, S. K., Banavar, J. R., and Sharma, M. G. (1992). Properties of a three-dimensional Poisson–Voronoi tessellation: a Monte Carlo study. *J. Stat. Phys.* **67**, 523–52.

Kumar, S. and Singh, R. N. (1995). Thermal conductivity of polycrystalline material. *J. Amer. Ceram. Soc.* **78**, 728–36.

Kuo, W., Kim, K. O., and Kim, T. (2006). Modeling and analyzing yield, burn-in and reliability for semiconductor manufacturing: overview. In Pham, H., ed., *Springer Handbook of Engineering Statistics*, chapter 9, pp. 153–69. Springer-Verlag, London.

Kutoyants, Y. A. (1998). *Statistical Inference for Spatial Poisson Processes*. Lecture Notes in Statistics **134**. Springer-Verlag, New York.

Land, S. and Wilkison, M. H. F. (2009). Comparison of spatial pattern spectra. In Wilkinson, M. H. F. and Roerdink, J. B. T. M., eds, *Mathematical Morphology and Its Application to Signal and Image Processing*, Lecture Notes in Computer Science **5720**, pp. 92–103. Springer-Verlag, Berlin.

Lantuéjoul, C. (1978a). Computation of the histograms of the number of edges and neighbours of cells in a tessellation. In Miles, R. E. and Serra, J., eds, *Geometrical Probability and Biological Structures: Buffon's 200th Anniversary*, Lecture Notes in Biomathematics **23**, pp. 323–9. Springer-Verlag, Berlin.

Lantuéjoul, C. (1978b). *La squelettisation et son application aux mesures topologiques des mosaïques polycristallines*. Thèse de Docteur-Ingénieur, École des Mines de Paris.

Lantuéjoul, C. (2002). *Geostatistical Simulation: Models and Algorithms*. Springer-Verlag, Berlin.

Laslett, G. M. (1982a). Censoring and edge effects in areal and line transect sampling of rock joint traces. *Math. Geol.* **14**, 125–49.

Laslett, G. M. (1982b). The survival curve under monotone density constraints with application to two-dimensional line segment processes. *Biometrika* **69**, 153–60.

Laslett, G. M., Kendall, W. S., Gleadow, A. J. W., and Duddy, I. R. (1982). Bias in measurement of fission track length distributions. *Nuclear Tracks* **6**, 79–85.

Last, G. (2006). Stationary partitions and Palm probabilities. *Adv. Appl. Prob.* **38**, 602–20.

Last, G. (2010). Modern random measures: Palm theory and related models. In Kendall, W. S. and Molchanov, I., eds, *New Perspectives in Stochastic Geometry*, pp. 77–110. Oxford University Press, Oxford.

Last, G. and Brandt, A. (1995). *Marked Point Processes on the Real Line: The Dynamic Approach*. Springer-Verlag, New York.

Last, G. and Holtmann, M. (1999). On the empty space function of some germ–grain models. *Pattern Recognition* **32**, 1587–600.

Last, G. and Penrose, M. D. (2013). Percolation and limit theory for the Poisson lilypond model. *Random Structures Algorithms* **42**, 226–49.

Last, G. and Schassberger, R. (1996). A flow conservation law for surface processes. *Adv. Appl. Prob.* **28**, 13–28.

Last, G. and Schassberger, R. (1998). On the distribution of the spherical contact vector of stationary germ–grain models. *Adv. Appl. Prob.* **30**, 36–52.

Last, G. and Schassberger, R. (2000). On stationary stochastic flows and Palm probabilities of surface processes. *Ann. Appl. Probab.* **10**, 463–92.

Last, G. and Schassberger, R. (2001). On the second derivative of the spherical contact distribution function of smooth grain models. *Probab. Theory Related Fields* **121**, 49–72.

Last, G. and Szekli, R. (2011). Comparisons and asymptotics for empty space hazard functions of germ–grain models. *Adv. Appl. Prob.* **43**, 943–62.

Lauschmann, H. and Mrkvička, T. (2009). Perimeter in material images: Comparison of continuous representation and global stereological estimation methods. *Mater. Char.* **60**, 1133–8.

Lautensack, C. (2007). *Random Laguerre Tessellations.* PhD thesis, Universität Karlsruhe (TH), Germany.

Lautensack, C. (2008). Fitting three-dimensional Laguerre tessellations to foam structures. *J. Appl. Stat.* **35**, 985–95.

Lautensack, C., Giertzsch, M., Godehardt, M., and Schladitz, K. (2008). Modelling a ceramic foam using locally adaptable morphology. *J. Microsc.* **230**, 396–404.

Lautensack, C. and Sych, T. (2006). 3D image analysis of open foams using random tessellations. *Image Anal. Stereol.* **25**, 87–93.

Lautensack, C. and Zuyev, S. (2008). Random Laguerre tessellations. *Adv. Appl. Prob.* **40**, 630–50.

Leistritz, L. and Zähle, M. (1992). Topological mean value relations for random cell complexes. *Math. Nachr.* **155**, 57–72.

León, C. A., Massé, J.-C., and Rivest, L.-P. (2006). A statistical model for random rotations. *J. Multivariate Anal.* **97**, 412–30.

Levitz, P. and Tchoubar, D. (1992). Disordered porous solids: from chord distributions to small angle scattering. *Journal de Physique I* **2**, 771–90.

Lewis, F. T. (1931). A comparison between the mosaic of polygons in a film of artifical emulsion and in cucumber epidermis and human amnion. *Anat. Rec.* **50**, 235–65.

Lewis, F. T. (1946). The shape of cells as a mathematical problem. *Amer. Scientist* **34**, 359–69.

Lewis, H. D., Walthers, K. L., and Johnson, K. A. (1973). Particle size distributions by area analysis: Modifications of the Saltykov method. *Metallography* **6**, 93–101.

Lewis, P. A. W. and Shedler, G. S. (1979). Simulation of non-homogeneous Poisson processes by thinning. *Naval Res. Logist. Quart.* **26**, 403–13.

Li, D. S., Tschopp, M. A., Khaleel, M., and Sun, X. (2012). Comparison of reconstructed spatial microstructure images using different statistical descriptors. *Comput. Mat. Sci.* **51**, 437–44.

Li, S., Ogura, Y., and Kreinovich, V. (2002). *Limit Theorems and Application of Set-valued and Fuzzy Set-valued Random Variables.* Kluwer Academic Publishers, Dordrecht.

Liemant, A., Matthes, K., and Wakolbinger, A. (1988). *Equilibrium Distributions of Branching Processes.* Mathematical Research **42**. Akademie-Verlag, Berlin.

Likeš, J. (1963). On the problem of particle number and size determination in opaque bodies. *Acta Tech. Acad. Sci. Hung.* **42**, 325–41.

Lindgren, G. and Rychlik, I. (1991). Slepian models and regression approximations in crossing and extreme value theory. *Int. Statist. Rev.* **59**, 195–225.

Lipskij, J. N., Nikonov, V. A., Skobeleva, T. I., and Kazimirov, D. A. (1977). *Catalogue of the Craters of the Mars and Statistics of the Craters of the Mars, Mercury and Moon.* Acad. Sciences USSR, Geol. Inst., Moscow. (The analysis is based on: Shaded relief map of Mars, 1:25000000. U. S. Geol. Survey 1972).

Liu, G., Yu, H., and Li, W. (1994). Efficient and unbiased evaluation of size and topology of space-filling grains. *Acta Stereol.* **13**, 281–6.

Lochmann, K., Anikeenko, A., Elsner, A., Medvedev, N., and Stoyan, D. (2006). Statistical verification of crystallization in hard sphere packings under densification. *Eur. Phys. J. B* **53**, 67–76.

Loosmore, N. B. and Ford, E. D. (2006). Statistical inference using the *G* or *K* point pattern spatial statistics. *Ecology* **87**, 1925–31.

Lord, G. W. and Willis, T. F. (1951). Calculation of air bubble distribution from results of a Rosiwal traverse of aerated concrete. *ASTM Bull.* **56**, 177–87.

Lorenz, C. D. and Ziff, R. M. (2001). Precise determination of the critical percolation threshold for the three-dimensional "Swiss cheese" model using a growth algorithm. *J. Chem. Phys.* **114**, 3659–61.

Lorenz, T. (2010). *Mutational Analysis. A Joint Framework for Cauchy Problems in and beyond Vector Spaces.* Lecture Notes in Mathematics **1996**. Springer-Verlag, Berlin.

Lorz, U. (1990). Cell-area distributions of planar sections of spatial Voronoi mosaics. *Mater. Char.* **3**, 297–311.

Lorz, U. (1995). Statistics for the spatial Poisson–Voronoi tessellation. In Titterington, D. M., ed., *Complex Stochastic Systems and Engineering*, pp. 141–53. Clarendon Press, Oxford.

Lorz, U. and Hahn, U. (1993). Geometric characteristics of spatial Voronoi tessellations and planar sections. Preprint 93-05, TU Bergakademie Freiberg.

Lotwick, H. W. (1984). Some models for multitype spatial point processes, with remarks on analysing multitype patterns. *J. Appl. Prob.* **21**, 575–82.

Lotwick, H. W. and Silverman, B. W. (1981). Convergence of spatial birth-and-death processes. *Math. Proc. Camb. Phil. Soc* **90**, 155–65.

Louis, A. K., Riplinger, M., Spiess, M., and Spodarev, E. (2011). Inversion algorithms for the spherical Radon and cosine transform. *Inverse Problems* **27**, 035015.

Lowry, M. I. and Miller, C. T. (1995). Pore-scale modeling of nonwetting-phase residual in porous media. *Water Resour. Res.* **31**, 455–73.

Lu, B. and Torquato, S. (1992). Lineal-path function for random heterogeneous materials. *Phys. Rev. A* **45**, 922–9.

Lück, S., Fichtl, A., Sailer, M., Joos, H., Brenner, R. E., Walther, P., and Schmidt, V. (2013). Statistical analysis of the intermediate filament network in cells of mesenchymal lineage by greyvalue-oriented image segmentation. *Comput. Stat.* **28**, 139–60.

Lück, S., Sailer, M., Schmidt, V., and Walther, P. (2010). Three-dimensional analysis of intermediate filament networks using sem tomography. *J. Microsc.* **239**, 1–16.

Lücke, T. and Tittel, R. (1993). An improved description of non woven materials by an assembly of straight lines. *Chem. Biochem. Eng. Q.* **7**, 169–75.

Mack, C. (1954). The expected number of clumps formed when convex laminae or bodies are placed at random and with random orientation on a plane area. *Proc. Camb. Phil. Soc.* **50**, 581–5.

Mack, C. (1956). On clumps formed when convex laminae or bodies are placed at random in two or three dimensions. *Proc. Camb. Phil. Soc.* **52**, 246–50.

Macosko, C. W. and Miller, D. R. (1991). Calculation of average molecular properties during nonlinear, living copolymerization. *Makromol. Chem.* **192**, 377–404.

Mahin, W. K., Hanson, K., and Morris, J. W. (1980). Comparative analysis of the cellular and Johnson–Mehl microstructures through computer simulation. *Acta Metallurgica* **28**, 443–53.

Maier, R., Mayer, J., and Schmidt, V. (2004). Distributional properties of the typical cell of stationary iterated tessellations. *Math. Methods Oper. Res.* **59**, 287–302.

Maier, R. and Schmidt, V. (2003). Stationary iterated tessellations. *Adv. Appl. Prob.* **35**, 337–53.

Malin, M. C., Edgett, K. S., Posiolova, L. V., McColley, S. M., and Dobrea, E. Z. N. (2006). Present-day impact cratering rate and contemporary gully activity on Mars. *Science* **314**, 1573–7.

Malinowski, M. T. and Michta, M. (2010). Stochastic set differential equations. *Nonlinear Anal.* **72**, 1247–56.

Månsson, M. and Rudemo, M. (2002). Random patterns of nonoverlapping convex grains. *Adv. Appl. Prob.* **34**, 718–38.

Manwart, C., Torquato, S., and Hilfer, R. (2000). Stochastic reconstruction of sandstones. *Phys. Rev. E* **62**, 893–9.

Marchant, J. C. and Dillon, P. L. P. (1961). Correlation between random-dot samples and the photographic emulsion. *J. Opt. Soc. Amer.* **51**, 641–4.

Marcus, A. (1972). Some point process models of lunar and planetary surfaces. In Lewis, P. A. W., ed., *Stochastic Point Processes: Statistical Analysis, Theory, and Applications*, pp. 682–99. John Wiley & Sons, Inc., New York.

Mardia, K. V., Edwards, R., and Puri, M. L. (1977). Analysis of central place theory. *Bull. Int. Statist. Inst.* **47**, 93–110.

Mardia, K. V. and Jupp, P. E. (2000). *Directional Statistics*. John Wiley & Sons, Ltd, Chichester.

Marriott, F. H. C. (1971). Buffon's problem for non-random directions. *Biometrics* **27**, 233.

Martínez, V. J., Arnalte-Mur, P., Saar, E., de la Cruz, P., Pons-Borderia, M. J., Paredes, S., Fernadez-Soto, A., and Tempel, E. (2009). Reliability of the detection of the baryon acoustic peak. *Astrophys. J.* **696**, L93–7. (Erratum: 2009, vol. **703**, p. L184).

Martínez, V. J. and Saar, E. (2002). *Statistics of the Galaxy Distribution*. Chapman & Hall/CRC, Boca Raton.

Mase, S. (1985). On the possible form of size distributions for Gibbsian processes of mutually non-intersecting discs. *J. Appl. Prob.* **23**, 646–59.

Mase, S. (1990). Mean characteristics of Gibbsian point processes. *Ann. Inst. Statist. Math.* **42**, 203–20.

Mase, S. (1992). Uniform LAN condition of planar Gibbsian point processes and optimality of maximum-likelihood estimators of soft-core potential functions. *Probab. Theory Related Fields* **92**, 51–67.

Mase, S. (1995). Stereological estimation of particle size distributions. *Adv. Appl. Prob* **27**, 350–66.

Matérn, B. (1960). Spatial Variation. *Meddelanden fran Statens Skogsforskningsinstitut* **49**, 1–144. See also Matérn (1986).

Matérn, B. (1971). Doubly stochastic Poisson processes in the plane. In Patil, G., ed., *Statistical Ecology, Volume 1*, pp. 195–213. Pennsylvania State University Press, University Park.

Matérn, B. (1986). *Spatial Variation*. Lecture Notes in Statistics **36**. Springer-Verlag, Berlin.

Matheron, G. (1967). *Elements pour une theorie des mileux poreux*. Masson, Paris.

Matheron, G. (1971). *The Theory of Regionalized Variables and its Applications*. École national supérieure des mines, Paris.

Matheron, G. (1975). *Random Sets and Integral Geometry*. John Wiley & Sons, Inc., New York.

Matheron, G. (1978). La formule de Steiner pour les erosions. *J. Appl. Prob.* **15**, 126–35.

Matheron, G. (1989). *Estimating and Choosing*. Springer-Verlag, Berlin.

Mathieu, O., Cruz-Orive, L. M., Hoppeler, H., and Weibel, E. R. (1983). Estimating length density and quantifying anisotropy in skeletal muscle capillaries. *J. Microsc.* **131**, 131–46.

Matos, I. (2009). *Limited Range Coverage Problems*. PhD thesis, Departamento de Matemática, Universidade de Aveiro, Portugal.

Matoušek, J. and Nešetřil, J. (2008). *Invitation to Discrete Mathematics*. Oxford University Press, Oxford.

Mattfeldt, T. and Mall, G. (1984). Estimation of length and surface of anisotropic capillaries. *J. Microsc.* **135**, 181–90.

Mattfeldt, T. and Stoyan, D. (2000). Improved estimation of the pair correlation function of random sets. *J. Microsc.* **200**, 158–73.

Matthes, K. (1963). Stationäre zufällige Punktfolgen. *Jahresbericht Deutsche Math. Verein.* **66**, 66–79.

Matthes, K., Kerstan, J., and Mecke, J. (1978). *Infinitely Divisible Point Processes*. John Wiley & Sons, Ltd, Chichester.

Matthes, K., Warmuth, W., and Mecke, J. (1979). Bemerkungen zu einer Arbeit von Nguyen Xuan Xanh und Hans Zessin. *Math. Nachr.* **88**, 117–27.

Mayer, J. (2004). A time-optimal algorithm for the estimation of contact distribution functions of random sets. *Image Anal. Stereol.* **23**, 177–83.

McLachlan, G. J. and Krishnan, T. (2008). *The EM Algorithm and Extensions*. John Wiley & Sons, Inc., New York, 2nd edition.

Mecke, J. (1967). Stationäre zufällige Maße auf lokalkompakten Abelschen Gruppen. *Z. Wahrscheinlichkeitsth. verw. Geb.* **9**, 36–58.

Mecke, J. (1968). Eine charakteristische Eigenschaft der doppelt stochastischen Poissonschen Prozesse. *Z. Wahrscheinlichkeitsth. verw. Geb.* **11**, 74–81.

Mecke, J. (1972). Zufällige Maße auf lokalkompakten Hausdorffschen Räumen. *Beiträge zur Analysis* **3**, 7–30.

Mecke, J. (1975). Invarianzeigenschaften allgemeiner Palmscher Maße. *Math. Nachr.* **65**, 335–44.

Mecke, J. (1979). An explicit description of Kallenberg's lattice-type point process. *Math. Nachr.* **89**, 185–95.

Mecke, J. (1980). Palm methods for stationary random mosaics. In Ambartzumian, R. V., ed., *Combinatorial Principles in Stochastic Geometry*, pp. 124–32. Armenian Academy of Sciences Publ., Erevan.

Mecke, J. (1981a). Formulas for stationary planar fibre processes III – Intersection with fibre systems. *Math. Operationsf. Statist., Ser. Statistics* **12**, 201–10.

Mecke, J. (1981b). Stereological formulas for manifold processes. *Prob. Math. Statist.* **2**, 31–5.

Mecke, J. (1983). Inequalities for intersection densities of superpositions of stationary Poisson hyperplane processes. In Jensen, E. B. and Gundersen, H. J. G., eds, *Second International Workshop on Stereology and Stochastic Geometry*, pp. 115–24. Memoirs No. 6. Institute of Mathematics, Department of Theoretical Statistics, University of Aarhus.

Mecke, J. (1984a). Isoperimetric properties of stationary random mosaics. *Math. Nachr.* **117**, 75–82.

Mecke, J. (1984b). Parametric representation of mean values for stationary random mosaics. *Math. Operationsf. Statist., Ser. Statistics* **15**, 437–42.

Mecke, J. (1984c). Random tessellations generated by hyperplanes. In Ambartzumian, R. V. and Weil, W., eds, *Stochastic Geometry, Geometric Statistics, Stereology*, Teubner-Texte zur Mathematik **65**, pp. 104–9. B. G. Teubner Verlagsgesellschaft, Leipzig.

Mecke, J. (1986). On some inequalities for Poisson networks. *Math. Nachr.* **128**, 81–6.

Mecke, J. (1988a). An extremal property of random flats. *J. Microsc.* **151**, 205–9.

Mecke, J. (1988b). Random r-flats meeting a ball. *Arch. Math.* **51**, 378–84.

Mecke, J. (1991). On the intersection density of flat processes. *Math. Nachr.* **151**, 69–74.

Mecke, J. (1995). Inequalities for the anisotropic Poisson polytope. *Adv. Appl. Prob* **27**, 56–62.

Mecke, J. (1999). On the relationship between the 0-cell and the typical cell of a stationary random tessellation. *Pattern Recognition* **32**, 1645–8.

Mecke, J. (2010). Inhomogeneous random planar tessellations generated by lines. *J. Contemp. Math. Anal., Armen.* **45**, 357–67.

Mecke, J. and Muche, L. (1995). The Poisson Voronoi tessellation, I: A basic identity. *Math. Nachr.* **176**, 199–208.

Mecke, J. and Nagel, W. (1980). Stationäre räumliche Faserprozesse und ihre Schnittzahlrosen. *Elektron. Informationsverarb. Kyb.* **16**, 475–83.

Mecke, J., Nagel, W., and Weiss, V. (2008). A global construction of homogeneous random planar tessellations that are stable under iteration. *Stochastics* **80,** 51–67.

Mecke, J., Schneider, R., Stoyan, D., and Weil, W. (1990). *Stochastische Geometrie.* Birkhäuser, Basel.

Mecke, J. and Stoyan, D. (1980a). Formulas for stationary planar fibre processes I — general theory. *Math. Operationsf. Statist., Ser. Statistics* **12,** 267–79.

Mecke, J. and Stoyan, D. (1980b). Stereological problems for spherical particles. *Math. Nachr.* **96,** 311–17.

Mecke, J. and Stoyan, D. (2001). The specific connectivity number of random networks. *Adv. Appl. Prob.* **33,** 576–85.

Mecke, J. and Thomas, C. (1986). On an extreme value problem for flat processes. *Stochastic Models* **2,** 273–80.

Mecke, K. (1994). *Integralgeometrie in der Statistischen Physik: Perkolation, komplexe Flüssigkeiten und die Struktur des Universums.* Reihe Physik **25.** Verlag Harry Deutsch, Frankfurt.

Mecke, K. R. (1996). Morphological characterization of patterns in reaction-diffusion systems. *Phys. Rev. E* **53,** 4794–800.

Mecke, K. R. (2000). Additivity, convexity, and beyond: applications of Minkowski Functionals in statistical physics. In Mecke, K. R. and Stoyan, D., eds, *Statistical Physics and Spatial Statistics: the Art of Analyzing and Modeling Spatial Structures and Pattern Formation,* Lecture Notes in Physics **554,** pp. 111–84. Springer-Verlag, Berlin.

Mecke, K. R. and Seyfried, A. (2002). Strong dependence of percolation thresholds on polydispersity. *Europhys. Lett.* **58,** 28–34.

Mecke, K. R. and Wagner, H. (1991). Euler characteristic and related measures for random geometric sets. *J. Stat. Phys.* **64,** 843–50.

Medvedev, N. N. (2000). *The Voronoi–Delaunay Method for Non-crystal Structures* (in Russian). SB Russian Academy of Sciences, Novosibirsk.

Medvedev, N. N., Voloshin, V. P., Luchnikov, V. A., and Gavrilova, M. L. (2006). An algorithm for three-dimensional Voronoi S-network. *J. Comput. Chem.* **27,** 1676–92.

Meester, R. and Roy, R. (1994). Uniqueness of unbounded occupied and vacant components in Boolean models. *Ann. Appl. Probab* **4,** 933–51.

Meester, R. and Roy, R. (1996). *Continuum Percolation.* Cambridge University Press, Cambridge.

Meester, R., Roy, R., and Sarkar, A. (1994). Nonuniversality and continuity of the critical covered volume fraction in continuum percolation. *J. Stat. Phys.* **75,** 123–34.

Meijering, J. L. (1953). Interface area, edge length and number of vertices in crystal aggregates with random nucleation. *Philips Res. Rep.* **8,** 270–90.

Metropolis, N., Rosenbluth, A. W., Rosenbluth, M. N., Teller, A. H., and Teller, E. (1953). Equation of state calculations by fast computing machines. *J. Chem. Phys.* **21,** 1087–92.

Michalke, W., Lang, M., Kreitmeier, S., and Göritz, D. (2001). Simulations on the number of entanglements of a polymer network using knot theory. *Phys. Rev. E* **64,** 6300–7.

Michel, J. and Paroux, K. (2003). Local convergence of the Boolean shell model towards the thick Poisson hyperplane process in the Euclidean space. *Adv. Appl. Prob.* **35,** 354–61.

Michel, J. and Paroux, K. (2007). Empirical polygon simulation and central limit theorems for the homogenous Poisson line process. *Methodol. Comput. Appl. Probab.* **9,** 541–56.

Michell, J. (1767). An inquiry into the probable parallax, and magnitude, of the fixed stars, from the quantity of light which they afford us, and the particular circumstances of their situation. *Phil. Trans. R. Soc. London* **57,** 234–64.

Miles, R. E. (1964a). Random polygons determined by random lines in a plane. *Proc. Nat. Acad. Sci. (USA)* **52**, 901–7.

Miles, R. E. (1964b). Random polygons determined by random lines in a plane, II. *Proc. Nat. Acad. Sci. (USA)* **52**, 1157–60.

Miles, R. E. (1970). On the homogeneous planar Poisson point process. *Math. Biosci.* **6**, 85–127.

Miles, R. E. (1971a). Poisson flats in Euclidean spaces. Part II: Homogeneous Poisson flats and the complementary theorem. *Adv. Appl. Prob.* **3**, 1–43.

Miles, R. E. (1971b). Random points, sets and tessellations on the surface of a sphere. *Sankhyā A* **33**, 145–74.

Miles, R. E. (1972a). Multi-dimensional perspectives on stereology. *J. Microsc.* **95**, 181–95.

Miles, R. E. (1972b). The random division of space. *Suppl. Adv. Appl. Prob.* **4**, 243–66.

Miles, R. E. (1973). The various aggregates of random polygons determined by random lines in a plane. *Adv. Math.* **10**, 256–90.

Miles, R. E. (1974a). A synopsis of 'Poisson Flats in Euclidean Spaces'. In Harding, E. F. and Kendall, D. G., eds, *Stochastic Geometry*, pp. 202–27. John Wiley & Sons, Ltd, London.

Miles, R. E. (1974b). On the elimination of edge-effects in planar sampling. In Harding, E. F. and Kendall, D. G., eds, *Stochastic Geometry*, pp. 228–47. John Wiley & Sons, Ltd, London.

Miles, R. E. (1976). Estimating aggregate and overall characteristics from thick sections by transmission microscopy. *J. Microsc.* **107**, 227–33.

Miles, R. E. (1983). Stereology for embedded aggregates of not necessarily convex particles. In Jensen, E. B. and Gundersen, H. J. G., eds, *Second International Workshop on Stereology and Stochastic Geometry*, pp. 127–47. Memoirs No. 6. Institute of Mathematics, Department of Theoretical Statistics, University of Aarhus.

Miles, R. E. (1984). A comprehensive set of stereological formulae for embedded aggregates of not-necessarily-convex particles. *J. Microsc.* **134**, 127–36.

Miles, R. E. (1985). A comprehensive set of stereological formulae for embedded aggregates of not-necessarily-convex particles. *J. Microsc* **138**, 115–25.

Miles, R. E. (1986). Spatial tessellations and their stereology. In Ishizaka, S., ed., *Science on Form: Proceedings of the First International Symposium for Science on Form*, pp. 147–55. KTK Scientific Publishers, Tokyo.

Miles, R. E. (1987). Dihedral angle distributions. *Acta Stereol.* **6**, 19–24.

Miles, R. E. (1995). A heuristic proof of a long-standing conjecture of D. G. Kendall concerning the shapes of certain large random polygons. *Adv. Appl. Prob* **27**, 397–417.

Miles, R. E. and Maillardet, R. J. (1982). The basis structures of Voronoi and generalized Voronoi polygons. *J. Appl. Prob.* **19A**, 97–112.

Milne, R. K. and Westcott, M. (1972). Further results for Gauss-Poisson processes. *Adv. Appl. Prob.* **4**, 151–76.

Minlos, R. A. (1968). Lectures on statistical physics. *Usp. Mat. Nauk* **23**, 133–90.

Molchanov, I. S. (1992). Handling with spatial censored observations in statistics of Boolean models of random sets. *Biometrical J.* **34**, 617–31.

Molchanov, I. S. (1993). *Limit Theorems for Unions of Random Sets*. Lecture Notes in Mathematics **1561**. Springer-Verlag, Berlin.

Molchanov, I. S. (1994). On statistical analysis of Boolean models with non-random grains. *Scand. J. Statist.* **21**, 73–82.

Molchanov, I. S. (1996). A limit theorem for scaled vacancies of the Boolean model. *Stoch. Stoch. Rep.* **58**, 45–65.

Molchanov, I. S. (1997). *Statistics of the Boolean Model for Practitioners and Mathematicians*. John Wiley & Sons, Inc., New York.

Molchanov, I. S. (2005). *Theory of Random Sets*. Springer-Verlag, London.

Molchanov, I. S. and Chiu, S. N. (2000). Smoothing techniques and estimation methods for nonstationary boolean models with applications to coverage. *Biometrika* **87**, 265–83.

Molchanov, I. S. and Stoyan, D. (1994). Asymptotic properties of estimators for parameters of the Boolean model. *Adv. Appl. Prob.* **26**, 301–23.

Molchanov, I. S. and Stoyan, D. (1995). Statistics of compact sets and random polygons. *Stochastic Models* **12**, 199–214.

Molchanov, I. S., Stoyan, D., and Fyodorov, K. (1993). Directional analysis of planar fibre networks: application to cardboard microstructure. *J. Microsc.* **172**, 257–61.

Molek, H., Pohlmann, S., Reuter, F., and Stoyan, D. (1981). Entwicklung eines komplexen Durchtrennungsgrades von Gesteinsverbänden mit Hilfe stereologischer Methoden. *Neue Bergbautechnik* **11**, 221–4.

Møller, J. (1989). Random tessellations in $\mathbb{R}^d$. *Adv. Appl. Prob.* **24**, 37–73.

Møller, J. (1992). Random Johnson–Mehl tessellations. *Adv. Appl. Prob.* **24**, 814–44.

Møller, J. (1994). *Lectures on Random Voronoi Tessellations*. Lecture Notes in Statistics **87**. Springer-Verlag, Berlin.

Møller, J. (1995). Generation of Johnson–Mehl crystals and comparative analysis of models for random nucleation. *Adv. Appl. Prob.* **27**, 367–83.

Møller, J. (1999). Markov chain Monte Carlo and spatial point processes. In Barndorff-Nielsen, O. E., Kendall, W. S., and van Lieshout, M. N. M., eds, *Stochastic Geometry: Likelihood and Computation*, pp. 141–72. Chapman & Hall/CRC, Boca Raton.

Møller, J. (2001). A review of perfect simulation in stochastic geometry. In Basawa, I. V., Heyde, C. C., and Taylor, R. L., eds, *Selected Proceedings of the Symposium on Inference for Stochastic Processes*, IMS Lecture Notes Monograph Series **37**, pp. 333–56. Institute of Mathematical Statistics, Beachwood, Ohio.

Møller, J. and Helisová, K. (2010). Likelihood inference for unions of interacting discs. *Scand. J. Statist.* **37**, 365–81.

Møller, J., Huber, M. L., and Wolpert, R. L. (2010). Perfect simulation and moment properties for the Matérn type III process. *Stoch. Process. Appl.* **120**, 2142–58.

Møller, J. and Stoyan, D. (2014). Stochastic geometry and random tessellations. In van de Weygaert, R., Vegter, G., Icke, V., and Ritzerveld, J., eds, *Tessellations in the Sciences: Virtues, Techniques and Applications of Geometric Tilings*. Forthcoming.

Møller, J., Syversveen, R. A., and Waagepetersen, R. P. (1998). Log Gaussian Cox processes. *Scand. J. Statist.* **25**, 451–82.

Møller, J. and Waagepetersen, R. P. (2004). *Statistical Inference and Simulation for Spatial Point Processes*. Chapman & Hall/CRC, Boca Raton.

Møller, J. and Zuyev, S. (1996). Gamma-type results and other related properties of Poisson processes. *Adv. Appl. Prob.* **28**, 662–73.

Möller, O. (1989). A fast statistical procedure solving Wicksell's corpuscle problem. *Elektron. Informationsverarb. Kyb.* **25**, 581–5.

Mollison, D. (1977). Spatial contact models for ecological and epidemic spread (with discussion). *J. Roy. Statist. Soc. B* **39**, 283–326.

Molloy, M. and Reed, B. (1995). A critical point for random graphs with a given degree sequence. *Random Structures Algorithms* **6**, 161–79.

Moran, P. A. P. (1972). The probabilistic basis of stereology. *Supp. Adv. Appl. Prob.* **4**, 69–91.

Moran, P. A. P. (1976). Another quasi-Poisson planar point process. *Probab. Theory Related Fields* **33**, 269–72.

Morgan, F. (2009). *Geometric Measure Theory. A Beginner's Guide*. Academic Press, Burlington, MA, 4th edition.

Mörters, P. (2010). Random fractals. In Kendall, W. S. and Molchanov, I., eds, *New Perspectives in Stochastic Geometry*, pp. 275–304. Oxford University Press, Oxford.

Morton, R. R. A. (1966). The expected number and angle of intersection between random curves in a plane. *J. Appl. Prob.* **3**, 559–62.

Mouton, P. R. (2011). *Unbiased Stereology: a Concise Guide*. Johns Hopkins University Press, Baltimore.

Mrkvička, T. and Mattfeldt, T. (2011). Testing histological images of mammary tissues on compatibility with the Boolean model of random sets. *Image Anal. Stereol.* **30**, 11–18.

Mrkvička, T. and Rataj, J. (2008). On the estimation of intrinsic volume densities of stationary random closed sets. *Stoch. Process. Appl.* **118**, 213–31.

Muche, L. (1993). An incomplete Voronoi tessellation. *Appl. Mathematicae* **22**, 45–53.

Muche, L. (1996a). Distributional properties of the three-dimensional Poisson Delaunay cell. *J. Stat. Phys.* **84**, 147–67.

Muche, L. (1996b). The Poisson–Voronoi tessellation II. Edge length distribution functions. *Math. Nachr.* **178**, 271–83.

Muche, L. (1998). The Poisson Voronoi tessellation III. Miles' formula. *Math. Nachr.* **191**, 247–67.

Muche, L. (2005). The Poisson–Voronoi tessellation: relationships for edges. *Adv. Appl. Prob.* **37**, 279–96.

Muche, L. (2010). Contact and chord length distribution functions of the Poisson–Voronoi tessellation in high dimensions. *Adv. Appl. Prob.* **42**, 48–68.

Muche, L. and Ballani, F. (2011). The second volume moment of the typical cell and higher moments of edge lengths of the spatial Poisson–Voronoi tessellation. *Monatsh. Math.* **163**, 71–80.

Muche, L. and Stoyan, D. (1992). Contact and chord length distributions of the Poisson–Voronoi tessellation. *J. Appl. Prob.* **29**, 467–71.

Müller, A. and Stoyan, D. (2002). *Comparison Methods for Stochastic Models and Risks*. John Wiley & Sons, Ltd, Chichester.

Myles, J. P., Flenley, E. C., Fieller, N. R. J., Alkinson, H. V., and Jones, H. (1995). Statistical tests for clustering of second phases in composite materials. *Philos. Mag.* **72**, 515–28.

Nagel, W. (1983). Dünne Schnitte von stationären räumlichen Faserprozessen. *Math. Operationsf. Statist., Ser. Statistics* **14**, 569–76.

Nagel, W. (2010). Stereology. In Kendall, W. S. and Molchanov, I., eds, *New Perspectives in Stochastic Geometry*, pp. 451–75. Oxford University Press, Oxford.

Nagel, W. and Weiss, V. (2003). Limits of sequences of stationary planar tessellations. *Adv. Appl. Prob.* **35**, 123–38.

Nagel, W. and Weiss, V. (2004). Crack STIT tessellations – existence and uniqueness of tessellations that are stable with respect to iterations. *Isvest. Nac. Akad. Nauk Armenii (Mat.)* **39**, 84–114.

Nagel, W. and Weiss, V. (2005). Crack STIT tessellations: characterization of stationary random tessellations stable with respect to iteration. *Adv. Appl. Prob.* **37**, 859–83.

Nagel, W. and Weiss, V. (2006). STIT tessellations in the plane. *Rend. Circ. Mat. Palermo (2) Suppl.* **77**, 441–58.

Nagel, W. and Weiss, V. (2008). Mean values for homogeneous STIT tessellations in 3D. *Image Anal. Stereol.* **27**, 29–37.

Nemat-Nasser, S. and Hori, M. (1999). *Micromechanics: Overall Properties of Heterogeneous Materials.* Elsevier, Amsterdam, 2nd revised edition.

Newman, M. E. J. (2003). The structure and function of complex networks. *SIAM Review* **45**, 167–256.

Newman, M. E. J., Watts, D. J., and Barabási, A. L. (2006). *The Structure and Dynamics of Networks.* Princeton University Press, Princeton, New Jersey.

Neyman, J. and Scott, E. L. (1958). Statistical approach to problems of cosmology. *J. Roy. Statist. Soc. B* **20**, 1–43.

Neyman, J. and Scott, E. L. (1972). Processes of clustering and applications. In Lewis, P. A. W., ed., *Stochastic Point Processes: Statistical Analysis, Theory and Applications*, pp. 646–81. John Wiley & Sons, Inc., New York.

Nguyen, H. T. (2005). Fuzzy and random sets. *Fuzzy Sets and Systems* **156**, 349–56.

Nguyen, H. T. (2006). *An Introduction to Random Sets.* Chapman & Hall/CRC, Boca Raton.

Nguyen, H. T. and Wu, B. (2006). *Fundamentals of Statistics with Fuzzy Data.* Studies in Fuzziness and Soft Computing **198**. Springer-Verlag, New York.

Nguyen, X. X. (1979). Ergodic theorems for subadditive spatial processes. *Probab. Theory Related Fields* **48**, 159–76.

Nguyen, X. X. and Zessin, H. (1979a). Ergodic theorems for spatial processes. *Probab. Theory Related Fields* **48**, 133–58.

Nguyen, X. X. and Zessin, H. (1979b). Integral and differential characterizations of Gibbs processes. *Math. Nachr.* **88**, 105–15.

Nicholson, W. L. (1970). Estimation of linear properties of particle size distributions. *Biometrika* **57**, 273–97.

Nicholson, W. L. (1976). Estimation of linear functionals by maximum likelihood. *J. Microsc.* **107**, 323–36.

Nicholson, W. L. (1978). Application of statistical methods in quantitative microscopy. *J. Microsc.* **113**, 223–336.

Niskanen, K., Kajanto, I., and Pakarinen, P. (1998). Paper structure. In Niskanen, K. J., ed., *Paper Physics*, Papermaking Science and Technology Series **16**, pp. 14–53. Fapet Oy, Helsinki.

Norberg, T. (1984). Convergence and existence of random set functions. *Ann. Prob.* **12**, 726–32.

Norberg, T. (1989). Existence theorems for measures on continuous posets, with applications to random set theory. *Math. Scand.* **64**, 15–51.

Norberg, T. (1992a). An ordered random set coupling. *Theory Probab. Appl.* **37**, 161–3.

Norberg, T. (1992b). On the existence of ordered couplings of random sets–with applications. *Isr. J. Math.* **77**, 241–64.

Norros, I. and Reittu, H. (2006). On a conditionally Poissonian graph process. *Adv. Appl. Prob.* **38**, 59–75.

Nott, D. J. and Rydén, T. (1999). Pairwise likelihood methods for inference in image models. *Biometrika* **86**, 661–76.

Nott, D. J. and Wilson, R. J. (1996). Size distributions for excursion sets. In Jeulin, D., ed., *Advances in Theory and Applications of Random Sets*, pp. 176–96. World Scientific, Singapore.

Nott, D. J. and Wilson, R. J. (1997). Parameter estimation for excursion set texture models. *Signal Processing* **63**, 199–210.

Nuske, R. S., Sprauer, S., and Saborowski, J. (2009). Adapting the pair-correlation function for analysing the spatial distribution of canopy gaps. *Forest Ecol. Manag.* **259**, 107–16.

Nutting, P. G. (1913). On the absorption of light in heterogeneous media. *London, Edinb. & Dublin Philos. Mag. Ser. 6* **26,** 423–6.

Nychka, D. (1983a). Smooth non-parametric estimates of particle size distributions. *Acta Stereol.* **2,** 25–8.

Nychka, D. (1983b). Solving integral equations with noisy data: An application of smoothing splines in pathology. In Heiner, K. W., Sacher, R. S., and Wilkinson, J. W., eds, *Computer Science and Statistics: Proceedings of the 14th Symposium on the Interface,* pp. 157–63. Springer-Verlag, New York.

Nychka, D., Wahba, G., Goldfarb, S., and Pugh, T. (1984). Cross-validated spline methods for the estimation of three-dimensional tumor-size distributions from observations on two-dimensional cross-sections. *J. Amer. Statist. Assoc.* **79,** 832–46.

Ogata, Y. (1981). On Lewis' simulation method for point processes. *IEEE Trans. Inf. Theor.* **IT-27,** 23–31.

Ogata, Y. and Tanemura, M. (1981). Estimation of interaction potentials of spatial point-patterns through the maximum-likelihood procedure. *Ann. Inst. Statist. Math.* **33,** 315–38.

Ogata, Y. and Tanemura, M. (1984). Likelihood analysis of spatial point-patterns. *J. Roy. Statist. Soc. B* **46,** 496–518.

Ogawa, T. and Tanemura, M. (1974). Geometrical considerations on hard-core problems. *Progress Theor. Phys.* **51,** 399–417.

Ohanian, V. K. (1973). On random Markovian colouring of the plane with two colours (in Russian). *Dokl. Akad. Nauk. Armenian SSR* **58,** 193–8.

Ohanian, V. K. (1978). Combinatorial principles in stochastic geometry of random segment processes. In Ambartzumian, R. V., ed., *Combinatorial Principles in Stochastic Geometry,* pp. 81–106. Publ. Akad. Sci. Armen. SSR.

Ohser, J. (1980). On statistical analysis of the Boolean model. *Elektron. Inf.-Verarb. Kyb.* **16,** 651–53.

Ohser, J. (1981). A remark on the estimation of the rose of directions of fibre processes. *Math. Operationsf. Statist., Ser. Statistics* **12,** 581–5.

Ohser, J. (1983). On estimators for the reduced second-moment measure of point processes. *Math. Operationsf. Statist., Ser. Statistics* **14,** 63–71.

Ohser, J. and Lorz, U. (1994). *Quantitative Gefügeanalyse. Theoretische Grundlagen und Anwendungen.* Freiberger Forschungshefte **B276.** Deutscher Verlag für Grundstoffindustrie, Leipzig.

Ohser, J. and Mücklich, F. (1995). Stereology for some classes of polyhedrons. *Adv. Appl. Prob.* **27,** 384–96.

Ohser, J. and Mücklich, F. (2000). *Statistical Analysis of Microstructures in Materials Science.* John Wiley & Sons, Ltd, Chichester.

Ohser, J. and Schladitz, K. (2009). *3D Images of Material Structures. Processing and Analysis.* Wiley-VCH, Weinheim.

Ohser, J. and Stoyan, D. (1980). Zur Beschreibung gewisser zufälliger Muster in der Geologie. *Z. angew. Geol.* **26,** 209–12.

Ohser, J. and Stoyan, D. (1981). On the second-order and orientation analysis of planar stationary point processes. *Biometrical J.* **23,** 523–33.

Ohser, J. and Tscherny, H. (1988). *Grundlagen der quantitativen Gefügeanalyse.* Freiberger Forschungshefte **B264.** Deutscher Verlag für Grundstoffindustrie, Leipzig.

Okabe, A., Boots, B., Sugihara, K., and Chiu, S. N. (2000). *Spatial Tessellations – Concepts and Applications of Voronoi Diagrams.* John Wiley & Sons, Ltd, Chichester, 2nd edition.

Okabe, A. and Sugihara, K. (2012). *Spatial Analysis Along Networks: Statistical and Computataional Methods.* John Wiley & Sons, Inc., New York.

Olsbo, V. (2007). On the correlation between the volumes of the typical Poisson–Voronoi cell and the typical Stienen sphere. *Adv. Appl. Prob.* **39**, 883–92.

Ong, M. S., Kuang, Y. C., and Ooi, M. P.-L. (2012). Statistical measures of two dimensional point set uniformity. *Comput. Statist. Data Anal.* **56**, 2159–81.

Øren, P.-E. and Bakke, S. (2002). Process based reconstruction of sandstones and prediction of transport properties. *Trans. Porous Med.* **46**, 311–43.

Øren, P.-E. and Bakke, S. (2003). Reconstruction of Berea sandstone and pore-scale modelling of wettability effects. *J. Petrol. Sci. Eng.* **39**, 177–99.

Ornstein, L. S. and Zernike, F. (1914). Accidental deviations of density and opalescence at the critical point of a single substance. *Proc. R. Neth. Acad. Arts Sci.* **17**, 793–806.

Osher, S. and Fedkiw, R. (2003). *Level Set Methods and Dynamic Implicit Surfaces*. Springer-Verlag, New York.

Oualkacha, K. and Rivest, L.-P. (2009). A new statistical model for random unit vectors. *J. Multivariate Anal.* **100**, 70–80.

Overby, D. R. and Johnson, M. (2005). Studies on depth-of-field effects in microscopy supported by numerical simulations. *J. Microsc.* **220**, 176–89.

Palm, C. (1943). Intensitätsschwankungen im Fernsprechverkehr. *Ericsson Technics* **44**, 1–189.

Pelikan, K., Saxl, I., and Poнížil, P. (1994). Germ–grain model of short-fibre composites. In Wojnar, L., ed., *Stermath'94 Proc. 4th Int. Conf. Stereology and Image Analysis in Materials Science*, pp. 389–96. Fotobit-Design, Krakow.

Penrose, M. (2003). *Random Geometric Graphs*. Oxford Studies in Probability **5**. Oxford University Press, Oxford.

Penrose, M. D. (1997). The longest edge of the random minimal spanning tree. *Ann. Appl. Probab.* **7**, 340–61.

Penrose, M. D. (2007). Laws of large numbers in stochastic geometry with statistical applications. *Bernoulli* **13**, 1124–50.

Penrose, M. D. and Wade, A. R. (2010). Random directed and on-line networks. In Kendall, W. S. and Molchanov, I., eds, *New Perspectives in Stochastic Geometry*, pp. 248–74. Oxford University Press, Oxford.

Penttinen, A. (1984). *Modelling interactions in spatial point-patterns: parameter estimation by the maximum-likelihood method*. Jyväskyla Studies in Computer Science, Economics and Statistics **7**, Jyväskyla.

Penttinen, A. and Niemi, A. (2007). On statistical inference for the random set generated Cox process with set-marking. *Biometrical J.* **49**, 197–213.

Penttinen, A. and Stoyan, D. (1989). Statistical analysis for a class of line segment processes. *Scand. J. Statist.* **16**, 153–61.

Penttinen, A., Stoyan, D., and Henttonen, H. M. (1992). Marked point processes in forest statistics. *Forest Sci.* **38**, 806–24.

Peshkin, M. A., Strandburg, K. J., and Rivier, N. (1991). Entropic predictions for cellular networks. *Phys. Rev. Lett.* **7**, 1803–6.

Peyrega, C., Jeulin, D., Delisée, C., and Malvestio, J. (2009). 3D morphological modelling of a random fibrous network. *Imaga Anal. Stereol.* **28**, 129–41.

Picard, N. and Bar-Hen, A. (2000). Estimation of the envelope of a point set with loose boundaries. *Appl. Math. Lett.* **13**, 13–8.

Pielou, E. C. (1977). *Mathematical Ecology*. John Wiley & Sons, Inc., New York.

Piterbarg, V. I. (1996). *Asymptotic Methods in the Theory of Gaussian Processes and Fields*. Translation of Mathematical Monographs **148**. American Mathematical Society, Providence, RI.

Pohlmann, S., Mecke, J., and Stoyan, D. (1981). Stereological formulas for stationary surface processes. *Math. Operationsf. Statist., Ser. Statistics* **12**, 429–40.

Pollard, D. (2002). *A User's Guide to Measure Theoretic Probability*. Cambridge University Press, Cambridge.

Pólya, G. (1918). Zahlentheoretisches und Wahrscheinlichkeitstheoretisches über die Sichtweite im Walde. *Arch. Math. Phys.* **27**, 135–42.

Poole, D. C., Batra, S., Mathieu-Costello, O., and Rakusan, K. (1992). Capillary geometrical changes with fiber shortening in rat myocardium. *Circ. Res.* **70**, 697–706.

Porod, G. (1951). Die Röntgenkleinwinkelstreuung von dichtgepackten kolloiden Systemen I. *Kolloid Zeitschrift* **124**, 83–114.

Porod, G. (1952). Die Röntgenkleinwinkelstreuung von dichtgepackten kolloiden Systemen II. *Kolloid Zeitschrift* **125**, 51–7.

Pothuaud, L., Porion, P., Lespessailles, E., Benhamou, C. L., and Levitz, P. (2000). A new method for three-dimensional skeleton graph analysis of porous media: application to trabecular bone microarchitecture. *J. Microsc.* **199**, 149–61.

Poupon, A. (2004). Voronoi and Voronoi-related tessellations in studies of protein structure and interaction. *Curr. Opin. Struct. Biol.* **2**, 233–41.

Prager, S. (1969). Improved variational bounds on some bulk properties of a two-phase medium. *J. Chem. Phys.* **50**, 4305–12.

Preparata, J. P. and Shamos, M. I. (1985). *Computational Geometry, An Introduction*. Springer-Verlag, New York.

Preston, C. J. (1974). *Gibbs States on Countable Sets*. Cambridge University Press, Cambridge.

Preston, C. J. (1976). *Random Fields*. Lecture Notes in Mathematics **534**. Springer-Verlag, Berlin.

Preston, C. J. (1977). Spatial birth-and-death processes. *Bull. Int. Statist. Inst.* **46**, 371–91.

Priolo, A., Jaeger, H. M., Dammers, A. J., and Radelaar, S. (1992). Conductance of two-dimensional disordered Voronoi networks. *Phys. Rev. B* **46**, 14889–92.

Prokešová, M. (2003). Bayesian MCMC estimation of the rose of directions. *Kybernetika* **39**, 701–17.

Provatas, N., Haataja, M., Asikainen, J., Majaniemi, S., Alava, M., and Ala-Nissila, T. (2000). Fiber deposition models in two and three spatial dimensions. *Colloids Surfaces A: Physicochem. Eng. Aspects* **165**, 209–29.

Quine, M. P. and Robinson, J. (1992). Estimation for a linear growth model. *Statist. Prob. Lett.* **15**, 293–7.

Quine, M. P. and Watson, D. F. (1984). Radial simulation of n-dimensional Poisson processes. *J. Appl. Prob.* **21**, 548–57.

Quintanilla, J. A. and Ziff, R. M. (2007). Asymmetry in the percolation thresholds of fully penetrable disks with two different radii. *Phys. Rev. E* **76**, 051115.

Radecke, W. (1980). Some mean value relations on stationary random mosaics in the space. *Math. Nachr.* **97**, 203–10.

Rahman, A. (1966). Liquid structure and self-diffusion. *J. Chem. Phys.* **45**, 2584–92.

Räisänen, V. I., Heyden, S., Gustafsson, P.-J., Alava, M. J., and Niskanen, K. J. (1997). Simulation of the effect of a reinforcement fiber on network mechanics. *Nordic Pulp Paper Res. J.* **12**, 162–6.

Rancoita, P. M. V., Giusti, A., and Micheletti, A. (2011). Intensity estimation of stationary fibre processes from digital images with a learned detector. *Image Anal. Stereol.* **30**, 167–78.

Rao, M. M. (1993). *Conditional Measures and Applications*. Marcel Dekker, New York.

Rasson, J. P. and Hermans, M. (1988). On a connection between Davidson's entropy and a test of randomness for point and line process in the plane. *Atti Accad. Peloritana Pericolanti, Cl. Sci, Fis. Mat. Nat.* **65**, 337–46.

Rataj, J. (1993). Random distances and edge-correction. *Statistics* **24**, 377–85.

Rataj, J. (1996). Estimation of oriented direction distirbution of a planar body. *Adv. Appl. Prob.* **28**, 394–404.

Rataj, J. and Saxl, I. (1989). Analysis of planar anisotropy by means of the Steiner compact. *J. Appl. Prob.* **26**, 490–502.

Rathbun, S. L. (1996). Estimation of Poisson intensity using partially observed concomitant variables. *Biometrics* **52**, 226–42.

Rathie, P. N. (1992). On the volume of the typical Poisson-Delaunay cell. *J. Appl. Prob.* **29**, 740–4.

Rau, C. and Chiu, S. N. (2011). Grain rotations and distortions in the asymptotic variance of vacancy of the Boolean model. *J. Math. Anal. Appl.* **384**, 647–57.

Redenbach, C. (2009a). Microstructure models for cellular materials. *Comput. Mat. Sci.* **44**, 1397–407.

Redenbach, C. (2009b). Modelling foam structures using random tessellations. In Capasso, V., Aletti, G., and Micheletti, A., eds, *Stereology and Image Analysis: ECS10 – 10th European Congress of ISS*, The MIRIAM Project Series **4**. Societa Editrice Esculapio–Progetto Leonardo, Bologna.

Redenbach, C. (2011). On the dilated facets of a Poisson-Voronoi tessellation. *Image Anal. Stereol.* **30**, 31–8.

Redenbach, C., Shklyar, I., and Andrä, H. (2012). Laguerre tessellations for elastic stiffness simulations of closed foams with strongly varying cell sizes. *Int. J. Eng. Sci.* **50**, 70–8.

Reiss, R.-D. (1993). *A Course on Point Processes*. Springer-Verlag, Berlin.

Reitzner, M. (2010). Random polytopes. In Kendall, W. S. and Molchanov, I., eds, *New Perspectives in Stochastic Geometry*, pp. 45–76. Oxford University Press, Oxford.

Reitzner, M., Spodarev, E., and Zaporozhets, D. (2012). Set reconstruction by Voronoi cells. *Adv. Appl. Prob.* **44**, 938–53.

Rényi, A. (1967). Remarks on the Poisson process. *Studia Sci. Math. Hung.* **2**, 119–23.

Rhines, F. N. (1986). *Microstructology. Behaviour and Microstructure of Materials*. Dr. Riederer-Verlag, Stuttgart.

Rice, S. O. (1944, 1945). Mathematical analysis of random noise. *Bell Syst. Tech. J.* **23**, 282–332; **24**, 45–146.

Richards, F. M. (1974). The interpretation of protein structures: total volume, group volume distributions and packing density. *J. Mol. Biol.* **82**, 1–14.

Richeson, D. S. (2008). *Euler's Gem. The Polyhedron Formula and the Birth of Topology*. Princeton University Press, Princeton, New Jersey.

Ripley, B. D. (1976). The second-order analysis of stationary point processes. *J. Appl. Prob.* **13**, 255–66.

Ripley, B. D. (1977). Modelling spatial patterns (with discussion). *J. Roy. Statist. Soc. B* **39**, 172–212.

Ripley, B. D. (1979). Test of 'randomness' for spatial point-patterns. *J. Roy. Statist. Soc. B* **41**, 368–74.

Ripley, B. D. (1981). *Spatial Statistics*. John Wiley & Sons, Inc., New York. (Reprinted in 2004.)

Ripley, B. D. (1982). Edge effects in spatial stochastic processes. In Ranneby, B., ed., *Statistics in Theory and Practice. Essays in Honour of Bertil Matérn*, pp. 242–62. Swedish University of Agricultural Sciences, Umea.

Ripley, B. D. (1988). *Statistical Inference for Spatial Processes*. Cambridge University Press, Cambridge.

Ripley, B. D. and Silverman, B. W. (1978). Quick tests for spatial interaction. *Biometrika* **65**, 641–2.

Rivier, N. and Lissowski, A. (1982). On the correlation between sizes and shapes of cells in epithelial mosaics. *J. Phys. A: Math. Gen.* **15**, L143–8.

Robbins, H. E. (1944). On the measure of a random set I. *Ann. Math. Statist.* **15**, 70–4.

Robbins, H. E. (1945). On the measure of a random set II. *Ann. Math. Statist.* **16**, 342–7.

Robins, V. (2002). Computational topology for point data: Betti numbers of α-shapes. In Mecke, K. R. and Stoyan, D., eds, *Morphology of Condensed Matter. Physics and Geometry of Spatially Complex Systems*, Lecture Notes in Physics **600**, pp. 261–74. Springer-Verlag, Berlin.

Rodrigez-Itube, I., Cox, D. R., and Isham, V. (1987). Some models for rainfall based on stochastic point processes. *Proc. Roy. Soc. London A* **410**, 269–88.

Rodrigez-Itube, I., Cox, D. R., and Isham, V. (1988). A point process model for rainfall: further developments. *Proc. Roy. Soc. London A* **417**, 283–98.

Rosenthal, J. S. (2006). *A First Look at Rigorous Probability Theory*. World Scientific, Singapore, 2nd edition.

Rosiwal, A. (1898). Über geometrische Gesteinsanalysen. Ein einfacher Weg zur ziffernmäßigen Feststellung des Quantitatsverhältnisses der Mineralbestandteile gemengter Gesteine. *Verh. K. K. Geol. Reichsanst.* **5/6**, 143–75.

Rother, W. and Zähle, M. (1990). A short proof of a principal kinematic formula and extensions. *Trans. Amer. Math. Soc.* **321**, 547–58.

Rother, W. and Zähle, M. (1992). Absolute curvature measures, II. *Geom. Dedicata* **41**, 229–40.

Royall, C. P., Dzubiella, J., Schmidt, M., and van Blöaderen, A. (2007). Nonequilibrium sedimentation of colloids on the particle scale. *Phys. Rev. Lett.* **98**, 188304.

Ruelle, D. (1969). *Statistical Mechanics*. John Wiley & Sons, Inc., New York.

Ruelle, D. (1970). Superstable interaction in classical mechanics. *Commun. Math. Phys.* **18**, 127–59.

Rychlik, I. and Lindgren, G. (1993). CROSSREG — A technique for first passage and wave density analysis. *Probab. Engrg. Inform. Sci.* **7**, 125–48.

Rysz, J. and Wiencek, K. (1980). Stereology of spherical carbide particles in steels. *Arch. Nauki o Materialach* **I**, 151–68.

Sahimi, M. (2003). *Heterogeneous Materials I: Linear Transport and Optical Properties*. Springer-Verlag, New York.

Salinetti, G. and Wets, R. J.-B. (1986). On the convergence in distribution of measurable multifunctions (random sets), normal integrands, stochastic processes and stochastic infima. *Math. Oper. Res.* **11**, 385–419.

Saltykov, S. A. (1945). *Stereometric Metallography* (in Russian). State Publishing House for Metals Sciences, Moscow.

Saltykov, S. A. (1950). *Introduction to Sterometric Metallography* (in Russian). Publishing House of Academy of Science of Armenian SSR, Erevan.

Saltykov, S. A. (1974). *Stereometrische Metallographie*. Deutscher Verlag für Grundstoffindustrie, Leipzig.

Sampson, W. W. (2001). The structural characterisation of fibre networks in papermaking processes a review. In Baker, C. F., ed., *The Science of Papermaking. Transactions of the 12th Fundamental Research Symposium*, pp. 1205–88. FRC The Pulp & Paper Fundamental Research Society.

Sampson, W. W. (2004). A model for fibre contact in planar random fibre networks. *J. Mater. Sci.* **39**, 2775–81.

Sandau, K. (1993). An estimation procedure for the joint distribution of spatial direction and thickness of flat bodies using vertical sections. Part I: Theoretical considerations. *Biometrical J.* **35**, 649–60.

Sandau, K. and Vogel, H.-J. (1993). An estimation procedure for the joint distribution of spatial direction and thickness of flat bodies using vertical sections. Part II: An application in soil micromorphology. *Biometrical J.* **35**, 661–75.

Santaló, L. (1976). *Integral Geometry and Geometric Probability.* Addison-Wesley, Reading, MA.

Santaló, L. (1980). Random lines and tessellations in a plane. *Stochastica* **4**, 3–13.

Santaló, L. (1984). Mixed random mosaics. *Math. Nachr.* **117**, 129–33.

Savary, L., Jeulin, D., and Thorel, A. (1999). Morphological analysis of carbon-polymer composite materials from thick sections. *Acta Stereol.* **18**, 297–303.

Saxl, I. (1989). *Stereology of Objects with Internal Structure.* Academia, Prague.

Saxl, I. and Ponížil, P. (2001). Grain size estimation: w–s diagram. *Mater. Char.* **46**, 113–18.

Saxl, I. and Ponížil, P. (2002). Bernoulli cluster field: Voronoi tessellations. *Appl. Math.* **47**, 157–67.

Saxl, I., Ponížil, P., and Sülleiova, K. (2003). Stereology and simulation of heterogeneous crystalline media. *Int. J. Mater. Prod. Tech.* **18**, 1–25.

Schaap, W. E. and van de Weygaert, R. (2000). Continuous fields and discrete samples. Reconstruction through Delaunay tessellations. *Astron. Astrophys.* **363**, L29–32.

Schack-Kirchner, H., Wilpert, K. V., and Hildebrand, E. E. (2000). The spatial distribution of soil hyphae in structured spruce-forest soils. *Plant Soil* **224**, 195–205.

Scheidegger, A. E. (1979). Beziehungen zwischen Orientierungsstruktur der Tallagen und der Kluftstellungen in Österreich. *Mitt. Österr. Geograph. Ges.* **121**, 187–95.

Scheike, T. H. (1994). Anisotropic growth of Voronoi cells. *Adv. Appl. Prob.* **26**, 43–53.

Schlather, M. (1999). Introduction to positive definite functions and to unconditional simulation of random fields. Technical Report ST 99-10, Lancaster University.

Schlather, M. (2000). A formula for the edge length distribution function of the Poisson Voronoi tessellation. *Math. Nachr.* **214**, 113–9.

Schlather, M. (2001a). On the second-order characteristics of marked point processes. *Bernoulli* **7**, 99–117.

Schlather, M. (2001b). Simulation and analysis of random fields. *R News* **1**, 10–20.

Schlather, M. and Stoyan, D. (1997). The covariance of the Steinen model. In Jeulin, D., ed., *Advances in Theory and Applications of Random Sets*, pp. 157–74. World Scientific, Singapore.

Schmidt, V. (1985). Poisson bounds for moments of shot noise processes. *Statistics* **16**, 253–62.

Schmidt, V. and Spodarev, E. (2005). Joint estimators for the specific intrinsic volumes of stationary random sets. *Stoch. Process. Appl.* **115**, 959–81.

Schmitt, M. (1991). Estimation of the density in a stationary Boolean model. *J. Appl. Prob.* **28**, 702–8.

Schneider, R. (1979a). Bestimmung konvexer Körper durch Krümmungsmasse. *Comment. Math. Helvet.* **54**, 42–60.

Schneider, R. (1979b). Boundary structure and curvature of convex bodies. In Tolke, J. and Wills, J. M., eds, *Contributions to Geometry. Proc. Geometrie-Symp. Siegen 1978*, pp. 13–59. Birkhäuser, Basel.

Schneider, R. (1980). Parallelmengen mit Vielfachheit und Steiner-Formeln. *Geom. Ded.* **9**, 111–27.

Schneider, R. (1987). Geometric inequalities for Poisson processes of convex bodies and cylinders. *Result Math.* **11**, 165–85.

Schneider, R. (1993). *Convex Bodies: The Brunn–Minkowski Theory.* Encyclopedia of Mathematics and Its Applications **44**. Cambridge University Press, Cambridge.

Schneider, R. and Weil, W. (1992). *Integralgeometrie*. B. G. Teubner, Stuttgart.

Schneider, R. and Weil, W. (2000). *Stochastische Geometrie*. B. G. Teubner, Stuttgart.

Schneider, R. and Weil, W. (2008). *Stochastic and Integral Geometry*. Springer-Verlag, Berlin.

Schneider, R. and Weil, W. (2010). Classical stochastic geometry. In Kendall, W. S. and Molchanov, I., eds, *New Perspectives in Stochastic Geometry*, pp. 1–42. Oxford University Press, Oxford.

Schreiber, T. (2010). Limit theorems in stochstic geometry. In Kendall, W. S. and Molchanov, I., eds, *New Perspectives in Stochastic Geometry*, pp. 111–44. Oxford University Press, Oxford.

Schulz, G. E. W., Schwan, L. O., and Franke, P. (1993). The mean normalized Euler characteristic of a simultaneously starting and growing 2D Voronoi tessellation with Poisson distributed nuclei. *J. Mater. Sci.* **28**, 2076–714.

Schüth, F., Sing, K. S. W., and Weitkamp, J., eds (2002). *Handbook of Porous Solids*. Wiley-VCH, Weinheim.

Schwandtke, A. (1985). Distributional analysis of dihedral angles in single-phase polycrystalline structures – comments on a paper by J. Rys and A. Kasprzyk. *Metalurgie i Odlewnictwo* **11**, 171–82.

Schwandtke, A. (1988). Second-order quantities for stationary weighted fibre processes. *Math. Nachr.* **139**, 321–34.

Schwandtke, A., Ohser, J., and Stoyan, D. (1987). Improved estimation in planar sampling. *Acta Stereol.* **6**, 325–34.

Schwandtke, A., Stoyan, D., and Schmidt, V. (1988). Some remarks on the stereological estimation of particle characteristics. *Acta Stereol.* **7**, 143–53.

Schwertel, J. and Stamm, H. (1997). Analysis and modelling of tessellations by means of image analysis methods. *J. Microsc.* **186**, 198–209.

Scott, G. D. (1960). Packing of spheres. *Nature* **188**, 908–9.

Serra, J. (1975). *Anisotropy fast characterization*. Fascicules de morphologie mathématique appliquée **8**. École de Mines de Paris, Fontainebleau.

Serra, J. (1982). *Image Analysis and Mathematical Morphology*. Academic Press, London.

Serra, J. (1987). Boolean random functions. *Acta Stereol.* **6**, 325–30.

Serra, J. (1988). *Image Analysis and Mathematical Morphology. Volume 2*. Academic Press, London.

Serra, J. (1989). Boolean random functions. *J. Microsc.* **156**, 41–63.

Serra, J. (2009). The random spread model. In Passare, M., ed., *Complex Analysis and Digital Geometry: Proceedings from the Kiselmanfest 2006*, pp. 283–310. Uppsala Universitet.

Shannon, C. E. (1949). Communication in the presence of noise. *Proc. IRE* **37**, 10–21.

Shepilov, M. P., Kalmykov, A. E., and Sycheva, G. A. (2006). Ordering effects in spatial arrangement of particles in phase separated sodium borosilicate glass. *Phys. Chem. Glass. Eur. J. Glass Sci. Technol. B* **47**, 339–43.

Sherman, M. (2011). *Spatial Statistics and Spatio-Temporal Data: Covariance Functions and Directional Properties*. John Wiley & Sons, Ltd, Chichester.

Shimatani, I. K. (2010). Spatially explicit neutral models for population genetics and community ecology: Extensions of the Neyman–Scott clustering process. *Theor. Popul. Biol.* **77**, 32–41.

Shimatani, K. (2002). Point processes for fine scale spatial genetics and molecular ecology. *Biometrical J.* **44**, 325–52.

Sibson, R. (1980). The Dirichlet tessellation as an aid in data analysis. *Scand. J. Statist.* **7**, 14–20.

Sibson, R. (1981). A brief description of natural neighbour interpolation. In Barnett, V., ed., *Interpreting Multivariate Data*, pp. 21–36. John Wiley & Sons, Ltd, Chichester.

Sigman, K. (1995). *Stationary Marked Point Processes: An Intuitive Approach*. Chapman & Hall, New York.

Silverman, B. W., Jones, M. C., Wilson, J. D., and Nychka, D. W. (1990). A smoothed EM approach to indirect estimation problems, with particular reference to stereology and emission tomography. *J. Roy. Statist. Soc. B* **52**, 271–324.

Singh, H., Gokhale, A. M., Mao, Y., and Spowart, J. E. (2006). Computer simulations of realistic microstructures of discontinuously reinforced aluminum alloy (DRA) composites. *Acta Materialia* **54**, 2131–43.

Sivakumar, K. and Goutsias, J. (1997a). Discrete morphological size distributions and densities: estimation techniques and applications. *J. Electron. Imaging* **6**, 31–53.

Sivakumar, K. and Goutsias, J. (1997b). Morphologically constrained discrete random sets. In Jeulin, D., ed., *Advances in Theory and Applications of Random Sets*, pp. 49–66. World Scientific, Singapore.

Sivakumar, K. and Goutsias, J. (1997c). On the discretization of morphological operators. *J. Vis. Commun. Image R.* **8**, 39–49.

Sivakumar, K. and Goutsias, J. (1999). Morphologically constrained GRFs: Applications to texture synthesis and analysis. *IEEE Trans. Pattern Anal. Mach. Intell.* **21**, 99–113.

Skare, Ø., Møller, J., and Jensen, E. B. V. (2007). Bayesian analysis of spatial point processes in the neighbourhood of Voronoi networks. *Statist. Comput.* **17**, 369–79.

SKM95 = Stoyan, Kendall and Mecke (1995).

Small, C. G. (1996). *The Statistical Theory of Shape*. Springer-Verlag, New York.

Smalley, I. J. (1966). Contraction crack networks in basalt flows. *Geol. Mag.* **103**, 110–4.

Smith, C. and Guttman, L. (1953). Measurement of internal boundaries in three-dimensional structures by random sectioning. *Trans. AIME* **197**, 81–7.

Snethlage, M., Martínez, V. J., Stoyan, D., and Saar, E. (2002). Point field models for the galaxy point pattern: Modelling the singularity of the two-point correlation function. *Astron. Astrophys.* **388**, 758–65.

Snyder, D. L. (1975). *Random Point Processes*. John Wiley & Sons, Inc., New York.

Snyder, D. L. and Miller, M. I. (1991). *Random Point Processes in Time and Space*. Springer-Verlag, New York, 2nd edition.

Soille, P. (1999). *Morphological Image Analysis*. Springer-Verlag, Berlin.

Soille, P. (2003). *Morphological Image Analysis: Principle and Applications*. Springer-Verlag, Berlin, 2nd edition.

Sok, R. M., Knackstedt, M. A., Sheppard, A. P., Pinczewski, W. V., Lindquist, W. B., Venkatarangan, A., and Paterson, L. (2002). Direct and stochastic generation of network models from tomographic images: Effect of topology on residual saturations. *Trans. Porous Med.* **46**, 345–71.

Solomon, H. (1978). *Geometric Probability*. SIAM, Philadephia.

Solomonoff, R. and Rapoport, A. (1951). Connectivity of random nets. *Bull. Math. Biophys.* **13**, 107–17.

Song, C., Wang, P., and Makse, H. A. (2008). A phase diagram for jammed matter. *Nature* **453**, 629–32.

Sonntag, U., Stoyan, D., and Hermann, H. (1981). Random set models in the interpretation of small-angle scattering data. *Phys. Stat. Sol. (a)* **68**, 281–8.

Spektor, A. G. (1950). Analysis of distribution of spherical particles in non-transparent structures. *Zavod. Lab.* **16**, 173–7.

Spiess, M. and Spodarev, E. (2011). Anisotropic Poisson processes of cylinders. *Methodol. Comput. Appl. Probab.* **13**, 801–19.

Srinivasan, S. K. (1974). *Stochastic Point Processes and Their Applications*. Hafner Press, New York.

Srinivasan, S. K. and Vijayakumar, A., eds (2003a). *Point Prcesses and Product Densities*. Narosa, New Delhi.

Srinivasan, S. K. and Vijayakumar, A., eds (2003b). *Stochastic Point Processes*. Narosa, New Delhi.

Stachurski, Z. H. (2011). On structure and properties of amorphous materials. *Materials* **4**, 1564–98.

Stapper, C. H., McLaren, A., and Dreckmann, M. (1980). Yield model for productivity optimization of VLSI memory chips with redundancy and partially good product. *IBM J. Res. Dev.* **24**, 398–409.

Stephan, H. (1975). *Allocortex*. Handbuch der mikroskopischen Anatomie des Menschen, Band IV, Nervensystem, Teil 9. Springer-Verlag, Berlin.

Sterio, D. G. (1984). The unbiased estimation of numbers and sizes of arbitrary particles using the disector. *J. Microsc.* **134**, 127–36.

Stiny, J. (1929). *Technische Gesteinskunde für Bauingenieure, Kulturtechniker, Land- und Forstwirte*. Springer-Verlag, Wien, 2nd edition.

Stoica, R. S., Martínez, V. J., Mateu, J., and Saar, E. (2005). Detection of cosmic filaments using the Candy model. *Astron. Astrophys.* **434**, 423–32.

Stoica, R. S., Martínez, V. J., and Saar, E. (2010). Filaments in observed and mock galaxy catalogues. *Astron. Astrophys.* **510**, A38.

Stoyan, D. (1979a). Interrupted point processes. *Biometrical J.* **21**, 607–10.

Stoyan, D. (1979b). On the accuracy of lineal analysis. *Biometrical J.* **21**, 439–49.

Stoyan, D. (1979c). Proof of some fundamental formulas of stereology for non-Poisson grain models. *Math. Operationsf. Statist., Ser. Optimization* **10**, 573–81.

Stoyan, D. (1981). On the second-order analysis of stationary planar fibre processes. *Math. Nachr.* **102**, 189–99.

Stoyan, D. (1982). Stereological formulae for size distributions through marked point processes. *Prob. Math. Statist.* **2**, 161–6.

Stoyan, D. (1983). Inequalities and bounds for variances of point processes and fibre processes. *Math. Operationsf. Statist., Ser. Statistics* **14**, 409–19.

Stoyan, D. (1984a). Estimating the volume density from thin sections. *Biometrical J.* **27**, 427–30.

Stoyan, D. (1984b). Further stereological formulae for spatial fibre processes. *Math. Operationsf. Statist., Ser. Statistics* **15**, 421–8.

Stoyan, D. (1984c). On correlations of marked point processes. *Math. Nachr.* **116**, 197–207.

Stoyan, D. (1984d). Weighted fibres and surfaces in stereology. In Ambartzumian, R. V. and Weil, W., eds, *Stochastic Geometry, Geometric Statistics, Stereology*, Teubner-Texte zur Mathematik **65**, pp. 188–96. B. G. Teubner Verlagsgesellschaft, Leipzig.

Stoyan, D. (1985a). Practicable methods for the determination of the pair correlation function of fibre processes. In Nagel, W., ed., *Geobild '85: Workshop on Geometrical Problems of Image Processing, Georgenthal (GDR), January 14–18, 1985: Proceedings*, pp. 131–40. Wissenschaftliche Beiträge der Friedrich-Schiller-Universität Jena.

Stoyan, D. (1985b). Stereological determination of orientations, second-order quantities and correlations for random fibre systems. *Biom J.* **27**, 411–25.

Stoyan, D. (1986). On generalized planar tessellations. *Math. Nachr.* **128**, 215–9.

Stoyan, D. (1987). Statistical analysis of spatial point processes: a soft-core model and cross-correlation of marks. *Biometrical J.* **29**, 971–80.

Stoyan, D. (1988). Thinnings of point processes and their use in the statistical analysis of a settlement pattern with deserted villages. *Statistics* **19**, 45–56.

Stoyan, D. (1989). Statistical inference for a Gibbs point process of mutually non-intersecting discs. *Biometrical J.* **31**, 153–61.

Stoyan, D. (1990a). Stereological formulae for a random system of non-overlapping spheres. *Statistics* **21**, 131–6.

Stoyan, D. (1990b). Stereology and stochastic geometry. *Int. Statist. Rev.* **58**, 227–42.

Stoyan, D. (1991). Describing the anisotropy of marked planar point processes. *Statistics* **22**, 449–62.

Stoyan, D. (1993). A spatial statistical analysis of a work of art: did Hans Arp make a "truly random" collage? *Statistics* **24**, 71–80.

Stoyan, D. (1994). Caution with "fractal" point-patterns! *Statistics* **25**, 267–70.

Stoyan, D. (1998). Random sets: Models and statistics. *Int. Statist. Rev.* **66**, 1–27.

Stoyan, D. and Beneš, V. (1991). Anisotropy analysis for particle systems. *J. Microsc.* **164**, 159–68.

Stoyan, D., Davtyan, A., and Turetayev, D. (2002). Shape statistics for random domains and particles. In Mecke, K. R. and Stoyan, D., eds, *Morphology of Condensed Matter. Physics and Geometry of Spatially Complex Systems*, Lecture Notes in Physics 66, pp. 299–316. Springer-Verlag, Berlin.

Stoyan, D. and Gerlach, W. (1987). Stereological determination of curvature distributions of spatial fibre systems. *J. Microsc.* **148**, 297–305.

Stoyan, D. and Gloaguen, R. (2011). Nucleation and growth of geological faults. *Nonlin. Processes Geophys.* **18**, 529–36.

Stoyan, D. and Grabarnik, P. (1991). Second-order characteristics for stochastic structures connected with Gibbs point processes. *Math. Nachr.* **151**, 95–100.

Stoyan, D., Kendall, W. S., and Mecke, J. (1995). *Stochastic Geometry and its Applications*. John Wiley & Sons, Ltd, Chichester, 2nd edition. (Reprinted in 2008.)

Stoyan, D. and Mecke, J. (1983a). *Stochastische Geometrie*. Akademie-Verlag, Berlin.

Stoyan, D. and Mecke, J. (1983b). *Stochastische Geometrie Eine Einführung*. Akademie-Verlag, Berlin.

Stoyan, D., Mecke, J., and Pohlmann, S. (1980). Formulas for stationary planar fibre processes II — Partially oriented fibre systems. *Math. Operationsf. Statist., Ser. Statistics* **11**, 281–6.

Stoyan, D. and Mecke, K. (2005). The Boolean model: from Matheron till today. In Bilodeau, M., Meyer, F., and Schmitt, M., eds, *Space, Structure and Randomness*, Lecture Notes in Statistics **183**, pp. 151–81. Springer-Verlag, New York.

Stoyan, D. and Molchanov, I. S. (1997). Set-valued means of random particles. *J. Math. Imaging Vis.* **7**, 111–21.

Stoyan, D. and Ohser, J. (1982). Correlations between planar random structures (with an ecological example). *Biometrical J.* **24**, 631–47.

Stoyan, D. and Ohser, J. (1984). Cross-correlation measure of weighted random measures and their estimation. *Teor. Verojatn. Primen.* **29**, 338–47.

Stoyan, D. and Penttinen, A. (2000). Recent applications of point process methods in forestry statistics. *Statist. Sci.* **15**, 61–78.

Stoyan, D. and Schlather, M. (2000). Random sequential adsorption: relationship to dead leaves and characterization of variability. *J. Stat. Phys.* **100**, 969–79.

Stoyan, D. and Steyer, H.-L. (1979). Zur Genauigkeit der Linearanalyse. *Neue Hütte* **24**, 303–7.

Stoyan, D. and Stoyan, H. (1980a). Gedanken zur Entstehung der Säulenformen bei Basalten. *Z. Geol. Wiss.* **8**, 1529–37.

Stoyan, D. and Stoyan, H. (1980b). On some partial orderings of random closed sets. *Math. Operationsf. Statist., Ser. Optimization* **11**, 145–54.

Stoyan, D. and Stoyan, H. (1983). Über eine Methode zur Quantifizierung von Korrelationen zwischen geologischen Liniensystemen. *Z. angew. Geologie* **29**, 512–17.

Stoyan, D. and Stoyan, H. (1985). On one of Matérn's hard-core point process models. *Math. Nachr.* **122**, 205–14.

Stoyan, D. and Stoyan, H. (1986). Simple stochastic models for the analysis of dislocation distributions. *Phys. Status Solidi A* **97**, 163–72.

Stoyan, D. and Stoyan, H. (1994). *Fractals, Random Shapes and Point Fields*. John Wiley & Sons, Ltd, Chichester.

Stoyan, D. and Stoyan, H. (1996). Estimating pair-correlation functions of planar cluster processes. *Biometrical J.* **38**, 259–71.

Stoyan, D. and Stoyan, H. (2000). Improving ratio estimators of second order point process characteristics. *Scand. J. Statist.* **27**, 641–56.

Stoyan, D., Stoyan, H., Tscheschel, A., and Mattfeldt, T. (2001). On the estimation of distance distribution functions for point processes and random sets. *Image Anal. Stereol.* **20**, 65–9.

Stoyan, D., von Wolfersdorf, L., and Ohser, J. (1990). Stereological problems for spherical particles–second-order theory. *Math. Nachr.* **146**, 33–46.

Stoyan, D., Wagner, A., Hermann, H., and Elsner, A. (2011). Statistical characterization of the pore space of random systems of hard spheres. *J. Non-Cryst. Solids* **357**, 1508–15.

Streit, R. L. (2010). *Poisson Point Processes: Imaging, Tracking and Sensing*. Springer-Verlag, New York.

Stroeven, M., Askes, H., and Sluys, L. J. (2004). Numerical determination of representative volumes for granular materials. *Comput. Methods Appl. Mech. Engrg.* **193**, 3221–38.

Stroeven, P. (2000). A stereological approach to roughness of fracture surfaces and tortuosity of transport paths in concrete. *Cement Concrete Comp.* **22**, 331–41.

Stroeven, P., Le, N. L. B., Sluys, L. J., and He, H. (2012). Porosimetry by random node structuring in virtual concrete. *Image Anal. Stereol.* **31**, 79–87.

Sukiasian, G. S. (1978). Random triangles on the plane (in Russian). *Akad. Nauk Armjan. SSR Dokl.* **66**, 150–5.

Sukiasian, G. S. (1980). Processes of chords on lines intersecting random circle fields on a plane (in Russian). *Akad. Nauk Armyan. SSR Dokl.* **70**, 297–300.

Sukiasian, G. S. (1982). Random sections of polyhedra (in Russian). *Dokl. Akad. Nauk SSSR* **263**, 809–12.

Sukiasian, G. S. (1987). Randomizable point systems. *Acta Appl. Math.* **9**, 83–95.

Sulanke, R. (1961). Die Verteilung der Sehnenlängen an ebenen und räumlichen Figuren. *Math. Nachr.* **23**, 51–74.

Suwa, N., Takahashi, T., Saito, K., and Sawai, T. (1976). Morphometrical method to estimate the parameters of distribution functions assumed for spherical bodies from measurements on a random section. *Tohoku J. Exp. Med.* **118**, 101–18.

Szekli, R. (1995). *Stochastic Ordering and Dependence in Applied Probability*. Lecture Notes in Statistics **97**. Springer-Verlag, New York.

Talbot, J., Tarjus, G., van Tassel, P. R., and Viot, P. (2000). From car parking to protein adsorption: an overview of sequential adsorption processes. *Colloids Surfaces A: Physicochem. Eng. Aspects* **165**, 287–324.

Talukdar, M. S., Torsaeter, O., and Ioannidis, M. A. (2002). Stochastic reconstruction of particulate media from two-dimensional images. *J. Colloid Interface Sci.* **248**, 419–28.

Talukdar, M. S., Torsaeter, O., Ioannidis, M. A., and Howard, J. J. (2002a). Stochastic reconstruction, 3D characterization and network modeling of chalk. *J. Petrol. Sci. Eng.* **35**, 1–21.

Talukdar, M. S., Torsaeter, O., Ioannidis, M. A., and Howard, J. J. (2002b). Stochastic reconstruction of chalk from 2D images. *Trans. Porous Med.* **48**, 101–23.

Tanaka, U., Ogata, Y., and Stoyan, D. (2008). Parameter estimation and model selection for Neyman-Scott point processes. *Biometrical J.* **50**, 43–57.

Tanemura, M. (1988). Random packing and random tessellation in relation to the dimension of space. *J. Microsc.* **151**, 247–55.

Tanner, J. C. (1983a). Polygons formed by random lines in a plane: some further results. *J. Appl. Prob.* **20**, 778–87.

Tanner, J. C. (1983b). The proportion of quadrilaterals formed by random lines in a plane. *J. Appl. Prob.* **20**, 400–4.

Taylor, C. C. (1983). A new method for unfolding sphere-size distributions. *J. Microsc.* **132**, 57–66.

Teichmann, J., Ballani, F., and van den Boogaart, K. G. (2013). Generalizations of Matérn's hard-core point processes. *Spatial Statistics* **3**, 33–53.

Thäle, C. and Weiss, V. (2010). New mean values for homogeneous spatial tessellations that are stable under iteration. *Image Anal. Stereol.* **29**, 143–57.

Thall, P. F. (1983). A theorem on regular infinitely divisible Cox processes. *Stoch. Process. Appl.* **16**, 205–10.

Thiedmann, R., Hartnig, C., Manke, I., Schmidt, V., and Lehnert, W. (2009). Local structural characteristics of pore space in GDLs of PEM fuel cells based on geometric 3D graphs. *J. Electrochem. Soc.* **156**, B1339–47.

Thiedmann, R., Spettl, A., Stenzel, O., Zeibig, T., Hindson, J. C., Saghi, Z., Greenham, N. C., Midgley, P. A., and Schmidt, V. (2012). Networks of nanoparticles in organic-inorganic composites: algorithmic extraction and statistical analysis. *Image Anal. Stereol.* **31**, 27–42.

Thiessen, A. H. and Alter, J. C. (1911). Climatological data for July, 1911. District No. 11, Great Basin. *Monthly Weather Review* **39**, 1082–4.

Thomas, C. (1984). Extremum properties of the intersection densities of stationary Poisson hyperplane processes. *Math. Operationsf. Statist., Ser. Statistics* **15**, 443–50.

Thompson, E. (1930). Quantitative microscopic analysis. *J. Geol.* **38**, 193–222.

Thönnes, E. (2001). The conditional Boolean model revisited. *Markov Process. Related Fields* **7**, 77–96.

Thorisson, H. (2000). *Coupling, Stationarity, and Regeneration.* Springer-Verlag, New York.

Thovert, J.-F. and Adler, P. M. (2004). Trace analysis for fracture networks of any convex shape. *Geophys. Res. Lett.* **31**, L22502.

Thovert, J.-F. and Adler, P. M. (2011). Grain reconstruction of porous media: Application to a Bentheim sandstone. *Phys. Rev. E* **83**, 056116.

Thovert, J.-F., Yousefian, F., Spanne, P., Jacquin, C. G., and Adler, P. M. (2001). Grain reconstruction of porous media: Application to a low-porosity Fontainebleau sandstone. *Phys. Rev. E* **63**, 061307.

Tong, C. S., Choy, S. K., Chiu, S. N., Zhao, Z. Z., and Liang, Z. T. (2008). Characterization of shapes for use in classification of starch grains images. *Microsc. Res. Tech.* **71**, 651–8.

Torquato, S. (1991). Random heterogeneous media: Microstructure and improved bounds on effective properties. *Appl. Mech. Rev.* **44**, 37–76.

Torquato, S. (2002). *Random Heterogeneous Materials: Microstructure and Macroscopic Properties.* Springer-Verlag, New York.

Torquato, S. (2006). Necessary conditions on realizable two-point correlation functions of random media. *Ind. Eng. Chem. Res.* **45**, 6923–8.

Torquato, S. (2010). Optimal design of heterogeneous materials. *Annu. Rev. Mater. Res.* **40**, 101–29.

Torquato, S. (2012). Effect of dimensionality on the continuum percolation of overlapping hyperspheres and hypercubes. *J. Chem. Phys.* **136**, 054106.

Torquato, S. and Lu, B. (1993). Chord-length distribution function for two-phase random media. *Phys. Rev. E* **47**, 2950–3.

Torquato, S. and Stillinger, F. H. (2010). Jammed hard-particle packings: From Kepler to Bernal and beyond. *Rev. Mod. Phys.* **82**, 2633–72.

Tscheschel, A. and Chiu, S. N. (2008). Quasi-plus sampling edge correction for spatial point patterns. *Comput. Statist. Data Anal.* **52,** 5287–95.

Tscheschel, A. and Stoyan, D. (2003). On the estimation variance for the specific Euler--Poincaré characteristic of random networks. *J. Microsc.* **211,** 80–8.

Tscheschel, A. and Stoyan, D. (2006). Statistical reconstruction of random point patterns. *Comput. Statist. Data Anal.* **51,** 859–71.

Underwood, E. E. (1970). *Quantitative Stereology.* Addison-Wesley, Reading, MA.

Underwood, E. E. (1976). Appendix Basic Stereology. In *Proc Fourth Int. Cong. for Stereology 1975,* pp. 509–13. National Bureau of Standards Special Publ. 431. NBS, Washington, DC.

van Dalen, G., Nootenboom, P., van Vliet, L. J., Voortman, L., and Esveld, E. (2007). 3D imaging, analysis and modelling of porous cereal products using X-ray microtomography. *Image Anal. Sterol.* **26,** 169–77.

van de Laan, M. J. (1995). Efficiency of the NPMLE in the line-segment problem. *Scand. J. Statist.* **23,** 527–50.

van de Weygaert, R. (1994). Fragmenting the universe III. The construction and statistics of 3-D Voronoi tessellations. *Astron. Astrophys.* **283,** 361–406.

van de Weygaert, R. (2007). Voronoi tessellations and the cosmic web: spatial patterns and clustering across the universe. In Gold, C. M., ed., *4th International Symposium on Voronoi Diagrams in Science and Engineering,* pp. 230–9. IEEE Computer Society, Los Alamitos, CA.

van de Weygaert, R. and Icke, V. (1989). Fragmenting the universe II. Voronoi vertices as Abell clusters. *Astron. Astrophys.* **213,** 1–9.

van de Weygaert, R., Vegter, G., Ritzerveld, J., and Icke, V., eds (2014). *Tessellations in the Sciences: Virtues, Techniques and Applications of Geometric Tilings.* Forthcoming.

van der Hofstad, R. (2010a). Percolation and random graphs. In Kendall, W. S. and Molchanov, I., eds, *New Perspectives in Stochastic Geometry,* pp. 173–247. Oxford University Press, Oxford.

van der Hofstad, R. (2010b). *Random Graphs and Complex Networks.* http://www.win.tue.nl/ rhofstad/NotesRGCN2010.pdf.

van Es, B. (1991). *Aspects of Nonparametric Density Estimation.* CWI Tracts **77,** Centrum voor Wiskunde en Informatica, Amsterdam.

van Es, B. and Hoogendoorn, A. W. (1990). Kernel estimation in Wicksell's corpuscle problem. *Biometrika* **77,** 139–45.

van Lieshout, M. N. M. (1995). *Stochastic Geometry Models in Image Analysis and Spatial Statistics.* CWI Tracts **108,** Centrum voor Wiskunde en Informatica, Amsterdam.

van Lieshout, M. N. M. (1999). Size-biased random closed sets. *Pattern Recognition* **32,** 1631–44.

van Lieshout, M. N. M. (2000). *Markov Point Processes and Their Applications.* Imperial College Press, London.

van Lieshout, M. N. M. (2006). Maximum likelihood estimation for random sequential adsorption. *Adv. Appl. Prob.* **38,** 889–98.

van Lieshout, M. N. M. (2012). An introduction to planar random tessellation models. *Spatial Statistics* **1,** 40–9.

van Lieshout, M. N. M, and Baddeley, A. J. (1996). A non-parametric measure of spatial interaction in point patterns. *Statist. Neerl.* **50,** 344–61.

van Lieshout, M. N. M. and Stoica, R. S. (2003). The Candy model: properties and inference. *Stat. Neerl.* **57,** 177–206.

van Zwet, E. W. (2004). Laslett's line segment problem. *Bernoulli* **10,** 377–96.

Vanderbei, R. P. and Shepp, L. A. (1988). A probabilistic model for the time to unravel a strand of DNA. *Stochastic Models* **4,** 299–314.

Vanmarcke, E. (2010). *Random Fields: Analysis and Synthesis*. World Scientific, Singapore, revised and expanded edition.

Villa, E. and Rios, P. R. (2010). Transformation kinetics for surface and bulk nucleation. *Acta Materialia* **58**, 2752–68.

Vogel, H.-J. (2002). Topological characterization of porous media. In Mecke, K. R. and Stoyan, D., eds, *Morphology of Condensed Matter. Physics and Geometry of Spatially Complex Systems*, Lecture Notes in Physics **600**, pp. 75–92. Springer-Verlag, Berlin.

Voloshin, V. P., Anikeenko, A. V., Medvedev, N. N., Geiger, A., and Stoyan, D. (2010). Hydration shells in Voronoi tessellations. In Mostafavi, M. A., ed., *Seventh International Symposium on Voronoi Diagrams in Science and Engineerng*, pp. 254–9. IEEE Computer Society, Los Alamitos, CA.

von Economo, C. F. and Koskinas, G. N. (1925). *Die Cytoarchitektonik der Hirnrinde des erwachsenen Menschen*. Springer-Verlag, Wien.

Voronoi, G. (1908). Nouvelles applications des parametres continus a la theorie des formes quadratiques. *J. Reine angew. Math.* **134**, 198–287.

Voss, F., Gloaguen, C., Fleischer, F., and Schmidt, V. (2011). Densities of shortest path lengths in spatial stochastic networks. *Stochastic Models* **27**, 141–67.

Voss, F., Gloaguen, C., and Schmidt, V. (2009). Capacity distributions in spatial stochastic models for telecommunication networks. *Image Anal. Stereol.* **28**, 155–63.

Voss, F., Gloaguen, C., and Schmidt, V. (2010). Scaling limits for shortest path lengths along the edges of stationary tessellations. *Adv. Appl. Prob.* **42**, 936–52.

Voss, K. (1980). Exakte stereologische Formeln und Näherungslösungen für konvexe Körper. *J. Inf. Process. Cybern.* **16**, 485–91.

Voss, K. (1982). Frequencies of n-polygons in planar sections of polyhedra. *J. Microsc.* **128**, 111–20.

Voss, K. and Schubert, W. (1975). Zur numerischen Auswertung von Schnittflächenverteilungen. *Biometrical J.* **17**, 189–95.

Voss, K. and Stoyan, D. (1985). On the stereological estimation of numerical density of particle systems by an object counting method. *Biometrical J.* **27**, 919–24.

Warren, W. G. (1971). The centre satellite concept as a basis for ecological sampling. In Patil, G., Pielou, E. C., and Waters, W. E., eds, *Statistical Ecology, Volume 2*, pp. 87–118. Pennsylvania State University Press, University Park.

Watson, G. S. (1971). Estimation functionals of particle size distributions. *Biometrika* **58**, 483–90.

Watts, D. J. and Strogatz, S. H. (1998). Collective dynamics of 'small-world' networks. *Nature* **393**, 440–2.

Weaire, D. (1974). Some remarks on the arrangement of grains in a polycrystal. *Metallography* **7**, 157–60.

Weber, W. (1977). Zur Methode der Lokalisierung und Charakterisierung tiefer Bruchstrukturen für minerogenetische Untersuchungen. In Weber, W. and Korcemagin, V. A., eds, *Tiefenbruchstrukturen und postmagmatische Mineralization*. Freiberger Forschungshefte C **329**, pp. 9–52. Deutscher Verlag für Grundstoffindustrie, Leipzig.

Weese, J. (1995). Density estimation and regularization at the example of Wicksell's corpuscle problem. Technical report, Materialforschungszentrum Freiburg.

Weese, J., Korat, E., Maier, D., and Honerkamp, J. (1997). Unfolding sphere size distributions with a density estimator based on Tikhonov regularization. *J. Comput. Phys.* **138**, 331–53.

Weibel, E. R. (1980). *Stereological Methods. Volume 2: Theoretical Foundations*. Academic Press, London.

Weil, W. (1982a). An application of the central limit theorem for Banach-space-valued random variables to the theory of random sets. *Probab. Theory Related Fields* **60**, 203–8.

Weil, W. (1982b). Inner contact probabilities for convex bodies. *Adv. Appl. Prob.* **14,** 582–99.

Weil, W. (1984). Densities of quermassintegrals for stationary random sets. In Ambartzumian, R. V. and Weil, W., eds, *Stochastic Geometry, Geometric Statistics, Stereology,* Teubner-Texte zur Mathematik **65,** pp. 233–47. B. G. Teubner Verlagsgesellschaft, Leipzig.

Weil, W. (1988). Expectation formulas and isoperimetric properties for non-isotropic Boolean models. *J. Microsc.* **151,** 235–45.

Weil, W. (1995). The estimation of mean shape and mean particle number in overlapping particle systems in the plane. *Adv. Appl. Prob.* **27,** 102–19.

Weil, W. (1997). The mean normal distribution of stationary random sets and particle processes. In Jeulin, D., ed., *Advances in Theory and Applications of Random Sets,* pp. 21–33. World Scientific, Singapore.

Weil, W. (2007). Random sets (in particular Boolean models). In Weil, W., ed., *Stochastic Geometry,* Lecture Notes in Mathematics **1892,** pp. 185–245. Springer-Verlag, Berlin.

Weil, W. and Wieacker, J. A. (1987). A representation theorem for random sets. *Prob. Math. Statist.* **9,** 147–51.

Weiss, V. (1995). Second-order quantities for random tessellations of $\mathbb{R}^d$. *Stoch. Stoch. Rep.* **55,** 195–205.

Weiss, V. and Cowan, R. (2011). Topological relationships in spatial tessellations. *Adv. Appl. Prob.* **43,** 963–84.

Weiss, V. and Nagel, W. (1994). Second-order stereology for planar fibre processes. *Adv. Appl. Prob.* **26,** 906–18.

Weiss, V., Ohser, J., and Nagel, W. (2010). Second moment measure and K-function for planar STIT tessellations. *Image Anal. Stereol.* **29,** 121–31.

Weiss, V. and Zähle, M. (1988). Geometric measures for random curved mosaics of $\mathbb{R}^d$. *Math. Nachr.* **138,** 313–26.

Wicksell, S. D. (1925). The corpuscle problem I. *Biometrika* **17,** 84–9.

Wicksell, S. D. (1926). The corpuscle problem II. *Biometrika* **18,** 152–72.

Widom, J. S. and Rowlinson, B. (1970). New model for the study of liquid–vapour phase transitions. *J. Chem. Phys.* **52,** 1670–84.

Wieacker, J. A. (1985). Intersections of random hypersurfaces and visibility. *Prob. Theory Related Fields* **71,** 405–33.

Wieacker, J. A. (1989). Geometric inequalities for random surfaces. *Math. Nachr.* **142,** 73–106.

Wiencek, K. and Stoyan, D. (1993). Spatial correlations in metal structures and their analysis, II: The covariance. *Mater. Char.* **31,** 47–53.

Wijers, B. J. (1995). Consistent non-parametric estimation for a one-dimensional line segment process observed in an interval. *Scand. J. Statist.* **22,** 335–60.

Wilder, R. L. (1963). *Topology of Manifolds.* American Mathematical Society, New York.

Willis, J. R. (1978). Variational principles and bounds for the overall properties of composites. In Provan, J. W., ed., *Continuum Models of Discrete Systems,* pp. 185–215. University of Waterloo Press, Waterloo.

Wilson, J. D. (1987). A smoothed EM algorithm for the solution of Wicksell's corpuscle problem. *J. Statist. Comput. Simul.* **3,** 195–221.

Winkler, G. (2003). *Image Analysis, Random Fields and Markov Chain Monte Carlo Methods: a Mathematical Introduction.* Springer-Verlag, New York.

Wirjadi, O., Schladitz, K., Rack, A., and Breuel, T. (2009). Applications of anisotropic image filters for computing 2D and 3D-fiber orientations. In Capasso, V., Aletti, G., and Micheletti, A., eds, *Stereology and Image Analysis: ECS10 – 10th European Congress of ISS,* The MIRIAM Project Series **4.** Societa Editrice Esculapio–Progetto Leonardo, Bologna.

Worsley, K. J. (1994). Local maxima and the expected Euler characteristic of excursion sets of χ^2, F and t fields. *Adv. Appl. Prob.* **26**, 13–42.

Worsley, K. J. (1997). The geometry of random images. *Chance* **9**, 27–40.

Wu, H.-I. and Schmidt, P. W. (1973). Intersect distributions and small-angle X-ray scattering theory III. The intersect distribution for an ellipsoid. *J. Appl. Cryst.* **6**, 66–72.

Wu, H.-I., Sharpe, P. J. H., Walker, J., and Penridge, L. K. (1985). Ecological field theory: a spatial analysis of resource interference among plants. *Ecol. Modelling* **29**, 215–43.

Yadin, M. and Zacks, S. (1985). The visibilty of stationary and moving targets in the plane subject to shadowing elements. *J. Appl. Prob.* **22**, 776–86.

Yadin, M. and Zacks, S. (1988). Visibility probabilities on line segments in three-dimensional spaces subject to random Poisson fields of obscuring spheres. *Naval Res. Logist. Quart.* **35**, 558–69.

Yeh, J. (2006). *Real Analysis. Theory of Measure and Integration*. World Scientific, Singapore, 2nd edition.

Yeong, C. L. Y. and Torquato, S. (1998a). Reconstructing random media. *Phys. Rev. E* **57**, 495–506.

Yeong, C. L. Y. and Torquato, S. (1998b). Reconstructing random media. II. Three-dimensional media from two-dimensional cuts. *Phys. Rev. E* **58**, 224–33.

Zachary, C. E. and Torquato, S. (2011). Improved reconstructions of random media using dilation and erosion processes. *Phys. Rev. E* **84**, 056102.

Zacks, S. (1994). *Stochastic Visibility in Random Fields*. Lecture Notes in Statistics **95**. Springer-Verlag, Berlin.

Zähle, M. (1982). Random processes of Hausdorff-rectifiable closed sets. *Math. Nachr.* **108**, 49–72.

Zähle, M. (1983). Random set processes in homogeneous Riemannian spaces. *Math. Nachr.* **110**, 179–93.

Zähle, M. (1984a). Curvature measures and random sets I. *Math. Nachr.* **119**, 327–39.

Zähle, M. (1984b). Properties of signed curvature measures. In Ambartzumian, R. V. and Weil, W., eds, *Stochastic Geometry, Geometric Statistics, Stereology*, Teubner-Texte zur Mathematik **65**, pp. 256–66. B. G. Teubner Verlagsgesellschaft, Leipzig.

Zähle, M. (1984c). Thick section stereology for random fibres. *Math. Operationsf. Statist., Ser. Statistics* **15**, 429–36.

Zähle, M. (1986). Curvature measures and random sets II. *Probab. Theory Related Fields* **71**, 37–58.

Zähle, M. (1987a). Curvatures and currents for unions of sets of positive reach. *Geom. Dedicata* **23**, 155–71.

Zähle, M. (1987b). Polyhedron theorems for non-smooth cell complexes. *Math. Nachr.* **131**, 299–310.

Zähle, M. (1988). Random cell complexes and generalized sets. *Ann. Prob.* **16**, 1742–66.

Zähle, M. (1989). Absolute curvature measures. *Math. Nachr.* **140**, 83–90.

Zähle, M. (1990). A kinematic formula and moment measures of random sets. *Math. Nachr.* **149**, 325–40.

Zähle, U. (1984d). Local interpretation of Palm distributions of surface measures. *Math. Nachr.* **119**, 341–56.

Zaninetti, L. (2006). On the large-scale structure of the universe as given by the Voronoi diagrams. *Chin. J. Astron. Astrophys.* **6**, 387–95.

Zessin, H. (1983). The method of moments for random measures. *Probab. Theory Related Fields* **62**, 395–409.

Zhang, C., Zhang, Y., and Fang, Y. (2006). Localized coverage boundary detection for wireless sensor networks. In *Proceedings of the 3rd International Conference on Quality Service in Heterogeneous Wired/wireless Networks*, Waterloo, Ontario, Canada. Session: Algorithms in sensor networks, Article number: 12.

Zhang, C., Zhang, Y., and Fang, Y. (2009). Localized algorithms for coverage boundary detection in wireless sensor networks. *Wireless Networks* **15,** 3–20.

Ziegel, J. and Kiderlen, M. (2010a). Estimation of surface area and surface area measure of three-dimensional sets from digitizations. *Image Vision Comput.* **28,** 64–77.

Ziegel, J. and Kiderlen, M. (2010b). Stereolgical estimation of surface area from digital images. *Image Anal. Stereol.* **29,** 99–110.

Zuyev, S. (1999). Stopping sets: Gamma-type results and hitting properties. *Adv. Appl. Prob.* **31,** 355–66.

Zuyev, S. (2010). Stochastic geometry and telecommunications networks. In Kendall, W. S. and Molchanov, I., eds, *New Perspectives in Stochastic Geometry*, pp. 520–54. Oxford University Press, Oxford.

Zuyev, S. and Quintanilla, J. (2003). Estimation of percolation thresholds via percolation in inhomogeneous media. *J. Math. Phys.* **44,** 6040–6.

Zuyev, S. A. and Sidorenko, A. F. (1985a). Continuous models of percolation theory. I. *Teoret. Mat. Fiz.* **62,** 76–86. (English translation: *Theor. Math. Phys.* **62,** 51–8).

Zuyev, S. A. and Sidorenko, A. F. (1985b). Continuous models of percolation theory. II. *Teoret. Mat. Fiz.* **62,** 253–262. (English translation: *Theor. Math. Phys.* **62,** 171–7).

Author index

Subject index

When a term is referred to more than one page, number(s) in **bold** refer to the page(s) containing the definition or technical details.

WILEY SERIES IN PROBABILITY AND STATISTICS

The *Wiley Series in Probability and Statistics* is well established and authoritative. It covers many topics of current research interest in both pure and applied statistics and probability theory. Written by leading statisticians and institutions, the titles span both state-of-the-art developments in the field and classical methods.

Reflecting the wide range of current research in statistics, the series encompasses applied, methodological and theoretical statistics, ranging from applications and new techniques made possible by advances in computerized practice to rigorous treatment of theoretical approaches.

This series provides essential and invaluable reading for all statisticians, whether in academia, industry, government, or research.

*Now available in a lower priced paperback edition in the Wiley Classics Library.
†Now available in a lower priced paperback edition in the Wiley-Interscience Paperback Series.

BARTOSZYNSKI and NIEWIADOMSKA-BUGAJ · Probability and Statistical Inference, *Second Edition*

BASILEVSKY · Statistical Factor Analysis and Related Methods: Theory and Applications

BATES and WATTS · Nonlinear Regression Analysis and Its Applications

BECHHOFER, SANTNER, and GOLDSMAN · Design and Analysis of Experiments for Statistical Selection, Screening, and Multiple Comparisons

BEIRLANT, GOEGEBEUR, SEGERS, TEUGELS, and DE WAAL · Statistics of Extremes: Theory and Applications

BELSLEY · Conditioning Diagnostics: Collinearity and Weak Data in Regression

† BELSLEY, KUH, and WELSCH · Regression Diagnostics: Identifying Influential Data and Sources of Collinearity

BENDAT and PIERSOL · Random Data: Analysis and Measurement Procedures, *Fourth Edition*

BERNARDO and SMITH · Bayesian Theory

BHAT and MILLER · Elements of Applied Stochastic Processes, *Third Edition*

BHATTACHARYA and WAYMIRE · Stochastic Processes with Applications

BIEMER, GROVES, LYBERG, MATHIOWETZ, and SUDMAN · Measurement Errors in Surveys

BILLINGSLEY · Convergence of Probability Measures, *Second Edition*

BILLINGSLEY · Probability and Measure, *Anniversary Edition*

BIRKES and DODGE · Alternative Methods of Regression

BISGAARD and KULAHCI · Time Series Analysis and Forecasting by Example

BISWAS, DATTA, FINE, and SEGAL · Statistical Advances in the Biomedical Sciences: Clinical Trials, Epidemiology, Survival Analysis, and Bioinformatics

BLISCHKE and MURTHY (editors) · Case Studies in Reliability and Maintenance

BLISCHKE and MURTHY · Reliability: Modeling, Prediction, and Optimization

BLOOMFIELD · Fourier Analysis of Time Series: An Introduction, *Second Edition*

BOLLEN · Structural Equations with Latent Variables

BOLLEN and CURRAN · Latent Curve Models: A Structural Equation Perspective

BOROVKOV · Ergodicity and Stability of Stochastic Processes

BOSQ and BLANKE · Inference and Prediction in Large Dimensions

BOULEAU · Numerical Methods for Stochastic Processes

* BOX · Bayesian Inference in Statistical Analysis

BOX · Improving Almost Anything, *Revised Edition*

* BOX and DRAPER · Evolutionary Operation: A Statistical Method for Process Improvement

BOX and DRAPER · Response Surfaces, Mixtures, and Ridge Analyses, *Second Edition*

BOX, HUNTER, and HUNTER · Statistics for Experimenters: Design, Innovation, and Discovery, *Second Editon*

BOX, JENKINS, and REINSEL · Time Series Analysis: Forcasting and Control, *Fourth Edition*

BOX, LUCEÑO, and PANIAGUA-QUIÑONES · Statistical Control by Monitoring and Adjustment, *Second Edition*

* BROWN and HOLLANDER · Statistics: A Biomedical Introduction

CAIROLI and DALANG · Sequential Stochastic Optimization

*Now available in a lower priced paperback edition in the Wiley Classics Library.
†Now available in a lower priced paperback edition in the Wiley-Interscience Paperback Series.

CASTILLO, HADI, BALAKRISHNAN, and SARABIA · Extreme Value and Related Models with Applications in Engineering and Science

CHAN · Time Series: Applications to Finance with R and S-Plus®, *Second Edition*

CHARALAMBIDES · Combinatorial Methods in Discrete Distributions

CHATTERJEE and HADI · Regression Analysis by Example, *Fourth Edition*

CHATTERJEE and HADI · Sensitivity Analysis in Linear Regression

CHERNICK · Bootstrap Methods: A Guide for Practitioners and Researchers, *Second Edition*

CHERNICK and FRIIS · Introductory Biostatistics for the Health Sciences

CHILES and DELFINER · Geostatistics: Modeling Spatial Uncertainty, *Second Edition*

CHIU, STOYAN, KENDALL and MECKE · Stochastic Geometry and its Applications, *Third Edition*

CHOW and LIU · Design and Analysis of Clinical Trials: Concepts and Methodologies, *Second Edition*

CLARKE · Linear Models: The Theory and Application of Analysis of Variance

CLARKE and DISNEY · Probability and Random Processes: A First Course with Applications, *Second Edition*

* COCHRAN and COX · Experimental Designs, *Second Edition*

COLLINS and LANZA · Latent Class and Latent Transition Analysis: With Applications in the Social, Behavioral, and Health Sciences

CONGDON · Applied Bayesian Modelling

CONGDON · Bayesian Models for Categorical Data

CONGDON · Bayesian Statistical Modelling, *Second Edition*

CONOVER · Practical Nonparametric Statistics, *Third Edition*

COOK · Regression Graphics

COOK and WEISBERG · An Introduction to Regression Graphics

COOK and WEISBERG · Applied Regression Including Computing and Graphics

CORNELL · A Primer on Experiments with Mixtures

CORNELL · Experiments with Mixtures, Designs, Models, and the Analysis of Mixture Data, *Third Edition*

COX · A Handbook of Introductory Statistical Methods

CRESSIE · Statistics for Spatial Data, *Revised Edition*

CRESSIE and WIKLE · Statistics for Spatio-Temporal Data

CSÖRGŐ and HORVÁTH · Limit Theorems in Change Point Analysis

DAGPUNAR · Simulation and Monte Carlo: With Applications in Finance and MCMC

DANIEL · Applications of Statistics to Industrial Experimentation

DANIEL · Biostatistics: A Foundation for Analysis in the Health Sciences, *Eighth Edition*

* DANIEL · Fitting Equations to Data: Computer Analysis of Multifactor Data, *Second Edition*

DASU and JOHNSON · Exploratory Data Mining and Data Cleaning

DAVID and NAGARAJA · Order Statistics, *Third Edition*

* DEGROOT, FIENBERG, and KADANE · Statistics and the Law

DEL CASTILLO · Statistical Process Adjustment for Quality Control

DeMARIS · Regression with Social Data: Modeling Continuous and Limited Response Variables

---

*Now available in a lower priced paperback edition in the Wiley Classics Library.

†Now available in a lower priced paperback edition in the Wiley-Interscience Paperback Series.

DEMIDENKO · Mixed Models: Theory and Applications

DENISON, HOLMES, MALLICK and SMITH · Bayesian Methods for Nonlinear Classification and Regression

DETTE and STUDDEN · The Theory of Canonical Moments with Applications in Statistics, Probability, and Analysis

DEY and MUKERJEE · Fractional Factorial Plans

DE ROCQUIGNY · Modelling Under Risk and Uncertainty: An Introduction to Statistical, Phenomenological and Computational Models

DILLON and GOLDSTEIN · Multivariate Analysis: Methods and Applications

* DODGE and ROMIG · Sampling Inspection Tables, *Second Edition*

* DOOB · Stochastic Processes

DOWDY, WEARDEN, and CHILKO · Statistics for Research, *Third Edition*

DRAPER and SMITH · Applied Regression Analysis, *Third Edition*

DRYDEN and MARDIA · Statistical Shape Analysis

DUDEWICZ and MISHRA · Modern Mathematical Statistics

DUNN and CLARK · Basic Statistics: A Primer for the Biomedical Sciences, *Fourth Edition*

DUPUIS and ELLIS · A Weak Convergence Approach to the Theory of Large Deviations

EDLER and KITSOS · Recent Advances in Quantitative Methods in Cancer and Human Health Risk Assessment

* ELANDT-JOHNSON and JOHNSON · Survival Models and Data Analysis

ENDERS · Applied Econometric Time Series, *Third Edition*

† ETHIER and KURTZ · Markov Processes: Characterization and Convergence

EVANS, HASTINGS, and PEACOCK · Statistical Distributions, *Third Edition*

EVERITT, LANDAU, LEESE, and STAHL · Cluster Analysis, *Fifth Edition*

FEDERER and KING · Variations on Split Plot and Split Block Experiment Designs

FELLER · An Introduction to Probability Theory and Its Applications, Volume I, *Third Edition,* Revised; Volume II, *Second Edition*

FITZMAURICE, LAIRD, and WARE · Applied Longitudinal Analysis, *Second Edition*

* FLEISS · The Design and Analysis of Clinical Experiments

FLEISS · Statistical Methods for Rates and Proportions, *Third Edition*

† FLEMING and HARRINGTON · Counting Processes and Survival Analysis

FUJIKOSHI, ULYANOV, and SHIMIZU · Multivariate Statistics: High-Dimensional and Large-Sample Approximations

FULLER · Introduction to Statistical Time Series, *Second Edition*

† FULLER · Measurement Error Models

GALLANT · Nonlinear Statistical Models

GEISSER · Modes of Parametric Statistical Inference

GELMAN and MENG · Applied Bayesian Modeling and Causal Inference from Incomplete-Data Perspectives

GEWEKE · Contemporary Bayesian Econometrics and Statistics

GHOSH, MUKHOPADHYAY, and SEN · Sequential Estimation

GIESBRECHT and GUMPERTZ · Planning, Construction, and Statistical Analysis of Comparative Experiments

GIFI · Nonlinear Multivariate Analysis

GIVENS and HOETING · Computational Statistics

GLASSERMAN and YAO · Monotone Structure in Discrete-Event Systems

GNANADESIKAN · Methods for Statistical Data Analysis of Multivariate Observations, *Second Edition*

GOLDSTEIN · Multilevel Statistical Models, *Fourth Edition*

GOLDSTEIN and LEWIS · Assessment: Problems, Development, and Statistical Issues

GOLDSTEIN and WOOFF · Bayes Linear Statistics

GREENWOOD and NIKULIN · A Guide to Chi-Squared Testing

GROSS, SHORTLE, THOMPSON, and HARRIS · Fundamentals of Queueing Theory, *Fourth Edition*

GROSS, SHORTLE, THOMPSON, and HARRIS · Solutions Manual to Accompany Fundamentals of Queueing Theory, *Fourth Edition*

* HAHN and SHAPIRO · Statistical Models in Engineering

HAHN and MEEKER · Statistical Intervals: A Guide for Practitioners

HALD · A History of Probability and Statistics and their Applications Before 1750

† HAMPEL · Robust Statistics: The Approach Based on Influence Functions

HARTUNG, KNAPP, and SINHA · Statistical Meta-Analysis with Applications

HEIBERGER · Computation for the Analysis of Designed Experiments

HEDAYAT and SINHA · Design and Inference in Finite Population Sampling

HEDEKER and GIBBONS · Longitudinal Data Analysis

HELLER · MACSYMA for Statisticians

HERITIER, CANTONI, COPT, and VICTORIA-FESER · Robust Methods in Biostatistics

HINKELMANN and KEMPTHORNE · Design and Analysis of Experiments, Volume 1: Introduction to Experimental Design, *Second Edition*

HINKELMANN and KEMPTHORNE · Design and Analysis of Experiments, Volume 2: Advanced Experimental Design

HINKELMANN (editor) · Design and Analysis of Experiments, Volume 3: Special Designs and Applications

* HOAGLIN, MOSTELLER, and TUKEY · Fundamentals of Exploratory Analysis of Variance

* HOAGLIN, MOSTELLER, and TUKEY · Exploring Data Tables, Trends and Shapes

* HOAGLIN, MOSTELLER, and TUKEY · Understanding Robust and Exploratory Data Analysis

HOCHBERG and TAMHANE · Multiple Comparison Procedures

HOCKING · Methods and Applications of Linear Models: Regression and the Analysis of Variance, *Second Edition*

HOEL · Introduction to Mathematical Statistics, *Fifth Edition*

HOGG and KLUGMAN · Loss Distributions

HOLLANDER and WOLFE · Nonparametric Statistical Methods, *Second Edition*

HOSMER and LEMESHOW · Applied Logistic Regression, *Second Edition*

HOSMER, LEMESHOW, and MAY · Applied Survival Analysis: Regression Modeling of Time-to-Event Data, *Second Edition*

*Now available in a lower priced paperback edition in the Wiley Classics Library.

†Now available in a lower priced paperback edition in the Wiley-Interscience Paperback Series.

---

*Now available in a lower priced paperback edition in the Wiley Classics Library.

†Now available in a lower priced paperback edition in the Wiley-Interscience Paperback Series.

RYAN · Sample Size Determination and Power

RYAN · Statistical Methods for Quality Improvement, *Third Edition*

SALEH · Theory of Preliminary Test and Stein-Type Estimation with Applications

SALTELLI, CHAN, and SCOTT (editors) · Sensitivity Analysis

SCHERER · Batch Effects and Noise in Microarray Experiments: Sources and Solutions

* SCHEFFE · The Analysis of Variance

SCHIMEK · Smoothing and Regression: Approaches, Computation, and Application

SCHOTT · Matrix Analysis for Statistics, *Second Edition*

SCHOUTENS · Levy Processes in Finance: Pricing Financial Derivatives

SCOTT · Multivariate Density Estimation: Theory, Practice, and Visualization

* SEARLE · Linear Models

† SEARLE · Linear Models for Unbalanced Data

† SEARLE · Matrix Algebra Useful for Statistics

† SEARLE, CASELLA, and McCULLOCH · Variance Components

SEARLE and WILLETT · Matrix Algebra for Applied Economics

SEBER · A Matrix Handbook For Statisticians

† SEBER · Multivariate Observations

SEBER and LEE · Linear Regression Analysis, *Second Edition*

† SEBER and WILD · Nonlinear Regression

SENNOTT · Stochastic Dynamic Programming and the Control of Queueing Systems

* SERFLING · Approximation Theorems of Mathematical Statistics

SHAFER and VOVK · Probability and Finance: It's Only a Game!

SHERMAN · Spatial Statistics and Spatio-Temporal Data: Covariance Functions and Directional Properties

SILVAPULLE and SEN · Constrained Statistical Inference: Inequality, Order, and Shape Restrictions

SINGPURWALLA · Reliability and Risk: A Bayesian Perspective

SMALL and McLEISH · Hilbert Space Methods in Probability and Statistical Inference

SRIVASTAVA · Methods of Multivariate Statistics

STAPLETON · Linear Statistical Models, *Second Edition*

STAPLETON · Models for Probability and Statistical Inference: Theory and Applications

STAUDTE and SHEATHER · Robust Estimation and Testing

STOYAN · Counterexamples in Probability, *Second Edition*

STOYAN, KENDALL, and MECKE · Stochastic Geometry and Its Applications, *Second Edition*

STOYAN and STOYAN · Fractals, Random Shapes and Point Fields: Methods of Geometrical Statistics

STREET and BURGESS · The Construction of Optimal Stated Choice Experiments: Theory and Methods

STYAN · The Collected Papers of T. W. Anderson: 1943–1985

SUTTON, ABRAMS, JONES, SHELDON, and SONG · Methods for Meta-Analysis in Medical Research

TAKEZAWA · Introduction to Nonparametric Regression

*Now available in a lower priced paperback edition in the Wiley Classics Library.

†Now available in a lower priced paperback edition in the Wiley-Interscience Paperback Series.

Printed and bound by CPI Group (UK) Ltd, Croydon, CR0 4YY

12/01/2025

14624503-0005